AF353036

Nanomedicine Formulations Based on PLGA Nanoparticles for Diagnosis, Monitoring and Treatment of Disease: From Bench to Bedside

Nanomedicine Formulations Based on PLGA Nanoparticles for Diagnosis, Monitoring and Treatment of Disease: From Bench to Bedside

Editor

Oya Tagit

MDPI • Basel • Beijing • Wuhan • Barcelona • Belgrade • Manchester • Tokyo • Cluj • Tianjin

Editor
Oya Tagit
BioInterfaces
FHNW
Muttenz
Switzerland

Editorial Office
MDPI
St. Alban-Anlage 66
4052 Basel, Switzerland

This is a reprint of articles from the Special Issue published online in the open access journal *Pharmaceutics* (ISSN 1999-4923) (available at: www.mdpi.com/journal/pharmaceutics/special_issues/nanomedicine_formulations).

For citation purposes, cite each article independently as indicated on the article page online and as indicated below:

LastName, A.A.; LastName, B.B.; LastName, C.C. Article Title. *Journal Name* **Year**, *Volume Number*, Page Range.

ISBN 978-3-0365-4490-8 (Hbk)
ISBN 978-3-0365-4489-2 (PDF)

Contents

About the Editor

Oya Tagit

Prof. Dr. Oya Tagit is a full professor of BioInterfaces at the University of Applied Science and Arts Northwest Switzerland. In her research, she aims to address currently unmet biomedical needs in the diagnosis, monitoring, and treatment of disease with engineered nanoparticles tailored for the desired biological targets and functions. Interdisciplinary approaches at the interface between materials chemistry, nanotechnology, and biology are utilized for bench-to-bedside development of optical and magnetic detection probes, (pre-)clinical imaging agents, and drug delivery systems based on inorganic and polymeric nanoparticles.

Preface to "Nanomedicine Formulations Based on PLGA Nanoparticles for Diagnosis, Monitoring and Treatment of Disease: From Bench to Bedside"

Nanomedicine is among the most promising emerging fields that can provide innovative and radical solutions to unmet needs in pharmaceutical formulation development. Encapsulation of active pharmaceutical ingredients within nano-size carriers offers several benefits, namely, protection of the therapeutic agents from degradation, their increased solubility and bioavailability, improved pharmacokinetics, reduced toxicity, enhanced therapeutic efficacy, decreased drug immunogenicity, targeted delivery, and simultaneous imaging and treatment options with a single system.

Poly(lactide-co-glycolide) (PLGA) is one of the most commonly used polymers in nanomedicine formulations due to its excellent biocompatibility, tunable degradation characteristics, and high versatility. Furthermore, PLGA is approved by the European Medicines Agency (EMA) and the Food and Drug Administration (FDA) for use in pharmaceutical products. Nanomedicines based on PLGA nanoparticles can offer tremendous opportunities in the diagnosis, monitoring, and treatment of various diseases.

This Special Issue aims to focus on the bench-to-bedside development of PLGA nanoparticles including (but not limited to) design, development, physicochemical characterization, scale-up production, efficacy and safety assessment, and biodistribution studies of these nanomedicine formulations.

Oya Tagit
Editor

pharmaceutics

Article

Effect of Solvents, Stabilizers and the Concentration of Stabilizers on the Physical Properties of Poly(D,L-lactide-*co*-glycolide) Nanoparticles: Encapsulation, In Vitro Release of Indomethacin and Cytotoxicity against HepG2-Cell

Musaed Alkholief [1], Mohd Abul Kalam [1], Md Khalid Anwer [2] and Aws Alshamsan [1,*]

1 Nanobiotechnology Unit, Department of Pharmaceutics, College of Pharmacy, King Saud University, P.O. Box 2457, Riyadh 11451, Saudi Arabia; malkholief@ksu.edu.sa (M.A.); makalam@ksu.edu.sa (M.A.K.)
2 Department of Pharmaceutics, College of Pharmacy, Prince Sattam Bin Abdulaziz University, Al-Kharj 11942, Saudi Arabia; m.anwer@psau.edu.sa
* Correspondence: aalshamsan@ksu.edu.sa

Citation: Alkholief, M.; Kalam, M.A.; Anwer, M.K.; Alshamsan, A. Effect of Solvents, Stabilizers and the Concentration of Stabilizers on the Physical Properties of Poly(D,L-lactide-*co*-glycolide) Nanoparticles: Encapsulation, In Vitro Release of Indomethacin and Cytotoxicity against HepG2-Cell. *Pharmaceutics* 2022, 14, 870. https://doi.org/10.3390/pharmaceutics14040870

Academic Editors: Juan José Torrado, Bruno Sarmento and Oya Tagit

Received: 6 February 2022
Accepted: 12 April 2022
Published: 15 April 2022

Abstract: A biocompatible, biodegradable and FDA-approved polymer [Poly lactic-*co*-glycolic acid (PLGA)] was used to prepare the nanoparticles (NPs) to observe the effect of solvents, stabilizers and their concentrations on the physical properties of the PLGA-NPs, following the encapsulation and in vitro release of Indomethacin (IND). PLGA-NPs were prepared by the single-emulsion solvent evaporation technique using dichloromethane (DCM)/chloroform as the organic phase with Polyvinyl-alcohol (PVA)/Polyvinylpyrrolidone (PVP) as stabilizers to encapsulate IND. The effects of different proportions of PVA/PVP with DCM/chloroform on the physiochemical properties (particle size, the polydispersity index, the zeta potential by Malvern Zetasizer and morphology by SEM) of the NPs were investigated. DSC was used to check the physical state, the possible complexation of PLGA with stabilizer(s) and the crystallinity of the encapsulated drug. Stabilizers at all concentrations produced spherical, regular-shaped, smooth-surfaced discrete NPs. Average size of 273.2–563.9 nm was obtained when PVA (stabilizer) with DCM, whereas it ranged from 317.6 to 588.1 nm with chloroform. The particle size was 273.2–563.9 nm when PVP was the stabilizer with DCM, while it was 381.4–466.6 nm with chloroform. The zeta potentials of PVA-stabilized NPs were low and negative (−0.62 mV) while they were comparatively higher and positive for PVP-stabilized NPs (+17.73 mV). Finally, drug-loaded optimal NPs were composed of PLGA (40 mg) and IND (4 mg) in 1 mL DCM/chloroform with PVA/PVP (1–3%), which resulted in sufficient encapsulation (54.94–74.86%) and drug loading (4.99–6.81%). No endothermic peak of PVA/PVP appeared in the optimized formulation, which indicated the amorphous state of IND in the core of the PLGA-NPs. The in vitro release study indicated a sustained release of IND (32.83–52.16%) from the PLGA-NPs till 72 h and primarily followed the Higuchi matrix release kinetics followed by Korsmeyer–Peppas models. The cell proliferation assay clearly established that the organic solvents used to prepare PLGA-NPs had evaporated. The PLGA-NPs did not show any particular toxicity in the HepG2 cells within the dose range of IND (250–500 μg/mL) and at an equivalent concentration of PLGA-NPs (3571.4–7142.7 μg/mL). The cytotoxicity of the hepatotoxic drug (IND) was reduced by its encapsulation into PLGA-NPs. The outcomes of this investigation could be implemented to prepare PLGA-NPs of acceptable properties for the encapsulation of low/high molecular weight drugs. It would be useful for further in vitro and in vivo applications to use this delivery system.

Keywords: PLGA; Indomethacin; nanoparticles; solvents; stabilizers; morphology; particle-size; encapsulation; drug release; cytotoxicity

1. Introduction

Natural and synthetic polymers have been extensively used by researchers for drug delivery applications for decades [1–5]. Natural polymers (e.g., bovine serum albumin) have fallen out of favor due to impurity and cost effectiveness issues [6,7]. Synthetic polymers (e.g., poly D,L-Lactide-co-glycolic Acid, PLGA), on the other hand, have become more preferable to formulation scientists as drug delivery carriers, due to their superior biodegradability and biocompatibility properties [8,9]. A wide variety of synthetic polymers have been studied, with PLGA being one of the most used. PLGA is biocompatible and biodegradable, and has been explored in many drug delivery applications, such as controlled delivery carriers for peptide, proteins, antigens, genes, vaccines, growth factors and many other macromolecular therapeutics [8,10–19]. The frequent application of PLGA for drug delivery is due to its attractive mechanical and handling properties [20]. In addition, researchers prefer using PLGA because it has been approved by World Health Organization and Food Drug Administration (FDA) for human use since 1989 [5,7,9,21–24].

PLGA-NPs can be synthesized using different techniques, including emulsion solvent evaporation or diffusion, solvent displacement, salting-out, nanoprecipitation and many other techniques [8,25–29]. Of these methods, the single (o/w) or double ($w/o/w$) emulsion solvent evaporation are the oldest and most extensively used techniques to encapsulate the hydrophilic and hydrophobic drugs into the PLGA-based micro- or nanoparticles [6,9,22]. In this method, PLGA and Indomethacin (IND) are dissolved in organic solvent (chloroform/dichloromethane/ethyl acetate/acetone etc.) and then emulsified by the stabilizer present in the aqueous phase via homogenization followed by sonication [28]. The nanodroplets obtained are further treated by evaporating the organic solvent(s) before purifying and collecting the PLGA-NPs by washing with purified water (ultracentrifugation) for its further characterization [21,25]. The morphological and physical/physicochemical characteristics (particle size, its distribution, physical and stability, release profile of the encapsulated drug, etc.) of the NPs can be greatly influenced by the formulation parameters [30,31]. The above characteristics of PLGA-NPs might change by changing the type of organic solvent(s) used to solubilize the polymer and the type of emulsifier/stabilizer, and their respective concentrations. Other factors that play a role include the mechanical forces (such as homogenization/probe-sonication, magnetic stirring rate, etc.) used to emulsify the mixture of organic and solvents, and the duration for the complete evaporation of the organic solvent [27,30].

Indomethacin (IND) is a nonsteroidal anti-inflammatory drug (NSAID) used to relieve fever, pain, swelling and stiffness of joints in moderate to severe arthritic conditions such as rheumatoid, osteo and gouty arthritis or spondylitis. It is also used for other purposes including in the shoulder pain affected by tendinitis or bursitis. It acts by the reversible inhibition of cyclooxygenase enzymes (COX-I and COX-II), also known as prostaglandin-endoperoxide synthase. The COX-1 catalyzes the synthesis of prostaglandins and thromboxane, while COX-2 is articulated in response to inflammation and injury. Thus, the antipyretic, analgesic and anti-inflammatory actions of IND take place as a result of decreased prostaglandin synthesis. Unlike other NSAIDs, IND also inhibits the enzyme phospholipase-A2 (causes the release of arachidonic acid from phospholipids). The antipyretic effect of IND might be due to its action on the hypothalamus causing vasodilation, increased peripheral blood flow and thus the dissipation of heat [32]. The poor aqueous solubility of IND (0.937 mg/L at 25 °C) leads to poor bioavailability [33]. Thus, taking high doses of IND is needed for some conditions. High doses of this drug for prolonged use may cause intestinal bleeding and increase the risk of stroke or fatal heart attack or myocardial infarction [34]. The encapsulation of IND into nanoparticulate drug carriers is needed to improve its bioavailability and reduce its dose and dosing frequency.

In the present study, we investigated the effect of different formulation parameters on the physiochemical properties of the PLGA-NPs. We prepared four batches of PLGA-NPs using two different organic solvents, namely chloroform and dichloromethane (DCM), with two commonly used stabilizers, namely Polyvinyl alcohol (PVA) and Polyvinylpyrrolidone

(PVP) and studied their influence on the particle size, its distribution and their morphological characteristics. We also investigated the effects of different concentrations and viscosities of the aqueous solution of the stabilizers on the PLGA-NPs. In addition, we utilized differential scanning calorimetry to check any polymer–stabilizer or drug interactions and the crystallinity of the encapsulated drug. To our knowledge, most of the literature has focused on the release behavior of the payload from the PLGA-NPs system. Thus, here we synthesized drug-loaded PLGA-NPs by the single emulsion solvent evaporation technique using IND as a model lipophilic drug. The physicochemical properties including the drug encapsulation, its loading and in vitro drug release and release kinetics were investigated in detail. Moreover, the toxicity of PLGA-NPs (with and without IND) was examined by performing the cytotoxicity using HepG2 cells.

2. Materials and Methods

2.1. Materials

Poly (D,L-lactic-*co*-glycolic acid) (50:50) with molecular weight 40,000–75,000 and Polyvinylpyrrolidone (PVP; Av. Mw ~ 55,000) were purchased from Sigma Aldrich Co. (St. Louis, MO, USA). Indomethacin (IND) was procured from Winlab, UK. Polyvinyl alcohol (PVA) with 17,200 Mw was purchased from AVONCHEM Ltd. (Macclesfield, Cheshire, UK). Dichloromethane (DCM) and chloroform were purchased from PANREAC QUIMICA SA, (Barcelona, Spain) and MERK (Darmstadt, Germany), respectively. Purified water was obtained through Milli-Q water purifier (Millipore, France) system.

2.2. Chromatographic Analysis of Indomethacin

The reverse phase (RP) HPLC-UV method was used for the analysis of IND at 241 nm wavelength [35–37]. "The HPLC system (Waters®, Model-1500-series controller, Milford, MA, USA) equipped with UV-detector (Waters®, Model-2489, dual absorbance detector, USA), binary pump (Waters®, Model-1525, USA), automated sampling system (Waters®, Model-2707 Autosampler, USA). The system was run by "Breeze software" (Breeze™ HPLC Systems From Waters Corporation, Milford, MA, USA). A C_{18} analytical column (Macherey-Nagel 150 × 4.6 mm, 5 µm) was used at room temperature. The chromatography was performed by isocratic elution of the mobile phase composed of 75:25, *v/v* of Acetonitrile and Milli-Q® water (the pH of water was adjusted to 3.2 by Ortho-Phosphoric acid). The flow rate was 1 mL/min and the volume of injection was 30 µL. Total run time was 5 min and retention time (R_t) for IND was 3.5 min. The calibration curve was constructed in the concentration range 0.1, 0.2, 0.4, 0.8, 1.6, 2.5, 5.0 and 10 µgmL^{-1}. The straight-line equation obtained was $y = 150{,}383x - 6548.7$ with a correlation coefficient (R^2) of 0.9991.

2.3. Preparation of PLGA Nanoparticles

The PLGA-NPs were prepared by the single emulsion solvent evaporation technique. Here, the PLGA-NPs were prepared using the oil-in-water (*o/w*) single emulsion solvent evaporation technique [38]. In brief, 40 mg of PLGA (50:50; Mol. Wt. of 40–75 kDa) and 4 mg of IND were dissolved into 1 mL of the organic phase consisting of either chloroform or dichloromethane (DCM). These organic phases were then emulsified after the addition of 5 mL of varying concentrations of (1%, 3%, 6% or 9%, *w/v*) of polyvinyl alcohol (PVA) or polyvinyl pyrrolidone (PVP) by pulse sonication using a probe sonicator at 40% power for 40 s (4 cycles, 10 s each) using a probe sonicator (Sonics & Materials, Inc., Newtown, PA, USA) on an ice bath. The organic solvent was evaporated by magnetic stirring at 800 rpm at room temperature for overnight to obtain the suspension of NPs. The suspended PLGA-NPs were purified after washing with Milli-Q® water to remove the extra stabilizers and collected by ultracentrifugation at 15,000 rpm for 30 min at 4 °C (Preparative ultracentrifuge, WX-series by Hitachi Koki, Hitachinaka, Japan). Finally, the drug-loaded PLGA-NPs were also prepared by the same procedure, where 4 mg of IND was dissolved into the organic phase containing PLGA. The washing process was performed three times for the purification and collection of PLGA-NPs. The obtained NPs were then dispersed in 10 mL of Milli-Q water, freeze-dried and were stored at $-80\,^\circ$C [39] for further experiments.

2.4. Particle Size and Zeta Potential Measurement

The hydrodynamic diameter (mean particle size), size distribution and zeta potential of the developed PLGA-NPs were checked by Dynamic Light Scattering (DLS) measurement by Zetasizer Nano-Series (Nano-ZS, Malvern Instruments Limited, Worcestershire, UK) as per the reported methods [28,40]. Before starting the measurements, the suspensions of the NPs were suitably diluted with Milli-Q® water (Millipore, France). For zeta potential, by considering the dielectric constant of the dispersant (i.e., 78.5 for water), the electrophoretic mobility was calculated at 25 °C by the software. The electrophoretic values were utilized to estimate the zeta potential by the software DTS V-4.1 (Malvern, Worcestershire, UK) installed in the system. All the measurements were performed in triplicate.

2.5. Morphological Characterization of PLGA-NPs

The shape and surface morphology of the PLGA-NPs was observed and determined by micrographs of the NPs obtained by Scanning Electron Microscopy (SEM) (Zeiss EVO LS10; Cambridge, UK) using the gold sputter technique. The freeze-dried PLGA-NPs were coated with gold using an ion-sputter (Q150R Sputter unit) from Quorum Technologies Ltd. (East Sussex, UK) in an argon atmosphere at 20 mA for 1 min. Observation and imaging was completed at 10 to 20 kV accelerating voltage and 8.5 mm of working distance. The magnification for the SEM images was kept at around 10 to 15 KX [28].

2.6. FTIR Spectral Analysis

Attenuated total reflectance (ATR) FTIR spectrum of solid samples (of pure drug (IND), PLGA 50:50, Empty PLGA-NPs and IND-loaded PLGA-NPs-5) were recorded using Bruker ALPHA spectrometer (Bruker Optics, Rosenheim, Germany). The spectrometer was equipped with a diamond prism with single bounce reflection ATR accessory. A thin layer of the solid sample was kept over the ATR accessory and the spectra were recorded from 4000–450 cm^{-1} wavenumber range using OPUS software (V-7.8), Bruker Optics, Rosenheim, Germany.

2.7. Viscosity Evaluation of the Surfactant Solutions

Viscosity study of the different concentrations of the stabilizers (PVA and PVP) was conducted to explore if there was a direct or indirect relationship between the viscosity and the characterization parameters [41]. The viscosity of the stabilizer solutions was evaluated by "Brookfield Viscometer (Brookfield Engineering Laboratories, Middleboro, MA, USA)" following as reported [42,43]. The viscosities (Pa.s) were determined at different shear rates (300–500 s^{-1} for PVA and 360–460 for PVP s^{-1}) at $25 \pm 1\,^\circ$C temperature.

2.8. Encapsulation Efficiency and Drug-Loading Capacity

For the determination of the encapsulation efficiency (%EE) drug loading (%DL), 10 mg of drug-loaded PLGA-NPs were dispersed in 5 mL of Milli-Q water. The dispersion was centrifuged at 13,500 rpm at 10 °C for 15 min (PRISM-R, Labnet International Inc., Edison, NJ, USA). The supernatant was removed and the sediments of PLGA-NPs (precipitant) were washed by Milli-Q water in triplicate with 2 min of vortexing. Thereafter, 10 mL 1:1 (*v*/*v*) mixture of CHCl$_3$ and DCM (to dissolve the PLGA and solubilize the IND) was added in the precipitant or pellet and vortexed. For complete extraction of the drug (Indomethacin, IND) from NPs, the mixture was subjected to ultrasonication (Model-3510, Branson Ultrasonic Corporation) for 20 min. The resulting suspension was further centrifuged for 10 min at 13,500 rpm. The supernatant was collected and the concentration of the drug was analyzed by the reported HPLC-UV method at 241 nm wavelength [35–37].

The %EE and %DL were calculated in triplicate according to the following equations (Equations (1) and (2)).

$$\%EE = \left(\frac{\text{Amount of IND in NPs precipitant (mg)}}{\text{Initial amount of IND added (mg)}} \right) \times 100 \tag{1}$$

$$\%DL = \left(\frac{\text{Amount of IND in NPs precipitant (mg)}}{\text{Total mount of NPs (mg)}} \right) \times 100 \tag{2}$$

2.9. In Vitro Release and Kinetics Study

Based on the physicochemical features, two formulations from each batch were selected for the comparative in vitro release profile of IND from different PLGA-NPs. The in vitro release of IND from the selected and optimized PLGA-NPs was performed in in triplicate by the dialysis tubing method [44–46]. Phosphate buffer saline (PBS, pH 7.4) with 0.25% (w/v) of sodium lauryl sulphate (SLS) was used as the release media. An in vitro release study was also performed using dilute HCl solution (pH 1.2) with 0.25% (w/v) of SLS as release medium. A total of 10 mg of freeze-dried drug-loaded NPs was put in 1 mL of PBS mixed by vortexing and transferred into dialysis tubing. Both the ends of the dialysis tubing were tied and put into 100 mL capacity beakers containing 50 mL of the release medium. Beakers were placed in a shaking water bath (100 opm) where the temperature of water bath was maintained at $37 \pm 0.5\,^\circ$C. At stipulated time points 1 mL of the sample was taken out from each beaker and centrifuged at 13,500 rpm for 10 min. The supernatants was collected and 30 µL of each was injected into the HPLC-UV system for the analysis of released drug [37]. The cumulative amount drug released (%DR) was calculated by the following equation (Equation (3)), where DF stands for dilution factor.

$$\%DR = \frac{\text{Conc. (µg/mL)} \times \text{DF} \times \text{Volume of release medium (mL)}}{\text{Initial amount of IND in NPs (µg)}} \times 100 \tag{3}$$

To check the kinetics and mechanism of IND release from the PLGA-NPs, the in vitro release data were fitted into different kinetic models [47,48]. The applied kinetic models include the zero-order (%DR versus time, Equation (4)), first-order (Log %D remaining versus time, Equation (5)), Higuchi matrix (%DR versus square root of time, Equation (6)), Korsmeyer–Peppas (Log fraction DR versus Log time, Equation (7)) and Hixson–Crowell (cube root of %D remaining in the polymeric matrix versus time, Equation (8)). From the values of slopes and co-efficient of correlation (R^2) of kinetic plots obtained in different model equations, the release/diffusion-exponent (n-value) was calculated. The determination of n-values could provide an idea about the mechanism of drug release from the NPs [36,49,50].

$$A_t = k_0 t \tag{4}$$

$$Log\,A_0 - Log\,A_t = k_1 t / 2.303 \tag{5}$$

$$A_t = k_{HM}\sqrt{t} \tag{6}$$

$$A_t = k_{KP}\,t^n \tag{7}$$

$$\sqrt[3]{A_0} - \sqrt[3]{A_t} = k_{HC}t \tag{8}$$

In the above equations, A_t is the amount of released IND at time 't', A_0 is the initial amount of IND in the PLGA-NPs. The expression 'k' is the rate constants (k_0 for zero-order, k_1 for first order, k_{HM} for Higuchi matrix, k_{KP} for Korsmeyer–Peppas, and K_{HC} for the Hixson–Crowell model). The n-value in the Korsmeyer–Peppas equation is known as the diffusion or release exponent.

2.10. Differential Scanning Calorimetry (DSC)

The analysis of the physical state and possible complexation of the PLGA with the used stabilizers as well as the crystallinity of the encapsulated drug (IND) was checked through the differential scanning calorimetry technique through a DSC-8000 (Perkin Elmer Instruments, Shelton, WA, USA) at the heating rate of 10 °C per min. Around 4–5 mg of PLGA-NPs was weighed and put in aluminum pans and the lids were crimped. The pans were hermetically sealed and placed in the sample cell of the calibrated DSC instrument to check the melting temperature of the materials [40,51]. Both the reference and sample cells were continuously purged with N_2 gas at the rate of 20 mL per min. The results were analyzed by the software Pyris™ (Version-11, PerkinElmer Inc., Houston, TX, USA) attached to the system. The thermal behavior of the materials used in the PLGA-NPs was examined at 10 °C per min of heating rate within 20 to 240 °C temperature range.

2.11. Stability of IND-Loaded PLGA-NPs

A short-term stability on drug-loaded freeze-dried PLGA-NPs (IND-PLGA-NPs-5) was performed [28,52,53]. The drug-loaded NPs (10 mg) were crammed into glass containers and stored at 4 ± 0.5 °C, 30 ± 1 °C (as per the Saudi Arabian climatic zone (IVa)) and 37 ± 1 °C for 1-month. The alteration in the particle size, polydispersity index, zeta potential, encapsulation efficiency (%EE) and drug-loading capacity (%DL) with storage time was determined at the 15th and 30th day to recognize the stability of the NPs. All the measurements were performed in triplicate.

2.12. In Vitro Cytotoxicity Studies

2.12.1. Maintenance and Growth of Cell Line

The human hepatic cancer cell (HepG2) lines were cultured under CO_2 (5%) at 37 °C using Dulbecco's Modified Eagle's Medium (DMEM), procured from UFC Biotech, Riyadh, Saudi Arabia). The media were accompanied with 1% mixture of "Penicillin-Streptomycin" (Thermo Fischer Scientific, Waltham, MA, USA), 10% Fetal Bovine Serum (FBS) and 1% L-Glutamine obtained from Alpha Chemika, Maharashtra, India and BioWest, Riverside, MO, USA, respectively, while the MTT was obtained from "Sigma Aldrich", St. Louis, MO, USA.

2.12.2. MTT Assay

The in vitro cytotoxic activities of the free drug (IND solution) and the IND-loaded PLGA-NPs were investigated as compared to the empty PLGA-NPs-5 using HepG2 cells by assessing their viability through an MTT assay [54]. The cell cultures in the phase of exponential growth were "Trypsinized" and diluted in DMEM to obtain a total cell count of 5×10^5 cells/mL. The suspension of cells were then transferred into flat-bottomed 96 well microplate in 100 µL medium and left overnight to attach. The medium was changed with 100 µL of fresh medium containing IND, IND-loaded NPs and empty PLGA-NPs-5 using the equivalent amount of NPs to obtain IND concentrations (5–1000 µg/mL) in the same media as the solvent. An equivalent amount of empty NPs was also used. The free IND was dissolved in DMSO and diluted with the media, keeping the concentration of DMSO not more than 1%, (v/v). Three wells were used as a negative control for cell viability, where only cells (without treatment) were suspended in the FBS free DMEM. Each product dilution was evaluated in triplicate. After 24 h, 48 h and 72 h of incubation, the cell suspensions were removed and the wells were washed with PBS. Then, 20 µL of MTT solution (at 5 mg/mL concentration in PBS) and 80 µL of the medium was added and incubated at 37 °C for 4 h. Subsequently, the culture fluid (containing media and MTT) was discarded leaving the precipitate of formazan crystals. The crystals were dissolved in 100 µL mixture (99.4 mL/0.6 mL/10 g) of DMSO, acetic acid and sodium lauryl sulfate at room temperature for 10–15 min. The microplates were observed and absorbance was measured at 570 nm using spectrophotometric microplate reader (Synergy HT, BioTek Inst., Winooski, VT, USA). The results were normalized as compared to the absorbance obtained for viable control cells (by considering as 100%) and expressed as plots of cells viability versus log concentration of IND. The IC_{50} values were determined for the employed cells against each product including the empty NPs using GraphPad Prism 5, San Diego, CA, USA.

2.13. Statistical Analysis

All the experiments were performed in triplicate ($n = 3$). The data were expressed as mean with standard deviation (SD). Data were analyzed and compared by the Paired *t*-test using GraphPad Prism 5, USA. The *p*-value, less than ($p < 0.05$) was considered as statistically significant.

3. Results and Discussion

3.1. Preparation of PLGA-NPs

For the preparation of NPs, we used PLGA as the main polymer mixed with either PVP or PVA, while either chloroform or DCM were used as the organic solvents. We employed the single emulsion solvent evaporation method which is helpful in optimizing the design of the NPs (size, morphology and surface properties) making them better candidates for drug delivery applications. In this method, the emulsification of the two phases and the stabilization (droplet protection) of the NPs are prime factors, which are greatly influenced by the type of stabilizers used and their quantity. PVA and PVP were chosen as stabilizers because they facilitate the formulation of smaller PLGA-NPs with a spherical shape and a narrow granulometric distribution even at higher concentrations, in addition to their ability to prevent the coalescence of particles to aggregate [55–58]. Although PVP-stabilized PLGA-NPs have shown similar features to PVA-stabilized NPs in terms of the encapsulation and loading of low molecular weight drugs [59], under similar conditions PVP-stabilized PLGA-NPs had higher efficiencies compared to PVA-stabilized NPs for the encapsulation and loading of proteins [59–61]. As for the type of organic solvents used, chloroform and DCM are the most commonly used to prepare the PLGA-NPs by the emulsification solvent evaporation method. However, reports have indicated that other solvents can be used, but mostly in combination, such as DCM-acetone that is used to dissolve rifampicin and DCM-acetone-ethanol used to dissolve Estradiol valerate [62]. Additionally, propylene carbonate has been utilized to reduce the toxic effect of benzyl alcohol to encapsulate estrogen into PLGA-NPs [57,58]. The solvent evaporation process during the preparation of PLGA-NPs is a critical step to encapsulate the drug(s) as it decreases the diffusion of drug(s) to the external aqueous phase, in addition to obtaining smaller particles [63,64]. The chloroform and DCM as organic solvents for PLGA are easily evaporated in a short duration even at a reduced pressure (vacuum rotational evaporator) or with magnetic stirring at normal atmospheric pressure [55,63]. The smaller particles are obtained due to the higher solvent-front kinetic energy that leads to the higher rate of diffusion of the organic phase at the interface and a higher rate of droplet dispersion, which results in smaller NPs [63]. Solvent evaporation causes a volume reduction in the dispersion phase which subsequently increases the density and viscosity of the dispersed NPs. This phenomenon prevents the coalescence and agglomeration of the NPs during the evaporation of organic solvents [64].

Using the proper preparation technique, such as single emulsion or double emulsion, is critical as the chosen method affects the physicochemical properties of NPs and the drug-loading capacity [65,66]. For example, the encapsulation and loading of lipophilic agents through the single emulsion technique supports the formation of smaller NPs than that of the double emulsion technique [17,67,68]. The employed emulsion solvent evaporation technique, also called the ultrasonication solvent evaporation technique, was first introduced by Vanderhoff et al. and has become a very popular method for the synthesis of polymer-based nanoparticles [7,62,65,69]. The process parameters in this technique, such as the volume of the aqueous phase containing stabilizer/emulsifier, magnetic stirring, homogenization speed and drug/PLGA ratio (in the case of drug-loaded NPs) can easily be controlled and optimized to obtain the NPs with an improved surface morphology and the desired size distribution [55,65]. This method has been used frequently to encapsulate low and high molecular weight drugs [27,28,70–72]. Moreover, smaller NPs could be obtained using a small quantity of PLGA and a large volume of a stabilizer-containing aqueous phase at a high homogenization speed while larger particles can be obtained at higher PLGA concentrations [55,56]. A high amount of polymer caused an increased viscosity of the internal phase, leading to its insufficient and non-homogenous dispersion into the aqueous phase during emulsification and resulting in larger NPs [55].

3.2. Effect of Formulation Factors on Particle Size and Zeta Potential

The PLGA-NPs prepared by the single emulsion solvent evaporation method produced particles with an average size in the range of 273–563 nm when PVA was used as the stabilizer and DCM was used as the organic solvent, whereas it was in the range of 317–588 nm when chloroform was used as the solvent. As for PVP, the mean particle size was in the range of 287–684 nm when DCM was used as the solvent, while it was in the range of 381–466 nm when chloroform was the solvent (Table 1).

Table 1. PLGA-NPs prepared by the single-emulsion solvent evaporation technique.

Formulations	Organic Solvents *	Type of Stabilizer, Its Concentration *	Mean ± SD *, $n = 3$		
			Particle Size (nm)	Polydispersity Index	Zeta-Potential (mV)
Batch 1					
PLGA-NPs-1	$CHCl_3$	PVA (1%, w/v)	328.1 ± 24.4	0.096 ± 0.079	−1.90 ± 0.91
PLGA-NPs-2	$CHCl_3$	PVA (3%, w/v)	345.2 ± 8.3	0.061 ± 0.036	−1.12 ± 0.06
PLGA-NPs-3	$CHCl_3$	PVA (6%, w/v)	588.1 ± 23.9	0.096 ± 0.028	−0.74 ± 0.12
PLGA-NPs-4	$CHCl_3$	PVA (9%, w/v)	317.6 ± 2.7	0.285 ± 0.088	−0.68 ± 0.11
Batch 2					
PLGA-NPs-5	DCM	PVA (1%, w/v)	273.2 ± 3.3	0.087 ± 0.064	−0.92 ± 0.21
PLGA-NPs-6	DCM	PVA (3%, w/v)	406.3 ± 8.5	0.213 ± 0.044	−0.89 ± 0.28
PLGA-NPs-7	DCM	PVA (6%, w/v)	554.4 ± 23.8	0.239 ± 0.045	−0.76 ± 0.06
PLGA-NPs-8	DCM	PVA (9%, w/v)	563.9 ± 54.4	0.170 ± 0.015	−0.62 ± 0.58
Batch 3					
PLGA-NPs-9	$CHCl_3$	PVP (1%, w/v)	456.9 ± 61.4	0.152 ± 0.069	17.73 ± 3.45
PLGA-NPs-10	$CHCl_3$	PVP (3%, w/v)	448.8 ± 22.4	0.142 ± 0.035	14.05 ± 2.05
PLGA-NPs-11	$CHCl_3$	PVP (6%, w/v)	466.6 ± 90.5	0.256 ± 0.181	3.86 ± 1.26
PLGA-NPs-12	$CHCl_3$	PVP (9%, w/v)	381.4 ± 83.5	0.255 ± 0.179	2.67 ± 1.41
Batch 4					
PLGA-NPs-13	DCM	PVP (1%, w/v)	287.8 ± 12.0	0.035 ± 0.029	10.24 ± 3.93
PLGA-NPs-14	DCM	PVP (3%, w/v)	566.9 ± 64.1	0.268 ± 0.042	5.82 ± 2.65
PLGA-NPs-15	DCM	PVP (6%, w/v)	609.3 ± 26.9	0.512 ± 0.055	5.37 ± 1.06
PLGA-NPs-16	DCM	PVP (9%, w/v)	684.2 ± 43.1	0.846 ± 0.016	4.91 ± 2.37

* PVA = Polyvinyl alcohol; PVP = Polyvinyl pyrrolidone; $CHCl_3$ = Chloroform; DCM = Dichloromethane; SD = Standard deviation.

Overall, the results of the particle size measurement indicate that PLGA-NPs prepared using PVA as stabilizer were smaller compared to PVP when DCM was used to solubilize the PLGA, with an exception at 1% PVP. This might be attributed to the good stabilizing property of PVP at this concentration (having optimum viscosity) with DCM. At the varying concentrations of PVA (1%, 3%, 6% and 9%, w/v) and using DCM as the organic solvent by keeping the other formulations factors constant (such as volume of solvents, stirring speed, amount of PLGA), an increase in the size of the NPs was found from 273–563 nm (Table 1). Similarly, an increase in particle size was observed from 317–588 nm as the concentration of PVA was increased and chloroform was used as the organic solvent. Contrary to this observation, the particle size was smallest (317 nm) at the highest (9%, w/v) concentration of PVA when chloroform was used, which might be due to the strong anchoring of the hydrophobic segment of PVA with the matrix of PLGA which remained closely associated with polymer surface [73]. The mechanistic effect of stabilizers on the size of PLGA-NPs might be due to the interpenetration of PVA/PVP and PLGA molecules during the formation of NPs [74,75]. These stabilizers might develop a uniform coating network throughout the surface of PLGA, where PVA cannot shield completely the negative surface charge (due to –COOH functional groups of PLGA) as compared to PVP. Thus, PVA caused a weak negative zeta potential [15], while PVP showed a relatively high positive zeta potential. In spite of the weak zeta potential, the PLGA-NPs were stabilized by the varying layers of PVA/PVP which surrounded the NPs by mechanism of steric hindrance [76].

The particle size and zeta potential distributions of the four NPs prepared with 1% stabilizers are represented in Figure 1. The values of polydispersity index and zeta potential are also summarized in Table 1. The zeta potentials of the NPs prepared with PVA showed low negatively charged surfaces, while most of the NPs prepared with PVP were found to have comparatively higher and positive surface charges. To check the effect of stabilizer on surface charges, the zeta potentials of 1% PVA and PVP solutions were also checked. The PVA solution showed slightly negative (-2.26 ± 0.49 mV) zeta potentials while it was positive ($+5.06 \pm 1.58$ mV) for the PVP solution. Thus, we could state that the positive zeta potential of PVP solution might be the reason for the slightly higher zeta potentials of the PLGA-NPs developed using PVP as the stabilizer. The low polydispersity values in all cases of NPs (except in the cases where 6% and 9% PVP, which were 0.512 and 0.846, respectively) indicated the unimodal distribution of the particles in the dispersion medium.

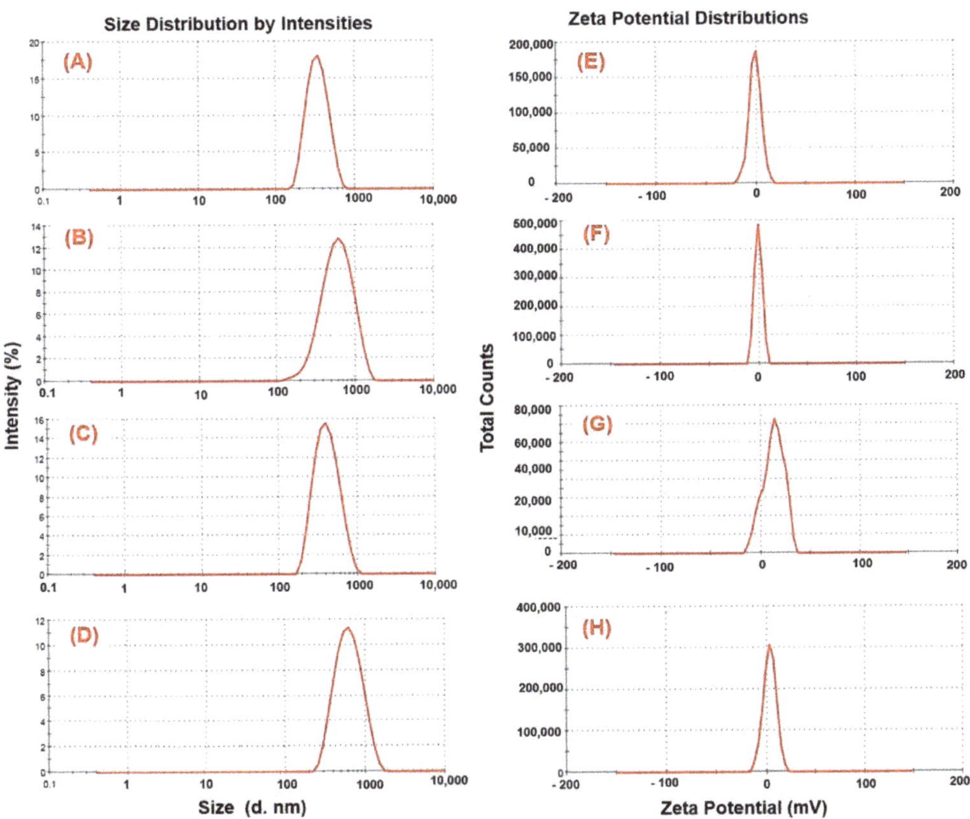

Figure 1. Particle size and zeta potential distribution curves of PLGA-NPs prepared with: (**A**) 1% PVA and chloroform; (**B**) 1% PVA and DCM (**C**) 1% PVP with chloroform (**D**) 1% PVP with DCM and (**E–H**) are their zeta potentials, respectively.

As for the type of organic solvent, we observed that using DCM as the organic phase resulted in comparatively good-sized particles as compared to chloroform when NPs were prepared with the PVA at 1%, 3%, 6% and 9%, *w/v* concentrations. DCM produced uniform and smaller PLGA-NPs than chloroform (stabilized by PVA) which might be due to the comparatively low solubility/miscibility of DCM with water. On the other hand, when chloroform was used as an organic phase, it produced larger NPs. This contrasts with the previous work, where they reported that relatively smaller NPs were produced when they employed vitamin-E-TPGS as the stabilizer [22,27,40,77]. This can be explained by the degree of stabilization that occurs either from the stabilizer itself, or other formulation parameters (concentration of PLGA in organic phase, speed of magnetic stirring and homogenization and sonication power, etc.) [55,57,58]. While PLGA-NPs with larger particles sizes are generally unwanted, they still might have an advantage in cases where the high amount of lipophilic drugs (relative to the amount of PLGA) need to be encapsulated as compared to smaller particles (subject to the solubility of the stabilizer in aqueous phase and its miscibility with the organic phase). However, this should be pursued carefully as the release kinetics of the encapsulated drug(s) may be influenced [22,78,79]. Although both organic solvents have limited solubility in water, the higher solubility of chloroform in the aqueous phase as compared to DCM [80] facilitates its rapid diffusion from oil droplets towards the outer aqueous phase during the emulsification process, which causes the rapid precipitation of PLGA, consequently obstructing the formation of smaller sized NPs [15].

Overall, PLGA-NPs prepared by the single emulsion solvent evaporation technique produced smaller NPs with a smooth surface and reliable properties when the most commonly used stabilizer (PVA) was employed, even at higher concentrations with DCM as the organic solvent, which is in agreement with previous reports [81,82].

3.3. Effect of the Two Stabilizers on the Morphology of PLGA-NPs

The effect of PVA and PVP at different concentrations on the morphology of PLGA-NPs was observed in two different organic solvents, chloroform and DCM. An obvious effect in the morphological structure of the NPs was found, which can be observed in most of the SEM images as shown in the Figures 2 and 3. At lower concentrations (1% and 3%, *w/v*) of PVA, NPs obtained had a spherical and regular shape with a smooth surface (Figure 2A,B, respectively) with no aggregation. This confirmed the relevance and suitability of the selected formulation parameters to prepare the PLGA-NPs. At higher concentrations (6% and 9%, *w/v*) of PVA, however, NPs tended to aggregate and form irregular shapes (Figure 2C,D, respectively). These aggregates might have been formed due to residual stabilizers, which in turn increase the viscosity of the aqueous phase, requiring extra washing of the NPs during ultracentrifugation [76]. Likewise, we observed a similar outcome when PVP was used as a stabilizer. At 1% and 3% *w/v* concentrations of PVP, we obtained NPs that had a smooth surface and spherical, regular shapes (Figure 3A,B, respectively). Whereas, the aggregated–irregular-shaped particles were found with higher concentrations (6% and 9%, *w/v*) of PVP (Figure 3C,D, respectively). On the other hand, when DCM was used as an organic solvent, the prepared PLGA-NPs with either PVA or PVP as stabilizer produced regular spherical-shaped NPs with smooth surfaces at all the four concentrations employed (Figures 4 and 5). It is noteworthy that the population of NPs were very low but discrete and spherical at all concentrations of PVA with equal amounts of PLGA, which might be due to the strong emulsifying property of PVA even at low concentrations and its good surface stabilizing feature towards PLGA-NPs. The presence of PVA at the interface prevented the coalescence of nano-droplets and the aggregation of NPs, which might be due to reduction in the overall energy of the two phases by PVA [57,83]. This might be attributed to the high HLB value of PVA (>18) than that of PVP (14 to 16) [77,84] and might be due to the low viscosity of PVP as compared to PVA, even at the same concentrations of the two stabilizers. Overall, employing DCM as an organic solvent seems to produce more regular and spherical nanoparticles compared to chloroform, which is advantageous in terms of the shelf-stability of the PLGA-NPs due to the narrow, uniform and unimodal granulometric distribution of the particles [55–58]. Based on these results, NPs prepared with 1% of either PVA or PVP were used for further analysis.

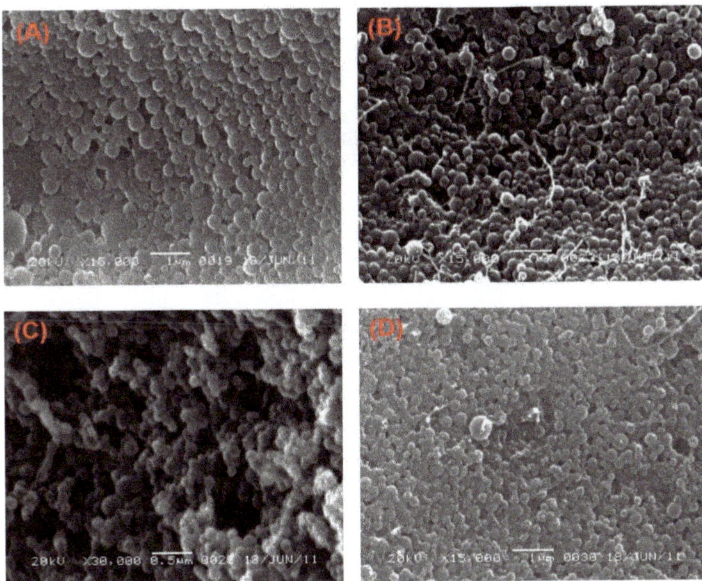

Figure 2. SEM images of the PLGA-NPs prepared with chloroform as organic solvent with different concentrations of PVA in the aqueous phase. (**A**) 1%, *w/v* PVA; (**B**) 3%, *w/v* PVA; (**C**) 6%, *w/v* PVA and (**D**) 9%, *w/v* PVA.

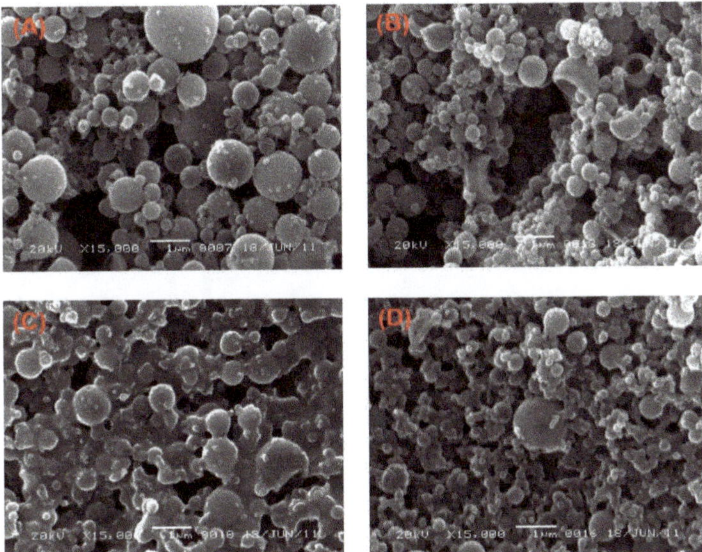

Figure 3. SEM images of the PLGA-NPs prepared with chloroform as organic solvent with different concentrations of PVP in the aqueous phase. (**A**) 1%, *w/v* PVP; (**B**) 3%, *w/v* PVP; (**C**) 6%, *w/v* PVP and (**D**) 9%, *w/v* PVP.

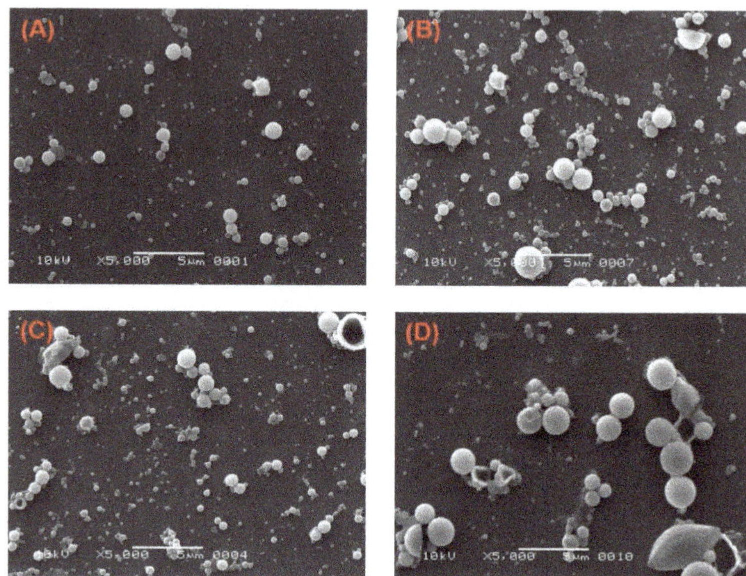

Figure 4. SEM images of the PLGA-NPs prepared with DCM as organic solvent with different concentrations of PVA in the aqueous phase. (**A**) 1%, *w/v* PVA; (**B**) 3%, *w/v* PVA; (**C**) 6%, *w/v* PVA and (**D**) 9%, *w/v* PVA.

Figure 5. SEM images of the PLGA-NPs prepared with DCM as organic solvent with different concentrations of PVP in the aqueous phase. (**A**) 1%, *w/v* PVP; (**B**) 3%, *w/v* PVP; (**C**) 6%, *w/v* PVP and (**D**) 9%, *w/v* PVP.

3.4. Fourier Transform Infrared (FTIR) Analysis

The FTIR spectra were recorded to investigate the possible interaction (if any) between PLGA and IND during the encapsulation of IND into the PLGA-NPs. The FTIR spectra of PLGA, IND, empty PLGA-NPs and IND-PLGA-NPs-5 are presented in Figure 6.

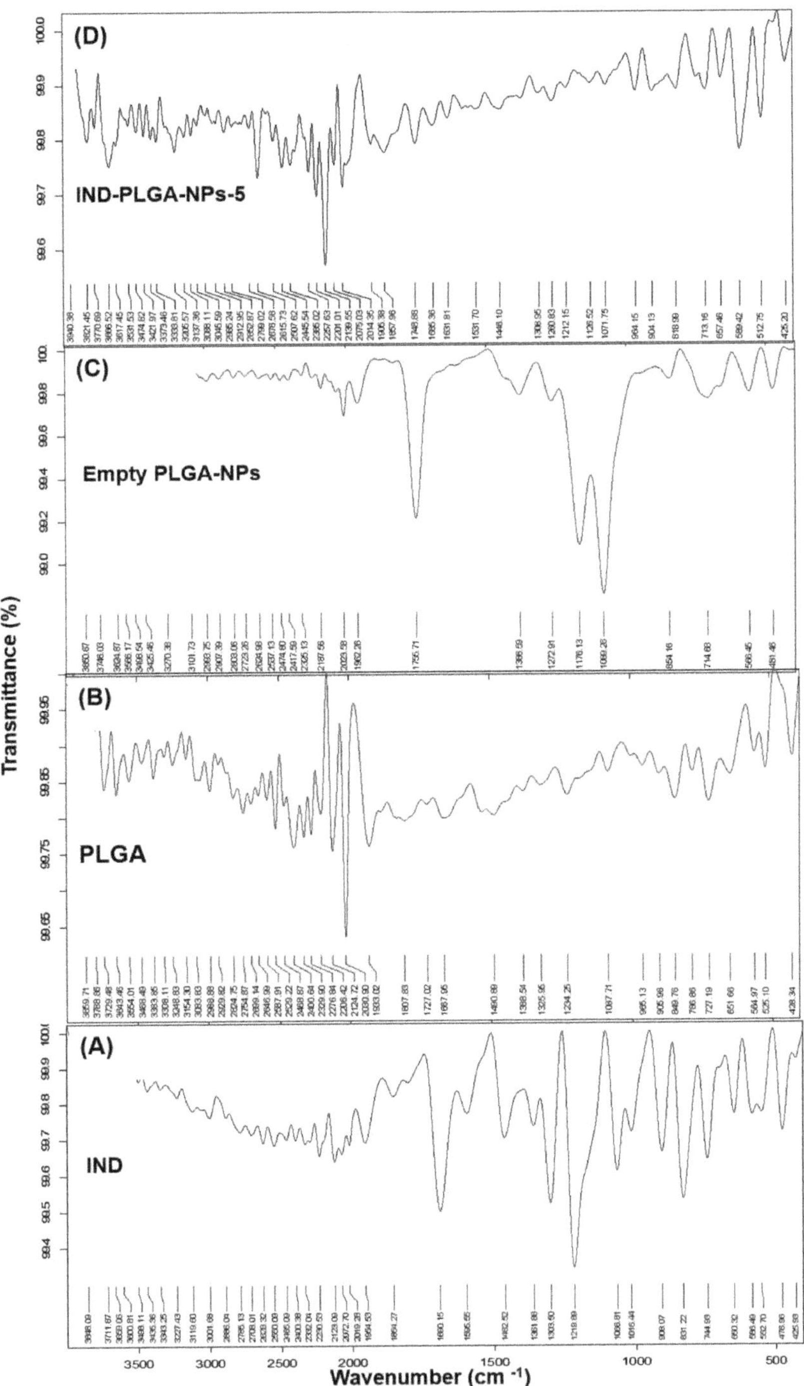

Figure 6. FT-IR spectra of pure IND (**A**), PLGA (**B**), empty PLGA-NPs (**C**) and IND-loaded PLGA-NPs-5 (**D**).

The sample of pure IND demonstrated the characteristic bands at 1585 cm^{-1} (aromatic C=C stretching and 1680 cm^{-1} (carboxyl stretching). The C–Cl stretching at 831 cm^{-1}, C-O stretching ether at 1219 cm^{-1} and 1066 cm^{-1}. Low intensity peaks at 2123 cm^{-1} and 2230 cm^{-1} wavenumbers were possibly indicate the O–H (carboxy) stretching [85]. From pure PLGA the band at 1933 cm^{-1} was the characteristic peak of –C=O– stretching of aliphatic polyesters and also at 2030 cm^{-1} and 2124 cm^{-1} C=O bond stretching was observed. Moreover, the peaks at 727 cm^{-1} and 846 cm^{-1} were due to the C-H bending (vib). The IR scan of empty PLGA-NPs demonstrated characteristic bands at 1089 cm^{-1} and 1176 cm^{-1} which were due to the C-O stretching (ether) and peak at 1755 cm^{-1} was due to the aromatic C=C stretching. A less intense band appeared at 2023 cm^{-1} might be due to the –C=O– stretching of aliphatic polyesters (characteristic of PLGA) [86–88]. The IND-loaded PLGA-NPs has shown a prominent IR band at 2139 cm^{-1} (–C=O– stretching of aliphatic polyesters and a less prominent band at 2014 cm^{-1}, are characteristic bands for PLGA). The characteristic bands of IND did not appear in the drug-loaded NPs-5, which might suggest the proper encapsulation of IND into core of the NPs. Moreover, from this IR spectral analysis, it was concluded that IND was compatible with the PLGA (main excipient) used to prepare the NPs. However, the broadening and less intense bands appeared in the IR spectrum of IND-PLGA-NPs-5, also suggesting the absence of a chemical interaction between the drug (IND) and the polymer (PLGA).

3.5. Viscosity-Shear Rate Relationship of Different Concentrations of PVA and PVP

In this study, the viscosity versus concentration relationship for PVA and PVP solutions at four different concentrations (1%, 3%, 6% and 9%, w/v) of each formulation, was investigated using Milli-Q water as the solvent [89,90]. An increment in the viscosity of aqueous solutions was noticed which was proportional to the concentration of the stabilizers (Figure 7). The concentration dependence on the viscosity was more pronounced in the case of PVA (Figure 7A) as compared to PVP (Figure 7B) at the same concentrations of both the stabilizer. Each solution showed pseudoplastic behavior, which was due to the concentration dependency of PVA and PVP.

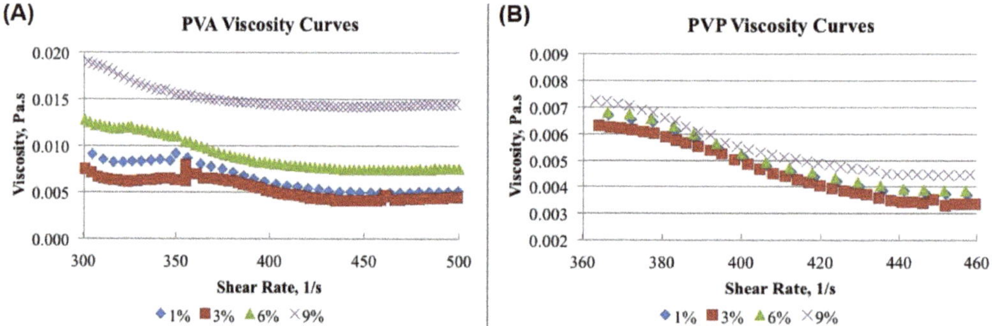

Figure 7. Relationship between viscosity and shear rate in the aqueous solutions of PVA (**A**) and PVP (**B**).

The viscosity (Pa.s) of each solution was higher at the low shear rate (s^{-1}) and the viscosity of aqueous solutions of PVA and PVP increased with increasing their concentrations, exceptionally in case of 3% PVA as compared to 1% PVA. This might be due to the fact that viscosity is highly time and temperature-dependent. The decreased viscosity of 3% PVA solution might be attributed to the unwanted delayed processing in viscosity determination, also the temperature of the solution might be increased [91]. Here, we observed that the viscosity of the PVA–aqueous solutions decreased less rapidly with increasing the shear rate compared to PVP solutions. This indicated that PVA solutions were more viscous than the PVP solutions at same concentrations. Although these are well known phenomena, in this study we tried to correlate the effect of viscosity of the stabilizers on the size and morphology of PLGA-NPs. The measurements of viscosity of polymer(s) manifests the molecular interactions between the polymer (here, PVA and PVP) and the solvents, and from this the degree or magnitude of the interaction could be predicted [41,92]. The relatively smaller sized PLGA-NPs were obtained at lower concentrations of PVA and PVP which may have resulted from the low viscous solution of the stabilizers. Due to the low viscosity of the aqueous phase, their good interaction with the PLGA in organic phase led to the stress-free formation of PLGA-NPs during the organic solvent evaporation [15]. Viscosity of PVP aqueous solutions exhibited Newtonian flow at low shear-rate and lower concentrations, and shear-thinning at high shear-rate and higher concentrations (Figure 7B). The findings related to the effect of shear rate on the viscosity of aqueous solutions of stabilizer could be useful in understanding their effect in the development of PLGA-NPs [93,94].

3.6. Loading of Drug (IND) and In Vitro Release

Considering the facts of the above findings about the empty PLGA-NPs, a few drug-loaded formulations were developed to encapsulate a model lipophilic drug (indomethacin, IND). Where, PVA/PVP (1% and 3% w/v), PLGA (40 mg) and IND (4 mg), DCM and CHCl$_3$ were used as organic solvent(s) to dissolve the polymer and the drug. The developed formulations were further characterized.

3.6.1. Encapsulation Efficiency and Drug Loading Capacity

The encapsulation efficiencies and drug loading capacities of the selected PLGA-NPs of the four batches (Batch-1 to Batch-4) are summarized in Table 2. The physicochemical properties indicated the acceptable size of PLGA-NPs, good polydispersity with high encapsulation of IND at 1% PVA (stabilizer) and DCM as organic solvent. The mean size of PLGA-NPs-5, polydispersity and zeta-potential were 275.4 ± 8.5 nm, -0.157 ± 0.044 and -1.13 ± 0.29 mV respectively, with $72.96 \pm 4.59\%$ encapsulation, $6.63 \pm 0.41\%$ drug loading and highest amount ($52.16 \pm 3.83\%$) of cumulative drug release at 72 h as compared to other formulations. Although the highest encapsulation and drug loading ($74.86 \pm 4.78\%$ and $6.81 \pm 0.45\%$) was found with PLGA-NPs-6 (3% PVA and DCM) but the cumulative release was poor ($37.88 \pm 2.11\%$). Similarly, the size of PLGA-NPs-13 (286.9 ± 13.2 nm) was comparable to PLGA-NPs-5 and slightly high positive surface charge (10.4 ± 3.94 mV) but other parameters including the drug release were not sufficiently good. Overall, at similar preparation conditions the PLGA-NPs-5 showed satisfactory physiochemical characteristics; thus, it was assumed as the best one amongst the eight optimized PLGA-NPs of the four batches.

Table 2. Physicochemical characteristics of some selected formulations after the encapsulation of Indomethacin (IND). Data are presented as Mean $\pm$ SD, $n = 3$.

Formulations	Particle Size (nm)	Polydispersity Index	Zeta Potential (mV)	Encapsulation Efficiency (%)	Drug Loading (%)	Cumulative Drug Release (%) at pH 7.4		Cumulative Drug Release (%) at pH 1.2	
						At 12 h	At 72 h	At 12 h	At 72 h
PVA Group formulations									
Batch-1									
1%, PVA with CHCl$_3$ (PLGA-NPs-1)	331.8 $\pm$ 8.4	0.174 $\pm$ 0.035	-2.89 ± 0.95	68.74 $\pm$ 5.47	6.24 $\pm$ 0.51	16.46 $\pm$ 0.63	43.94 $\pm$ 2.43	41.28 $\pm$ 3.09	42. 16 $\pm$ 2.49
3%, PVA with CHCl$_3$ (PLGA-NPs-2)	547.4 $\pm$ 10.3	0.288 $\pm$ 0.078	-0.81 ± 0.14	69.59 $\pm$ 6.15	6.32 $\pm$ 0.56	13.09 $\pm$ 0.78	32.83 $\pm$ 1.76	30.38 $\pm$ 2.01	32.89 $\pm$ 3.88
Batch-2									
1%, PVA with DCM (PLGA-NPs-5)	275.4 $\pm$ 8.5	0.157 $\pm$ 0.048	-1.13 ± 0.29	72.96 $\pm$ 4.59	6.63 $\pm$ 0.42	18.82 $\pm$ 1.38	52.16 $\pm$ 3.83	31.70 $\pm$ 2.03	40.06 $\pm$ 1.32
3%, PVA with DCM (PLGA-NPs-6)	410.9 $\pm$ 11.3	0.205 $\pm$ 0.061	-1.14 ± 0.58	74.86 $\pm$ 4.78	6.81 $\pm$ 0.45	14.58 $\pm$ 0.81	37.88 $\pm$ 2.11	37.04 $\pm$ 1.07	39. 17 $\pm$ 1.49
PVP Group formulations									
Batch-3									
1%, PVP with CHCl$_3$ (PLGA-NPs-9)	439.3 $\pm$ 15.1	0.179 $\pm$ 0.033	16.1 $\pm$ 3.5	54.94 $\pm$ 6.34	4.99 $\pm$ 0.58	15.76 $\pm$ 0.86	39.90 $\pm$ 2.28	34.05 $\pm$ 1.27	38.14 $\pm$ 1.83
3%, PVP with CHCl$_3$ (PLGA-NPs-10)	379.5 $\pm$ 10.5	0.177 $\pm$ 0.065	15.5 $\pm$ 2.6	58.91 $\pm$ 4.84	5.36 $\pm$ 0.44	16.83 $\pm$ 0.95	43.64 $\pm$ 2.53	26.46 $\pm$ 2.28	32.48 $\pm$ 2.12
Batch-4									
1%, PVP with DCM (PLGA-NPs-13)	286.9 $\pm$ 13.2	0.066 $\pm$ 0.002	10.4 $\pm$ 3.94	63.33 $\pm$ 6.77	5.76 $\pm$ 0.62	19.42 $\pm$ 3.47	48.07 $\pm$ 2.47	33.13 $\pm$ 1.18	35.81 $\pm$ 2.08
3%, PVP with DCM (PLGA-NPs-14)	569.6 $\pm$ 15.6	0.276 $\pm$ 0.055	7.6 $\pm$ 2.37	65.37 $\pm$ 5.36	5.94 $\pm$ 0.49	19.05 $\pm$ 1.08	51.68 $\pm$ 2.99	39.89 $\pm$ 1.86	43.01 $\pm$ 1.17

3.6.2. In Vitro Drug Release and Release Kinetics

A sustained release of IND from all the optimized PLGA-NPs was found at pH 7.4, as shown in Figure 8). The in vitro release profiles of IND from the selected formulations of the four batches (PLGA-NPs-1 with 1% PVA and PLGA-NPs-2 with 3% PVA from Batch-1, PLGA-NPs-5 with 1% PVA and PLGA-NPs-6 with 3% PVA (Figure 8A) from Batch-2, PLGA-NPs-9 with 1% PVP and PLGA-NPs-10 with 3% PVP from Batch-3 and PLGA-NPs-13 with 1% PVP and PLGA-NPs-14 with 3% PVP from Batch-4 (Figure 8B) were summarized in Table 2. For the comparative release assessment, two time points, i.e., 12 h and 72 h were chosen for all the formulations. The cumulative release of the drug from PVA group PLGA-NPs at 12 h were almost same without any significant difference ($p < 0.05$), a highest amount $52.16 \pm 3.83\%$ of drug was released from PLGA-NPs-5 (1% PVA), which was significantly ($p < 0.05$) higher from the rest of the three formulations of PVA group. The use of DCM as an organic solvent to prepare this formulation provided a good encapsulation of IND which might have resulted in its higher and sustained release in PBS with 0.25% SLS.

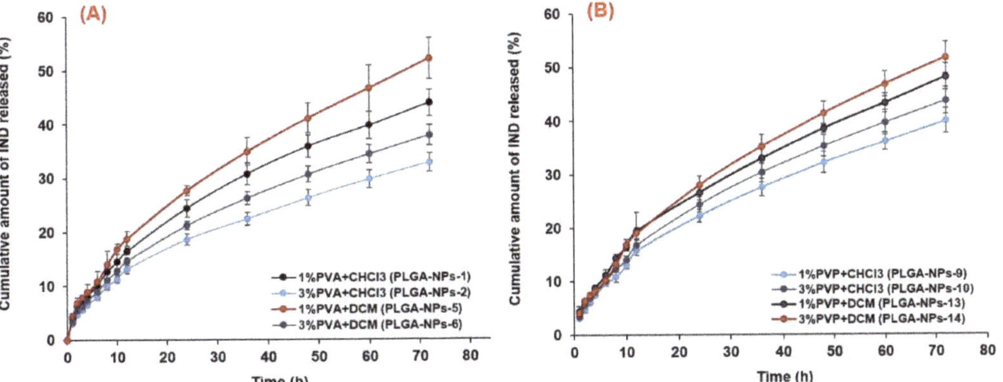

Figure 8. In vitro release profiles of IND from different PLGA-NPs in PBS (pH 7.4). Prepared with different concentrations of stabilizers, (**A**) PVA and (**B**) PVP.

At 12 h, the release of IND from PVP group formulations prepared with $CHCl_3$ were almost same (15–16%) with no significant difference ($p < 0.05$), while a slightly improved release (19.4%) was found with the formulations prepared with DCM. However, the highest amount 51.6% of drug was released from PLGA-NPs-14 (3% PVP), which was slightly higher from the rest of the three formulations of PVP group. The higher concentration of PVP (3%, w/v), provided better stabilization to this formulation, which might be the reason for improved and sustained release of IND in the selected release medium.

The in vitro release profiles of IND in acidic conditions (pH 1.2) from the selected PVA group and PVP group formulations of the four batches are illustrated in Figure 9A, B, respectively, while the cumulative amount of the drug released at 12 h and 72 h is also summarized in Table 2. The highest amount (42.16%) of drug was released at 72 h from PLGA-NPs-1 (1% PVA with $CHCl_3$), slightly less amount (40.06%) was released from PLGA-NPs-5 (1% PVA with DCM) and 39.2% was from PLGA-NPs-6 (3%PVA with DCM) which were significantly ($p < 0.05$) higher from the PLGA-NPs-2 (3%PVA with $CHCl_3$) of PVA group formulations. Similarly, for PVP group formulations, the highest amount (43.16%) of drug was released at 72 h from PLGA-NPs-41 (3% PVP with DCM), slightly less (38.14%) was released from PLGA-NPs-9 (1% PVP with $CHCl_3$) and an almost equivalent amount (32.5% and 35.81%) was released from PLGA-NPs-10 and PLGA-NPs-13, respectively.

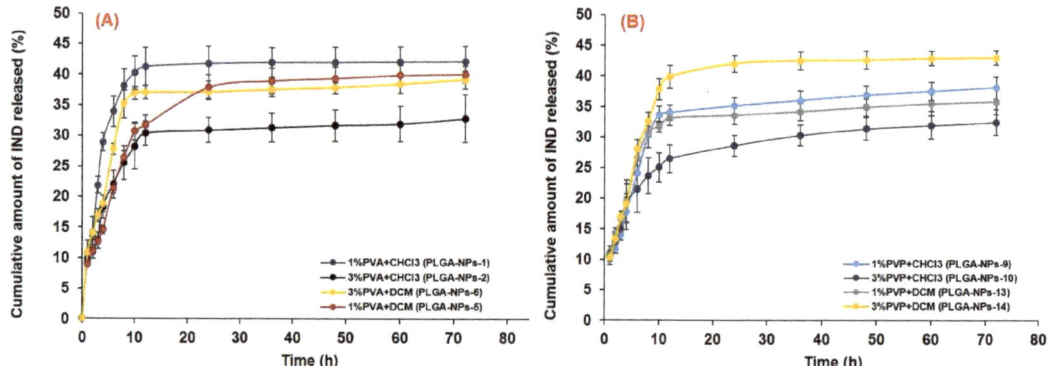

Figure 9. In vitro release profiles of IND from different PLGA-NPs in acidic pH (1.2). Prepared with different concentrations of stabilizers, (**A**) PVA and (**B**) PVP.

Overall, the cumulative amount of drug released from most of the PVA and PVP group PLGA-NPs in acidic release medium followed a biphasic release pattern as illustrated in Figure 9A,B. Initially, a burst release was found from all the formulations followed by a sustained release of the drug, while a sustained release pattern was observed throughout the experiment performed at pH 7.4. Around 21.3–33.9% and 21.4–28.01% of drugs were released within 6 h from PVA and PVP group formulations, respectively. At 12 h, 30.38–41.27% and 26.46–39.89% drugs were release from PVA and PVP group formulations, respectively, while a maximum 42.01% (from PVA group) and 42.61% (from PVP group) were released in 48 h. After 12 h, clearly the prolonged and sustained drug release occurred till 72 h. An initial burst release might be beneficial for the use of IND in terms of acute inflammatory conditions, as it would assist the quick achievement of the therapeutic level, followed by a sustained release to maintain the therapeutic concentration of IND. The initial burst release of a drug might be due to loosely bound or surface-adsorbed drugs and the fast dissolution of IND in an acidic environment (pH 1.2). Secondly the sustained release could be due to the diffusion of the drug across the PLGA matrix, following the drug dissolution and polymer erosion.

The release of the drug at 12 h was very high at an acidic pH as compared to pH 7.4 as shown in Table 2. The release of the drug at acidic and basic conditions are almost equal from PLGA-NPs-1, PLGA-NPs-2 and PLGA-NPs-6 at 72 h study. While from other formulations the drug release was comparatively higher at pH 7.4 at 72 h, which could be attributed to high drug release during the initial hours at acidic condition, might saturated the release even after maintaining the sink condition. Thus, further release from the formulation could not have occurred, so the release in acidic condition was low in some formulations. Moreover, the degradation mechanism of PLGA at pH 7.4 takes place "inside-out" and at acidic pH it occurs as "outside-in". At pH 7.4, in some PLGA-NPs the pore formation and surface pitting might occurred. Whereas, at an acidic pH the PLGA-NPs maintained their smooth surfaces during the process of polymer degradation and rupturing. The rupturing of the PLGA-NPs was accredited to the crystallization of the oligomeric degradation of PLGA due to its low solubility at acidic pH. Moreover, Zolnik and Burgess [95], reported that the degradation of PLGA taken place in an extra homogeneous pattern at acidic pH as compared to pH 7.4. This might be the result of entire NPs experienced a close-to-uniform pH at acidic condition. However, at pH 7.4, the local micro environmental pH within the PLGA-NPs was varied significantly due to accumulation of acid oligomers. Thus, a heterogeneous degradation resulted in the random creation of pores within PLGA-NPs degraded at pH 7.4 which was not occurred during their degradation acidic pH.

The sustained release of IND from the PLGA-NPs was obtained from initial to 72 h, when PBS (pH 7.4) with 0.25% (v/v) SLS was used as release medium. Thus, only the release data obtained from this experiment was subjected to different kinetic models to check the mechanism of drug release from PLGA-NPs. The plots obtained in the applied models were represented in the supplementary data (Figures S1–S4). The calculated values of kinetic parameters including the rate constants of different reactions (k_0 for zero-order, k_1 for first order, k_{HM} for Higuchi-Matrix, k_{KP} for Korsmeyer–Peppas and K_{HC} for Hixson-Crowell models) were summarized in Table 3. Based on the best fit model with highest correlation coefficient (R^2) value, it was found that the selected NPs from the four batches primarily followed the Higuchi matrix release model, indicating the diffusion-controlled drug release from the PLGA-NPs, which was further substantiated by the mathematical calculation of the release-exponents (Table 3). The Higuchi-Matrix mathematical model in this investigation recommended that the initial drug concentration in the PLGA-NPs was much higher as compared to the solubilization of PLGA, even after uninterrupted maintenance of the sink condition throughout the experiment. Moreover, the diffusion of drug (IND) remained persistent and the swelling of polymer-matrix of the NPs was not significant, which was also suggested in some previous reports [49,96].

Table 3. Fitting of the in vitro release data in different release kinetic models.

Formulations	Release Kinetic Models														
	Zero Order			First Order			Higuchi Matrix			Korsmeyer–Peppas			Hixon–Crowell		
	R^2	$*k_0$ (10^{-2})	n-Value	R^2	$*k_1$ (10^{-1})	n-Value	R^2	$*k_{HM}$ (10^{-2})	n-Value	R^2	$*k_{KP}$ (10^{-1})	n-Value	R^2	$*k_{HC}$ (10^{-0})	n-Value
PLGA-NPs-1	0.966	1.66	0.0024	0.985	9.31	0.0014	0.998	4.64	0.2366	0.996	1.78	0.2461	0.979	1.38	0.0009
PLGA-NPs-2	0.964	1.28	0.0018	0.979	9.35	0.0009	0.999	3.54	0.0173	0.998	1.35	0.2412	0.974	1.39	0.0006
PLGA-NPs-5	0.968	1.85	0.0029	0.989	9.28	0.0018	0.997	5.26	0.0282	0.994	2.08	0.2563	0.984	1.44	0.0012
PLGA-NPs-6	0.965	1.44	0.0021	0.982	9.33	0.0012	0.999	4.02	0.0203	0.998	1.58	0.2462	0.976	1.42	0.0008
PLGA-NPs-9	0.963	1.43	0.0022	0.982	9.33	0.0013	0.998	4.12	0.0216	0.997	1.65	0.2579	0.976	1.27	0.0009
PLGA-NPs-10	0.964	1.58	0.0024	0.984	9.32	0.0014	0.841	4.51	0.0236	0.998	1.81	0.2562	0.978	1.30	0.0009
PLGA-NPs-13	0.936	1.81	0.0077	0.958	9.29	0.0058	0.991	5.07	0.0822	0.990	2.01	0.0947	0.951	1.34	0.0037
PLGA-NPs-14	0.967	1.76	0.0029	0.959	9.30	0.0022	0.997	5.11	0.0284	0.994	2.02	0.2649	0.983	1.36	0.0012

$*k$ = Rate constants (k_0 for zero-order, k_1 for first-order, k_{HM} for Higuchi-Matrix, k_{KP} for Korsmeyer-Peppas and K_{HC} for Hixson-Crowell model) and n-value = Diffusion exponent.

Apart from the Higuchi-Matrix model, the second-best fit model was the Korsmeyer–Peppas with second highest R^2-values (Table 3). Where, linearity in the release curves of IND till 72 h (Log fraction drug released versus log time) with the higher R^2 values were found (Supplementary Data, Figures S1–S4). This also indicated the sustained release feature of the PLGA-NPs. The diffusion-exponent values were fallen between 0 and 0.5 for all the formulations of four batches (Table 3), indicated a Fickian diffusion mechanism of IND release. Among the applied release kinetic models, for all the PLGA-NPs from the four batches, the Higuchi-matrix model was appeared as the best fitted one followed by Korsmeyer–Peppas model. This model described the progressive loss of PLGA-NPs with time which was due to polymer erosion, and the release of IND from PLGA-NPs was occurred due to the diffusion of release medium into the matrix of PLGA. Concurrently, swelling and erosion of polymer matrix occurred, which might be the reason of sustained release of the encapsulated IND in the present investigation, which was also reported in previous report [97].

For the preferred PLGA-NPs-5, the R^2 and n values were 0.999 and 0.0282 (Higuchi matrix model); 0.994 and 0.2563 (Korsmeyer–Peppas model) as shown in Table 3. In both the model, though the n-values were lower than 0.5, yet, these values still indicated the diffusion-controlled release mechanism for IND from the PLGA-NPs [49,98]. As, the n-values were <0.45 in the present case of IND-loaded spherical PLGA-NPs, a Fickian-diffusion release mechanism was directed [97,99]. The non-linear curve fitting of in vitro release data obtained for PLGA-NPs-5, using the Higuchi matrix model (primarily) and

Korsmeyer–Peppas model (secondary), shows that the release constant (k, indicating the rate of drug release) was directly proportional to the diffusion exponent. Therefore, the drug release was depended on the structural and physical properties of the IND and the PLGA-matrix. The highest values of "k" (k_{HP}, 2.08×10^{-1} and k_{HM}, 5.26×10^{-2} h^{-1}) were found with PLGA-NPs-5 (Table 3). These findings were in agreement with the previous reports for the release of vancomycin from chitosan-alginate microparticles [100] and from the microporous calcium phosphate ceramics [101].

3.7. DSC Analysis

Differential Scanning Calorimetry was conducted on the optimized lyophilized NPs (PLGA-NPs-5) to determine any possible complexation of the polymer (PLGA) with the stabilizers (PVA and PVP), and to check the crystallinity of the encapsulated model drug (IND). The overlay scans of DSC were represented in Figure 10. A sharp endothermic peak at 157 °C for IND was appeared as shown in Figure 10A, indicates the crystalline characteristic of the drug. The reported melting point of indomethacin is 158 °C, a slight variation in the melting temperature indicated the purity of the drug. The endothermic peaks of PLGA (Figure 10B), PVA (Figure 10C) and PVP (Figure 10D) appeared at 60 °C, 202 °C and 152 °C respectively. The peak of IND was almost diminished in the PLGA-NPs suggesting that the drug was encapsulated into the core of the NPs. The appearance of a very low intensity melting peak (at around 158 °C) in case of PLGA-NPs-5 (Figure 10E), might be due to the presence of surface adsorbed drug in the crystalline state.

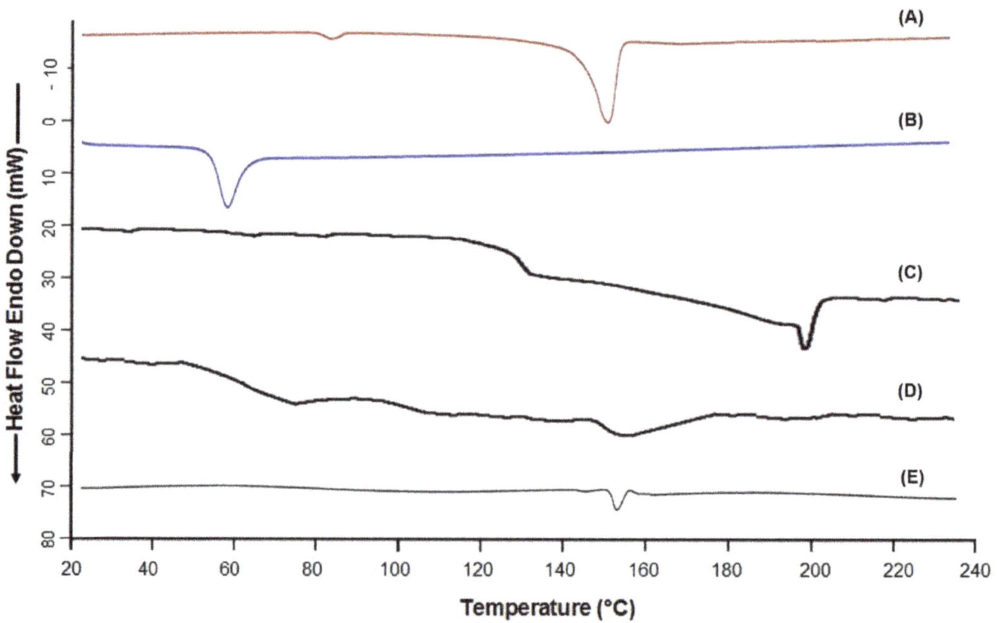

Figure 10. Overlay DSC thermograms of Indomethacin. Indomethacin, IND alone (**A**); PLGA alone (**B**); PVA alone (**C**); PVP alone (**D**) and PLGA-NPs-5 from Batch-2 (**E**).

Although, no endothermic peak of stabilizer was noticed in the DSC scan of the lyophilized PLGA-NPs, indicating that the stabilizer was completely washed out during the purification steps by ultracentrifugation or the traces of the stabilizer (if remained), was not in the crystalline state. Furthermore, non-appearance of any new peaks throughout the scanning of the NPs, indicating the good compatibility among all the excipients used to develop the NPs.

3.8. Stability Studies

The results of the stability study (Table 4) indicate that the freeze-dried PLGA-NPs-5 were stable at different temperatures for 1 month. Almost no changes in the mentioned characteristic parameters for the PLGA-NPs were found stored at 4 °C. A slight variation in particle sizes were noted in the samples store at 37 °C, and such changes were non-significant. No obvious change in the particle size was associated with the stabilizer's protection effect on the PLGA-NPs shell, which was also reported in the previous short term stability on drug loaded PLGA-NPs [28,52].

Table 4. Storage stability of IND-loaded PLGA-NPs (IND-PLGA-NPs-5). Results are represented as mean ± SD of triple measurements.

Characterization Parameters	Storage Time (Days)		
	Initial (0 Day)	At 15th Day	At 30th Day
At 4 °C			
Particle Size (nm)	275.4 ± 8.5	275.8 ± 8.67	276.9 ± 7.81
Polydispersity-index	0.157 ± 0.048	0.158 ± 0.051	0.163 ± 0.054
Zeta-Potential (mV)	−1.13 ± 0.29	−1.12 ± 0.27	−1.11 ± 0.28
Encapsulation efficiency (%)	72.96 ± 4.59	72.54 ± 4.51	71.02 ± 4.63
Drug loading (%)	6.63 ± 0.42	6.50 ± 0.43	6.46 ± 0.45
At 30 °C			
Particle Size (nm)	275.4 ± 8.5	277.2 ± 8.65	279.0 ± 8.81
Polydispersity-index	0.157 ± 0.048	0.158 ± 0.049	0.159 ± 0.047
Zeta-Potential (mV)	−1.13 ± 0.29	−1.12 ± 0.27	−1.11 ± 0.31
Encapsulation efficiency (%)	72.96 ± 4.59	71.87 ± 4.54	71.0 ± 4.64
Drug loading (%)	6.63 ± 0.42	6.53 ± 0.41	6.46 ± 0.42
At 37 °C			
Particle Size (nm)	275.4 ± 8.5	277.8 ± 9.36	279.8 ± 6.95
Polydispersity-index	0.157 ± 0.048	0.159 ± 0.049	0.161 ± 0.048
Zeta-Potential (mV)	−1.13 ± 0.29	−1.12 ± 0.28	−1.11 ± 0.29
Encapsulation efficiency (%)	72.96 ± 4.59	71.38 ± 4.72	70.28 ± 4.22
Drug loading (%)	6.63 ± 0.42	6.44 ± 0.39	6.42 ± 0.41

3.9. Cytotoxicity Study by MTT Assay

Drug encapsulation into PLGA-NPs provides a good solution to minimize the potential toxicity while retaining the therapeutic efficacy of numerous loaded drugs including chemotherapeutics, analgesics and NSAIDs [102]. Here, we sought to encapsulate IND into PLGA-NPs to reduce the toxicities associated with its free delivery and due to the organic solvents used in its preparation, by evaluating the cell survival rate during MTT assay. The cytotoxicity of IND, empty PLGA-NPs and IND-loaded NPs as well as PLGA-only were evaluated after an incubation with the HepG2 cells for 24 h, 48 h and 72 h. The results for IC_{50} were represented in Figure 11, which indicated a decrease in the cell viabilities with increasing concentrations of the tests.

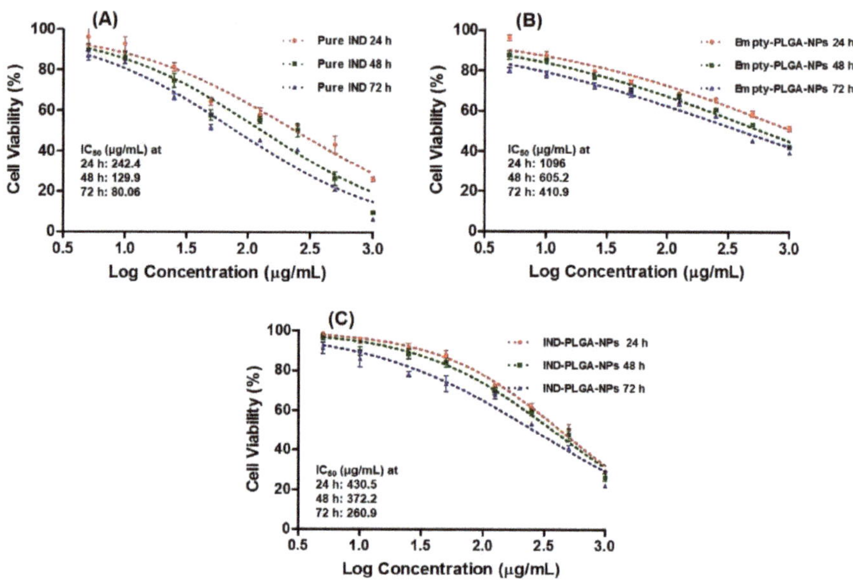

Figure 11. Cytotoxicity after 24 h, 48 h and 72 h of incubation in HepG2 cells: pure IND (**A**); empty-PLGA-NPs (**B**) and IND-PLGA-NPs-5 (**C**) with IC_{50} values (μg/mL).

The histograms of the percentage cell viability against varying concentrations of IND and the NPs were represented in Figure 12. As shown in Figure 12, cell proliferation of IND was examined at 5–1000 μg/mL concentrations, where the observed IC_{50} were 242.4, 129.9 and 80.06 μg/mL at 24 h, 48 h and 72 h, respectively. A report of Xiong et al., 2013, indicated that the PLGA-NPs was not significantly toxic up to 300 μg/mL against RAW264.7 and BEAS-2B cells [103]. Likewise, in the present investigation, we found that the IND-loaded PLGA-NPs-5, has shown increased cell viability at an equivalent drug concentration (5–1000 μg/mL) and the observed IC_{50} were 430.5, 372.2 and 260.9 μg/mL at 24 h, 48 h and 72 h, respectively. Moreover, the cell proliferation assay conducted using empty-PLGA-NPs at equivalent concentration of IND (i.e., the amount of empty-PLGA-NPs those containing the same amount of IND in the drug-loaded NPs) has shown IC_{50} values of 1096, 605.2 and 410.9 μg/mL at 24 h, 48 h and 72 h, respectively. Thus, even at these high concentrations of PLGA, the cells were viable at IC_{50} of 410.9 μg/mL when the duration of treatment was prolonged (72 h). This might be due to the time-dependent effect of PLGA-NPs on the cytotoxicity.

The MTT assay results clearly demonstrating the reduction in toxicity of IND by 66% in case of formulation (IND-PLGANPs) and empty PLGA-NPs with IC_{50} of 7142.7 μg/mL. Thus, we can assume that even at this very high concentration of PLGA (empty PLGA-NPs/IND-PLGA-NPs-5), the product was non-toxic as even at these high concentrations of PLGA, the cells were viable and the IC_{50} was significantly reduced as the duration of treatment was prolonged (i.e., 410.9 μg/mL at 72 h), indicating the time-dependent effect of PLGA-NPs on the cytotoxicity.

At high concentrations the PLGA-NPs was toxic, this might be associated with the physicochemical properties including the particle size, zeta-potential and selective targeting of the PLGA-NPs. These play an important role in terms of its efficacy in biological systems. In previous reports, it was found that the increasing the particle sizes, the IC_{50} of cytotoxicity was also increased, which suggests that small sized NPs encouraged higher cytotoxicity at low IC_{50} values [104,105]. The cytotoxicity results show that PLGA-NPs induced an anti-proliferative effect at an average particle size of 189 nm, while in the present investigation the size of the tested NPs was 275.4 nm, this might be one reason for its toxic nature at some concentrations. Even though our formulations did not exhibit any potential toxicity of the PLGA-NPs against HepG2 cells, might be due to its non-targeting and high zeta-potential properties [103–105].

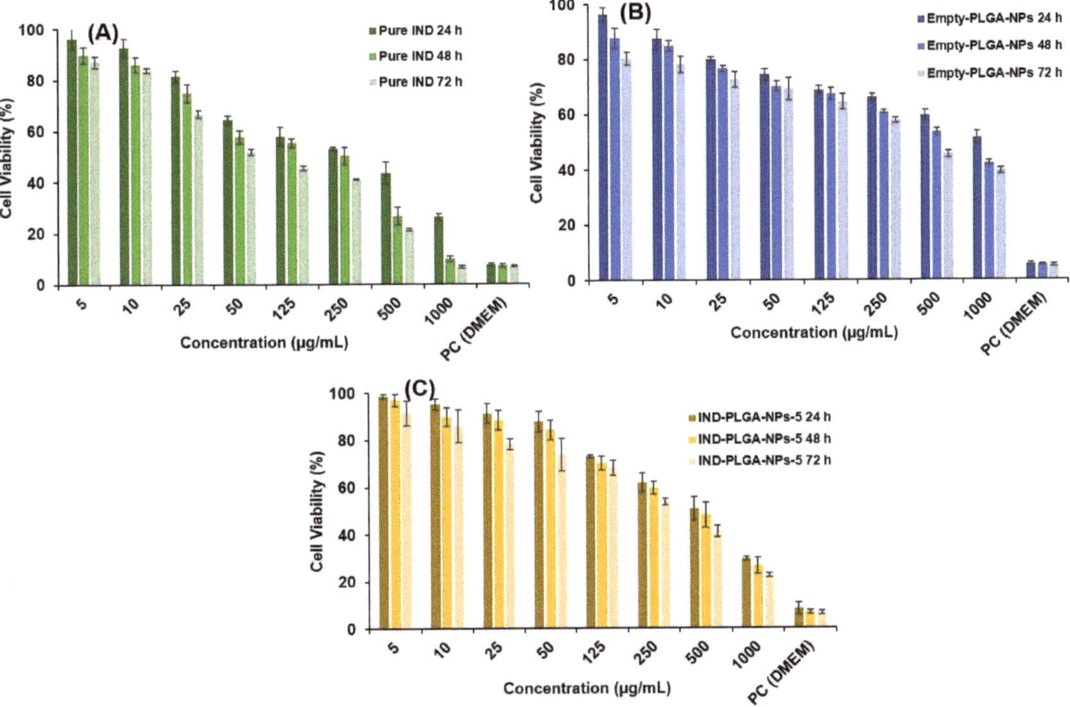

Figure 12. Histograms of the percentage cell viability of HepG2 after 24 h, 48 h and 72 h of incubation with varying concentrations (5–1000 µg/mL) of pure IND (**A**) and its equivalent concentration of empty-PLGA-NPs (**B**) and IND-loaded PLGA-NPs (**C**). Data were presented as the mean of three independent experiments with error bars as standard deviation (SD).

The HepG2 are liver-derived human cell lines those display abundant parenchymal cell functions containing the metabolism of polycyclic aromatic hydrocarbons. The observed proliferation rate of HepG2 (i.e., doubling time) was measured, and it was found that these cells were doubled after 48.95 h at 5×10^5 cells/well, cell density of HepG2 cells. The highest live cells were observed at 72 h, and these results were similar to Miura et al., 1999, who seeded 4×10^5 cells/well in 6-well plates, and they believed that this was a more appropriate cell number for 6-well plates [106]. The doubling time of HepG2 cells was 48.95 h in the present investigation, which was consistent with other studies, where the doubling time of HepG2 was around 2.05 days (49.2 h) [107–109]. The growth rate at 24 h, 48 h, and 72 h were 53%, 52%, and 49%, respectively, and there was no major change in the growth rate with cell viability of more than 94.27% at 72 h. Thereby, no significant change in doubling time was found.

Overall, the PLGA-NPs in the present investigation did not show any particular toxicity at most of the concentrations and could not influence the normal growth of HepG2 cells within the IND dose range (250–500 µg/mL) and at equivalent amount of PLGA-NPs (3571.4–7142.7 µg/mL). Conclusively, an ideal size, zeta-potential, dose, duration of treatment and targeting ability of PLGA-NPs to deliver the encapsulated drugs at desired site are essential to develop an efficient and safe drug carrier.

A significantly ($p < 0.05$) high IC_{50} value for the PLGA-NPs indicated that the FDA-approved polymer used to prepare the NPs was non-toxic. The above findings also indicated that the complete evaporation of the organic solvent (DCM) used to prepare the NPs. Moreoverthe cytotoxicity of hepatotoxic drug (IND) can be reduced by its encapsulation into PLGA-NPs, which was in agreement with previous reports [102,110].

4. Conclusions

We have acknowledged some process variables that are decisive to obtain proper sized, uniformly distributed PLGA-NPs with a smooth surface and a solid dense structure discrete. This study has shed light on the influence of preparation condition on physicochemical properties of PLGA-NPs, following the single-emulsion solvent-evaporation method to prepare the NPs. The PVA at lower concentrations was proved to be a good stabilizer to obtain an optimized PLGA-NPs with excellent size and surface properties, when DCM was used as organic solvent to dissolve the polymer. Controlling the particle size and morphology of the NPs may have a potential impact in tailoring the drug delivery applications of such nano-carriers. The developed and characterized PLGA-based NPs as controlled delivery system in the present investigation was stable and would have good potential applications for the loading of hydrophilic and poorly soluble lipophilic therapeutic agents. A short-term storage physical stability indicated that the optimized PLGA-NPs was stable at different temperatures for 30 days. A prolonged release of the IND from the optimized NPs suggested the sustained and controlled release property of PLGA. A significantly ($p < 0.05$) high IC_{50} value for the NPs during the cytotoxicity study, indicated that the developed PLGA-NPs was non-toxic. Moreover, the adverse effect associated with the cytotoxic drugs such as IND (a hepatotoxic drug) could be overcome by their encapsulation into the PLGA-NPs. Nonetheless, further studies are needed to substantiate the current outcomes, which we are currently conducting in our lab that may further extend the utility of PLGA-NPs.

Supplementary Materials: The following supporting information can be downloaded at: https://www.mdpi.com/article/10.3390/pharmaceutics14040870/s1, Figure S1: Different kinetic model plots for PLGA-NPs-1 (prepared with 1%, *w/v* PVA as stabilizer and $CHCl_3$ as organic solvent) (A) and PLGA-NPs-2 (prepared with 3%, *w/v* PVA as stabilizer and $CHCl_3$ as organic solvent (B); Figure S2: Different kinetic model plots for PLGA-NPs-5 (prepared with 1%, *w/v* PVA as stabilizer and DCM as organic solvent) (A) and PLGA-NPs-6 (prepared with 3%, *w/v* PVA as stabilizer and DCM as organic solvent (B); Figure S3: Different kinetic model plots for PLGA-NPs-9 (prepared with 1%, *w/v* PVP as stabilizer and $CHCl_3$ as organic solvent) (A) and PLGA-NPs-10 (prepared with 3%, *w/v* PVP as stabilizer and $CHCl_3$ as organic solvent (B); Figure S4: Different kinetic model plots for PLGA-NPs-13 (prepared with 1%, *w/v* PVP as stabilizer and DCM as organic solvent) (A) and PLGA-NPs-14 (prepared with 3%, *w/v* PVP as stabilizer and DCM as organic solvent (B).

Author Contributions: Conceptualization, M.A.K.; Formal analysis, M.A.; Funding acquisition, A.A.; Investigation, M.A.K.; Methodology, M.A., M.A.K. and M.K.A.; Project administration, A.A.; Resources, M.A., M.A.K. and M.K.A.; Software, M.A.K.; Supervision, A.A.; Validation, M.A.; Visualization, M.A.; Writing—original draft, M.A.K. and M.K.A.; Writing—review and editing, M.A. and A.A. All authors have read and agreed to the published version of the manuscript.

Funding: This research received no external funding.

Institutional Review Board Statement: Not applicable.

Informed Consent Statement: Not applicable.

Data Availability Statement: The data presented in this study are available on request from the corresponding author.

Acknowledgments: The authors are very grateful to the Deanship of Scientific Research and Research Centre, College of Pharmacy, King Saud University, Riyadh, Saudi Arabia for their funding and support.

Conflicts of Interest: The authors declare no any conflict of interest.

References

1. Patra, J.K.; Das, G.; Fraceto, L.F.; Campos, E.V.R.; del Pilar Rodriguez-Torres, M.; Acosta-Torres, L.S.; Diaz-Torres, L.A.; Grillo, R.; Swamy, M.K.; Sharma, S.; et al. Nano based drug delivery systems: Recent developmentsi and future prospects. *J. Nanobiotechnol.* **2018**, *16*, 71. [CrossRef] [PubMed]
2. Sung, Y.K.; Kim, S.W. Recent advances in polymeric drug delivery systems. *Biomater. Res.* **2020**, *24*, 12. [CrossRef] [PubMed]
3. Wu, Q.-X.; Lin, D.-Q.; Yao, S.-J. Design of Chitosan and Its Water Soluble Derivatives-Based Drug Carriers with Polyelectrolyte Complexes. *Mar. Drugs* **2014**, *12*, 6236–6253. [CrossRef] [PubMed]
4. Martinho, N.; Damgé, C.; Reis, C.P. Recent Advances in Drug Delivery Systems. *J. Biomater. Nanobiotechnol.* **2011**, *2*, 510–526. [CrossRef]
5. Dinarvand, R.; Sepehri, N.; Manouchehri, S.; Rouhani, H.; Atyabi, F. Polylactide-*co*-glycolide nanoparticles for controlled delivery of anticancer agents. *Int. J. Nanomed.* **2011**, *6*, 877–895. [CrossRef]
6. Astete, C.E.; Sabliov, C.M. Synthesis and characterization of PLGA nanoparticles. *J. Biomater. Sci. Polym. Ed.* **2006**, *17*, 247–289. [CrossRef]
7. Jain, R.A. The manufacturing techniques of various drug loaded biodegradable poly(lactide-*co*-glycolide) (PLGA) devices. *Biomaterials* **2000**, *21*, 2475–2490. [CrossRef]
8. Makadia, H.K.; Siegel, S.J. Poly lactic-co-glycolic acid (PLGA) As biodegradable controlled drug delivery carrier. *Polymers* **2011**, *3*, 1377–1397. [CrossRef]
9. Tulla-Puche, J.; Albericio, F. Polymers and drug delivery systems. *Curr. Drug Deliv.* **2012**, *9*, 367–394. [CrossRef]
10. Fathi-Achachelouei, M.; Keskin, D.; Bat, E.; Vrana, N.E.; Tezcaner, A. Dual growth factor delivery using PLGA nanoparticles in silk fibroin/PEGDMA hydrogels for articular cartilage tissue engineering. *J. Biomed. Mater. Res. Part B Appl. Biomater.* **2020**, *108*, 2041–2062. [CrossRef]
11. Gu, P.; Wusiman, A.; Zhang, Y.; Liu, Z.; Bo, R.; Hu, Y.; Liu, J.; Wang, D. Rational Design of PLGA Nanoparticle Vaccine Delivery Systems To Improve Immune Responses. *Mol. Pharm.* **2019**, *16*, 5000–5012. [CrossRef] [PubMed]
12. Liang, G.F.; Zhu, Y.L.; Sun, B.; Hu, F.H.; Tian, T.; Li, S.C.; Xiao, Z.D. PLGA-based gene delivering nanoparticle enhance suppression effect of miRNA in HePG2 cells. *Nanoscale Res. Lett.* **2011**, *6*, 447. [CrossRef] [PubMed]
13. Ramezani, M.; Ebrahimian, M.; Hashemi, M. Current Strategies in the Modification of PLGA-based Gene Delivery System. *Curr. Med. Chem.* **2017**, *24*, 728–739. [CrossRef]
14. Petrizzo, A.; Conte, C.; Tagliamonte, M.; Napolitano, M.; Bifulco, K.; Carriero, V.; De Stradis, A.; Tornesello, M.L.; Buonaguro, F.M.; Quaglia, F.; et al. Functional characterization of biodegradable nanoparticles as antigen delivery system. *J. Exp. Clin. Cancer Res.* **2015**, *34*, 114. [CrossRef]
15. Cegnar, M.; Kos, J.; Kristl, J. Cystatin incorporated in poly(lactide-*co*-glycolide) nanoparticles: Development and fundamental studies on preservation of its activity. *Eur. J. Pharm. Sci.* **2004**, *22*, 357–364. [CrossRef]
16. Bisht, R.; Rupenthal, I.D. PLGA nanoparticles for intravitreal peptide delivery: Statistical optimization, characterization and toxicity evaluation. *Pharm. Dev. Technol.* **2018**, *23*, 324–333. [CrossRef]
17. Mundargi, R.C.; Babu, V.R.; Rangaswamy, V.; Patel, P.; Aminabhavi, T.M. Nano/micro technologies for delivering macromolecular therapeutics using poly(d,l-lactide-*co*-glycolide) and its derivatives. *J. Control. Release* **2008**, *125*, 193–209. [CrossRef]
18. Anderson, J.M.; Shive, M.S. Biodegradation and biocompatibility of PLA and PLGA microspheres. *Adv. Drug Deliv. Rev.* **1997**, *28*, 5–24. [CrossRef]
19. Freitas, S.; Merkle, H.P.; Gander, B. Microencapsulation by solvent extraction/evaporation: Reviewing the state of the art of microsphere preparation process technology. *J. Control. Release* **2005**, *102*, 313–332. [CrossRef]
20. Martins, C.; Sousa, F.; Araújo, F.; Sarmento, B. Functionalizing PLGA and PLGA Derivatives for Drug Delivery and Tissue Regeneration Applications. *Adv. Health Mater.* **2018**, *7*, 1701035. [CrossRef]
21. Bala, I.; Hariharan, S.; Kumar, M.N.V.R. PLGA Nanoparticles in Drug Delivery: The State of the Art. *Crit. Rev. Ther. Drug Carr. Syst.* **2004**, *21*, 387–422. [CrossRef] [PubMed]
22. McCall, R.L.; Sirianni, R.W. PLGA nanoparticles formed by single- or double-emulsion with vitamin E-TPGS. *J. Vis. Exp.* **2013**, *82*, e51015. [CrossRef] [PubMed]
23. Park, K.; Skidmore, S.; Hadar, J.; Garner, J.; Park, H.; Otte, A.; Soh, B.K.; Yoon, G.; Yu, D.; Yun, Y.; et al. Injectable, long-acting PLGA formulations: Analyzing PLGA and understanding microparticle formation. *J. Control. Release* **2019**, *304*, 125–134. [CrossRef] [PubMed]
24. Uskoković, D.; Stevanovic, M. Poly(lactide-*co*-glycolide)-based Micro and Nanoparticles for the Controlled Drug Delivery of Vitamins. *Curr. Nanosci.* **2009**, *5*, 1–14. [CrossRef]
25. Alshamsan, A. Nanoprecipitation is more efficient than emulsion solvent evaporation method to encapsulate cucurbitacin I in PLGA nanoparticles. *Saudi Pharm. J.* **2014**, *22*, 219–222. [CrossRef]
26. McCarron, P.A.; Donnelly, R.F.; Marouf, W. Celecoxib-loaded poly(D,L-lactide-*co*-glycolide) nanoparticles prepared using a novel and controllable combination of diffusion and emulsification steps as part of the salting-out procedure. *J. Microencapsul.* **2006**, *23*, 480–498. [CrossRef]

27. Alkholief, M.; Albasit, H.; Alhowyan, A.; Alshehri, S.; Raish, M.; Kalam, A.; Alshamsan, A. Employing a PLGA-TPGS based nanoparticle to improve the ocular delivery of Acyclovir. *Saudi Pharm. J.* **2018**, *27*, 293–302. [CrossRef]

28. Kalam, A.; Alshamsan, A. Poly (d,l-lactide-*co*-glycolide) nanoparticles for sustained release of tacrolimus in rabbit eyes. *Biomed. Pharmacother.* **2017**, *94*, 402–411. [CrossRef]

29. Beck-Broichsitter, M.; Rytting, E.; Lebhardt, T.; Wang, X.; Kissel, T. Preparation of nanoparticles by solvent displacement for drug delivery: A shift in the "ouzo region" upon drug loading. *Eur. J. Pharm. Sci.* **2010**, *41*, 244–253. [CrossRef]

30. Sahana, D.; Mittal, G.; Bhardwaj, V.; Kumar, M. PLGA Nanoparticles for Oral Delivery of Hydrophobic Drugs: Influence of Organic Solvent on Nanoparticle Formation and Release Behavior In Vitro and In Vivo Using Estradiol as a Model Drug. *J. Pharm. Sci.* **2008**, *97*, 1530–1542. [CrossRef]

31. Caputo, F.; Clogston, J.; Calzolai, L.; Rösslein, M.; Prina-Mello, A. Measuring particle size distribution of nanoparticle enabled medicinal products, the joint view of EUNCL and NCI-NCL. A step by step approach combining orthogonal measurements with increasing complexity. *J. Control. Release* **2019**, *299*, 31–43. [CrossRef]

32. Lucas, S. The Pharmacology of Indomethacin. *Headache J. Head Face Pain* **2016**, *56*, 436–446. [CrossRef]

33. Dannenfelser, R.-M.; Yalkowsky, S.H. Data base of aqueous solubility for organic non-electrolytes. *Sci. Total Environ.* **1991**, *109–110*, 625–628. [CrossRef]

34. Williams, C.M. Using medications appropriately in older adults. *Am. Fam. Physician* **2002**, *66*, 1917.

35. Kalam, M.A.; Khan, A.A.; Khan, S.; Almalik, A.; Alshamsan, A. Optimizing indomethacin-loaded chitosan nanoparticle size, encapsulation, and release using Box–Behnken experimental design. *Int. J. Biol. Macromol.* **2016**, *87*, 329–340. [CrossRef]

36. Boon, V.; Glass, B.; Nimmo, A. High-Performance Liquid Chromatographic Assay of Indomethacin in Porcine Plasma with Applicability to Human Levels. *J. Chromatogr. Sci.* **2006**, *44*, 41–44. [CrossRef]

37. Novakova, L.; Matysová, L.; Havlíková, L.C.; Solich, P. Development and validation of HPLC method for determination of indomethacin and its two degradation products in topical gel. *J. Pharm. Biomed. Anal.* **2005**, *37*, 899–905. [CrossRef]

38. Kızılbey, K. Optimization of Rutin-Loaded PLGA Nanoparticles Synthesized by Single-Emulsion Solvent Evaporation Method. *ACS Omega* **2019**, *4*, 555–562. [CrossRef]

39. Niu, L.; Panyam, J. Freeze concentration-induced PLGA and polystyrene nanoparticle aggregation: Imaging and rational design of lyoprotection. *J. Control. Release* **2017**, *248*, 125–132. [CrossRef]

40. Alhowyan, A.A.; Altamimi, M.A.; Kalam, M.A.; Khan, A.A.; Badran, M.; Binkhathlan, Z.; Alkholief, M.; Alshamsan, A. Antifungal efficacy of Itraconazole loaded PLGA-nanoparticles stabilized by vitamin-E TPGS: In vitro and ex vivo studies. *J. Microbiol. Methods* **2019**, *161*, 87–95. [CrossRef]

41. Guettari, M.; Gomati, R. Influence of polyvinylpyrrolidone on the interaction between water and methanol. *Braz. J. Chem. Eng.* **2013**, *30*, 677–682. [CrossRef]

42. Kalam, M.A.; Raish, M.; Ahmed, A.; Alkharfy, K.M.; Mohsin, K.; Alshamsan, A.; Al-Jenoobi, F.I.; Al-Mohizea, A.M.; Shakeel, F. Oral bioavailability enhancement and hepatoprotective effects of thymoquinone by self-nanoemulsifying drug delivery system. *Mater. Sci. Eng. C Mater. Biol. Appl.* **2017**, *76*, 319–329. [CrossRef] [PubMed]

43. Shakeel, F.; Haq, N.; El-Badry, M.; Alanazi, F.K.; Alsarra, I. Ultra fine super self-nanoemulsifying drug delivery system (SNEDDS) enhanced solubility and dissolution of indomethacin. *J. Mol. Liq.* **2013**, *180*, 89–94. [CrossRef]

44. Ahnfelt, E.; Sjögren, E.; Hansson, P.; Lennernäs, H. In Vitro Release Mechanisms of Doxorubicin From a Clinical Bead Drug-Delivery System. *J. Pharm. Sci.* **2016**, *105*, 3387–3398. [CrossRef]

45. Alkholief, M. Optimization of Lecithin-Chitosan nanoparticles for simultaneous encapsulation of doxorubicin and piperine. *J. Drug Deliv. Sci. Technol.* **2019**, *52*, 204–214. [CrossRef]

46. Malinovskaya, Y.; Melnikov, P.; Baklaushev, V.; Gabashvili, A.; Osipova, N.; Mantrov, S.; Ermolenko, Y.; Maksimenko, O.; Gorshkova, M.; Balabanyan, V.; et al. Delivery of doxorubicin-loaded PLGA nanoparticles into U87 human glioblastoma cells. *Int. J. Pharm.* **2017**, *524*, 77–90. [CrossRef]

47. Chourasiya, V.; Bohrey, S.; Pandey, A. Formulation, optimization, characterization and in-vitro drug release kinetics of atenolol loaded PLGA nanoparticles using 3 3 factorial design for oral delivery. *Mater. Discov.* **2016**, *5*, 1–13. [CrossRef]

48. Dash, S.; Murthy, P.N.; Nath, L.; Chowdhury, P. Kinetic modeling on drug release from controlled drug delivery systems. *Acta Pol. Pharm.* **2010**, *67*, 217–223.

49. Kalam, M.A.; Humayun, M.; Parvez, N.; Yadav, S.; Garg, A.; Amin, S.; Sultana, Y.; Ali, A. Release kinetics of modified pharmaceutical dosage forms: A review. *Cont. J. Pharm. Sci.* **2007**, *1*, 30–35.

50. Wei, X.; Sun, N.; Wu, B.; Yin, C.; Wu, W. Sigmoidal release of indomethacin from pectin matrix tablets: Effect of in situ crosslinking by calcium cations. *Int. J. Pharm.* **2006**, *318*, 132–138. [CrossRef]

51. Altamimi, M.A.; Neau, S.H. Use of the Flory–Huggins theory to predict the solubility of nifedipine and sulfamethoxazole in the triblock, graft copolymer Soluplus. *Drug Dev. Ind. Pharm.* **2016**, *42*, 446–455. [CrossRef] [PubMed]

52. Li, Z.; Tao, W.; Zhang, D.; Wu, C.; Song, B.; Wang, S.; Wang, T.; Hu, M.; Liu, X.; Wang, Y.; et al. The studies of PLGA nanoparticles loading atorvastatin calcium for oral administration in vitro and in vivo. *Asian J. Pharm. Sci.* **2017**, *12*, 285–291. [CrossRef] [PubMed]

53. Raish, M.; Kalam, M.A.; Ahmad, A.; Shahid, M.; Ansari, M.A.; Ahad, A.; Ali, R.; Bin Jardan, Y.A.; Alshamsan, A.; Alkholief, M.; et al. Eudragit-Coated Sporopollenin Exine Microcapsules (SEMC) of *Phoenix dactylifera* L. of 5-Fluorouracil for Colon-Specific Drug Delivery. *Pharmaceutics* **2021**, *13*, 1921. [CrossRef] [PubMed]

54. Alshememry, A.; Kalam, M.A.; Almoghrabi, A.; Alzahrani, A.; Shahid, M.; Khan, A.A.; Haque, A.; Ali, R.; Alkholief, M.; Binkhathlan, Z.; et al. Chitosan-coated poly (lactic-*co*-glycolide) nanoparticles for dual delivery of doxorubicin and naringin against MCF-7 cells. *J. Drug Deliv. Sci. Technol.* **2022**, *68*, 103036. [CrossRef]

55. Mainardes, R.M.; Evangelista, R.C. PLGA nanoparticles containing praziquantel: Effect of formulation variables on size distribution. *Int. J. Pharm.* **2005**, *290*, 137–144. [CrossRef] [PubMed]

56. Murakami, H.; Kobayashi, M.; Takeuchi, H.; Kawashima, Y. Preparation of poly(dl-lactide-*co*-glycolide) nanoparticles by modified spontaneous emulsification solvent diffusion method. *Int. J. Pharm.* **1999**, *187*, 143–152. [CrossRef]

57. Kwon, H.-Y.; Lee, J.-Y.; Choi, S.-W.; Jang, Y.; Kim, J.-H. Preparation of PLGA nanoparticles containing estrogen by emulsification–diffusion method. *Colloids Surfaces A Physicochem. Eng. Asp.* **2001**, *182*, 123–130. [CrossRef]

58. Quintanar-Guerrero, D.; Fessi, H.; Allémann, E.; Doelker, E. Influence of stabilizing agents and preparative variables on the formation of poly (D, L-lactic acid) nanoparticles by an emulsification-diffusion technique. *Int. J. Pharm.* **1996**, *143*, 133–141. [CrossRef]

59. Feczkó, T.; Tóth, J.; Gyenis, J.J.C.; Physicochemical, S.A.; Aspects, E. Comparison of the preparation of PLGA–BSA nano-and microparticles by PVA, poloxamer and PVP. *Colloids Surf. A* **2008**, *319*, 188–195. [CrossRef]

60. Coombes, A.; Yeh, M.-K.; Lavelle, E.; Davis, S. The control of protein release from poly(dl-lactide *co*-glycolide) microparticles by variation of the external aqueous phase surfactant in the water-in oil-in water method. *J. Control. Release* **1998**, *52*, 311–320. [CrossRef]

61. Shakeri, S.; Roghanian, R.; Emtiazi, G.; Errico, C.; Chiellini, F.; Chiellini, E. Preparation of protein-loaded PLGA-PVP blend nanoparticles by nanoprecipitation method: Entrapment, Initial burst and drug release kinetic studies. *Nanomed. J.* **2015**, *2*, 175–186.

62. Esmaeili, F.; Atyabi, F.; Dinarvand, R. Preparation of PLGA nanoparticles using TPGS in the spontaneous emulsification solvent diffusion method. *J. Exp. Nanosci.* **2007**, *2*, 183–192. [CrossRef]

63. Jung, T.; Breitenbach, A.; Kissel, T. Sulfobutylated poly(vinyl alcohol)-graft-poly(lactide-*co*-glycolide)s facilitate the preparation of small negatively charged biodegradable nanospheres. *J. Control. Release* **2000**, *67*, 157–169. [CrossRef]

64. Lamprecht, A.; Ubrich, N.; Yamamoto, H.; Schäfer, U.; Takeuchi, H.; Lehr, C.-M.; Maincent, P.; Kawashima, Y. Design of rolipram-loaded nanoparticles: Comparison of two preparation methods. *J. Control. Release* **2001**, *71*, 297–306. [CrossRef]

65. Javadzadeh, Y.; Ahadi, F.; Davaran, S.; Mohammadi, G.; Sabzevari, A.; Adibkia, K. Preparation and physicochemical characterization of naproxen–PLGA nanoparticles. *Colloids Surf. B Biointerfaces* **2010**, *81*, 498–502. [CrossRef] [PubMed]

66. Auría-Soro, C.; Nesma, T.; Juanes-Velasco, P.; Landeira-Viñuela, A.; Fidalgo-Gomez, H.; Acebes-Fernandez, V.; Gongora, R.; Almendral Parra, M.J.; Manzano-Roman, R.; Fuentes, M. Interactions of Nanoparticles and Biosystems: Microenvironment of Nanoparticles and Biomolecules in Nanomedicine. *Nanomaterials* **2019**, *9*, 1365. [CrossRef] [PubMed]

67. Yang, Y.-Y.; Chia, H.-H.; Chung, T.-S. Effect of preparation temperature on the characteristics and release profiles of PLGA microspheres containing protein fabricated by double-emulsion solvent extraction/evaporation method. *J. Control. Release* **2000**, *69*, 81–96. [CrossRef]

68. Yang, Y.Y. Morphology, drug distribution, and in vitro release profiles of biodegradable polymeric microspheres containing protein fabricated by double-emulsion solvent extraction/evaporation method. *Biomaterials* **2001**, *22*, 231–241. [CrossRef]

69. Vanderhoff, J.W.; El-Aasser, M.S.; Ugelstad, J. Polymer Emulsification Process. U.S. Patent US4177177A, 4 December 1979.

70. Adibkia, K.; Omidi, Y.; Siahi, M.R.; Javadzadeh, A.R.; Barzegar-Jalali, M.; Barar, J.; Maleki, N.; Mohammadi, G.; Nokhodchi, A.; Javadzadeh, A. Inhibition of Endotoxin-Induced Uveitis by Methylprednisolone Acetate Nanosuspension in Rabbits. *J. Ocul. Pharmacol. Ther.* **2007**, *23*, 421–432. [CrossRef]

71. Alshamsan, A.; Binkhathlan, Z.; Kalam, M.A.; Qamar, W.; Kfouri, H.; Alghonaim, M.; Lavasanifar, A. Mitigation of Tacrolimus-Associated Nephrotoxicity by PLGA Nanoparticulate Delivery Following Multiple Dosing to Mice while Maintaining its Immunosuppressive Activity. *Sci. Rep.* **2020**, *10*, 6675. [CrossRef]

72. Adibkia, K.; Shadbad, M.R.S.; Nokhodchi, A.; Javadzedeh, A.; Barzegar-Jalali, M.; Barar, J.; Mohammadi, G.; Omidi, Y. Piroxicam nanoparticles for ocular delivery: Physicochemical characterization and implementation in endotoxin-induced uveitis. *J. Drug Target.* **2007**, *15*, 407–416. [CrossRef] [PubMed]

73. Prabha, S.; Labhasetwar, V. Critical determinants in PLGA/PLA nanoparticle-mediated gene expression. *Pharm. Res.* **2004**, *21*, 354–364. [CrossRef] [PubMed]

74. Nafee, N.; Taetz, S.; Schneider, M.; Schaefer, U.F.; Lehr, C.-M. Chitosan-coated PLGA nanoparticles for DNA/RNA delivery: Effect of the formulation parameters on complexation and transfection of antisense oligonucleotides. *Nanomed. Nanotechnol. Biol. Med.* **2007**, *3*, 173–183. [CrossRef] [PubMed]

75. Rezvantalab, S.; Drude, N.; Moraveji, M.K.; Güvener, N.; Koons, E.K.; Shi, Y.; Lammers, T.; Kiessling, F. PLGA-Based Nanoparticles in Cancer Treatment. *Front. Pharmacol.* **2018**, *9*, 1260. [CrossRef]

76. Mehrotra, A.; Pandit, J.K. Preparation and Characterization and Biodistribution Studies of Lomustine Loaded PLGA Nanoparticles by Interfacial Deposition Method. *J. Nanomed. Biother. Discov.* **2015**, *5*, 4. [CrossRef]

77. Turk, C.T.S.; Oz, U.C.; Serim, T.M.; Hascicek, C. Formulation and Optimization of Nonionic Surfactants Emulsified Nimesulide-Loaded PLGA-Based Nanoparticles by Design of Experiments. *AAPS PharmSciTech.* **2014**, *15*, 161–176. [CrossRef]

78. Panyam, J.; Dali, M.M.; Sahoo, S.K.; Ma, W.; Chakravarthi, S.S.; Amidon, G.L.; Levy, R.J.; Labhasetwar, V. Polymer degradation and in vitro release of a model protein from poly(d,l-lactide-*co*-glycolide) nano- and microparticles. *J. Control. Release* **2003**, *92*, 173–187. [CrossRef]
79. Siepmann, J.; Faisant, N.; Akiki, J.; Richard, J.; Benoit, J. Effect of the size of biodegradable microparticles on drug release: Experiment and theory. *J. Control. Release* **2004**, *96*, 123–134. [CrossRef]
80. Yeo, Y.; Basaran, O.A.; Park, K. A new process for making reservoir-type microcapsules using ink-jet technology and interfacial phase separation. *J. Control. Release* **2003**, *93*, 161–173. [CrossRef]
81. Cartiera, M.S.; Johnson, K.M.; Rajendran, V.; Caplan, M.J.; Saltzman, W.M. The uptake and intracellular fate of PLGA nanoparticles in epithelial cells. *Biomaterials* **2009**, *30*, 2790–2798. [CrossRef]
82. Wischke, C.; Schwendeman, S.P. Principles of encapsulating hydrophobic drugs in PLA/PLGA microparticles. *Int. J. Pharm.* **2008**, *364*, 298–327. [CrossRef] [PubMed]
83. Murakami, H.; Kawashima, Y.; Niwa, T.; Hino, T.; Takeuchi, H.; Kobayashi, M. Influence of the degrees of hydrolyzation and polymerization of poly(vinylalcohol) on the preparation and properties of poly(dl-lactide-*co*-glycolide) nanoparticle. *Int. J. Pharm.* **1997**, *149*, 43–49. [CrossRef]
84. Lv, C.; Su, Y.; Wang, Y.; Ma, X.; Sun, Q.; Jiang, Z. Enhanced permeation performance of cellulose acetate ultrafiltration membrane by incorporation of Pluronic F127. *J. Membr. Sci.* **2007**, *294*, 68–74. [CrossRef]
85. Jain, D.K.; Darwhekar, G.; Solanki, S.S.; Sharma, R. Osmotically regulated asymmetric capsular system for sustained delivery of indomethacin. *J. Pharm. Investig.* **2013**, *43*, 27–35. [CrossRef]
86. GÜMÜŞDERELIOĞLU, M.; Deniz, G. Synthesis, characterization and in vitro degradation of poly (dl-lactide)/poly (dl-lactide-*co*-glycolide) films. *Turk. J. Chem.* **1999**, *23*, 153–162.
87. Shams, T.; Parhizkar, M.; Illangakoon, U.; Orlu, M.; Edirisinghe, M. Core/shell microencapsulation of indomethacin/paracetamol by co-axial electrohydrodynamic atomization. *Mater. Des.* **2017**, *136*, 204–213. [CrossRef]
88. Singh, G.; Kaur, T.; Kaur, R.; Kaur, A. Recent biomedical applications and patents on biodegradable polymer-PLGA. *Int. J. Pharmacol. Pharm. Sci.* **2014**, *1*, 30–42.
89. Panda, S.; Mohanty, G.C.; Samal, R.; Roy, G.S. Viscosity studies of polyvinyl alcohol (PVA, Mw = 1,25,000) in solvent distilled water and aqueous solution of urea. *Mater. Sci. Res. India* **2010**, *7*, 443–448. [CrossRef]
90. Shimizu, Y.; Tanabe, T.; Yoshida, H.; Kasuya, M.; Matsunaga, T.; Haga, Y.; Kurihara, K.; Ohta, M. Viscosity measurement of Xanthan–Poly(vinyl alcohol) mixture and its effect on the mechanical properties of the hydrogel for 3D modeling. *Sci. Rep.* **2018**, *8*, 16538. [CrossRef]
91. Augusto, P.; Ibarz, A.; Cristianini, M. Effect of high pressure homogenization (HPH) on the rheological properties of tomato juice: Time-dependent and steady-state shear. *J. Food Eng.* **2012**, *111*, 570–579. [CrossRef]
92. Wu, C.-S.; Senak, L.; Bonilla, J.; Cullen, J. Comparison of relative viscosity measurement of polyvinylpyrrolidone in water by glass capillary viscometer and differential dual-capillary viscometer. *J. Appl. Polym. Sci.* **2002**, *86*, 1312–1315. [CrossRef]
93. Tóthová, J.; Lisý, V. Intrinsic viscosity of PVP polymers in extremely diluted solutions. *e-Polymers* **2013**, *13*, 22. [CrossRef]
94. Maggio, T.; Scales, C.; Romo-Uribe, A.; Liang, B. The shear rate and concentration dependence of viscosity in concentrated solutions of PVP and Bovine Mucin. *Contact Lens Anterior Eye* **2018**, *41*, S67. [CrossRef]
95. Zolnik, B.S.; Burgess, D.J. Effect of acidic pH on PLGA microsphere degradation and release. *J. Control. Release* **2007**, *122*, 338–344. [CrossRef]
96. Gouda, R.; Baishya, H.; Qing, Z. Application of mathematical models in drug release kinetics of carbidopa and levodopa ER tablets. *J. Dev. Drugs* **2017**, *6*, 1–8.
97. Supramaniam, J.; Adnan, R.; Kaus, N.H.M.; Bushra, R. Magnetic nanocellulose alginate hydrogel beads as potential drug delivery system. *Int. J. Biol. Macromol.* **2018**, *118*, 640–648. [CrossRef]
98. Singhvi, G.; Singh, M. Review: In vitro drug release characterization models. *Int. J. Pharm. Stud. Res.* **2011**, *2*, 77–84.
99. Jing, Z.; Dai, X.; Xian, X.; Du, X.; Liao, M.; Hong, P.; Li, Y. Tough, stretchable and compressive alginate-based hydrogels achieved by non-covalent interactions. *RSC Adv.* **2020**, *10*, 23592–23606. [CrossRef]
100. Gbureck, U.; Vorndran, E.; Barralet, J.E. Modeling vancomycin release kinetics from microporous calcium phosphate ceramics comparing static and dynamic immersion conditions. *Acta Biomater.* **2008**, *4*, 1480–1486. [CrossRef]
101. Unagolla, J.M.; Jayasuriya, A.C. Drug transport mechanisms and in vitro release kinetics of vancomycin encapsulated chitosan-alginate polyelectrolyte microparticles as a controlled drug delivery system. *Eur. J. Pharm. Sci.* **2018**, *114*, 199–209. [CrossRef]
102. Hung, H.-I.; Klein, O.J.; Peterson, S.W.; Rokosh, R.; Osseiran, S.; Nowell, N.H.; Evans, C.L. PLGA nanoparticle encapsulation reduces toxicity while retaining the therapeutic efficacy of EtNBS-PDT in vitro. *Sci. Rep.* **2016**, *6*, 33234. [CrossRef] [PubMed]
103. Xiong, S.; George, S.; Yu, H.; Damoiseaux, R.; France, B.; Ng, K.W.; Loo, J.S.-C. Size influences the cytotoxicity of poly (lactic-co-glycolic acid) (PLGA) and titanium dioxide (TiO$_2$) nanoparticles. *Arch. Toxicol.* **2012**, *87*, 1075–1086. [CrossRef] [PubMed]
104. Chiu, H.I.; Samad, N.A.; Fang, L.; Lim, V. Cytotoxicity of targeted PLGA nanoparticles: A systematic review. *RSC Adv.* **2021**, *11*, 9433–9449. [CrossRef]
105. Di-Wen, S.; Pan, G.-Z.; Hao, L.; Zhang, J.; Xue, Q.-Z.; Wang, P.; Yuan, Q.-Z. Improved antitumor activity of epirubicin-loaded CXCR4-targeted polymeric nanoparticles in liver cancers. *Int. J. Pharm.* **2016**, *500*, 54–61. [CrossRef]
106. Miura, N.; Matsumoto, Y.; Miyairi, S.; Nishiyama, S.; Naganuma, A. Protective Effects of Triterpene Compounds Against the Cytotoxicity of Cadmium in HepG2 Cells. *Mol. Pharmacol.* **1999**, *56*, 1324–1328. [CrossRef]

107. Louisa, M.; Suyatna, F.D.; Wanandi, S.I.; Asih, P.B.S.; Syafruddin, D. Differential expression of several drug transporter genes in HepG2 and Huh-7 cell lines. *Adv. Biomed. Res.* **2016**, *5*, 104. [CrossRef]
108. Norouzzadeh, M.; Kalikias, Y.; Mohamadpur, Z.; Sharifi, L.; Mahmoudi, M. Determining population doubling time and the appropriate number of HepG2 cells for culturing in 6-well plate. *Int. Res. Bas. Sci.* **2016**, *10*, 299–303.
109. Wen-Sheng, W.J.O. ERK signaling pathway is involved in p15INK4b/p16INK4a expression and HepG2 growth inhibition triggered by TPA and Saikosaponin a. *Oncogene* **2003**, *22*, 955–963. [CrossRef]
110. Sriuttha, P.; Sirichanchuen, B.; Permsuwan, U. Hepatotoxicity of Nonsteroidal Anti-Inflammatory Drugs: A Systematic Review of Randomized Controlled Trials. *Int. J. Hepatol.* **2018**, *2018*, 5253623. [CrossRef]

pharmaceutics

Article

Design, Development, Physicochemical Characterization, and In Vitro Drug Release of Formoterol PEGylated PLGA Polymeric Nanoparticles

Ernest L. Vallorz [1], David Encinas-Basurto [1], Rick G. Schnellmann [1,2,3] and Heidi M. Mansour [1,2,3,4,*]

1 Skaggs Pharmaceutical Sciences Center, The University of Arizona R. Ken Coit College of Pharmacy, 1703 E Mabel St., Tucson, AZ 85721, USA; vallorz@pharmacy.arizona.edu (E.L.V.); dencinas@pharmacy.arizona.edu (D.E.-B.); schnell@pharmacy.arizona.edu (R.G.S.)
2 Department of Medicine, The University of Arizona College of Medicine, 501 N Campbell Ave., Tucson, AZ 85724, USA
3 BIO5 Institute, The University of Arizona, 1657 E Helen St., Tucson, AZ 85719, USA
4 Center for Translational Science, Florida International University, Port St. Lucie, FL 34987, USA
* Correspondence: hmansour@fiu.edu; Tel.: +1-(772)-345-4731

Abstract: Polymeric nanoparticles' drug delivery systems represent a promising platform for targeted controlled release since they are capable of improving the bioavailability and tissue localization of drugs compared to traditional means of administration. Investigation of key parameters of nanoparticle preparation and their impact on performance, such as size, drug loading, and sustained release, is critical to understanding the synthesis parameters surrounding a given nanoparticle formulation. This comprehensive and systematic study reports for the first time and focuses on the development and characterization of formoterol polymeric nanoparticles that have potential application in a variety of acute and chronic diseases. Nanoparticles were prepared by a variety of solvent emulsion methods with varying modifications to the polymer and emulsion system with the aim of increasing drug loading and tuning particle size for renal localization and drug delivery. Maximal drug loading was achieved by amine modification of polyethylene glycol (PEG) conjugated to the poly(lactic-co-glycolic acid) (PLGA) backbone. The resulting formoterol PEGylated PLGA polymeric nanoparticles were successfully lyophilized without compromising size distribution by using either sucrose or trehalose as cryoprotectants. The physicochemical characteristics of the nanoparticles were examined comprehensively, including surface morphology, solid-state transitions, crystallinity, and residual water content. In vitro formoterol drug release characteristics from the PEGylated PLGA polymeric nanoparticles were also investigated as a function of both polymer and emulsion parameter selection, and release kinetics modeling was successfully applied.

Keywords: nanoparticle; solid-state characterization; in vitro; drug release kinetics modeling; PEGylation; PLGA diblock copolymer; biodegradable; biocompatible; amine; emulsion; polyvinyl alcohol (PVA); Pluronic triblock copolymer; trehalose; sucrose

Citation: Vallorz, E.L.; Encinas-Basurto, D.; Schnellmann, R.G.; Mansour, H.M. Design, Development, Physicochemical Characterization, and In Vitro Drug Release of Formoterol PEGylated PLGA Polymeric Nanoparticles. *Pharmaceutics* **2022**, *14*, 638. https://doi.org/10.3390/pharmaceutics14030638

Academic Editor: Oya Tagit

Received: 25 December 2021
Accepted: 9 March 2022
Published: 14 March 2022

Publisher's Note: MDPI stays neutral with regard to jurisdictional claims in published maps and institutional affiliations.

1. Introduction

Developments in the formulation and application of nanotechnology have advanced the administration of drugs to different organs for a wide array of diseases. Compared to carrier-independent conventional dosage forms, drug nanoparticles possess many key advantages. Nanoparticle drug delivery allows for improved solubility and stability of the drug, sustained drug release, improved patient compliance, and targeted delivery that can increase the therapeutic index of medicines [1–3].

Polymer-based particles are frequently studied and used as drug carriers in a wide variety of therapeutic applications since controlling their synthesis enables their physicochemical properties and drug release properties to be customized [4]. Thus, size distribution or side-chain composition of the polymer can be modified to improve nanoparticle performance. Among various polymers, the most widely used are aliphatic polyesters such as poly(lactic-co-glycolic acid) (PLGA) due to this particular polyester's favorable biodegradation characteristics and biocompatibility, and its success in FDA-approved sustained release injectable marketed pharmaceutical products [5]. Advances in the design of these particles has improved stability and circulation through PEGylation [6–8], biospecific targeting [9–12], and improved drug loading [2,13].

Formoterol fumarate dihydrate, a long-acting β_2 agonist (LABA), is a United States Food and Drug Administration (FDA)-approved agent for the treatment of asthma and obstructive pulmonary disease via inhalation. More recently, formoterol fumarate dihydrate has shown promise in treating mitochondrial dysfunction, which occurs in a variety of acute and chronic injuries [14–21]. Given that the β_2-adrenergic receptor is ubiquitously expressed, systemic delivery of formoterol risks potentially toxic side effects, particularly due to acute cardiovascular effects such as tachycardia and hypotension [22–25], as well as long-term cardiac remodeling [26–29]. Polymeric nanoparticles targeting the renal proximal tubules have shown promise at avoiding potential systemic toxicity using the FDA-approved diblock copolymer poly(lactic-co-glycolic) acid (PLGA) conjugated with FDA-approved polyethylene glycol (PEG) [10,30–32]. Thus, combining the potential renal targeting of polymeric nanoparticles with the mitochondrial biogenic and renoprotective effects of formoterol is likely to allow for enhanced biogenic effects and potentially improved recovery following renal injury while minimizing toxicity. In a proof-of-concept study, our laboratory has recently shown an ability to successfully deliver formoterol to the kidneys, providing renal drug targeting and sustained renal mitochondrial biogenesis while reducing the effect of the drug on the heart [33]. The purpose of this comprehensive and systematic study was to improve upon the previously synthesized nanoparticles, which were limited by formoterol drug loading and rapid release, by modifying the route of nanoparticle synthesis and through an evaluation of polymer modifications. Additionally, this study will characterize the physicochemical properties of the nanoparticles, quantify sustained release behavior, correlate the physicochemical properties with the sustained drug release properties, and mathematically model drug release with known mechanistic drug release models. To the authors' knowledge, this is the first study to report these findings.

2. Materials and Methods

2.1. Materials

Formoterol fumarate dihydrate [$C_{19}H_{24}N_2O_4\cdot0.5C_4H_4O_4\cdot H_2O$; 420.46 g/mol; 99.8% pure] was purchased from APAC Pharmaceuticals (Columbia, MD, USA). Poly(ethylene glycol) methyl ether-block-poly(lactide-co-glycolide) (PLGA-PEG), lactide:glycolide ratio 50:50, PLGA average M_n 55,000 g/mol, 30,000 g/mol, and 15,000 g/mol; PEG average M_n 5000 g/mol was purchased from Sigma-Aldrich (St. Louis, MO, USA). Poly(lactide-co-glycolide) methyl ether block- poly(ethylene glycol)-amine (PLGA-PEG-HN$_2$) PLGA average M_n 20,000 g/mol, PEG average M_n 5000 g/mol and poly(lactide-co-glycolide) methyl ether block- poly(ethylene glycol)-carboxylic acid (PLGA-PEG-COOH) PLGA average M_n 20,000 g/mol, and PEG average M_n 5000 g/mol were purchased from Nanosoft Polymers (Winston-Salem, NC, USA). Acetic acid (HPLC grade) was purchased from ThermoFisher (Waltham, MA, USA). Ethanol (99.9% purity, HPLC grade), hydrochloric acid 1 N (ACS grade), sodium hydroxide 1 N (ACS grade), poly(vinyl alcohol) (PVA) (molecular weight 89,000–98,000 g/mol, >99% hydrolyzed, reagent grade), Pluronic F127™ (molecular weight~12,600 g/mol, reagent grade), sodium choate hydrate (>99% purity), sodium deoxycholate (>99% purity) potassium phosphate monobasic (>99% purity), potassium phosphate dibasic (>98% purity), formic acid (97.5–98.5% purity), d-mannitol (ACS grade), and sucrose (99.5% purity) were purchased from Sigma-Aldrich (St. Louis, MO, USA). (+)-Trehalose dihydrate (387.32 g/mol) was purchased from Acros Organics (Fair Lawn, NJ, USA). Choloroform (ACS grade), anhydrous acetonitrile (LCMS grade), and methanol (LCMS grade) were purchased from Spectrum Chemical Mfg. Corp. (Gardena,

CA, USA). 0.45 µm polyvinylidene fluoride (PVDF) and 0.45 µm polytetrafluoroethylene (PTFE) membrane filters were purchased from MilliporeSigma (Burlington, MA, USA).

2.2. Methods

2.2.1. Solubility of Formoterol Fumarate Dihydrate in Aqueous and Organic Media

The solubility of formoterol fumarate dihydrate (APAC Pharmaceuticals, Columbia, MD, USA) was determined in various common aqueous and organic media to determine their suitability for use in the preparation of nanoparticles. The excess of formoterol fumarate dihydrate was added to a known volume of solvent. For variable pH samples, solution pH was adjusted with either hydrochloric acid or sodium hydroxide solutions (1 M Sigma-Aldrich, St. Louis, MO, USA). Vials were rotated gently for 24 h at 25 °C, as previously described [34]. Test solutions were filtered using 0.45 µm PVDF or PTFE membrane filters (MilliporeSigma, Burlington, MA, USA) for aqueous and organic solvents, respectively. Quantitative analysis of formoterol content was determined by high performance liquid chromatography (HPLC), as previously described [35]. Briefly, a C18-column (4.6 mm × 250 mm length, 5 µm pore size) (Phenomenex, Torrance, CA, USA) mobile phase consisting of methanol (Spectrum Chemical MFG Corp., Gardena, CA, USA) and 50 mM phosphoric acid (Sigma-Aldrich, St. Louis, MO, USA) buffer with 1% acetic acid (ThermoFisher, Waltham, MA, USA) at a ratio of 65:35, 1.0 mL/min flow rate and a column temperature of 40 °C was used to quantify formoterol content.

2.2.2. Preparation of Polymeric Nanoparticles

Nanoparticles were prepared by either single or double emulsion methods. For nanoparticles prepared by oil-in-water single emulsion, the polymer was dissolved in chloroform and formoterol fumarate dihydrate were dissolved in methanol before being added to the polymer solution. This was then emulsified in an aqueous solution of 3% PVA (Sigma-Aldrich, St. Louis, MO, USA) using a microtip probe sonicator (Qsonica, Newton, CT, USA) at 60 Watt of energy output for 3 min over ice and the organic solvent allowed to evaporate with stirring (700 rpm) at room temperature for at least 8 h. For nanoparticles prepared by water-in-oil-in-water double emulsion, formoterol fumarate dihydrate was dissolved in aqueous media before being added to a solution of polymer in chloroform (Spectrum Chemical Mfg. Corp., Gardena, CA, USA) at a 1:2 aqueous:organic ratio. An initial emulsion was formed by sonicating at 60 Watt using a microtip probe sonicator for 30 s before being added to a 3% PVA solution and sonicated again. The organic solvent was then allowed to evaporate with stirring at room temperature. For all nanoparticle syntheses, the particles were collected by centrifugation at 15,000 relative centrifugal force (rcf) and washed three times with distilled ultrapure water (18.2 MΩ.cm) (Milli-Q Plus, MilliporeSigma, Burlington, MA, USA). The samples were lyophilized at −80 °C under a vacuum < 0.133 mmHg (FreeZone 4.5 L, Labconco, Kansas City, MO, USA) with or without cryoprotectant and stored at −20 °C until use.

2.2.3. Effect of Polymeric Nanoparticle Synthesis Parameters on Formoterol Drug Loading

To determine drug loading, nanoparticles were dissolved in an acetonitrile solution and analyzed by HPLC method reported above. Drug loading was calculated, as previously reported [36], using Equation (1):

$$\text{DL (\%)} = \frac{\text{the amount of formoterol assayed}}{\text{the total amount of nanoparticles in the preparation}} \times 100 \qquad (1)$$

2.2.4. Effect of Polymeric Nanoparticle Synthesis Parameters on Particle Size

Nanoparticle size was determined immediately following washing. For the evaluation of the lyophilized particles, approximately 1 mg of nanoparticle was suspended in ultrapure water and centrifuged for 10 min at 15,000 rcf to remove the cryoprotectant. Nanoparticles were suspended at a concentration of ~1 mg/mL with ultrapure water, and hydrodynamic particle size was determined by photon correlation spectroscopy using the Zetasizer Nano ZS (Malvern Instruments Ltd., Malvern, UK) under previously reported conditions [37]. The suspended nanoparticles were evaluated using a scattering angle of 173° at a temperature of 25 °C in triplicate with a minimum of 10 measurements taken per replicate.

2.2.5. Impact of Nanoparticle Synthesis Parameters on Zeta Potential

Nanoparticle zeta potential (ζ) measurements were carried out in 0.1× normal saline solution at 25 °C and pH 7.2. The mean ζ was determined using the Zetasizer Nano ZS (Malvern Instruments Ltd., Malvern, UK) phase analysis light scattering technique.

2.2.6. Impact of Nanoparticle Synthesis Parameters on In Vitro Drug Release

Nanoparticles prepared as described above were dispersed in 10 mL of phosphate buffered saline (PBS) (pH 7.4) and incubated at 37 °C with gentle stirring, as previously described [38]. At determined intervals, an aliquot was taken and centrifuged at 15,000 rcf for 10 min. The supernatant was extracted and replaced with an equal volume of fresh PBS in order to maintain sink conditions. Formoterol content was chemically analyzed and quantified by HPLC, as described above. Modeling of formoterol in vitro drug release was carried out for three kinetic models, namely, the zero-order, first-order, and Korsmeyer–Peppas models. Zero-order kinetics were fitted to Equation (2):

$$Q_t - Q_0 = k_0 t \tag{2}$$

where Q_t is the amount of drug released after time t, Q_0 is the initial amount of drug in solution, and k_0 is the zero-order rate constant. First-order kinetics were fit to Equation (3):

$$\ln Q_t = \ln Q_0 - k_1 t \tag{3}$$

where Q_t is the amount of drug released after time t, Q_0 is the initial amount of drug in solution, and k_1 is the first-order rate constant. Finally, release kinetics were fit to the Korsmeyer–Peppas model, Equation (4):

$$Q_t = k t^n \tag{4}$$

where Q_t is the amount of drug released after time t, k is the rate constant, and n is the diffusion exponent for drug release. Nanoparticle release was determined in triplicated ($n = 3$) for each preparation. Data were plotted using Prism 9.0 (GraphPad® Software, San Diego, CA, USA).

2.2.7. Characterization of Nanoparticle Surface Morphology

Nanoparticle size and surface morphology was visualized using SEM (FEI Inspect S SEM, FEI Company, Hillsboro, OR, USA). Powders were deposited on double-sided carbon conductive adhesive tabs (Ted-Pella, Inc., Redding, CA, USA) attached to aluminum SEM stubs (Ted-Pella, Inc., Redding, CA, USA) and sputter-coated (Anatech Hummer 6.2, Union City, CA, USA) with gold for 90 s under argon plasma as previously reported [39].

2.2.8. X-Ray Powder Diffraction (XRPD)

The crystallinity of PLGA-PEG-NH$_2$, formoterol fumarate dihydrate, sucrose, trehalose, and lyophilized nanoparticles with and without cryoprotectant were examined using XRPD. The diffraction patterns of the samples were collected at room temperature scanning between 5.0° and 70.0° (2θ) at a rate of 2.00° per minute using a Philips PANalytical X'Pert PRO MPD (Malvern Panalytical, Malvern, UK) equipped with copper X-ray source (Kα radiation with $\lambda = 1.5406$ Å). The samples were loaded onto zero background single crystal silicon holders, as previously reported [37,39].

2.2.9. Thermal Analysis of Lyophilized Nanoparticles

Thermal analysis was performed by differential scanning calorimetry (DSC) and cross-polarized hot stage microscopy (HSM). DSC analysis was conducted as previously reported [40,41]. Thermal analysis and phase transition measurements for raw formoterol fumarate dihydrate, raw PLGA-PEG-NH$_2$ (20,000 MW PLGA, 5000 MW PEG), raw sucrose, raw trehalose, and lyophilized PLGA-PEG-NH$_2$ used with or without either sucrose or trehalose as a cryoprotectant were studied. Thermograms were acquired using the TA Q1000 differential scanning calorimeter with RSC090 colling accessory (TA Instruments, New Castle, DE, USA). A mass of between 1 and 5 mg of sample was weighed into anodized aluminum hermetic pans (TA Instruments) with an empty pan used as a reference. DSC measurements were performed at a heating rate of 10 °C/min from 0 to 350 °C. Ultrapure nitrogen gas was used as the purging gas at a rate of 50 mL/min. Analysis of thermograms was conducted using TA Universal Analysis (TA Instruments). All measurements were carried out in triplicate.

Solid-state phase transitions of the lyophilized nanoparticles were observed using cross-polarized light HSM similarly to previously reported [39]. Microscopy was conducted using a Leica DMLP cross-polarized microscope (Leica Mircosystems, Wetzlar, Germany) equipped with a Mettler FP 80 central processor and FP82 hot stage (Mettler Toldeo, Columbus, OH, USA). Lyophilized particles were mounted on a microscope slide and heated at a rate of 10 °C/min from 25 °C to 300 °C. The images were digitally captured using a Nikon Coolpix 8800 digital camera (Nikon, Tokyo, Japan) under 100× total magnification.

2.2.10. Residual Water Content Analysis by Karl Fischer Titration

The residual water content of lyophilized nanoparticles was quantified by Karl Fischer titration (KFT) colorimetric assay using a TitroLine® 7500 trace titrator (SI Analytics, Mainz, Germany). Around 3–7 mg of the sample was dissolved in 5 mL AQUA STAR anhydrous acetonitrile and injected into the titration cell. The measured moisture content was expressed in percentage as the result of the KFT. All measurements were completed in triplicate.

2.2.11. Statistical Analysis

Comparison of the difference between three groups was performed by one-way analysis of variance (ANOVA) with Tukey's *post hoc* test for comparisons (Prism 9.0, GraphPad Software, San Diego, CA, USA). In all cases, the *p* values of 0.05 or less were considered significant.

3. Results

3.1. Solubility of Formoterol Fumarate Dihydrate in Aqueous and Organic Media

The solubility of formoterol and its salt formoterol fumarate dihydrate has previously been described in water and some organic media; however, the solvents for which published literature exists are those most commonly used in inhalation drug development, such as various ionic and ethanol solutions, not the organic solvents most commonly used in the preparation of polymeric nanoparticles. The solubility of the fumarate salt of formoterol has a water solubility of 1.16 ± 0.02 mg/mL at 25 °C. Solubility is increased with the increasing volume fraction of low molecular weight alcohols such as ethanol and methanol (Figure 1A). Similar to previously reported studies, formoterol fumarate dihydrate likely forms a less soluble solvate with ethanol at volume fractions greater than 50% and sees a subsequent reduction in solubility with increasing cosolvent fraction. Similar solvent formation was not seen in water–methanol mixtures; however, this has been previously reported under different experimental conditions. Formoterol fumarate dihydrate sees increasing solubility in highly basic or acidic conditions (Figure 1B) as formoterol fumarate dihydrate contains both acidic and basic pKa(s) of around 8.6 and 9.8, respectively. Solubility of formoterol fumarate dihydrate in common non-ionic surfactants polyvinyl alcohol (PVA) and Pluronic® F127 remains unchanged at concentrations ranging from 0.1% to 5% in aqueous solution.

Solubilization is increased, however, in a concentration-dependent manner above the critical micelle concentration of sodium cholate (12 mM or 0.52%), an ionic surfactant (Figure 1C). Solubility in the common organic solvents dichloromethane (DCM), chloroform, acetonitrile, and acetate are also reported (Table 1), with acetone having the greatest solubility of 0.063 ± 0.004 mg/mL.

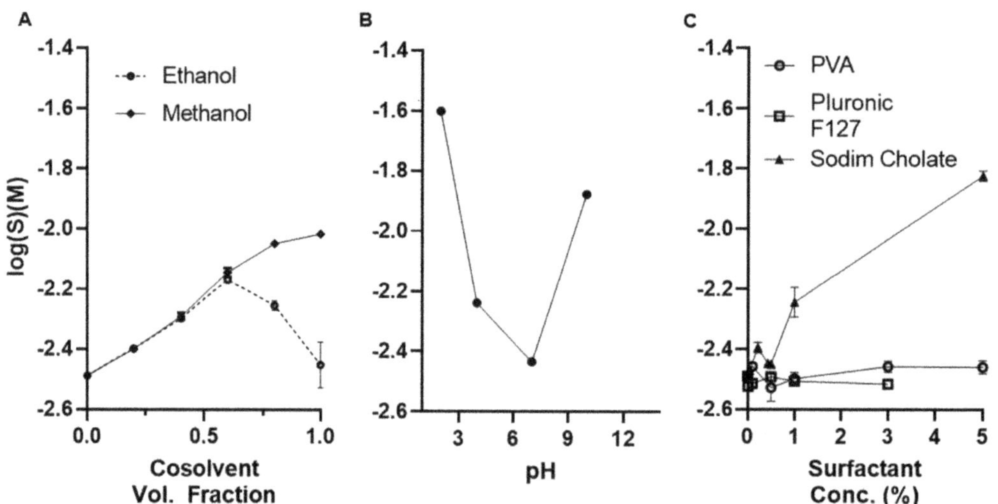

Figure 1. Solubility of formoterol determined in (**A**) ethanol and methanol, (**B**) water at various pH, and (**C**) aqueous solutions of various surfactants at 25 °C and gentle shaking for 24 h. Data presented are mean ($n = 3$) $\pm$ s.d.

Table 1. Solubility of formoterol in various organic solvents. DCM; dichloromethane, ACN; acetonitrile. Data are presented as the mean ($n = 3$) $\pm$ s.d.

Solvent	Formoterol Solubility (mg/mL)
DCM	0.001 ± 0.0004
Chloroform	0.002 ± 0.001
ACN	0.005 ± 0.001
Acetone	0.051 ± 0.004

3.2. Effect of Polymeric Nanoparticle Synthesis Parameters on Formoterol Drug Loading

Achieving significant drug loading of formoterol in PLGA nanoparticles is complicated by the low solubility in organic media, such as DCM and acetonitrile, where PLGA is freely soluble, compared to low molecular weight alcohols such as methanol, where PLGA and PLGA conjugates are practically insoluble. Drug loading of formoterol in PLGA-PEG nanoparticles prepared by single emulsion is improved with increasing PLGA molecular weight (35 mg/mL PLGA-PEG held constant), from 0.04% to 0.17% (Figure 2A), and increasing PLGA-PEG concentration (using 55,000 MW PLGA-PEG), up to 0.63% (Figure 2B). This was the maximum drug loading achievable by a single emulsion solvent evaporation method. While the higher molecular weight polymer (55,000 MW PLGA) achieved increased drug loading, the increases in drug loading are offset by increased nanoparticle size (data not shown) and reportedly increased degradation time in vivo [3–5], which could potentially lead to accumulation and toxicity. For these reasons, 20,000 MW PLGA polymers were selected for further study.

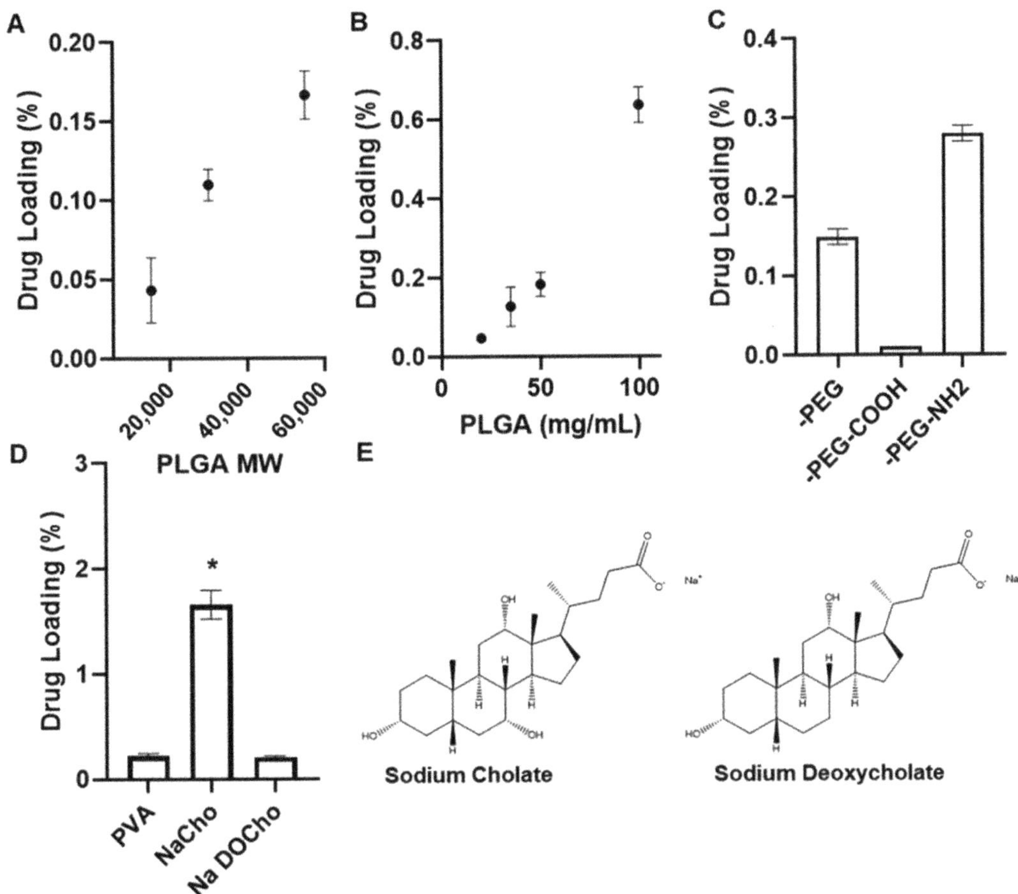

Figure 2. Formoterol drug loading of nanoparticles (**A**) prepared by single emulsion with increasing PLGA molecular weight, (**B**) prepared by single emulsion with increasing PLGA concentration, (**C**) prepared by double emulsion with modified PEG groups, (**D**) prepared by double emulsion with alterations to inner aqueous phase surfactant (**E**) structures of sodium cholate and sodium deoxycholate. PVA; 1% polyvinyl alcohol, NaCho; 12 mM sodium cholate, NaDOChol; 10 mM sodium deoxycholate. * indicates significance ($p < 0.05$). All data are presented as mean ($n = 3$) $\pm$ s.d.

Formoterol loading is further enhanced by changing from a single to double emulsion method and modification of the PEG terminus. Nanoparticles were prepared by water-in-oil-in-water double emulsion using PLGA-PEG, carboxylic acid modified PEG, or amine-modified PEG. PLGA-PEG-COOH reduced drug loading compared to methyl terminated PEG from 0.15% to 0.01%, whereas PLGA-PEG-NH$_2$ significantly increased drug loading to 1.39% (Figure 2C).

Interaction between the formoterol and the inner aqueous surfactant was evaluated using 10 mg/mL 20,000 MW PLGA-PEG-NH$_2$ and either 1% PVA, 12 mM sodium cholate or 10 mM sodium deoxycholate as the inner phase. The use of 10 mM sodium deoxycholate was required as stable nanoparticles did not form using 12 mM sodium deoxycholate. The use of the nonionic homopolymer surfactant PVA as the inner phase showed increased (0.22%) drug loading compared to their equivalent single emulsion prepared particles (Figure 2D). However, the use of the ionic surfactant sodium cholate showed a significant increase in drug loading, up to 1.66%. This increase in loading is completely nullified by the use of 10 mM sodium deoxycholate, which differs from sodium cholate by only the 7α-hydroxyl group (structures Figure 2E), suggesting this interaction is critical to the improved loading seen by sodium cholate.

3.3. Effect of Polymeric Nanoparticle Synthesis Parameters on Particle Size

The impact of sonication and PGLA concentration on particle size was evaluated with the goal of achieving particles with median hydrodynamic diameters between 300 and 500 nm. Sonication time was the first parameter to be evaluated for double emulsion-prepared particles (10 mg/mL PLGA-PEG-NH$_2$, 10 mM sodium cholate inner phase), increasing the secondary sonication from 30 to 600 s. Initially, there was a precipitous decrease in median diameter, from over 500 nm to 292 nm with 90 s of sonication. Further increases in sonication time resulted in no significant change in particle size (Figure 3A).

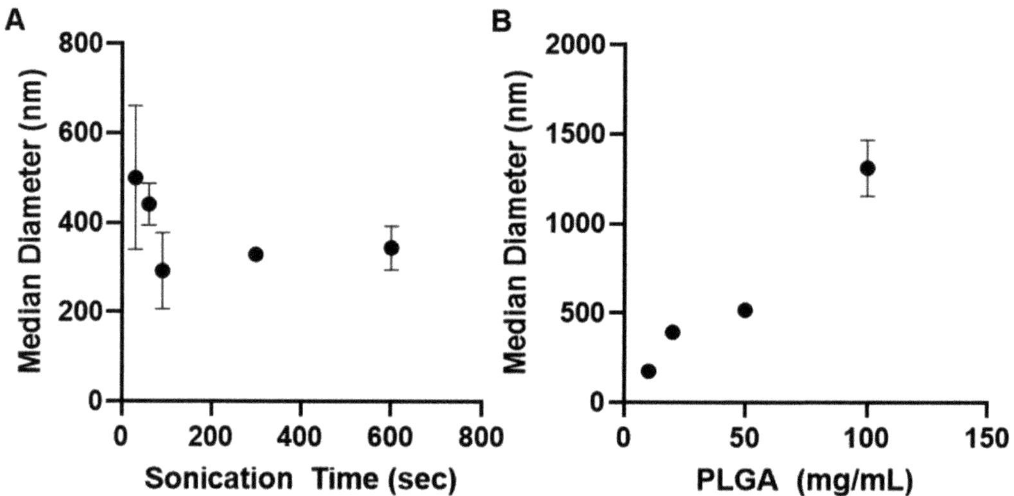

Figure 3. Median hydrodynamic diameter determined by zetasizer for nanoparticles prepared by double emulsion with (**A**) increasing 60 W ultrasonication time and (**B**) increasing polymer concentration in the organic phase. All data are presented as mean (n = 3) ± s.d.

The impact of polymer concentration was additionally determined at concentrations of PLGA-PEG-NH$_2$ ranging from 10 to 100 mg/mL. All particles were prepared by double-emulsion, used 12 mM sodium cholate as an inner phase, and were sonicated for 180 s during preparation of the secondary emulsion. Particle size proved to be extremely sensitive to increasing PLGA concentration, with particle size increasing proportionally to the increase in polymer (Figure 3B).

The impact of lyophilization was assessed following purification of the nanoparticles. To determine the impact of cryoprotectant selection and concentration on primary particle size, 1 mL of double-emulsion (12 mM sodium cholate inner phase, 180 s sonication) prepared PLGA-PEG-NH$_2$ nanoparticles at a concentration of 10 mg/mL were lyophilized in either 2.5%, 5%, or 10% of either sucrose, trehalose, or d-mannitol for 72 h. Resuspended nanoparticles lyophilized with mannitol as the cryoprotectant showed significantly increased particle size, over 1 μm (Figure 4). Conversely, both sucrose and trehalose cryoprotectants demonstrated decreased particle size growth post lyophilization with increasing cryoprotectant concentration, with sucrose performing slightly better than trehalose at all concentrations. For further characterization studies using nanoparticles lyophilized with cryoprotectants, 5% cryoprotectant concentration was used.

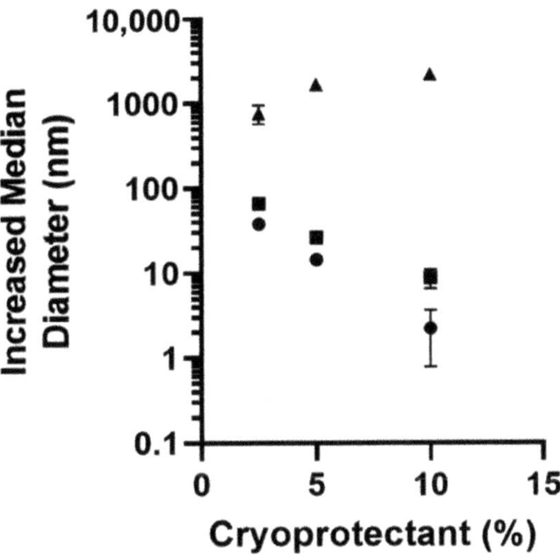

Figure 4. Median hydrodynamic diameter determined by zetasizer for PLGA-PEG-NH$_2$ nanoparticles prepared by double emulsion using either ▲ mannitol, ■ trehalose, or ● sucrose as cryoprotectants. All data are presented as mean ($n = 3$) $\pm$ s.d.

3.4. Impact of Nanoparticle Synthesis Parameters on Zeta Potential

Nanoparticle zeta potential following washing was determined in $0.1\times$ normal saline at 25 °C and neutral pH. Nanoparticle zeta potential was most strongly influenced by modification of PEG group (Figure 5). Methyl-endcapped PEG had a zeta potential of -0.792 ± 0.284 mV, whereas amine modified PEG had a positive zeta potential of 14.133 ± 0.404 mV and carboxylic acid modified PEG had a negative zeta potential of -34.87 ± 0.945 mV.

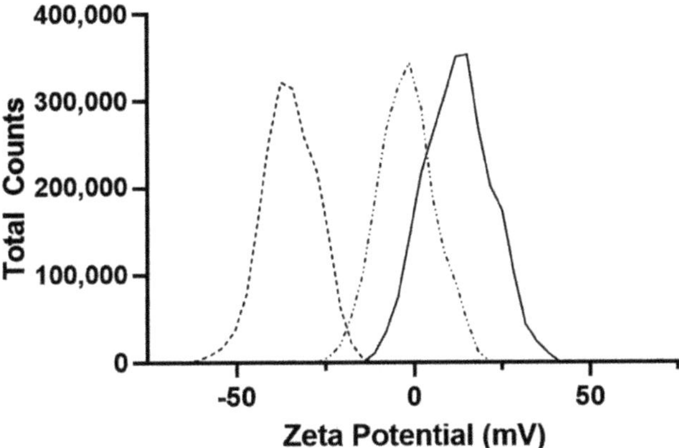

Figure 5. Zeta potential measurements of nanoparticles prepared with various PEG modifications immediately following synthesis after resuspension in $0.1\times$ normal saline solution at 25 °C and pH 7.2. ⋯⋯ PLGA-PEG-COOH, ⋯⋯ PLGA-PEG, —— PLGA-PEG-NH$_2$. All data are presented as mean ($n = 3$) $\pm$ s.d.

3.5. Impact of Nanoparticle Synthesis Parameters on Drug Release

Formoterol release is significantly altered by the method of synthesis. Single and double emulsion solvent evaporation methods were assessed using 50 mg/mL 55,000 MW

PLGA-PEG, 180 s sonication time and using 1% PVA as the inner aqueous phase. Release was measured out to 1 week (144 h) in PBS at 37 °C with constant stirring. Nanoparticles prepared by single emulsion showed significant burst release, with 80 and 90% of their entrapped drug released within the first 3 and 24 h, respectively. Comparatively, nanoparticles prepared by double emulsion demonstrated a slower initial phase of release, with 3-h release at 17% and 24-h release between 60 and 70% (Figure 6A). Analysis of the release kinetics showed non-fickian/anomalous diffusion (n > 0.43) for single emulsion-prepared nanoparticles and quasi-fickian diffusion for (n < 0.43) for double emulsion-prepared nanoparticles (Table 2).

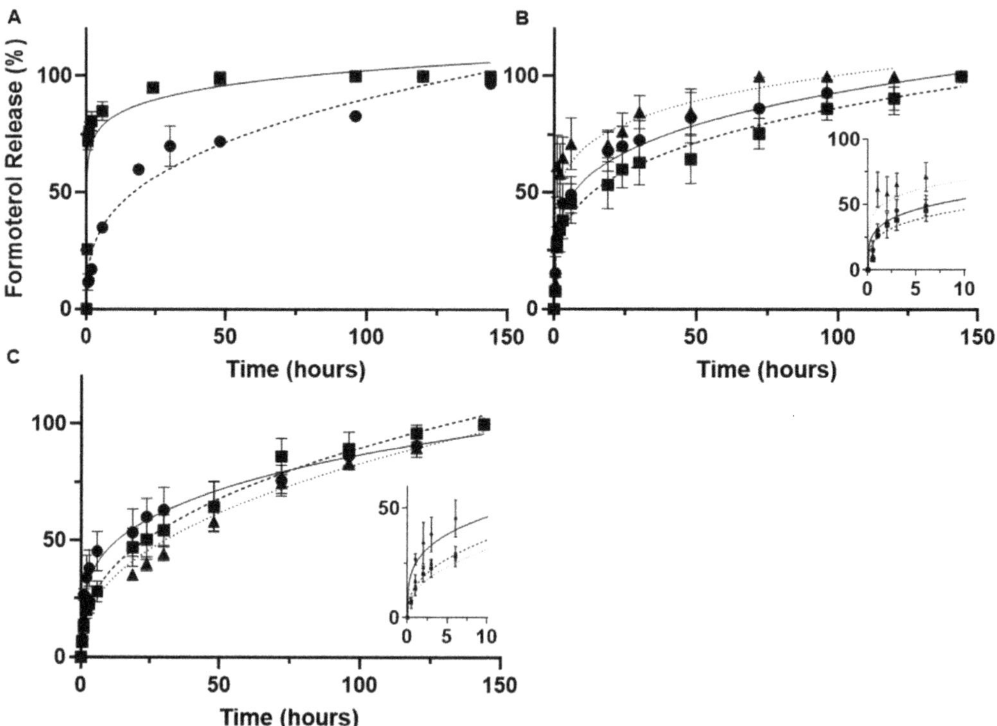

Figure 6. Cumulative formoterol release from nanoparticles at 37 °C and constant stirring (**A**) —■— oil-in-water single emulsion ··●·· water-in-oil-in-water double emulsion (**B**) nanoparticles prepared by double emulsion with —●— 1% PVA inner phase, ··■·· 12 mM sodium cholate inner phase, ··▲·· 10 mM sodium deoxycholate inner phase (**C**) nanoparticles prepared by double emulsion using —●— 10 mg/mL, ··■·· 20 mg/mL, ··▲·· 50 mg/mL PLGA-PEG-NH$_2$. Inserts of the first 10 h are included for (**B**,**C**). All data are presented as the mean of (n = 3) ± s.d.

Table 2. Nanoparticle release kinetics modeling. o/w; oil-in-water single emulsion w/o/w; water-in-oil-in-water double emulsion, PVA; nanoparticles prepared by double emulsion using 1% PVA inner phase, NaCholate; nanoparticles prepared by double emulsion using 12 mM sodium cholate inner phase, NaDOCholate; nanoparticles prepared by double emulsion using 10 mM sodium deoxycholate inner phase, 10/20/50 mg/mL; nanoparticles prepared by double emulsion using 10/20/50 mg/mL of polymer in the organic phase. All data are presented as the mean of ($n = 3$) release profiles.

	Zero Order	First Order	Korsmeyer-Peppas	
	R^2	R^2	R^2	n
o/w	0.33	0.94	0.88	0.56
w/o/w	0.77	0.97	0.97	0.34
PVA	0.65	0.89	0.94	0.23
NaCholate	0.75	0.86	0.94	0.27
NaDOCholate	0.49	0.81	0.84	0.17
10 mg/mL	0.75	0.86	0.94	0.27
20 mg/mL	0.85	0.96	0.97	0.40
50 mg/mL	0.92	0.96	0.98	0.42

Modification of the inner phase surfactant also impacted the initial release. Nanoparticle were prepared using 10 mg/mL 20,000 MW PLGA-PEG-NH$_2$ and by varying the inner phase between 1% PVA, 12 mM sodium cholate or 10 mM sodium deoxycholate. The PVA and sodium cholate-prepared particles showed slight differences in 3-h burse release (45 and 38%, respectively) and both were significantly lower than the sodium deoxycholate-prepared particles, which showed 65% drug release by 3-h (Figure 6B). Analysis of release kinetics showed no significant difference in release exponent between PVA and sodium cholate inner phases; however, there was a significant ($p < 0.05$) difference between those and sodium deoxycholate (Table 2). These differences were less apparent during the sustained release phase (time > 24 h). Finally, increasing concentration of PLGA-PEG-NH$_2$ during nanoparticle synthesis using 12 mM sodium cholate as the inner phase resulted in a decreased percentage of formoterol released within the first 3 h and increased rate of release beyond 24 h (Figure 6C). Increasing polymer concentration resulted in increasing the release exponent towards fickian (n = 0.43) release (Table 2).

3.6. Characterization of Nanoparticle Surface Morphology

The surface characteristics of nanoparticles prepared by double emulsion were assessed. Nanoparticles of PLGA-PEG with sodium cholate inner phase showed a high degree of surface roughness and size irregularity as well as a tendency to agglomerate (Figure 7A). Nanoparticles prepared by double-emulsion with PLGA-PEG-COOH and PLGA-PEG-NH$_2$ with sodium cholate inner phase were similarly sized, producing spherical particles that did not form aggregates and had smooth surface features (Figure 7B,C).

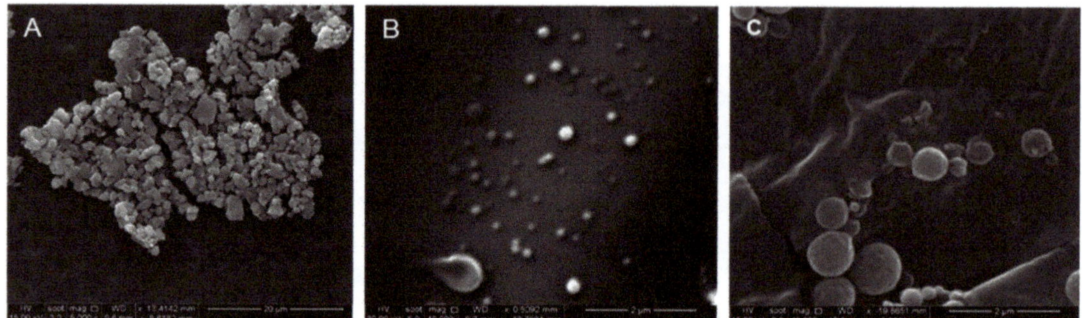

Figure 7. Scanning electron micrographs taken of double emulsion prepared nanoparticles of (**A**) PLGA-PEG, (**B**) PLGA-PEG-COOH, (**C**) PLGA-PEG-NH$_2$.

3.7. X-ray Powder Diffraction (XRPD)

X-ray diffractograms of nanoparticle raw materials (PLGA-PEG-NH$_2$, formoterol fumarate dihydrate and cryoprotectants) as well as lyophilized nanoparticles with and without cryoprotection were obtained (Figure 8). The diffraction pattern of raw materials formoterol fumarate dihydrate, sucrose, and trehalose showed multiple sharp peaks across the scanned range, indicating long range molecular order consistent with crystallinity (Figure 8A). Raw polymer samples and all lyophilized-prepared nanoparticles did not contain any sharp crystalline peaks (Figure 8A,B).

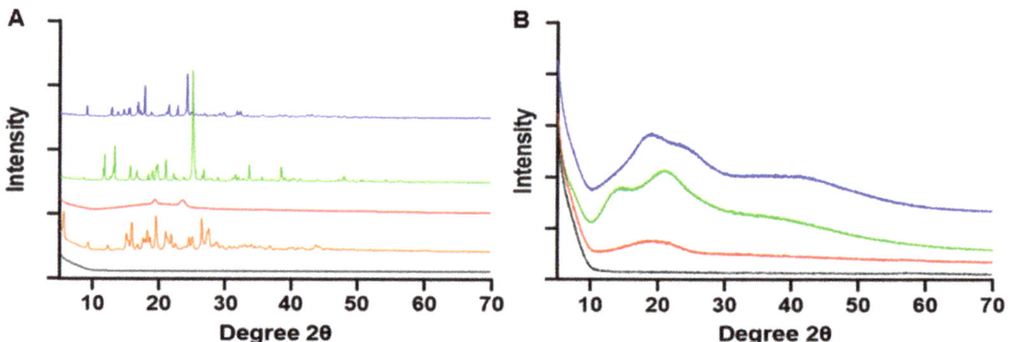

Figure 8. XRPD diffractograms (**A**) formoterol fumarate dihydrate (orange), PLGA-PEG-NH$_2$ (red), sucrose (green), and trehalose (blue) and (**B**) nanoparticles lyophilized without cryoprotectant (red), nanoparticles lyophilized with either sucrose (green) or trehalose (blue) cryoprotectants. Black is the blank background.

3.8. Thermal Analysis of Lyophilized Nanoparticles

Thermal analysis of nanoparticle components formoterol fumarate dihydrate, PLGA-PEG- NH$_2$, sucrose and trehalose as well as lyophilized nanoparticles with or without cryoprotectants are summarized in (Table 3).

Table 3. Differential scanning calorimetry (DSC) thermal analysis. T_g; glass transition temperature. Data presented are mean ($n = 3$) $\pm$ s.d.

Sample	Tg onset (°C)	Tg mid (°C)	Tg end (°C)	ΔCp (J/g °C)	Endotherm #1 onset (°C)	Endotherm #1 peak (°C)	Endotherm #1 enthalpy (J/g)	Endotherm #2 onset (°C)	Endotherm #2 peak (°C)	Endotherm #2 enthalpy (J/g)	Endotherm #3 onset (°C)	Endotherm #3 peak (°C)	Endotherm #3 enthalpy (J/g)	Endotherm #4 onset (°C)	Endotherm #4 peak (°C)	Endotherm #4 enthalpy (J/g)	Exotherm #1 onset (°C)	Exotherm #1 peak (°C)	Exotherm #1 enthalpy (J/g)
Raw Formoterol Fumarate Dihydrate					88.84 ± 1.87	104.62 ± 3.62	13.56 ± 1.04	116.28 ± 0.76	130.82 ± 2.73	136.7 ± 0.89									
Raw PLGA-PEG-NH2	41.56 ± 1.34	42.43 ± 1.51	43.34 ± 1.93	0.69 ± 0.44	43.32 ± 2.16	47.97 ± 4.72	4.95 ± 3.34												
Raw Sucrose											187.01 ± 0.59	189.43 ± 0.23	129.47 ± 0.4	219.35 ± 5.56	220.72 ± 4.48	150 ± 50.01			
Raw Trehalose								94.46 ± 0.07	95.79 ± 0.05	94.54 ± 2.28	190.98 ± 1.75	193.3 ± 3.24	92.98 ± 14.66						
Nanoparticle No Cryoprotectant	35.3 ± 1.25	36.12 ± 0.8	36.48 ± 0.69	0.3 ± 0.2	42.16 ± 6.75	43.99 ± 6.73	1.98 ± 1.14	82.62 ± 2.34	112.95 ± 9.6	8.67 ± 3.24									
Nanoparticle Sucrose Cryoprotectant	50.09 ± 1.1	54.3 ± 0.21	54.82 ± 0.43	1.34 ± 0.38	53.45 ± 0.32	57.15 ± 0.33	4.13 ± 0.43	70.93 ± 2.3	96.02 ± 1.58	10.11 ± 0.56	180.03 ± 0.36	184.74 ± 0.03	86.59 ± 2.38	213.08 ± 5.68	222.78 ± 2.12	119.77 ± 1.16	136.56 ± 1.93	148.26 ± 3.06	79.36 ± 5.03

Table 3. *Cont.*

Sample	Tg (°C)			ΔCp	Endotherm #1			Endotherm #2			Endotherm #3			Endotherm #4			Exotherm #1		
	onset (°C)	mid (°C)	end (°C)	J/g °C	onset (°C)	peak (°C)	enthalpy (J/g)	onset (°C)	peak (°C)	enthalpy (J/g)	onset (°C)	peak (°C)	enthalpy (J/g)	onset (°C)	peak (°C)	enthalpy (J/g)	onset (°C)	peak (°C)	enthalpy (J/g)
Nanoparticle a Trehalose Cryoprotectant	43.08 ± 3.39	44.68 ± 4.02	46.64 ± 2.94	1.21 ± 0.31	42.97 ± 1.61	50.17 ± 0.18	4.73 ± 3.73	58.99 ± 1.99	86.75 ± 3.45	16.89 ± 2.72									

For formoterol fumarate dihydrate, there was a bimodal endotherm with an initial peak of 104 °C and a main peak of 130 °C. Above 150 °C, thermal decomposition was seen in the form of a jagged baseline (Figure 9A). For PLGA-PEG-NH$_2$ polymer, there was a clear glass transition peak (T_g) from 1–43 °C combined with an endotherm at 48 °C and a broad decomposition starting at 250 °C (Figure 9B). Raw sucrose showed a sharp endotherm at 189 °C, followed by a broad decomposition endotherm at 220 °C (Figure 9C). Trehalose dihydrate showed a sharp endotherm at 95 °C, followed by a broad endotherm at 193 °C, followed by decomposition (Figure 9D). PLGA-PEG-NH$_2$ particles lyophilized without cryoprotectant showed slightly decreased T_g of 35–36 °C compared to the raw polymer and a similar first endotherm at 44 °C, followed by a broad endotherm starting from 82 °C and peaking at 112 °C (Figure 9E). Nanoparticles lyophilized with 5% sucrose as a cryoprotectant showed an increased T_g and first endotherm comparted to the raw polymer, 50–54 °C and 57 °C, respectively. Additionally, the broad endotherm at 96 °C transitioned into an exothermic peak at 148 °C, followed by an endotherm at 184 °C and secondary endotherm at 222 °C leading to decomposition (Figure 9F). Finally, nanoparticles lyophilized with trehalose as a cryoprotectant had a similar T_g to the raw polymer, 43–46 °C followed by a broad endotherm starting from 58 °C and peaking at 86 °C. No other endotherms were detected until decomposition started above 250 °C.

Lyophilized nanoparticles were also evaluated by cross-polarized HSM. All three particles (lyophilized without cryoprotectant, with sucrose cryoprotectant and with trehalose cryoprotectant) were dark and lacked birefringence at 25 °C and 37 °C. At 60 °C, the cryoprotectant free particles began melting, a process that continued until melting was fully completed at 130 °C (Figure 10A). Both formulations lyophilized with cryoprotectants did not have observable melts until closer to 100 °C (Figure 10B,C). Nanoparticles lyophilized with sucrose exhibited a liquid crystal transition, as noted by the marked diffuse birefringence at 130 °C (Figure 10B). Both lyophilates had completely melted by 160 °C and there were no observable transitions after that temperature.

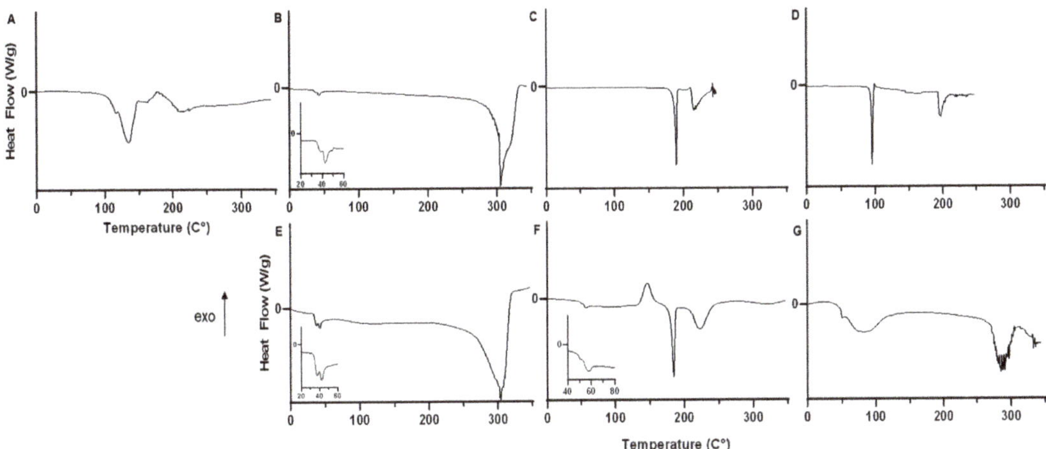

Figure 9. Differential scanning calorimetry thermograms of (**A**) formoterol fumarate dihydrate, (**B**) PLGA-PEG-NH$_2$ (**C**) Sucrose, (**D**) Trehalose, (**E**) PLGA-PEG-NH$_2$ nanoparticles lyophilized without cryoprotectant, (**F**) nanoparticles lyophilized with 5% sucrose cryoprotectant, (**G**) nanoparticles lyophilized with 5% trehalose cryoprotectant.

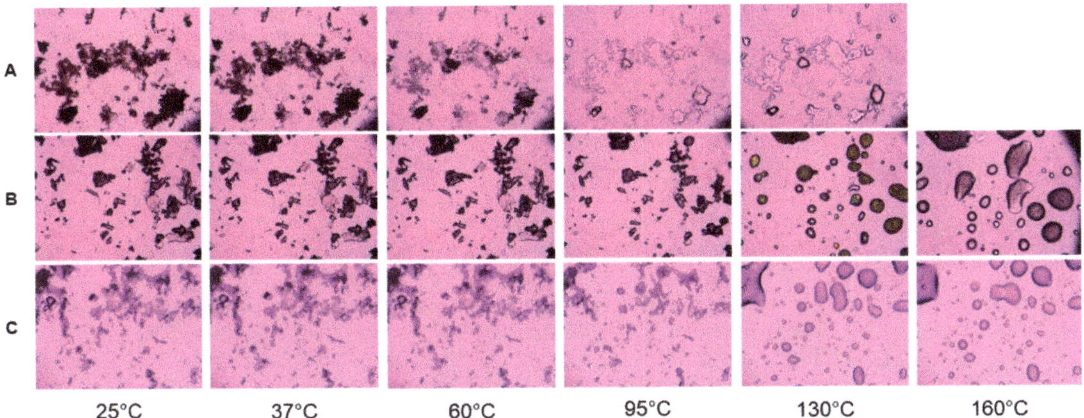

Figure 10. Representative hot stage microscopy images for (**A**) PLGA-PEG-NH$_2$ nanoparticles lyophilized without cryoprotectant, (**B**) nanoparticles lyophilized with 5% sucrose cryoprotectant, (**C**) nanoparticles lyophilized with 5% trehalose cryoprotectant.

3.9. Water Content Analysis by Karl Fischer Titration

Water content was determined for double emulsion-prepared lyophilates with and without cryoprotection (Table 4). For lyophilates of PLGA-PEG without cryoprotection, water content was 1.38 ± 0.20%. Comparatively, amine modification of the PEG group increased water content to 2.20 ± 0.61%. Lyophilization with 5% of either sucrose or trehalose resulted in significantly ($p < 0.05$) reduced water content, 0.78 ± 0.17% and 0.80 ± 0.19%, respectively.

Table 4. Residual water content of lyophilized nanoparticles with and without cryoprotectants. * indicates significantly different from other groups ($p < 0.05$) Data are presented as the mean ($n = 3$).

Nanoparticle	Water% (*w/w*)
PLGA-PEG (no cryoprotectant)	1.38 ± 0.20
PLGA-PEG-NH$_2$ (no cryoprotectant)	2.20 ± 0.61
PLGA-PEG-NH$_2$ (sucrose)	0.78 ± 0.17 *
PLGA-PEG-NH$_2$ (trehalose)	0.80 ± 0.19 *

4. Discussion

Formoterol fumarate dihydrate is an FDA-approved long-acting beta-2 adrenergic agonist that has been approved for the treatment of asthma and chronic obstructive pulmonary disease [42,43] and has shown promise in treating mitochondrial dysfunction in a variety of diseases [14–16,18,20,21]. As a raw material it exists as a crystalline powder that has been reported to be slightly soluble in water, soluble and sparingly soluble in methanol and ethanol, respectively, and practically insoluble in acetone and diethyl ether. The objective of this study was to develop a method of entrapping formoterol within a polymeric nanoparticle 300–500 nm in diameter for sustained drug release. Initially, an evaluation of the solubility of formoterol fumarate dihydrate (FFD) was required to determine optimal nanoparticle preparation, given FFD is only slightly soluble in water (1.16 ± 0.02 mg/mL) (Figure 1), which is in good agreement with the published literature [41]. Ethanol/Methanol and water mixtures show exponentially increasing solubilities of FFD with increasing alcohol content (Figure 1A). However, above 50% volume fraction of ethanol there is a decrease in solubility likely through the formation of an ehtanolate, which has been described previously [41]. Solubility in common organics used in nanoparticle syntheses was also established as low, with FFD being practically insoluble in dichloromethane (DCM), chloroform, acetonitrile, and acetone (Table 1). Solubility in nonionic surfactants including the homopolymer PVA and the triblock copolymer Pluronic F127 was unchanged; however, it increased significantly with the addition of ionic surfactant sodium cholate, with solubility increasing above the critical micelle concentration (12 mM or 0.5%) (Figure 1C). Given the apparent amphiphilic nature of FFD, both single (oil-in-water) and double (water-in-oil-in-water) methods of nanoparticle synthesis were evaluated.

Achieving significant drug loading by single emulsion solvent evaporation methods was challenging. Due to the low solubility of FFD in the common solvents DCM and chloroform, methanol, selected due to the decreased likelihood of forming solvates with formoterol, was added to allow sufficient FFD concentrations in the organic phase. Increased drug loading was observed with increased molecular weight of PLGA as well as increasing PLGA-PEG concentration in the organic phase (Figure 2A,B), both of which are in agreement with previously reported trends in loading of hydrophobic drugs [44–46]. Increasing PLGA concentration had a measured impact on nanoparticle size as well, with increasing polymer concentration resulting in increasing median hydrodynamic diameter (Figure 3B), which has also been previously reported for single emulsion prepared nanoparticles [46,47].

Sonication time was additionally optimized, with median particle diameter being reduced with increasing sonication time up to 90 s, and no significant change in median diameter with additional sonication time (Figure 3A).

Single emulsion nanoparticles were prepared; however, they exhibited significant burst release (Figure 6A), with over 90% of the entrapped drug being released within the first 24 h. This, along with first-order release kinetics (Table 2), suggests that in the PLGA-PEG single emulsion-prepared particles, the formoterol was not well entrapped within the polymer matrix but rather resided predominantly on the PLGA-PEG surface and the formoterol release was primarily diffusion limited.

In an effort to improve drug loading and the slow release of formoterol, a double emulsion solvent evaporation method of preparing nanoparticles was evaluated along with modifications to the polymer. By modifying the PEG group with either terminal acidic or amine residues. Nanoparticles prepared using 10 mg/mL polymer and a 1% PVA inner phase were compared and showed significantly improved ($p < 0.05$) drug loading with the amine modified polymer compared to both methyl and acidic residue terminated PEG. This was potentially due to the amine residues on the surface and microdomains, described by Rabanel et al. (2014) [48], within the polymer matrix interacting with formoterol and preventing its diffusion out of the organic phase as the polymer hardened. The modifications to the PEG surface moieties also had the effect of modifying the zeta potential of the prepared particles, with amine terminated PEG particles having a more positive zeta potential and carboxylic acid-terminated PEG particles having a more negative zeta potential (Figure 5). The impact of greater (more non-zero) surface potential is likely greater physical stability in suspension as the particles with higher surface charges are less likely to form agglomerates [49–51].

Further modification to the double emulsion method by modifying the inner phase was accomplished by exchanging the 1% PVA for a 0.5% solution of sodium cholate, which significantly increased drug loading (Figure 2D). The increase in loading is not likely due to any increased viscosity of the sodium cholate solution, which has a literature value (0.90–0.91 mPa s [52]) of approximately half that of a 1% PVA solution (1.6–2.5 mPa s [53]) at 25 °C. The improved drug loading was completely ameliorated when the sodium cholate was switched for sodium deoxycholate at the same concentration. The 7α-hydroxyl group of sodium cholate, critical in forming a hydrophilic axis in sodium cholate secondary structures [54,55], was not present in the sodium deoxycholate (Figure 2E), suggesting that interactions with this hydrophilic moiety were responsible for the increased loading. The addition of sodium cholate also appears to slow formoterol release. Nanoparticles prepared with sodium cholate have a significantly ($p < 0.05$) slower release than those made with sodium deoxycholate (Figure 6B, Table 2). However, there was no significant difference between particles prepared with either 0.5% sodium cholate or 1% PVA inner phases. Lastly, there was no significant impact on drug release kinetics from increasing amine-terminated PEG polymer concentration in double emulsion-prepared nanoparticles (Figure 6C). However, there is a clear trend between increasing Fickian release character and increasing polymer concentration, with n trending towards 0.43 (Table 2). Differences in drug loading, especially between single and double emulsion-prepared particles, would also impact release kinetics as greater concentration gradients would directly impact diffusion limited release mechanisms.

Selecting a suitable means of lyophilization is a critical and often overlooked aspect of nanoparticle synthesis methods, as the changes in solid state physicochemical properties can have significant impacts on particle size, stability, and drug release [56–58]. In this study, we assessed the impact of increasing concentrations of three cryoprotectants: sucrose, trehalose, and mannitol. Mannitol had a significantly detrimental impact on primary particle size, with median hydrodynamic diameter increasing ~1 μm over the range tested (Figure 4). Sucrose and trehalose cryoprotectants were much improved, with changes in nanoparticle size decreasing with increased cryoprotectant. Both sucrose and trehalose saw less than 30 nm increases in particle size at 5% cryoprotectant (Figure 4).

The impact of nanoparticle synthesis and lyophilization is apparent on powder analysis. With raw FFD, bound water is initially removed in an endotherm preceding the main melting endotherm at 130 °C (Figure 9A), which is in good agreement with the literature [41]. This melting is not apparent in any of the prepared nanoparticles (Figure 9E–G), potentially due to formoterol being in an amorphous state [40,41]. This is further supported by the lack of sharp peaks in the XRPD patterns (Figure 8) of the lyophilized nanoparticles, which indicates no long-range molecular order and is consistent with an amorphous particle. Comparing the glass transition temperatures (T_g) of raw PLGA-PEG-NH$_2$ polymer to that of lyophilized nanoparticles without cryoprotectant shows a ~6 °C decrease in T_g, which is potentially due to greater adsorbed water in the lyophilized particles (Table 4). Cryoprotection with either sucrose or trehalose showed no significant change in polymer glass transition (Figure 9F,G) and resulted in higher initial endotherm, which is attributed to the melting of PEG groups, and is visible in HSM micrographs (Figure 10). Cryoprotection of nanoparticles with sucrose results in an exotherm not present in the raw sucrose, a recrystallization peak at 148 °C (Figure 9F), which is confirmed by birefringency above 130 °C in HSM micrographs (Figure 10B), as has been previously described following sucrose lyophilization [59,60]. Lyophilization of nanoparticles with trehalose produces no similar amorphous-to-crystalline transition, only a broad melting endotherm between 60 and 120 °C (Figures 9G and 10C). Increased adsorbed water and decreased T_g, as indicated in nanoparticles lyophilized without cryoprotection, could negatively impact the solid-state stability of nanoparticles as well as impact the drug release [61–65], further reinforcing the importance of cryoprotection when lyophilizing nanoparticles. In these experiments, as stated in the Methods section, the prepared and lyophilized nanoparticles are stored at −20 °C until use. Under these conditions, it is highly unlikely that chemical degradation of the formoterol or polymer would occur. Additionally, the lowered molecular mobility of the polymer and entrapped drug at these temperatures makes spontaneous release of the drug highly unlikely as well.

5. Conclusions

This comprehensive and systematic study focused on the design, development, characterization, and in vitro drug release of PEGylated PLGA polymeric nanoparticles containing formoterol drug for sustained release drug delivery applications. Initial formoterol drug loading, nanoparticle size, and in vitro drug release kinetics of the polymeric nanoparticles were improved upon by modification of synthesis parameters such as polymer molecular weight, polymer concentration, PEG modification, surfactant selection, and sonication time. These changes resulted in significantly improved drug loading and sustained release over the course of 7 days. Solid state characterization of the nanoparticles lyophilates has also been reported and shows the importance of cryoprotectant selection in preserving nanoparticle characteristics.

Author Contributions: Conceptualization, E.L.V., R.G.S. and H.M.M.; methodology, E.L.V., D.E.-B., R.G.S. and H.M.M.; formal analysis, E.L.V., D.E.-B., R.G.S. and H.M.M.; investigation, E.L.V., D.E.-B., R.G.S. and H.M.M.; resources, R.G.S. and H.M.M.; data curation, E.L.V., D.E.-B., R.G.S. and H.M.M.; writing—original draft preparation, E.L.V., D.E.-B., R.G.S. and H.M.M.; writing—review and editing, E.L.V., D.E.-B., R.G.S. and H.M.M.; project administration, R.G.S. and H.M.M.; funding acquisition, R.G.S. and H.M.M.; supervision, R.G.S. and H.M.M. All authors have read and agreed to the published version of the manuscript.

Funding: All SEM images and data were collected in the W.M. Keck Center for Nano-Scale Imaging in the Department of Chemistry and Biochemistry at the University of Arizona with funding from the W.M. Keck Foundation Grant.

Data Availability Statement: The data presented in this study are available on request from the Corresponding Author.

Conflicts of Interest: The authors declare no conflict of interest.

References

1. Couvreur, P. Nanoparticles in drug delivery: Past, present and future. *Adv. Drug Deliv. Rev.* **2013**, *65*, 21–23. [CrossRef] [PubMed]
2. Farokhzad, O.C.; Langer, R. Impact of nanotechnology on drug delivery. *ACS Nano* **2009**, *3*, 16–20. [CrossRef] [PubMed]
3. Chenthamara, D.; Subramaniam, S.; Ramakrishnan, S.G.; Krishnaswamy, S.; Essa, M.M.; Lin, F.-H.; Qoronfleh, M.W. Therapeutic efficacy of nanoparticles and routes of administration. *Biomater. Res.* **2019**, *23*, 20. [CrossRef] [PubMed]
4. Mansour, H.M.; Sohn, M.; Al-Ghananeem, A.P.P. Materials for Pharmaceutical Dosage Forms: Molecular Pharmaceutics and Controlled Release Drug Delivery Aspects. *Int. J. Mol. Sci.* **2010**, *11*, 3298–3322. [CrossRef] [PubMed]
5. Rhee, Y.S.; Park, C.W.; DeLuca, P.P.; Mansour, H.M. Sustained-Release Injectable Drug Delivery Systems. *Pharm. Technol. Spec. Issue-Drug Deliv.* **2010**, *11*, 6–13.
6. Moghimi, S.M.; Hunter, A.C.; Murray, J.C. Long-circulating and target-specific nanoparticles: Theory to practice. *Pharmacol. Rev.* **2001**, *53*, 283–318. [PubMed]

7.   Williams, R.M.; Shah, J.; Tian, H.S.; Chen, X.; Geissmann, F.; Jaimes, E.A.; Heller, D.A. Selective Nanoparticle Targeting of the Renal Tubules. *Hypertension* **2018**, *71*, 87–94. [CrossRef]
8.   Muralidharan, P.; Mallory, E.; Malapit, M.; Hayes, D., Jr.; Mansour, H.M. Inhalable PEGylated Phospholipid Nanocarriers and PEGylated Therapeutics for Respiratory Delivery as Aerosolized Colloidal Dispersions and Dry Powder Inhalers. *Pharmaceutics* **2014**, *6*, 333–353. [CrossRef]
9.   Williams, R.M.; Shah, J.; Ng, B.D.; Minton, D.R.; Gudas, L.J.; Park, C.Y.; Heller, D.A. Mesoscale nanoparticles selectively target the renal proximal tubule epithelium. *Nano Lett.* **2015**, *15*, 2358–2364. [CrossRef] [PubMed]
10. Williams, R.M.; Jaimes, E.A.; Heller, D.A. Nanomedicines for kidney diseases. *Kidney Int.* **2016**, *90*, 740–745. [CrossRef]
11. Han, S.J.; Williams, R.M.; D'Agati, V.; Jaimes, E.A.; Heller, D.A.; Lee, H.T. Selective nanoparticle-mediated targeting of renal tubular Toll-like receptor 9 attenuates ischemic acute kidney injury. *Kidney Int.* **2020**, *98*, 76–87. [CrossRef]
12. Kim, S.H.; Jeong, J.H.; Chun, K.W.; Park, T.G. Target-specific cellular uptake of PLGA nanoparticles coated with poly(L-lysine)-poly(ethylene glycol)-folate conjugate. *Langmuir* **2005**, *21*, 8852–8857. [CrossRef] [PubMed]
13. Beck-Broichsitter, M.; Gauss, J.; Gessler, T.; Seeger, W.; Kissel, T.; Schmehl, T. Pulmonary targeting with biodegradable salbutamol-loaded nanoparticles. *J. Aerosol. Med. Pulm. Drug Deliv.* **2010**, *23*, 47–57. [CrossRef] [PubMed]
14. Jesinkey, S.R.; Funk, J.A.; Stallons, L.J.; Wills, L.P.; Megyesi, J.K.; Beeson, C.C.; Schnellmann, R.G. Formoterol restores mitochondrial and renal function after ischemia-reperfusion injury. *J. Am. Soc. Nephrol.* **2014**, *25*, 1157–1162. [CrossRef] [PubMed]
15. Cleveland, K.H.; Brosius, F.C.; Schnellmann, R.G. Regulation of mitochondrial dynamics and energetics in the diabetic renal proximal tubule by the β. *Am. J. Physiol. Renal. Physiol.* **2020**, *319*, F773–F779. [CrossRef] [PubMed]
16. Scholpa, N.E.; Williams, H.C.; Wang, W.; Corum, D.; Narang, A.; Tomlinson, S.; Sullivan, P.; Rabchevsky, A.S.; Schnellmann, R. Pharmacological Stimulation of Mitochondrial Biogenesis Using the Food and Drug Administration-Approved β. *J. Neurotrauma* **2019**, *36*, 962–972. [CrossRef] [PubMed]
17. Scholpa, N.E.; Simmons, E.C.; Crossman, J.D.; Schnellmann, R.G. Time-to-treatment window and cross-sex potential of Beta 2-adrenergic receptor-induced mitochondrial biogenesis-mediated recovery after spinal cord injury. *Toxicol. Appl. Pharmacol.* **2021**, *411*, 115366. [CrossRef] [PubMed]
18. Arif, E.; Solanki, A.K.; Srivastava, P.; Rahman, B.; Fitzgibbon, W.R.; Deng, P.; Budisavljevic, M.N.; Baicu, C.F.; Zile, M.R.; Megyesi, J.; et al. Mitochondrial biogenesis induced by the β2-adrenergic receptor agonist formoterol accelerates podocyte recovery from glomerular injury. *Kidney Int.* **2019**, *96*, 656–673. [CrossRef]
19. Vekaria, H.J.; Hubbard, W.B.; Scholpa, N.E.; Spry, M.L.; Gooch, J.L.; Prince, S.J.; Schnellmann, R.G.; Sullivan, P.G. Formoterol, a beta-2-adrenoreceptor agonist, induces mitochondrial biogenesis and promotes cognitive recovery after traumatic brain injury. *Neurobiol. Dis.* **2020**, *140*, 104866. [CrossRef]
20. Bhargava, P.; Schnellmann, R.G. Mitochondrial energetics in the kidney. *Nat. Rev. Nephrol.* **2017**, *13*, 629–646. [CrossRef]
21. Cameron, R.B.; Gibbs, W.S.; Miller, S.R.; Dupre, T.V.; Megyesi, J.; Beeson, C.C.; Schnellmann, R.G. Proximal Tubule Beta-2 Adrenergic Receptor Mediates Formoterol-Induced Recovery of Mitochondrial and Renal Function After Ischemia-Reperfusion Injury. *J. Pharmacol. Exp. Ther.* **2019**, *369*, 173–180. [CrossRef] [PubMed]
22. Levine, M.A.; Leenen, F.H. Role of beta 1-receptors and vagal tone in cardiac inotropic and chronotropic responses to a beta 2-agonist in humans. *Circulation* **1989**, *79*, 107–115. [CrossRef] [PubMed]
23. Brodde, O.E. Beta 1- and beta 2-adrenoceptors in the human heart: Properties, function, and alterations in chronic heart failure. *Pharmacol. Rev.* **1991**, *43*, 203–242.
24. Vyas, F.S.; Nelson, C.P.; Freeman, F.; Boocock, D.J.; Hargreaves, A.J.; Dickenson, J.M. β 2-adrenoceptor-induced modulation of transglutaminase 2 transamidase activity in cardiomyoblasts. *Eur. J. Pharmacol.* **2017**, *813*, 105–121. [CrossRef] [PubMed]
25. Koziczak-Holbro, M.; Rigel, D.F.; Dumotier, B.; Sykes, D.A.; Tsao, J.; Nguyen, N.; Bösch, J.; Jourdain, M.; Flotte, L.; Adachi, Y. Pharmacological Characterization of a Novel 5-Hydroxybenzothiazolone-Derived β 2-Adrenoceptor Agonist with Functional Selectivity for Anabolic Effects on Skeletal Muscle Resulting in a Wider Cardiovascular Safety Window in Preclinical Studies. *J. Pharmacol. Exp. Ther.* **2019**, *369*, 188–199. [CrossRef] [PubMed]
26. Molenaar, P.; Chen, L.; Parsonage, W.A. Cardiac implications for the use of beta2-adrenoceptor agonists for the management of muscle wasting. *Br. J. Pharmacol.* **2006**, *147*, 583–586. [CrossRef]
27. Yin, Q.; Yang, C.; Wu, J.; Lu, H.; Zheng, X.; Zhang, Y.; Lv, Z.; Zheng, X.; Li, Z. Downregulation of β-Adrenoceptors in Isoproterenol-Induced Cardiac Remodeling through HuR. *PLoS ONE* **2016**, *11*, e0152005. [CrossRef]
28. Dorn, G.W. Adrenergic pathways and left ventricular remodeling. *J. Card. Fail.* **2002**, *8*, S370–S373. [CrossRef]
29. Brouri, F.; Findji, L.; Mediani, O.; Mougenot, N.; Hanoun, N.; le Naour, G.; Hamon, M.; Lechat, P. Toxic cardiac effects of catecholamines: Role of beta-adrenoceptor downregulation. *Eur. J. Pharmacol.* **2002**, *456*, 69–75. [CrossRef]
30. He, J.; Chen, H.; Zhou, W.; Chen, M.; Yao, Y.; Zhang, Z.; Tan, N. Kidney targeted delivery of asiatic acid using a FITC labeled renal tubular-targeting peptide modified PLGA-PEG system. *Int. J. Pharm.* **2020**, *584*, 119455. [CrossRef]
31. Yu, H.; Lin, T.; Chen, W.; Cao, W.; Zhang, C.; Wang, T.; Ding, M.; Zhao, S.; Wei, H.; Guo, H.; et al. Size and temporal-dependent efficacy of oltipraz-loaded PLGA nanoparticles for treatment of acute kidney injury and fibrosis. *Biomaterials* **2019**, *219*, 119368. [CrossRef] [PubMed]
32. Nair, A.V.; Keliher, E.J.; Core, A.B.; Brown, D.; Weissleder, R. Characterizing the interactions of organic nanoparticles with renal epithelial cells in vivo. *ACS Nano* **2015**, *9*, 3641–3653. [CrossRef] [PubMed]

33. Vallorz, E.L.; Blohm-Mangone, K.; Schnellmann, R.G.; Mansour, H.M. Formoterol PLGA-PEG Nanoparticles Induce Mitochondrial Biogenesis in Renal Proximal Tubules. *AAPS J.* **2021**, *23*, 88. [CrossRef] [PubMed]
34. Aodah, A.; Pavlik, A.; Karlage, K.; Myrdal, P.B. Preformulation Studies on Piperlongumine. *PLoS ONE* **2016**, *11*, e0151707. [CrossRef]
35. Akapo, S.O.; Asif, M. Validation of a RP-HPLC method for the assay of formoterol and its related substances in formoterol fumarate dihydrate drug substance. *J. Pharm. Biomed. Anal.* **2003**, *33*, 935–945. [CrossRef]
36. Zhang, Z.; Feng, S.S. The drug encapsulation efficiency, in vitro drug release, cellular uptake and cytotoxicity of paclitaxel-loaded poly(lactide)-tocopheryl polyethylene glycol succinate nanoparticles. *Biomaterials* **2006**, *27*, 4025–4033. [CrossRef] [PubMed]
37. Duan, J.; Mansour, H.M.; Zhang, Y.; Deng, X.; Chen, Y.; Wang, J.; Pan, Y.; Zhao, J. Reversion of multidrug resistance by co-encapsulation of doxorubicin and curcumin in chitosan/poly(butyl cyanoacrylate) nanoparticles. *Int. J. Pharm.* **2012**, *426*, 193–201. [CrossRef] [PubMed]
38. Cheng, J.; Teply, B.A.; Sherifi, I.; Sung, J.; Luther, G.; Gu, F.X.; Levy-Nissenbaum, E.; Radovic-Moreno, A.F.; Langer, R.; Farokhzad, O.C. Formulation of functionalized PLGA-PEG nanoparticles for in vivo targeted drug delivery. *Biomaterials* **2007**, *28*, 869–876. [CrossRef] [PubMed]
39. Alabsi, W.; Al-Obeidi, F.A.; Polt, R.; Mansour, H.M. Organic Solution Advanced Spray-Dried Microparticulate/Nanoparticulate Dry Powders of Lactomorphin for Respiratory Delivery: Physicochemical Characterization, In Vitro Aerosol Dispersion, and Cellular Studies. *Pharmaceutics* **2020**, *13*, 26. [CrossRef] [PubMed]
40. Tajber, L.; Corrigan, D.O.; Corrigan, O.I.; Healy, A.M. Spray drying of budesonide, formoterol fumarate and their composites—I. Physicochemical characterisation. *Int. J. Pharm.* **2009**, *367*, 79–85. [CrossRef] [PubMed]
41. Jarring, K.; Larsson, T.; Stensland, B.; Ymén, I. Thermodynamic stability and crystal structures for polymorphs and solvates of formoterol fumarate. *J. Pharm. Sci.* **2006**, *95*, 1144–1161. [CrossRef]
42. Sharafkhaneh, A.; Mattewal, A.S.; Abraham, V.M.; Dronavalli, G.; Hanania, N.A. Budesonide/formoterol combination in COPD: A US perspective. *Int. J. Chron. Obstruct. Pulmon. Dis.* **2010**, *5*, 357–366. [CrossRef] [PubMed]
43. Saari, S.M.; Vidgren, M.T.; Herrala, J.; Turjanmaa, V.M.; Koskinen, M.O.; Nieminen, M.M. Possibilities of formoterol to enhance the peripheral lung deposition of the inhaled liposome corticosteroids. *Respir. Med.* **2002**, *96*, 999–1005. [CrossRef] [PubMed]
44. Jyothi, N.V.; Prasanna, P.M.; Sakarkar, S.N.; Prabha, K.S.; Ramaiah, P.S.; Srawan, G.Y. Microencapsulation techniques, factors influencing encapsulation efficiency. *J. Microencapsul.* **2010**, *27*, 187–197. [CrossRef] [PubMed]
45. Wischke, C.; Schwendeman, S.P. Principles of encapsulating hydrophobic drugs in PLA/PLGA microparticles. *Int. J. Pharm.* **2008**, *364*, 298–327. [CrossRef]
46. Ravi, S.; Peh, K.K.; Darwis, Y.; Murthy, B.K.; Singh, T.R.; Mallikarjun, C. Development and characterization of polymeric microspheres for controlled release protein loaded drug delivery system. *Indian J. Pharm. Sci.* **2008**, *70*, 303–309. [CrossRef]
47. Song, X.; Zhao, Y.; Hou, S.; Xu, F.; Zhao, R.; He, J.; Cai, Z.; Li, Y.; Chen, Q. Dual agents loaded PLGA nanoparticles: Systematic study of particle size and drug entrapment efficiency. *Eur. J. Pharm. Biopharm.* **2008**, *69*, 445–453. [CrossRef]
48. Rabanel, J.M.; Hildgen, P.; Banquy, X. Assessment of PEG on polymeric particles surface, a key step in drug carrier translation. *J. Control. Release* **2014**, *185*, 71–87. [CrossRef] [PubMed]
49. Jacobs, C.; Müller, R.H. Production and characterization of a budesonide nanosuspension for pulmonary administration. *Pharm. Res.* **2002**, *19*, 189–194. [CrossRef]
50. Feng, S.; Huang, G. Effects of emulsifiers on the controlled release of paclitaxel (Taxol) from nanospheres of biodegradable polymers. *J. Control. Release* **2001**, *71*, 53–69. [CrossRef]
51. Vega, E.; Gamisans, F.; García, M.L.; Chauvet, A.; Lacoulonche, F.; Egea, M.A. PLGA nanospheres for the ocular delivery of flurbiprofen: Drug release and interactions. *J. Pharm. Sci.* **2008**, *97*, 5306–5317. [CrossRef]
52. Ting, W.; Tong-Chun, B.; Wei, W.; Jian-Jun, Z.; Cheng-Wen, Z. Viscosity and activation parameters of viscous flow of sodium cholate aqueous solution. *J. Mol. Liq.* **2008**, *142*, 150–154.
53. Mohsen-Nia, M.; Modarress, H. Viscometric study of aqueous poly(vinyl alcohol) (PVA) solutions as a binder in adhesive formulations. *J. Adhes. Sci. Technol.* **2012**, *20*, 1273–1280. [CrossRef]
54. Santhanalakshmi, J.; Lakshmi, G.S.; Aswal, V.K.; Goyal, P.S. Small-angle neutron scattering study of sodium cholate and sodium deoxycholate interacting micelles in aqueous medium. *Proc. Indian Acad. Sci.* **2001**, *113*, 55–62. [CrossRef]
55. Maslova, V.A.; Kiselev, M.A. Structure of Sodium Cholate Micelles. *Crystallogr. Rep.* **2018**, *63*, 472–475. [CrossRef]
56. Abdelwahed, W.; Degobert, G.; Stainmesse, S.; Fessi, H. Freeze-drying of nanoparticles: Formulation, process and storage considerations. *Adv. Drug Deliv. Rev.* **2006**, *58*, 1688–1713. [CrossRef]
57. Fonte, P.; Soares, S.; Sousa, F.; Costa, A.; Seabra, V.; Reis, S.; Sarmento, B. Stability study perspective of the effect of freeze-drying using cryoprotectants on the structure of insulin loaded into PLGA nanoparticles. *Biomacromolecules* **2014**, *15*, 3753–3765. [CrossRef]
58. Holzer, M.; Vogel, V.; Mäntele, W.; Schwartz, D.; Haase, W.; Langer, K. Physico-chemical characterisation of PLGA nanoparticles after freeze-drying and storage. *Eur. J. Pharm. Biopharm.* **2009**, *72*, 428–437. [CrossRef]
59. Kedward, C.J.; MacNaughtan, W.; Mitchell, J.R. Isothermal and non-isothermal crystallization in amorphous sucrose and lactose at low moisture contents. *Carbohydr. Res.* **2000**, *329*, 423–430. [CrossRef]

60. van Eerdenbrugh, B.; Froyen, L.; Martens, J.A.; Blaton, N.; Augustijns, P.; Brewster, M.; van den Mooter, G. Characterization of physico-chemical properties and pharmaceutical performance of sucrose co-freeze-dried solid nanoparticulate powders of the anti-HIV agent loviride prepared by media milling. *Int. J. Pharm.* **2007**, *338*, 198–206. [CrossRef]
61. Alexis, F. Factors affecting the degradation and drug-release mechanism of poly(lactic acid) and poly[(lactic acid)-co-(glycolic acid)]. *Polym. Int.* **2004**, *54*, 36–46. [CrossRef]
62. Kranz, H.; Ubrich, N.; Maincent, P.; Bodmeier, R. Physicomechanical properties of biodegradable poly(D,L-lactide) and poly(D,L-lactide-co-glycolide) films in the dry and wet states. *J. Pharm. Sci.* **2000**, *89*, 1558–1566. [CrossRef]
63. Makadia, H.K.; Siegel, S.J. Poly Lactic-co-Glycolic Acid (PLGA) as Biodegradable Controlled Drug Delivery Carrier. *Polymers* **2011**, *3*, 1377–1397. [CrossRef] [PubMed]
64. Xu, Y.; Kim, C.S.; Saylor, D.M.; Koo, D. Polymer degradation and drug delivery in PLGA-based drug-polymer applications: A review of experiments and theories. *J. Biomed. Mater. Res. B Appl. Biomater.* **2017**, *105*, 1692–1716. [CrossRef]
65. Wang, B.; Tchessalov, S.; Cicerone, M.T.; Warne, N.W.; Pikal, M.J. Impact of sucrose level on storage stability of proteins in freeze-dried solids: II. Correlation of aggregation rate with protein structure and molecular mobility. *J. Pharm. Sci.* **2009**, *98*, 3145–3166. [CrossRef] [PubMed]

pharmaceutics

Article

Glycosylated Ang-(1-7) MasR Agonist Peptide Poly Lactic-co-Glycolic Acid (PLGA) Nanoparticles and Microparticles in Cognitive Impairment: Design, Particle Preparation, Physicochemical Characterization, and In Vitro Release

David Encinas-Basurto [1], John P. Konhilas [2], Robin Polt [3,4], Meredith Hay [5] and Heidi M. Mansour [1,4,6,7,*]

1   Skaggs Pharmaceutical Sciences Center, College of Pharmacy, The University of Arizona, Tucson, AZ 85721, USA; dencinas@pharmacy.arizona.edu
2   Department of Physiology and Sarver Heart Center, The University of Arizona, Tucson, AZ 85721, USA; konhilas@arizona.edu
3   Department of Chemistry & Biochemistry, The University of Arizona, Tucson, AZ 85721, USA; polt@email.arizona.edu
4   BIO5 Institute, The University of Arizona, Tucson, AZ 85721, USA
5   Department of Physiology and Evelyn F. McKnight, Brain Institute, The University of Arizona, Tucson, AZ 85721, USA; mhay@arizona.edu
6   Division of Translational and Regenerative Medicine, Department of Medicine, The University of Arizona College of Medicine, Tucson, AZ 85721, USA
7   Center for Translational Science, Florida International University, Port St. Lucie, FL 34987, USA
*   Correspondence: hmansour@fiu.edu; Tel.: +1-(772)-345-4731

Citation: Encinas-Basurto, D.; Konhilas, J.P.; Polt, R.; Hay, M.; Mansour, H.M. Glycosylated Ang-(1-7) MasR Agonist Peptide Poly Lactic-co-Glycolic Acid (PLGA) Nanoparticles and Microparticles in Cognitive Impairment: Design, Particle Preparation, Physicochemical Characterization, and In Vitro Release. *Pharmaceutics* **2022**, 14, 587. https://doi.org/10.3390/pharmaceutics14030587

Academic Editor: Oya Tagit

Received: 26 December 2021
Accepted: 5 March 2022
Published: 8 March 2022

Publisher's Note: MDPI stays neutral with regard to jurisdictional claims in published maps and institutional affiliations.

Abstract: Heart failure (HF) causes decreased brain perfusion in older adults, and increased brain and systemic inflammation increases the risk of cognitive impairment and Alzheimer's disease (AD). Glycosylated Ang-(1-7) MasR agonists (PNA5) has shown improved bioavailability, stability, and brain penetration compared to Ang-(1-7) native peptide. Despite promising results and numerous potential applications, clinical applications of PNA5 glycopeptide are limited by its short half-life, and frequent injections are required to ensure adequate treatment for cognitive impairment. Therefore, sustained-release injectable formulations of PNA5 glycopeptide are needed to improve its bioavailability, protect the peptide from degradation, and provide sustained drug release over a prolonged time to reduce injection administration frequency. Two types of poly(D,L-lactic-co-glycolic acid) (PLGA) were used in the synthesis to produce nanoparticles (≈0.769–0.35 μm) and microparticles (≈3.7–2.4 μm) loaded with PNA5 (ester and acid-end capped). Comprehensive physicochemical characterization including scanning electron microscopy, thermal analysis, molecular fingerprinting spectroscopy, particle sizing, drug loading, encapsulation efficiency, and in vitro drug release were conducted. The data shows that despite the differences in the size of the particles, sustained release of PNA5 was successfully achieved using PLGA R503H polymer with high drug loading (% DL) and high encapsulation efficiency (% EE) of >8% and >40%, respectively. While using the ester-end PLGA, NPs showed poor sustained release as after 72 h, nearly 100% of the peptide was released. Also, lower % EE and % DL values were observed (10.8 and 3.4, respectively). This is the first systematic and comprehensive study to report on the successful design, particle synthesis, physicochemical characterization, and in vitro glycopeptide drug release of PNA5 in PLGA nanoparticles and microparticles.

Keywords: PNA5 glycopeptide; mas receptor; angiotensin; PLGA diblock copolymer; ester and acid-end capped; double emulsion solvent evaporation; biocompatible; biodegradable; cardiovascular; neurodegenerative diseases

1. Introduction

Cases of dementia and heart failure (HF) both represent growing social, healthcare, and economic issues. It is estimated that more than 35 million people worldwide had dementia in 2010, with the number expected to double every 20 years [1,2]. HF causes decreased brain perfusion in older adults, and increased brain and systemic inflammation are known to increase the risk of cognitive impairment and Alzheimer's disease (AD) [3,4]. Decreased brain blood flow, increased reactive oxygen species (ROS) production, and proinflammatory mechanisms all appear to accelerate neurodegenerative disease progression, as seen in vascular contributions to cognitive impairment and dementia (VCID), AD, and related dementias [5] and, it is still an unmet condition that lacks effective treatment.

The renin-angiotensin system (RAS) is of importance in both cognitive function and HF. There is evidence that activating the ACE-AngII-AT1R pathway (in the RAS system) involves balancing ROS production and nitric oxide (NO) production in the brain. Xu, Sriramula [6] showed that the synthesis of Ang(1-7) by activating the ACE2 pathway correlated to decreased ROS production and increased nitric oxide synthase in the brain. Also, the G-protein-coupled receptor Mas (MasR) is activated by Ang(1-7) coupling. There is evidence that MasR is essential for standard object recognition processing, and the lack of the receptor impairs object recognition [7]. Our group hypothesized that Ang-(1-7) production and increased MasR activation might be beneficial for memory and cognitive function. In previous work, Hay, Vanderah [3] demonstrated that a systemic dose of Ang-(1-7) attenuated and even restored HF-induced cognitive impairment in mice, correlated to neuroprotective biomarkers.

However, it is known that low bioavailability and metabolic liability are observed after peptide or protein administration. High molecular weight, low lipophilicity, and charged functional groups in peptides affect their correct absorption. These features are the reason why most orally administered peptides have a low bioavailability (2%) and short half-life ($\approx$30 min) [8]. Although subcutaneous or intravenous dosing can help with absorption peptide opsonization, conformational changes, subunit protein dissociation, and systemic proteases remain a problem for biotherapeutics appropriate effect [8,9]. Hay, Polt [5] developed a novel Ang-(1-7) glycoside with MasR agonist activity and enhanced pharmacokinetics/pharmacodynamics properties. The Mas agonist Ang-1-6-O-Ser-Glc-NH_2 (PNA5) glycopeptides were rationally designed to improve bioavailability, stability, and brain uptake; PNA5 glycopeptide may be regarded as a new class of peptide drugs [10]. Previous in vitro work has shown that $t_{1/2}$ of PNA5 in serum samples is 1.06 ± 0.2 h. In vivo, after intravenous injection, bioavailability was improved compared to its native Ang-(1-7), but after 30 min neither PNA5 in serum nor CSF could be detected. Those observations were the motivation behind the encapsulation of PNA5 using FDA-approved polymer PLGA to get a slow and long sustained release to enhance its effects in the treatment and prevention of cognitive impairment.

Despite promising results and numerous potential PNA5 glycopeptide applications, limitations imposed by their short half-life required frequent injections to ensure adequate VCID treatment. Therefore, the proper formulation is needed for patients with this condition to achieve a long-term and consistent therapeutic effect [11–14]. Microparticles (MPs) and nanoparticles (NPs) have several advantages over the traditional application of the pure active pharmaceutical ingredient (API), such as increased drug bioavailability (and thus the possibility of reducing drug dosage and resultant side effects) [15]. Particle size has been reported to impact circulation time, cellular uptake, and release behavior. Apart from the significant physical differences between micro and nanoparticles, microparticles are thought to be able to entrap higher concentrations of medicines [16]. The FDA-approved diblock copolymer, pol(d,l-lactic-co-glycolic acid) (PLGA), is biocompatible and biodegradable [17] and has been extensively used for peptide/protein delivery and sustained-release formulations, including several successful pharmaceutical marketed sustained-release injectable products [18–21]. Depending on the drug type, it is feasible to alter the overall physical properties of the polymer-drug matrix by adjusting important parameters such as polymer molecular weight, lactide-to-glycolide ratio, and drug concentration to achieve a desired dosage and release interval; also, end-capped functional groups are a key component of PLGA [22]. These, in turn, affect peptide drug release from PLGA diblock copolymer and in vivo peptide drug pharmacokinetics [23].

In this comprehensive and systematic study, we successfully prepared PLGA NPs/MPs through a double emulsion solvent evaporation method to allow the successful and efficient encapsulation of PNA5 glycopeptide. Two types of poly(D,L-lactic-co-glycolic acid) (PLGA) were used in the synthesis to produce nanoparticles and microparticles loaded with PNA5 (ester and acid-end capped). Comprehensive physicochemical characterization including scanning electron microscopy, thermal analysis, molecular fingerprinting spectroscopy, particle sizing, drug loading, encapsulation efficiency, and in vitro drug release were conducted. To the authors' knowledge, this is the first systematic and comprehensive study to report on the successful design, particle synthesis, physicochemical characterization, and in vitro glycopeptide drug release of PNA5 PLGA nanoparticles and microparticles.

2. Materials and Methods

2.1. Materials

Various types of PLGA (50:50) diblock copolymers were purchased from Evonik (Essen, Germany) including 503H and 503 ($\approx$31 kDa, acid and ester end-capped, respectively). Polyvinyl Alcohol (PVA) (molecular weight 89,000-98,000 g/mol) was purchased from Sigma-Aldrich Chemical Co. (St. Louis, MO, USA). Dichloromethane and acetonitrile (HPLC grade) were obtained from Sigma-Aldrich (St. Louis, MO, USA). Glycosylated peptide (PNA5) was synthesized by the PolyPeptide Group, Torrence, CA via solid-phase synthesis.

2.2. Methods

2.2.1. Preparation of PNA5 PLGA NPs/MPs Using Double Emulsion Solvent Evaporation

PLGA NPs and MPs loaded with PNA5 peptides were prepared by the double-emulsification solvent evaporation method ($W_1/O/W_2$), with two esters and acid terminated PLGA 50:50 as a polymeric system (R503 and R503H, respectively), conditions and variables applied for the preparation of PNA5- micro/nanoparticles are shown in Table 1. Briefly, PLGA was dissolved in dichloromethane (DCM), forming the oil phase (O), then water phase in the first emulsion (W_1) was slowly added to the oil phase and sonicated at 50% amplitude for 180 s (Sonics Vibra-Cell vcx-500, Sonics & Materials, Inc., Newtown, CT, USA) to get the first emulsion (E_1), then E_1 was added to 2% w/v PVA solution (W_2) to form a second emulsion (E_2), and an E_2 was formed, either with homogenization (4500 rpm, Silverson L5M, Silverson Machines, Inc., East Longmeadow, MA, USA) or sonication (50% amplitude for 60 s) for MP and NPs, respectively. Furthermore, for PLGA MPs synthesis, E_2 was added to the W_3 phase to solidify the microspheres. After 3 h of stirring (oil phase solvent evaporation), MP and NPs were transferred to centrifuge tubes, washed three times (4000 rpm, 10 min, 10,000 rpm, 20 min, respectively) with distilled water. Before the freeze-drying process, particles were resuspended in a trehalose solution in a 1:2 weight ratio (PLGA:Trehalose).

Table 1. PNA5 loaded-PLGA microparticles (MPs) and nanoparticles (NPs): composition of formulations and manufacturing parameters.

Formulation	PLGA (mg)	Oil Phase (mL)	PNA5 (mg)	W_1 Phase (mL)	W_2 Phase (mL)	W_3 Phase (mL)	E_1 (min)	E_2 (min)
NPs	20	1	2	0.2	8	-	1	1
MPs	50	5	10	0.5	20	50	5	1

2.2.2. Characterization of the Prepared PNA5-Loaded PLGA MP/NPs

Particle Size, Size Distribution, and ζ Potential Measurements

The Z-average size and polydispersity index (PDI) of particles were measured by dynamic light scattering, using a Zetasizer (Nano ZS, Malvern Ltd., Malvern, UK). The zeta-potential, ζ, was measured by electrophoretic mobility using the same instrument. For that purpose, MP/NPs were diluted to $0.1 \times$ PBS, pH 7.4.

Scanning Electron Microscopy (SEM) Measurements

SEM was used to examine particle size and shape (FEI Inspect S microscope, FEI, Brno, Czech Republic). Double-side adhesive carbon tape was used to fix the samples to aluminum stubs (TedPella, Inc., Redding, CA, USA). An Anatech Hummer 6.2 sputtering system was used for sample gold-coating at 15 AC milliAmperes with about 7 kV of voltage for 90 s, getting 7 nm gold-thin film on the powder using a 9–12.5 mm working distance.

Encapsulation Efficiency (EE) and Drug loading (DL)

Peptide DL was determined by measuring the peptide content of dissolved particles by reversed-phase HPLC. For that purpose, 1 mg of dry particles was dissolved in 1 mL DMSO, and the solution was agitated at 37 °C for 60 min. Next, 10 mL MQ water was added, and the mixture was agitated at 37 °C for 60 min, to allow peptide extraction. A volume of 20 µL of the solution was filtered and injected into an HPLC system equipped with a C18 column. % EE was measured by HPLC injection of the first centrifuge supernatant of the NP/MPs. The drug-loading content (DLC %) and the drug entrapment efficiency (EE %) were calculated as shown in Equations (1) and (2), respectively:

$$\% \text{ DL} = (\text{Weight of Drug in NPs}/\text{Weight of NPs}) \times 100 \qquad (1)$$

$$\% \text{ EE} = (\text{Actual PNA5}/\text{Theoretical PNA5 Content}) \times 100 \qquad (2)$$

Chromatographic Equipment and Conditions

The analysis of PNA5 was performed by a reverse phase HPLC assay, using Phenomenex Prodigy 5 um ODS (3) 250 × 4.6 mm in an LC-2010HT next-generation HPLC (SHIMADZU, Tokyo, Japan). Mobile phase conditions were a gradient of acetonitrile (CH_3CN) in water (H_2O) with 0.1% trifluoroacetic acid (CF_3COOH) at a 1 mL/min flow rate. The start condition used was solvent A: solvent B in a ratio of 95:5; after 24 min, the ratio was 5:95. Solvent A composed 95% ACN and % 5 of the aqueous phase; solvent B was 20% ACN and 80% aqueous phase; the total run time was 24 min. Volume injection was 20 µL and the retention time for PNA5 was ~6 min.

2.2.3. Attenuated Total Reflectance (ATR) Fourier Transform Infrared Spectroscopy (FTIR)

A Nicolet Avatar 360 FTIR spectrometer (Varian Inc., Palo Alto, CA, USA) equipped with a DTGS detector and an attenuated total reflectance (ATR) accessory was used to record the spectra of raw PNA5 glycopeptide, PLGA diblock copolymer, and PNA5-loaded PLGA particles. Each spectrum was obtained across a wavenumber range of 4000–400 cm^{-1} for 32 scans at a spectral resolution of 2 cm^{-1}. Under the same experimental settings, a background spectrum was created. Data analysis was collected using the EZ OMNIC 9.1 software.

2.2.4. Differential Scanning Calorimeter (DSC)

Raw PNA5-PLGA formulations were investigated using thermal analysis and phase transition measurements. Thermograms were collected using a TA Q1000 differential scanning calorimeter (DSC) with T-Zero® technology, an automated computer-controlled RSC-90 cooling accessory, and an autosampler (TA Instruments, New Castle, DE, USA). In a hermetic anodized aluminum DSC pan, a mass of a 3–5 mg sample was accurately weighed. Using the T-Zero® hermetic press, the T-Zero® DSC pans were hermetically sealed (TA

Instruments). As a baseline, an empty hermetically sealed aluminum pan was employed. From 0.00 °C to 300 °C, DSC measurements were made at a rate of 5.00 °C/minute.

2.2.5. In Vitro Release of Peptide PLGA Nanoparticles and Microparticles

After resuspending 10 mg of freeze-dried MP/NPs in 10 mL of PBS buffer, the resulting particle suspension was incubated at 37 °C with magnetic stirring (200 rpm, n = 3). At certain time intervals, aliquots were taken, samples were centrifuged, and the supernatants were collected, followed by pellet resuspension in fresh PBS. The peptide concentration in the supernatant was determined with HPLC at 280 nm by comparing the concentration to a previously constructed standard calibration curve.

PNA5 in vitro drug release was fitted using four different kinetic models: zero-order, first-order, Higuchi, and Korsmeyer-Peppas.

Zero-order kinetics were fit to Equation (3):

$$Q_t - Q_0 = K_0 t \tag{3}$$

where Q_t is the amount of drug released after time t, Q_0 is the initial amount of drug in solution and K_0 is the zero-order rate constant.

First-order kinetics were fit to Equation (4):

$$\ln Q_t = \ln Q_0 - K_1 t \tag{4}$$

where Q_t is the amount of drug released after time t, Q_0 is the initial amount of drug in solution and K_1 is the first-order rate constant.

Higuchi kinetics were fit to Equation (5):

$$Q_t = K_H \times t^{1/2} \tag{5}$$

K_H is the Higuchi dissolving constant, and Q_t is the amount of drug released after time t.

Finally, release kinetics were fit to the Korsmeyer-Peppas model, Equation (6):

$$Q_t = K t^n \tag{6}$$

where Q_t is the amount of drug released after time t, K is the rate constant and n is the diffusion exponent for drug release.

3. Results

3.1. Characterization of the Prepared PNA5-Loaded PLGA MP/NPs

The influence of PLGA type (503H and 503) on the properties of PNA5-loaded particles was investigated in this study and for that, we generated two types of particles (NPs and MPs). As shown in Table 2, the diameter of the PLGA-based NPs, as determined by dynamic light scattering, was 0.76 μm and 0.35 μm for R503 and R503H, respectively. While the size of MPs was larger for R503 than 503H MP (3.7 μm versus 2.4 μm). Particles from both polymer types (R503 and R503H) and sizes (NP/MP) showed a negative surface charge higher for acid terminated PLGA of -30.4 ± 1.9 mV and 32.2 ± 2.5 mV for NP and MP, respectively. On the other hand, 503 polymers showed a ζpotential of -25 ± 1.5 and -24 ± 2.3 for NP and MP, respectively.

Table 2. Characteristics of nanoparticles (NPs) and microparticles (MPs) prepared from both PLGA types. EE = encapsulation efficiency; DL = drug loading; ζ = zeta potential; DI = polydispersity index. ($n = 3$).

Formulation		% EE	% DL	ζ Potential	Size (μm)	PDI	Span
NPs	503	10.8 ± 0.9	3.4 ± 0.2	−25 ± 1.5	0.769 ± 0.089	0.27 ± 0.04	-
	503H	42 ± 1.5	8.3 ± 0.4	−30.4 ± 1.9	0.35 ± 0.007	0.34 ± 0.02	-
MPs	503	13 ± 0.4	5.4 ± 0.5	−24 ± 2.3	3.7 ± 0.7	-	0.93 ± 0.01
	503H	55 ± 2.3	9.3 ± 0.5	−32 ± 2.5	2.4 ± 0.3	-	0.63 ± 0.04

To optimize the maximum amount of PNA5 encapsulated and to have a sustained release, we evaluated the two PLGA polymers (R503 and R503H) as NP and MPs. For NPs synthesis, an encapsulation efficiency (EE) of 10.8 and 42% was observed for R503 and R503H, respectively. For MPs, EE's of 13% and 55% were observed for 503 and 503H respectively. Regarding % drug loading (DL), MPs formulations yield a higher value than NPs. For example, when comparing the DL of 503H MPs to 503H NPs, the 503H MP showed a DL of 9.3% while NPs had a DL of 8.3%, with this trend being present in PLGA 503 particles.

The PNA5-loaded PLGA particles SEM micrographs are shown in Figure 1. Figure 1A,C are PLGA NPs using R503 and R503H. Figure 1B,D are for MPs of 503 and 503H, respectively. Since the optimized formula was uniform (according to low PDI), we chose a single image as a representation for each formulation. The results showed that PNA5-loaded PLGA NPs with a relatively smooth surface were spherical and had a homogenous distribution consistent with the abovementioned results showing low PDI values in all formulations.

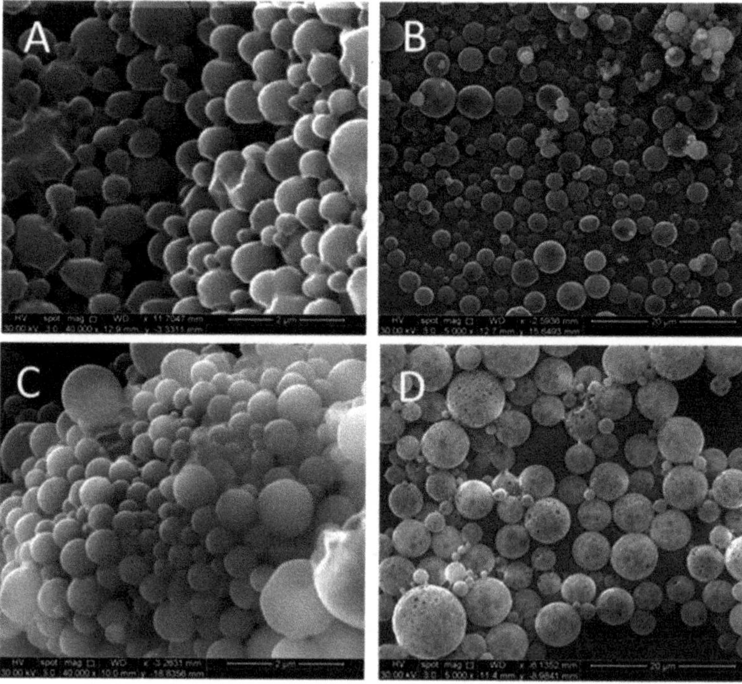

Figure 1. Scanning electron microscopy (SEM) images of microparticles (MPs) and nanoparticles (NPs) synthesized by double emulsion solvent evaporation for PLGA 503H (panels **A,B**) and PLGA 503 (panels **C,D**) diblock copolymers.

3.2. Attenuated Total Reflectance (ATR) Fourier Transform Infrared Spectroscopy (FTIR)

ATR FTIR molecular fingerprint spectra of pure PNA5 and PNA5-loaded particles are shown in Figure 2. The most typical spectra section for assessing peptide or protein-based drug secondary structure is the amide I area (1710–1590 cm^{-1}) [24]. A band at 1634 cm^{-1} in PNA5 spectra can be observed and represents the acidic carbonyl C = O stretching, and the spectra of PLGA MP of both polymers is also observed, however, in PLGA NPs formulations, that band is not present suggesting most of the peptide is entrapped inside the particle and is masked by PLGA bands. The intense band observed at 1762 cm^{-1} is attributed to the stretching vibration of the carbonyl groups present in the two monomers. The bands in the spectra of all formulations show a wide and sharp band of an -OH group, at 3150–3650 cm^{-1} [25].

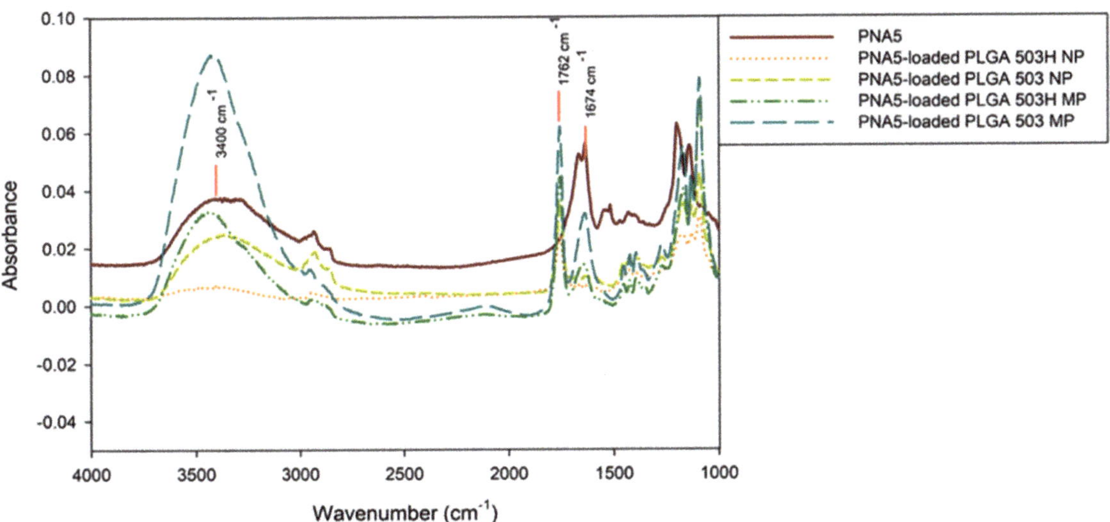

Figure 2. ATR FT-IR molecular fingerprinting spectra of Raw PNA5 glycopeptide and PNA5 PLGA polymeric nanoparticles and microparticles.

3.3. Differential Scanning Calorimetry (DSC)

DSC results of pure PNA5, PLGA NPs, and MPs using both polymers obtained in determining the physical state of the peptide are given in Figure 3. The DSC thermogram of PNA5 (Figure 3D) showed a sharp endothermic peak at 190 °C. On the other hand, two sets of peaks are observed in the DSC thermograms of different formulations as shown in Figure 3; a set of peaks at ~45 °C and an endotherm at ~190 °C. The first peak showed the T$_g$ of PLGA, while the second peak set showed the melting point, a first-order phase transition from the solid-state to the liquid state of PNA5.

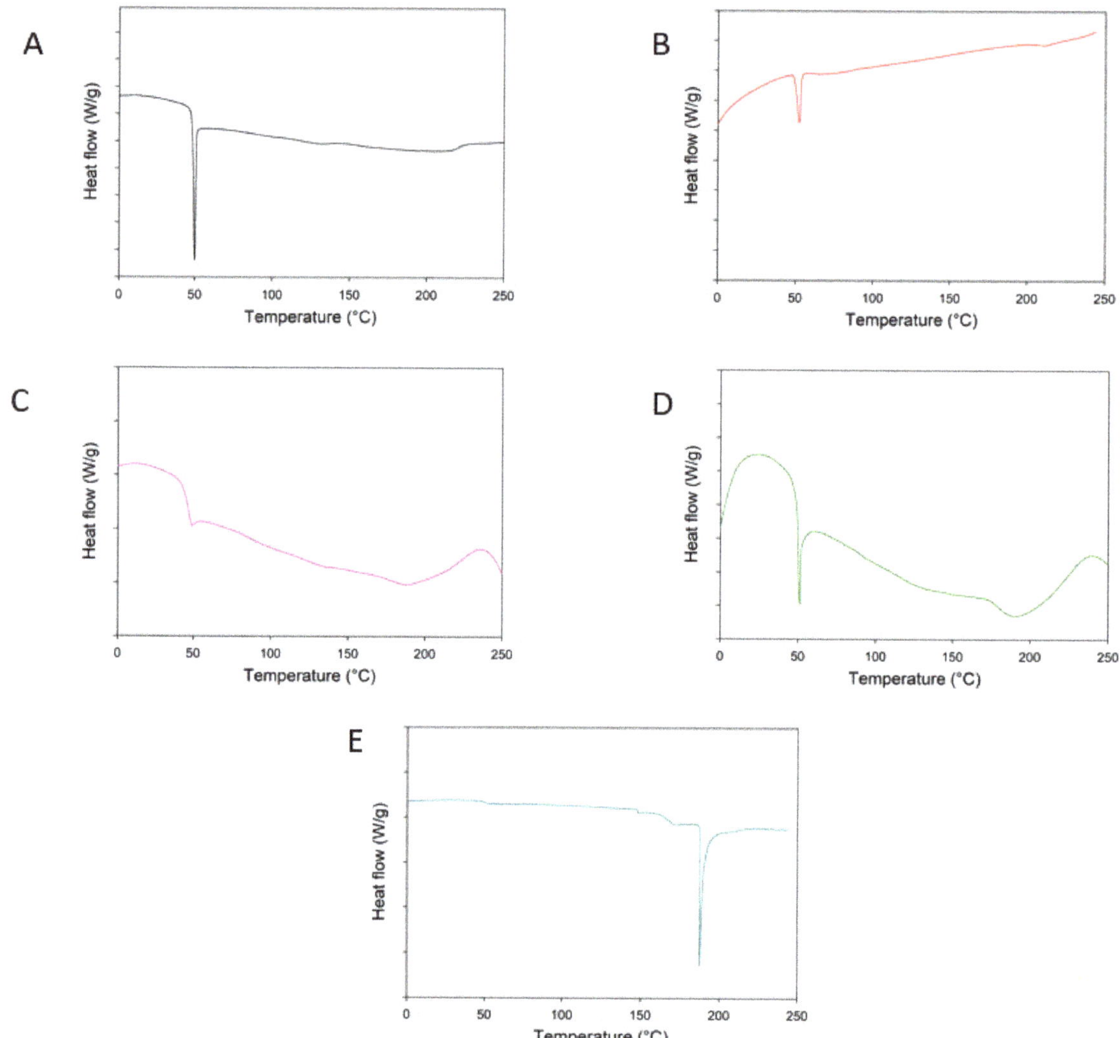

Figure 3. Differential scanning calorimetry (DSC) curves of Raw PNA5 glycopeptide and formulations: (**A**) 503 NPs, (**B**) 503H NPs, (**C**) 503 MPs, (**D**) 503H MPs, (**E**) Raw PNA5.

3.4. In Vitro Release of Peptide PLGA Nanoparticles and Microparticles

In vitro release profiles of PNA5 PLGA NP and MP with different PLGA types are shown in Figure 4. The burst release observed is influenced by its end group composition and size range. Lower burst release was observed for 503H in both NP and MP. For 503 H NPs, after 3 h, % released PNA5 was 12 ± 0.83, in contrast, ester terminated PLGA showed a % release of 56 ± 2.07. Regarding MPs, values of 11.19 ± 4.6 and 26.67 ± 1.8% release of PNA5 were observed for the 503H and 503 MPs, respectively. After burst release, PNA5 was released from PLGA particles in a sustained manner for all formulations, showing release for up to 14 days except for the 503 NPs. At the end of the experiment, for 503H formulation, release values of 51 ± 2.4% and 37 ± 4.5% were observed for PNA5. For ester-capped PLGA, the same trend was observed, NPs release much faster than MP formulation (100 versus 85.78 PNA5 % cumulative release).

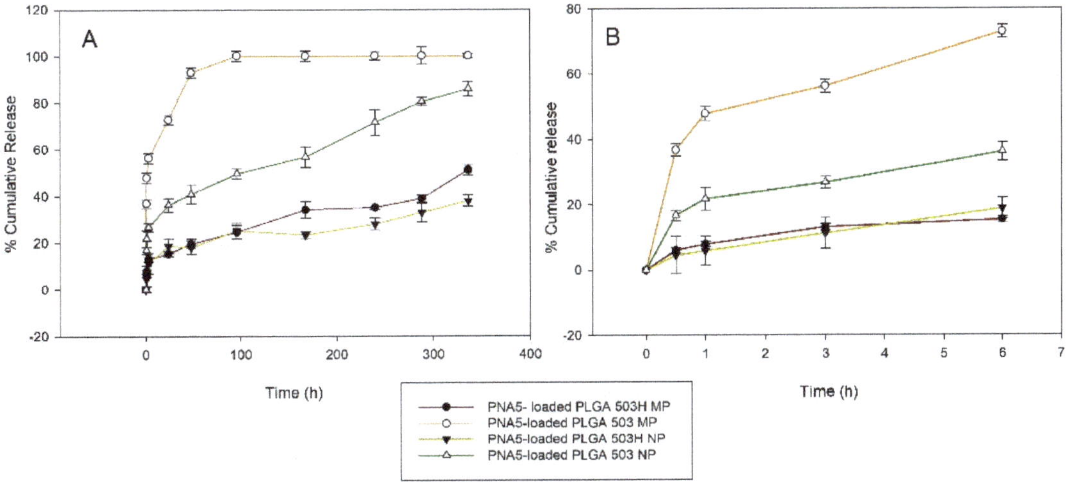

Figure 4. In vitro cumulative release of PNA5 glycopeptide over time from PLGA 503H and PLGA 503 polymeric nanoparticles and microparticles at pH = 7.4 and 37 °C. (**A**) 0–14 d; (**B**) 0–6 h.

4. Discussion

The main objective of the study was to prepare and evaluate the release of new MasR agonists PNA5 from PLGA NPs and MPs against cognitive impairment therapy. Previously in our group, PNA5 has shown a poor half-life of 1.0 ± 0.2 h after daily subcutaneous injections [5]. One potential means of addressing this problem is the encapsulation of polypeptide within a PLGA nano/micro particle carrier. These carriers offer the ability to encapsulate a wide variety of molecules, including nucleic acids, proteins and peptides, and small molecules, reducing dosing frequency and the ability of PLGA delivery systems to modify biological compounds' pharmacokinetics protecting the cargo from hydrolysis and subsequent degradation [22].

Both ester and acid terminated PLGA copolymers successfully synthesized both particle sizes by the w/o/w double emulsification solvent evaporation method. The double emulsion method was used to fabricate the PNA5 peptide-loaded PLGA NPs and MPs as this is the primary method that allows us to entrap highly hydrophilic molecules into a PLGA matrix. An important step in the preparation of particles was the formation of a stable homogenous primary emulsion, and proper energy input in the second emulsion to have the desired final size. The surface charge of particles showed high negative values, which means that the PLGA particles can be stable in suspension (Table 2) due to their strong electrostatic repulsive/attractive interactions. Further, PDI and span values for both sizes showed acceptable values.

The peptide % EE and % DL were highly influenced by the end-capped PLGA (acid or ester-terminated), suggesting that the acid-terminated part of the polymer can interact with any NH_2 group in the polypeptide chain [26]. Park, Lee [27], showed similar results when encapsulating drug molecules in PLGA; 503H particles showed a higher EE and DL than its ester-capped PLGA. The values of % DL in this article are comparable with commercial Lupron Depot®, which uses PLGA MPs and a Leuprolide acetate peptide with a DL of 8.5% [28]. It should be noted that only when the 503H is used for PLGA NP/MPs that this value can be achieved. Furthermore, the ionic interactions of peptides binding to polyesters can be explained by the observed behavior. Positively charged peptide moieties and negatively charged COOH end groups have been shown to interact via ionic forces, enhancing drug encapsulation and initial burst release [29]. This result could explain why acid carboxyl end-capped PLGA has a higher percentage of % EE and % DL than ester end-capped PLGA. More peptide molecules are attracted to polymers with more significant acid numbers because they have more negatively charged moieties.

DSC thermograms show the PNA5 endothermic peak as consistent with the literature [14]. The glass transition temperature of PLGA is responsible for the endothermic peak exhibited in all PLGA thermograms at about 48–50 °C [30]. The sharp endothermic PNA5 peak disappeared from the DSC curve of the PNA5-loaded PLGA particles formulation. The complete disappearance of the active substance peak could be attributed to either the formation of a homogeneous polymeric matrix or the dilution effect of the polymer [31]. The disappearance of the endothermic PNA5 peak in DSC curves of prepared particles indicated peptide incorporation, homogenous matrix formation, and amorphous structure [32].

Finally, different behavior was observed in the in vitro release study. A faster release of the polypeptide from nanoparticles was expected since the hydrolysis of the polymer is accelerated due to faster water penetration and acid-catalyzed degradation of the polyester matrix. Brauner, Schwarz [15], synthesized both PLGA NP and MP for trimethoprim for instillative treatments of urinary tract infections. They showed similar results to our study, where with NPs formulation of both polymers, the dissolution rate is higher than MPs. In their study, after 10 h of trimethoprim release, nearly 100% was completely released from the NP system and MPs formulation was around 75% using 503H polymer. Drug release from polymeric systems during the first phase, the active agent is released primarily through its diffusion through the polymer matrix, whereas during the second phase, the release is mediated by both drug diffusion and polymer matrix degradation. During polymer hydration, drug diffusion through the polymer matrix occurs slower [33]. To evaluate the drug release process and investigate the role of polymer in the dissolution profile, mathematical models were primarily used to predict the release of the encapsulated PNA5 as a function of time. As a result, the mathematical evaluation of our formulation's drug release kinetics was fitted against zero order, first order, Higuchi and Korsmeyer, and Peppas models (Table 3). This will ensure optimal pharmaceutical formulation(s) design as well as a better understanding of the release mechanism(s) through experimental verification [34].

Table 3. Mathematical models and kinetic constants.

Model	503H NP			503 NP			503H MP			503 MP		
	$K\ (h^{-1})$	R^2	n	$K\ (h^{-1})$	R^2	n	$K\ (h^{-1})$	R^2	n	$K\ (h^{-1})$	R^2	n
Zero-order	0.11	0.90	-	0.19	0.45	-	0.08	0.80	-	0.20	0.88	-
First-order	0.002	0.49	-	0.002	0.12	-	0.002	0.44	-	0.003	0.33	-
Higuchi	2.12	0.95	-	7.88	0.95	-	1.57	0.92	-	3.57	0.98	-
Korsmeyer and Peppas	5.40	0.96	0.35	6.66	0.91	0.27	20.2	0.98	0.19	44.8	0.97	0.17

Both PLGA 503H and PLGA 503 nano/microparticles correlated well to the Higuchi model (adjusted R^2 0.958 and 0.956 for NPs; 0.921 and 0.98, respectively) and Korsmeyer-Peppas (adjusted R^2 0.964 and 0.912 for NPs; 0.984 and 0.973, respectively) models. The diffusion exponent n of the Korsmeyer-Peppas model was between 0.17 and 0.35 suggesting a Fickian diffusion. The n value indicates the mechanisms to describe how the active compound is released from its matrix. In this case, the solvent diffusion is much greater than the process of polymeric chain relaxation. The kinetics of this phenomenon are characterized by diffusivity [35]. If n is less than 0.43 in the case of spherical encapsulation shape, then a Fickian diffusion release mechanism is implied [36]. The zero-order model describes drug release from a system in which the content is released at a constant rate independent of concentration. The first-order model describes a system in which the residual drug concentration uniquely determines the drug release rate. In addition, the Higuchi model hypothesizes that the system's edge effect must be negligible for drug diffusion and the dissolution and swelling of the system are irrelevant, showing a constant drug diffusivity. Korsmeyer-Peppas is another complete semi-empirical model that can be used to generalize the various release phenomena involving either diffusion or swelling [37,38].

Particles in the nanosized range have a greater surface area to volume ratio than particles in the micronized range, resulting in a faster dissolving rate, aided further by Brownian motion. This increased surface area of the nanoparticle should ideally allow for greater water contact/penetration into the particles while also allowing for faster degradation [39]. Due to the much faster drug release of 503 formulations, the nanometer and micrometer particles synthesized with the acid-capped PLGA seem to be better for PNA5 administration and improve its cognitive impairment effect in patients. Encapsulation peptides in PLGA diblock copolymer for sustained release injectable drug delivery continues to have significant clinical significance, as evidenced by multiple marketed pharmaceutical peptide PLGA products used clinically (e.g., leuprolide PLGA and octreotide PLGA) as sustained-release injectables [18] and continued research in this exciting area of peptide drug delivery [21,40,41].

5. Conclusions

This comprehensive and systematic study reports for the first time on the successful design and development of PNA5 loaded in the FDA-approved biocompatible biodegradable diblock copolymer PLGA as NPs and MPs by double emulsion solvent evaporation technique, comprehensive physicochemical characterization, and in vitro drug release. The formulations showed an acceptable PDI and span value for PLGA particles prepared by double emulsion, indicating monodisperse particles. The release of PNA5 from the particles was first with burst effect for both size and polymer used. The release was then slower and prolonged for up to 2 weeks for some formulations. The DSC thermal analysis and ATR FTIR molecular fingerprinting spectra analysis of the PNA5 PLGA particles revealed favorable intermolecular interaction between PNA5 and PLGA.

In this study, PLGA with the same molecular weight but different end-capping (acid or ester terminated) was used to compare the effects in formulation characterization. These end groups had an impact on properties, particularly the EE percent and release rate. The size of the particle also affected the release behavior due to the different surface-area ratios. The NP prepared with the polymer having the ester end-group released faster, whereas the NP prepared with only the acid end group released up to 14 days with 40% of PNA5 released. In terms of polymer selection, PLGA R503H provided several advantages over PLGA R503 in this study including higher drug loading, combined with a more sustained release over time, resulting in less burst release at the first sampling point.

Because the study of drug release kinetics provides critical information for better realization and optimization of nanoparticulate drug delivery, various methodologies for determining release kinetics from such formulations have been developed. The Higuchi model was the best-fitting mathematical model (among zero and first-order methods) for showing drug release patterns from PLGA particles in this study. The n values of Korsmeyer-Peppas are mostly less than 0.5, suggesting the release mechanism was governed by diffusion in all formulations. Regardless of the PLGA type used, both nanoparticles and microparticles demonstrated innovative PNA5 sustained-release formulations.

Author Contributions: D.E.-B.: Conceptualization, Investigation, Formal Analysis, Methodology, Writing—original, review, and editing; J.P.K.: Funding Acquisition, Supervision, Project Administration, Conceptualization, Methodology, Resources, Investigation, Writing—original, review, and editing; M.H.: Funding Acquisition, Supervision, Project Administration, Conceptualization, Methodology, Resources, Investigation, Writing—original, review, and editing; H.M.M.: Funding Acquisition, Supervision, Project Administration, Conceptualization, Investigation, Formal Analysis, Methodology, Resources, Investigation, Writing—original, review, and editing; R.P.: Conceptualization, Resources, Funding Acquisition, Writing—original, review, and editing. All authors have read and agreed to the published version of the manuscript.

Funding: The authors are grateful for grant funding from NIH NIA U01AG066623 (H.M.M., M.H., J.K. and R.P.). We are also grateful for funding provided by NIH NINDS R01NS091238 for the initial studies to produce PNA5 (R.P.).

Data Availability Statement: The data presented in this study are available on request from the corresponding author.

Acknowledgments: All SEM images and data were collected in the W.M. Keck Center for Nano-Scale Imaging in the Department of Chemistry and Biochemistry at the University of Arizona with funding from the W.M. Keck Foundation Grant. All FTIR spectra were collected in the W.M. Keck Center for Nano-Scale Imaging in the Department of Chemistry and Biochemistry at the University of Arizona. This instrument purchase was supported by Arizona Technology and Research Initiative Fund (A.R.S.§15-1648).

Conflicts of Interest: Meredith Hay is the founder and major share-holder in ProNeurogen, Inc., who has the exclusive license to all PNA5 technology. No funding from ProNeurogen was received in this study.

References

1. Cermakova, P.; Eriksdotter, M.; Lund, L.; Winblad, B.; Religa, P.; Religa, D. Heart failure and Alzheimer′s disease. *J. Intern. Med.* **2015**, *277*, 406–425. [CrossRef] [PubMed]
2. Prince, M.; Bryce, R.; Albanese, E.; Wimo, A.; Ribeiro, W.; Ferri, C.P. The global prevalence of dementia: A systematic review and metaanalysis. *Alzheimer's Dement.* **2013**, *9*, 63–75.e2. [CrossRef] [PubMed]
3. Hay, M.; Vanderah, T.W.; Samareh-Jahani, F.; Constantopoulos, E.; Uprety, A.R.; Barnes, C.A.; Konhilas, J. Cognitive impairment in heart failure: A protective role for angiotensin-(1-7). *Behav. Neurosci.* **2017**, *131*, 99. [CrossRef] [PubMed]
4. Suzuki, H.; Matsumoto, Y.; Ota, H.; Sugimura, K.; Takahashi, J.; Ito, K.; Miyata, S.; Furukawa, K.; Arai, H.; Fukumoto, Y. Hippocampal blood flow abnormality associated with depressive symptoms and cognitive impairment in patients with chronic heart failure. *Circ. J.* **2016**, *80*, 1773–1780. [CrossRef]
5. Hay, M.; Polt, R.; Heien, M.L.; Vanderah, T.W.; Largent-Milnes, T.M.; Rodgers, K.; Falk, T.; Bartlett, M.J.; Doyle, K.P.; Konhilas, J.P. A novel angiotensin-(1-7) glycosylated mas receptor agonist for treating vascular cognitive impairment and inflammation-related memory dysfunction. *J. Pharmacol. Exp. Ther.* **2019**, *369*, 9–25. [CrossRef]
6. Xu, P.; Sriramula, S.; Lazartigues, E. ACE2/ANG-(1–7)/Mas pathway in the brain: The axis of good. *Am. J. Physiol.-Regul. Integr. Comp. Physiol.* **2011**, *300*, R804–R817. [CrossRef]
7. Lazaroni, T.L.; Raslan, A.C.S.; Fontes, W.R.; de Oliveira, M.L.; Bader, M.; Alenina, N.; Moraes, M.F.; Santos, R.A.D.; Pereira, G.S. Angiotensin-(1–7)/Mas axis integrity is required for the expression of object recognition memory. *Neurobiol. Learn. Mem.* **2012**, *97*, 113–123. [CrossRef]
8. Bruno, B.J.; Miller, G.D.; Lim, C.S. Basics and recent advances in peptide and protein drug delivery. *Ther. Deliv.* **2013**, *4*, 1443–1467. [CrossRef]
9. Torchilin, V. Intracellular delivery of protein and peptide therapeutics. *Drug Discov. Today Technol.* **2008**, *5*, e95–e103. [CrossRef]
10. Apostol, C.R.; Hay, M.; Polt, R.J.P. Glycopeptide drugs: A pharmacological dimension between "Small Molecules" and "Biologics". *Peptides* **2020**, *131*, 170369. [CrossRef]
11. Khadka, P.; Ro, J.; Kim, H.; Kim, I.; Kim, J.T.; Kim, H.; Cho, J.M.; Yun, G.; Lee, J. Pharmaceutical particle technologies: An approach to improve drug solubility, dissolution and bioavailability. *Asian J. Pharm. Sci.* **2014**, *9*, 304–316. [CrossRef]
12. Sheth, P.; Sandhu, H.; Singhal, D.; Malick, W.; Shah, N.; Kislalioglu, M.S. Nanoparticles in the pharmaceutical industry and the use of supercritical fluid technologies for nanoparticle production. *Curr. Drug Deliv.* **2012**, *9*, 269–284. [CrossRef] [PubMed]
13. Mohammadi-Samani, S.; Taghipour, B. PLGA micro and nanoparticles in delivery of peptides and proteins; problems and approaches. *Pharm. Dev. Technol.* **2015**, *20*, 385–393. [CrossRef] [PubMed]
14. Alabsi, W.; Acosta, M.F.; Al-Obeidi, F.A.; Hay, M.; Polt, R.; Mansour, H.M. Synthesis, Physicochemical Characterization, In Vitro 2D/3D Human Cell Culture, and In Vitro Aerosol Dispersion Performance of Advanced Spray Dried and Co-Spray Dried Angiotensin (1–7) Peptide and PNA5 with Trehalose as Microparticles/Nanoparticles for Targeted Respiratory Delivery as Dry Powder Inhalers. *Pharmaceutics* **2021**, *13*, 1278.
15. Brauner, B.; Schwarz, P.; Wirth, M.; Gabor, F. Micro vs. nano: PLGA particles loaded with trimethoprim for instillative treatment of urinary tract infections. *Int. J. Pharm.* **2020**, *579*, 119158. [CrossRef]
16. Thundimadathil, J. Formulations of Microspheres and Nanoparticles for Peptide Delivery. In *Peptide Therapeutics: Strategy and Tactics for Chemistry, Manufacturing, and Controls*; The Royal Society of Chemistry: London, UK, 2019; pp. 503–530.
17. Mansour, H.M.; Sohn, M.; Al-Ghananeem, A.; DeLuca, P.P. Materials for Pharmaceutical Dosage Forms: Molecular Pharmaceutics and Controlled Release Drug Delivery Aspects. *Int. J. Mol. Sci.* **2010**, *11*, 3298–3322. [CrossRef]
18. Rhee, Y.S.; Park, C.W.; DeLuca, P.P.; Mansour, H.M. Sustained-Release Injectable Drug Delivery Systems. *Pharm. Technol. Spec. Issue-Drug Deliv.* **2010**, *34*, 6–13.

19. Zhang, C.; Yang, L.; Wan, F.; Bera, H.; Cun, D.; Rantanen, J.; Yang, M. Quality by design thinking in the development of long-acting injectable PLGA/PLA-based microspheres for peptide and protein drug delivery. *Int. J. Pharm.* **2020**, *585*, 119441. [CrossRef]
20. Wang, T.; Xue, P.; Wang, A.; Yin, M.; Han, J.; Tang, S.; Liang, R. Pore change during degradation of octreotide acetate-loaded PLGA microspheres: The effect of polymer blends. *Eur. J. Pharm. Sci.* **2019**, *138*, 104990. [CrossRef]
21. Zhang, C.; Wu, L.; Tao, A.; Bera, H.; Tang, X.; Cun, D.; Yang, M. Formulation and in vitro characterization of long-acting PLGA injectable microspheres encapsulating a peptide analog of LHRH. *J. Mater. Sci. Technol.* **2021**, *63*, 133–144. [CrossRef]
22. Makadia, H.K.; Siegel, S.J. Poly lactic-co-glycolic acid (PLGA) as biodegradable controlled drug delivery carrier. *Polymers* **2011**, *3*, 1377–1397. [CrossRef] [PubMed]
23. Rhee, Y.S.; Sohn, M.; Woo, B.H.; Thanoo, B.C.; DeLuca, P.P.; Mansour, H.M. Sustained-Release Delivery of Octreotide from Biodegradable Polymeric Microspheres. AAPS PharmSciTech: Special Theme—Sterile Products: Advances and Challenges in Formulation, Manufacturing. *Devices Regul. Asp.* **2011**, *12*, 1293–1301.
24. Ismail, R.; Sovány, T.; Gácsi, A.; Ambrus, R.; Katona, G.; Imre, N.; Csóka, I. Synthesis and statistical optimization of poly(lactic-co-glycolic acid) nanoparticles encapsulating GLP1 analog designed for oral delivery. *Pharm. Res.* **2019**, *36*, 99. [CrossRef] [PubMed]
25. Prabhuraj, R.; Bomb, K.; Srivastava, R.; Bandyopadhyaya, R. Dual drug delivery of curcumin and niclosamide using PLGA nanoparticles for improved therapeutic effect on breast cancer cells. *J. Polym. Res.* **2020**, *27*, 133.
26. Guo, W.; Quan, P.; Fang, L.; Cun, D.; Yang, M. Sustained release donepezil loaded PLGA microspheres for injection: Preparation, in vitro and in vivo study. *Asian J. Pharm. Sci.* **2015**, *10*, 405–414. [CrossRef]
27. Park, C.-W.; Lee, H.-J.; Oh, D.-W.; Kang, J.-H.; Han, C.-S.; Kim, D.-W. Preparation and in vitro/in vivo evaluation of PLGA microspheres containing norquetiapine for long-acting injection. *Drug Des. Dev. Ther.* **2018**, *12*, 711. [CrossRef]
28. Zhou, J.; Hirota, K.; Ackermann, R.; Walker, J.; Wang, Y.; Choi, S.; Schwendeman, A.; Schwendeman, S.P. Reverse engineering the 1-month Lupron Depot®. *AAPS J.* **2018**, *20*, 105. [CrossRef]
29. Sophocleous, A.M.; Desai, K.-G.H.; Mazzara, J.M.; Tong, L.; Cheng, J.-X.; Olsen, K.F.; Schwendeman, S.P. The nature of peptide interactions with acid end-group PLGAs and facile aqueous-based microencapsulation of therapeutic peptides. *J. Control. Release* **2013**, *172*, 662–670. [CrossRef]
30. Ayyoob, M.; Kim, Y.J. Effect of chemical composition variant and oxygen plasma treatments on the wettability of PLGA thin films, synthesized by direct copolycondensation. *Polymers* **2018**, *10*, 1132. [CrossRef]
31. Pang, J.; Luan, Y.; Li, F.; Cai, X.; Du, J.; Li, Z. Ibuprofen-loaded poly(lactic-co-glycolic acid) films for controlled drug release. *Int. J. Nanomed.* **2011**, *6*, 659.
32. Öztürk, A.A.; Kırımlıoğlu, G. Preparation and in vitro of characterization lamivudine loaded nanoparticles prepared by acid or ester terminated PLGA for effective oral antiretroviral therapy. *J. Res. Pharm.* **2019**, *23*, 897–913.
33. D'Souza, S.; Faraj, J.A.; Giovagnoli, S.; DeLuca, P.P. IVIVC from long acting olanzapine microspheres. *Int. J. Biomater.* **2014**, *2014*, 407065. [CrossRef] [PubMed]
34. Peppas, N.A.; Narasimhan, B.J.J.o.C.R. Mathematical models in drug delivery: How modeling has shaped the way we design new drug delivery systems. *J. Control. Release* **2014**, *190*, 75–81. [CrossRef] [PubMed]
35. Basak, S.C.; Kumar, K.S.; Ramalingam, M. Design and release characteristics of sustained release tablet containing metformin HCl. *Rev. Bras. Ciências Farm.* **2008**, *44*, 477–483. [CrossRef]
36. Singhvi, G.; Singh, M. In-vitro drug release characterization models. *Int. J. Pharm. Stud. Res.* **2011**, *2*, 77–84.
37. Paarakh, M.P.; Jose, P.A.; Setty, C.M.; Christoper, G.P. Release kinetics—concepts and applications. *Int. J. Pharm. Res. Technol.* **2018**, *8*, 12–20.
38. Jahromi, L.P.; Ghazali, M.; Ashrafi, H.; Azadi, A.J.H. A comparison of models for the analysis of the kinetics of drug release from PLGA-based nanoparticles. *Heliyon* **2020**, *6*, e03451. [CrossRef]
39. Panyam, J.; Dali, M.M.; Sahoo, S.K.; Ma, W.; Chakravarthi, S.S.; Amidon, G.L.; Levy, R.J.; Labhasetwar, V. Polymer degradation and in vitro release of a model protein from poly(D,L-lactide-co-glycolide) nano-and microparticles. *J. Control. Release* **2003**, *92*, 173–187. [CrossRef]
40. Du, X.; Xue, J.; Jiang, M.; Lin, S.; Huang, Y.; Deng, K.; Shu, L.; Xu, H.; Li, Z.; Yao, J. A multiepitope peptide, rOmp22, encapsulated in chitosan-PLGA nanoparticles as a candidate vaccine against Acinetobacter baumannii infection. *Int. J. Nanomed.* **2021**, *16*, 1819. [CrossRef]
41. Kostadinova, A.I.; Middelburg, J.; Ciulla, M.; Garssen, J.; Hennink, W.E.; Knippels, L.M.; van Nostrum, C.F.; Willemsen, L.E. PLGA nanoparticles loaded with beta-lactoglobulin-derived peptides modulate mucosal immunity and may facilitate cow's milk allergy prevention. *Eur. J. Pharmacol.* **2018**, *818*, 211–220. [CrossRef]

pharmaceutics

Article

Sustained and Long-Term Release of Doxorubicin from PLGA Nanoparticles for Eliciting Anti-Tumor Immune Responses

Jeongrae Kim [1,2,†], Yongwhan Choi [1,2,†], Suah Yang [1,2], Jaewan Lee [1,2], Jiwoong Choi [1,2], Yujeong Moon [2], Jinseong Kim [1,2], Nayeon Shim [1,2], Hanhee Cho [2], Man Kyu Shim [2], Sangmin Jeon [2], Dong-Kwon Lim [1], Hong Yeol Yoon [2] and Kwangmeyung Kim [1,2,*]

1   KU-KIST Graduate School of Converging Science and Technology, Korea University, 145 Anam-ro, Seongbuk-gu, Seoul 02841, Korea; 218312@kist.re.kr (J.K.); cyhwill@noxpharm.co.kr (Y.C.); haehwan@kist.re.kr (S.Y.); 220343@kist.re.kr (J.L.); 217802@kist.re.kr (J.C.); 218843@kist.re.kr (J.K.); sny3766@kist.re.kr (N.S.); dklim@korea.ac.kr (D.-K.L.)
2   Center for Theragnosis, Biomedical Research Institute, Korea Institute of Science and Technology (KIST), Seoul 02792, Korea; phoenix0310@kist.re.kr (Y.M.); ricky@kist.re.kr (H.C.); mks@kist.re.kr (M.K.S.); jeon@kist.re.kr (S.J.); seerou@kist.re.kr (H.Y.Y.)
*   Correspondence: kim@kist.re.kr; Tel.: +82-2-958-5916
†   Both authors contributed equally to this work.

Abstract: Immunogenic cell death (ICD) is a powerful trigger eliciting strong immune responses against tumors. However, traditional chemoimmunotherapy (CIT) does not last long enough to induce sufficient ICD, and also does not guarantee the safety of chemotherapeutics. To overcome the disadvantages of the conventional approach, we used doxorubicin (DOX) as an ICD inducer, and poly(lactic-co-glycolic acid) (PLGA)-based nanomedicine platform for controlled release of DOX. The diameter of 138.7 nm of DOX-loaded PLGA nanoparticles (DP-NPs) were stable for 14 days in phosphate-buffered saline (PBS, pH 7.4) at 37 °C. Furthermore, DOX was continuously released for 14 days, successfully inducing ICD and reducing cell viability in vitro. Directly injected DP-NPs enabled the remaining of DOX in the tumor site for 14 days. In addition, repeated local treatment of DP-NPs actually lasted long enough to maintain the enhanced antitumor immunity, leading to increased tumor growth inhibition with minimal toxicities. Notably, DP-NPs treated tumor tissues showed significantly increased maturated dendritic cells (DCs) and cytotoxic T lymphocytes (CTLs) population, showing enhanced antitumor immune responses. Finally, the therapeutic efficacy of DP-NPs was maximized in combination with an anti-programmed death-ligand 1 (PD-L1) antibody (Ab). Therefore, we expect therapeutic efficacies of cancer CIT can be maximized by the combination of DP-NPs with immune checkpoint blockade (ICB) by achieving proper therapeutic window and continuously inducing ICD, with minimal toxicities.

Keywords: cancer; chemoimmunotherapy; immunogenic cell death; nanomedicine; immune checkpoint blockade

Citation: Kim, J.; Choi, Y.; Yang, S.; Lee, J.; Choi, J.; Moon, Y.; Kim, J.; Shim, N.; Cho, H.; Shim, M.K.; et al. Sustained and Long-Term Release of Doxorubicin from PLGA Nanoparticles for Eliciting Anti-Tumor Immune Responses. *Pharmaceutics* **2022**, *14*, 474. https://doi.org/10.3390/pharmaceutics14030474

Academic Editor: Oya Tagit

Received: 28 January 2022
Accepted: 15 February 2022
Published: 22 February 2022

Publisher's Note: MDPI stays neutral with regard to jurisdictional claims in published maps and institutional affiliations.

1. Introduction

Immunogenic cell death (ICD) is representative of induced antitumor immunity, and in practice various cancer therapies could elicit immunogenicity by initiating ICD [1–3]. For instance, it has become known that several chemotherapeutics such as doxorubicin (DOX) kill cancer cells directly as well as cause successive antitumor immune responses by exposure to damage-associated molecular patterns (DAMPs) such as calreticulin (CRT), high mobility group box 1 (HMGB1), and adenosine triphosphate (ATP) [4–6]. Due to the ICD inducers, dendritic cells (DCs) subsequently act as a bridge between innate and adaptive immunity by presenting tumor antigens to T cells [7,8], leading to strong antitumor immunity based on the more immune-friendly reverted environment [9,10]. In spite of the key role of chemotherapeutics as potent inducers of ICD, their use has been severely limited for several reasons, such as poor pharmacokinetics (PK), high toxicity, and off-target effects [11]. Nanotechnology-based drug delivery systems (Nano-DDS) have been developed to overcome these traditional issues of cytotoxic drugs, and have enabled the prolonged circulation and increased tumor accumulation of drugs, and simultaneously reduced treatment-related toxicity [12–14]. Despite these promising signs, there still remain a few hurdles to be surmounted in nanomedicine, for example low delivery efficiency (<1%), heterogeneity of drug accumulation, and inadequate toxicokinetics causing side effects and immune adverse effects [15–18].

Nano-DDS-based local delivery could be a good alternative to solve most of these existing problems. Based on the nano-DDS, cytotoxic drugs could be relatively uniform and effective localized delivery at the target tumor sites, with minimal toxicities and controllable drug kinetics even elsewhere in the body [19–21]. Specifically, the effective and safe cancer therapy can be sufficiently provided by the optimal dose range of the nanomedicine inducing continuous ICD [22,23], especially based on the preserved tumor microenvironment (TME) with minimal histotoxicity and immunotoxicity [24–27]. Especially, immune checkpoint blockade (ICB) functions as a tumor suppressor by disrupting the existing interactions between immune cells and tumor cells [28]. ICB restores suppressed T cell-mediated antitumor immunity [29,30], and reverts tumor immune microenvironment (TIME) [31]. Therapeutic efficacies against cancer can be maximized in the responsive TIME [32]. There is no doubt of the promising potentials and therapeutic outcomes unleashing the strong antitumor immune responses in chemotherapy-based cancer immunotherapy [33–37].

In this regard, we hypothesized that ICD might be sufficiently induced during the whole treatment using poly(lactic-co-glycolic acid) (PLGA)-based nanomedicine by controlling the release kinetics of DOX. PLGA is a well-known biodegradable polymer, and has been widely used as a nanocarrier for controlled release of drug in the treatment of cancer [38,39]. The degradation rate and drug release kinetics of PLGA-based nanomedicine can be affected by its intrinsic polymer characteristics, such as the lactide:glycolide (L:G) ratio and the molecular weight [40–42]. Thus, PLGA platform is very suitable for controlling the efficacy and toxicity of drugs with the adequate drug loading and the desired controlled drug release profiles.

Tumor growth might be strongly inhibited, even in repeated treatment of DOX-loaded PLGA nanoparticles (DP-NPs) with reduced toxicities. The undevastated environment might fully promote antitumor immune responses, leading to high therapeutic performance, especially with the combination of ICB. To demonstrate our hypothesis, DOX was encapsulated with PLGA NPs using an oil-in-water (o/w) method, as we previously reported [43]. The morphology was examined using field emission scanning electron microscopy (FE-SEM). The efficiency of drug loading inside of NPs was measured using UV–vis spectroscopy. Size distribution, surface charge, and stability were determined using a zeta-sizer. Release kinetics of DOX were monitored under physiological conditions. Cellular uptake was observed using a confocal microscopy. Cytotoxicity was evaluated using cell counting kit-8 (CCK-8). ICD-associated DAMPs were investigated in cancer cell cultured system. All these features and behaviors of DP-NPs were characterized in vitro, and the other studies were performed in vivo. In vivo release kinetics of DOX from DP-NPs were monitored in CT26 tumor-bearing mice. Tumor growth inhibition by DP-NPs, also in combination therapy with anti-PD-L1 (aPD-L1) blockade, was monitored for 4 weeks. Finally, histology and immune cell analyses were performed simultaneously.

2. Materials and Methods

2.1. Materials

PLGA with acid endcaps (LG 50:50, Mn 5–10 kDa) was obtained from PolySciTech (West Lafayette, IN, USA). DOX hydrochloride (DOX-HCl) was obtained from FutureChem (Seoul, Korea). *N,N*-diisopropylethylamine (DIPEA), chloroform, dichloromethane (DCM), poly(vinyl alcohol) (PVA, hydrolyzed, Mw 30–70 kDa), dimethyl sulfoxide (DMSO), fluorescein amine, *N*-ethyl-*N'*-(3-(dimethylamino)propyl)carbodiimide (EDC), and *N*-

hydroxysuccinimide (NHS) were obtained from Sigma-Aldrich (St. Luis, MO, USA). Dialysis membranes (MWCO 3.5 kDa and 12–14 kDa) were purchased from Repligen Corporation (Waltham, MA, USA). CCK-8 was purchased from Dojindo (Rockville, MD, USA). ATP assay kit was purchased from Beyotime (Shanghai, China). Anti-CRT (mouse, AF647) mAb, anti-HMGB1 (mouse) pAb, and hematoxylin and eosin (H&E) staining were purchased from Abcam (Cambridge, UK). Terminal deoxynucleotidyl transferase (TdT) dUTP nick end labeling (TUNEL) assay kit was purchased from Promega (Madison, WI, USA). Amicon® centrifugal filter unit (MWCO 10 kDa) was purchased from Merk Millipore (Burlington, MA, USA). Tumor dissociation kit (mouse) was obtained from Miltenyi Biotec (Bergisch Gladbach, Germany). Anti-CD16/CD32 (mouse, Fc Block™) was purchased from BD Biosciences (San Jose, CA, USA). Anti-CD11c and anti-CD8a (mouse, APC), anti-CD40 (mouse, PE), anti-CD86 and anti-CD3 (mouse, FITC), and anti-CD45.2 (mouse, PE-Cy7) Abs were obtained from BioLegend (San Diego, CA, USA). aPD-L1 Ab (mouse) was purchased from BioXCell (Lebanon, NH, USA). CT26 (mouse colon carcinoma) was obtained from the Korean Cell Line Bank (Seoul, Korea). RPMI-1640 and fetal bovine serum (FBS) were obtained from Welgene Inc. (Gyeongsangbuk-do, Korea).

2.2. Preparation of DP-NPs

To prepare hydrophobic DOX, 100 mg of DOX-HCl was dissolved in 10 mL of deionized (DI) water and 200 μL of DIPEA was added additionally for 1 h. The desalted DOX was obtained after extraction with chloroform and subsequent evaporation. DP-NPs were prepared by the o/w single-emulsion solvent evaporation method. Briefly, 20 mg of PLGA was dissolved in 0.4 mL of DCM. Ten milligrams of desalted DOX dissolved in 0.4 mL of DCM was added into PLGA solution. For reference, to prepare PLGA NPs (P-NPs) as a vehicle, 0.4 mL of DCM without DOX was added into PLGA solution. Next, 3 mL of aqueous PVA solution (3 wt%) was slowly added into the previous mixture. This mixture was emulsified by sonication (180 watts) for 45 s in an ice bath, and stirred (800 rpm) to remove DCM for 3 h at room temperature (RT) after the addition of 6 mL of distilled water (DW). Then the suspension was centrifuged for 15 min at 15,000 rpm and washed twice with 30 mL of DW, subsequently lyophilized for 2 days after 0.45 μm filtration to obtain DP-NPs. To monitor cellular uptake of DP-NPs in vitro, fluorescein amine was chemically conjugated to carboxylic acid on the end of PLGA in the presence of EDC and NHS. In brief, 100 mg of PLGA was dissolved in 10 mL of DMSO, and mixed with 1 mL of DMSO containing 5.5 mg of EDC and 3.3 mg of NHS. Then, 1 mL of DMSO containing 10 mg of fluorescein amine was slowly added into the PLGA solution and stirred for 12 h at RT. The resulting solution was purified using a dialysis membrane (MWCO 3.5 kDa) against DMSO:DW (1:0, 1:1, 0:1 v/v%) for 3 days. The fluorescein-labeled PLGA (FITC-PLGA) was obtained after the lyophilization for 2 days. Finally, FITC-labeled DP-NPs were formulated as described above using FITC-PLGA and PLGA at 1:9 wt%. The DOX contents in DP-NPs and FITC-labeled DP-NPs were confirmed using a UV–vis spectrophotometer before use.

2.3. In Vitro Formulation and Characterization of DP-NPs

The morphology of DP-NPs was observed using FE-SEM (Regulus 8230, Hitachi, Tokyo, Japan). P-NPs and DP-NPs were dispersed in the DI water (0.2 mg/mL) for measurement of size distribution and surface charge. These features were measured using a zeta-sizer (Nano-ZS, Malvern Instruments, London, UK). The drug loading efficiency of DOX in DP-NPs was measured using UV–vis spectroscopy (Cary 60 UV–vis spectrophotometer, Agilent Technologies, Santa Clara, CA, USA), absorbance at 480 nm. The absorbance of DP-NPs (1 mg of DOX in 1 mL of DW) was measured at 480 nm, and calculated based on the calibration curve of DOX in various concentration and the baseline correction. The serum stability of P-NPs and DP-NPs were monitored in condition of 10% (v/v) of FBS-containing phosphate-buffered saline (PBS, pH 7.4) at 37 °C for 2 weeks. The size of both NPs was measured every 2 days using a zeta-sizer. Similarly, the release profile of DOX from DP-NPs, in 0.1% Tween 80-containing PBS (pH 7.4), was monitored in the water bath shaker (100 rpm) at 37 °C for 2 weeks. The DP-NPs (10 mg of DOX) initially within the dialysis membrane (MWCO 12–14 kDa) were dispersed in 10 mL of PBS, continually replaced as fresh PBS every other day. The amount of contained DOX in the harvested PBS was analyzed using UV–vis spectroscopy, absorbance at 480 nm.

2.4. In Vitro Cellular Uptake and Cytotoxicity

Cellular uptake imaging of FITC-labeled DP-NPs was performed in the CT26 tumor cells using a confocal microscopy (Leica TCS SP8, Wetzlar, Germany). Briefly, the CT26 tumor cells (1×10^5 cells) were seeded into dishes and incubated at 37 °C for 24 h. Then, the medium in dishes was replaced as 2 mL of new medium containing 2 µM of DOX-HCl or 10 µg of FITC-labeled DP-NPs (equiv. of 2 µM of DOX-HCl). The CT26 tumor cells were incubated for more 48 h, and they were washed with PBS (pH 7.4). Then, they were fixed with a paraformaldehyde (PFA) solution (4% in PBS) for 15 min. 4′,6-diamidino-2-phenylindole (DAPI) was utilized for nuclear staining at RT for 15 min. Cells were observed using a confocal microscopy equipped with diode, Ar, and HeNe lasers at 405, 488/514, and 633, respectively. Cell viability of the CT26 tumor was measured using CCK-8. The CT26 tumor cells (5×10^3 cells) were seeded into 96-well plates and incubated at 37 °C for 24 h. Cells were treated with various concentration (from 0.001 to 10 µM) of DOX-HCl and the equivalent DP-NPs, and further incubated for more 48 h. They were washed with PBS (pH 7.4) and incubated with a CCK-8 reagent-containing medium for 30 min. The absorbance at 450 nm of each medium was measured using a tunable microplate reader (VersaMax^{TM}, Molecular Devices, San Jose, CA, USA). For reference, the cell viability of the vehicle-treated CT26 tumor was also measured, in the same way as DP-NPs.

2.5. In Vitro Analysis of ICD-Associated DAMPs

To analyze the induced ICD in CT26 tumor cells, we observed several molecular markers of DAMPs such as CRT, HMGB1, and ATP. For reference, the analysis with the vehicle-treated group was also performed in each experiment, in the same way as DP-NPs. CRT exposed on the surface of CT26 tumor cells was observed using a confocal microscopy. Then, the medium in dishes was replaced as 2 mL of new medium containing 2 µM of DOX-HCl or 10 µg of DP-NPs (equiv. of 2 µM of DOX-HCl). The CT26 tumor cells were incubated for more 48 h, and they were washed with PBS (pH 7.4) and fixed with a PFA solution (4% in PBS) for 15 min. Cells were treated with the bovine serum albumin (BSA) blocking buffer (5% in PBS) to prevent the non-specific binding for 10 min. Then they were stained with CRT Ab for 1 h, and washed with PBS (pH 7.4). DAPI was utilized for nuclear staining at RT for 15 min. The stained CT26 tumor cells were observed using a confocal microscopy. Intensities of CRT was quantified using Image-Pro Plus software (Media Cybernetics, Silver Spring, MD, USA). The released HMGB1 and the secreted ATP in CT26 tumor cell culture medium was analyzed by Western blotting and ATP assay, respectively. Briefly, the CT26 tumor cells (1×10^6 cells) were seeded into culture dishes and incubated at 37 °C for 24 h. Then, the medium in dishes was replaced as the new medium (2 µM of DOX-HCl and the equivalent DP-NPs). The CT26 tumor cells were incubated for more 48 h. Their culture medium was collected and separated into two parts by centrifugation using a filter unit. The filtered sample (<10 kDa) was used to measure ATP levels, and the other sample (>10 kDa) was used to observe HMGB1 expressions. Band intensities of HMGB1 were quantified using ImageJ software (NIH, Bethesda, MD, USA).

2.6. In Vivo Drug Release Kinetics

All experiments with specific pathogen-free (SPF) mice were performed in compliance with the relevant laws and institutional guidelines of Korea Institute of Science and Technology (KIST), and Institutional Animal Care and Use Committee (IACUC) approved the experiments (approval number: KIST-2020-070, 27/05/2020). Six-week-old male BALB/c mice were purchased from OrientBio Inc. (Gyeonggi-do, Korea). The release of DOX from DP-NPs were observed by its fluorescence in CT26 tumor-bearing mice after the intratumoral injection of DP-NPs for 2 weeks. For reference, a DOX-HCl-treated mouse was also monitored as a control. Briefly, mice were subcutaneously injected with CT26 tumor cells (1×10^6 cells/60 µL) on their left flank. When the tumor volume reached 50–80 mm^3, 0.25 mg of DOX-HCl (10 mg/kg of DOX, DOX 10), 0.5 mg of DP-NPs (equiv. of 5 mg/kg of DOX, DP-NPs 5), 1 mg of DP-NPs (equiv. of 10 mg/kg of DOX, DP-NPs 10), and 1 mg of DP-NPs with two injections at intervals of 10 days (DP-NPs 10 × 2, equiv. of 10 mg/kg of DOX per injection) were intratumorally injected after dispersed using 30 µL of sterilized PBS (pH 7.4). Near-infrared fluorescence (NIRF) images were acquired by In Vivo Imaging System (IVIS, Lumina Series III, PerkinElmer, Waltham, MA, USA), and the fluorescence from these images were quantified using Living Image software (PerkinElmer, Waltham, MA, USA).

2.7. In Vivo Tumor Growth Inhibition and Toxicity

Mice were subcutaneously injected with CT26 tumor cells (1×10^6 cells/60 µL) on their left flank. When the tumor volume reached 50–80 mm^3, CT26 tumor-bearing mice were treated with PBS, P-NPs, DOX-HCl (10 mg/kg), DP-NPs (5, 10, 10 × 2 mg/kg), aPD-L1 Ab (10 × 5 mg/kg)-combined DOX-HCl (10 mg/kg) and DP-NPs (10 × 2 mg/kg). All samples were prepared using 30 µL of sterilized PBS (pH 7.4). Tumor growth, body weight change, and survival of each group were monitored for 4 weeks. Tumor volumes (calculated as $0.52 \times \text{Diameter}_{\text{Longitudinal}} \times \text{Diameter}_{\text{Transverse}}^2$) and body weight were measured every two days. We limited the maximum allowable size of subcutaneous tumors in control groups, and mice with a tumor size 3000 mm^3 were counted statistically dead, for survival. Photographs of spleens were obtained on the 14th day, and the weight of the spleen was also measured to examine the DOX-induced systemic toxicity.

2.8. In Vivo ICD-Associated DAMPs and Immune Cell Analysis

On the 14th day after the first injection, CT26 tumor tissues which were directly treated with PBS, P-NPs, DOX 10, and DP-NPs (5, 10, 10 × 2) and tumor-draining lymph nodes (TDLNs) were excised from the CT26 tumor-bearing mice to analyze ICD-associated DAMPs and immune cell population in the tumor tissues. The tumor tissues were dissociated with the tumor dissociation kit at 2% (v/v) of FBS-containing PBS, and centrifuged at 500 g for 10 min. The supernatant was harvested for Western blotting analysis of HMGB1, and the cells were collected for flow cytometric analysis of CTLs. The cells in TDLNs were also collected and centrifuged similarly for flow cytometric analysis of DCs. Red blood cell (RBC) lysis and single-cell suspension were performed for staining with the adequate Abs. FcR blocking reagent was used to eliminate non-specific staining. Then, the single cells from TDLNs were fluorescently stained with CD45, CD11c, CD40, and CD86 Abs for identifying matured DCs, and those from tumor tissues were stained with CD45, CD3, and CD8 Abs for identifying CTLs. After the incubation for 1 h, harvested cells were characterized by a flow cytometer (CytoFLEX, Beckman Coulter, CA, USA). These results were analyzed using FlowJo^{TM} software (BD Biosciences, NJ, USA).

2.9. TUNEL Assay and H&E Staining

Each sample was collected on the 14th day after the first treatment of PBS, P-NPs, DOX-HCl (10 mg/kg), and DP-NPs (5, 10, 10 × 2 mg/kg) in the previously stated in vivo experiments for TUNEL assay and H&E staining. TUNEL assay was performed to assess DOX-induced apoptotic changes in tumor tissues. The apoptotic cells were determined by TUNEL assay kit (Promega, Madison, WI, USA). Briefly, tumor tissues were deparaffinized and fixed with a PFA solution (4% in PBS) for 15 min. Proteinase K solution, equilibration and incubation buffers were sequentially added after the tissue permeabilization. Reaction was stopped in an hour, and the subsequent stains were performed with propidium iodide and DAPI. The localized green fluorescence of apoptotic cells in prepared samples were observed using a confocal microscopy. H&E staining was also performed in the tissues of tumor and major organs (spleen, liver, lung, kidney, and heart) to further investigate

DOX-induced in vivo toxicities. Briefly, the deparaffinized and rehydrated tumor tissues were sequentially stained with hematoxylin and eosin. After the dehydration, stained tumor tissues sliced into 10 μm sections were observed using an optical microscopy.

2.10. Statistical Analysis

In this study, one-way analysis of variance (ANOVA) was performed for a comparison of several experimental groups using Origin 2020 software (OriginLab Corporation, Northampton, MA, USA). Statistical significance was marked with an asterisk (* $p < 0.05$, ** $p < 0.01$, and *** $p < 0.001$) in the figures.

3. Results and Discussion

3.1. Characterization of the DP-NPs

In this study, we prepared P-NPs by the o/w single-emulsion solvent evaporation method. DP-NPs were also prepared in the same way as P-NPs with the hydrophobically desalted DOX. Their formulation showed a loading efficiency of 23.8% in that the amount of encapsulated DOX in 1 mg of DP-NPs was measured 238 ± 17 μg by UV–vis spectroscopy. We also confirmed each unit of DP-NPs was spherical, and they had a smooth particle surface similar to P-NPs from SEM images (Figure 1A). Unimodal size distributions were observed in DP-NPs as well as P-NPs, wherein the average particle size of P-NPs and DP-NPs was 138.7 ± 8.7 and 128.0 ± 3.9 nm, respectively (Figure 1B). The surface charge of the P-NPs and DP-NPs was −22.9 ± 6.5 and 11.8 ± 3.4 mV, respectively, and this difference was mainly because of a surface charge change resulting from the encapsulation of the hydrophobically modified DOX into the P-NPs. Actually, the changes in surface charges were directly affected by the hydrophobicity/hydrophilicity of DOX. The characteristics of the P-NPs and DP-NPs were presented in Figure 1C. The particle size of the P-NPs and DP-NPs was stable in PBS with 10% FBS for 2 weeks (Figure 1D). The release behavior of DOX from DP-NPs was investigated for 2 weeks. DP-NPs dispersed in PBS (pH 7.4) with 0.1% (v/v) tween 80 were kept in a shaking incubator while maintaining a constant temperature (37 °C). Under these physiological conditions, a total of 46.6 ± 2.0% of DOX was released from the DP-NPs for 14 days (Figure 1E). This in vitro release study showed the encapsulated DOX had a short-term initial burst release within 6 h followed by a sustained release manner for almost 2 weeks. On the other hand, over nine-tenths of DOX was freed from DOX control, as opposed to one half of DOX was released from the DP-NPs. In summary, DP-NPs were successfully nanoformulated and spherical in shape with unimodal size distribution. They also showed the excellent serum stability for 2 weeks, with sustained drug release-rate kinetics.

3.2. In Vitro Cellular Uptake, Cytotoxicity, and ICD of DP-NPs

We used CT26 tumor cells to identify in vitro cellular uptake and localization of DOX, and to perform integrative analyses of the follow-on effects such as DOX-induced cytotoxicity and ICD. DOX-HCl and FITC-labeled DP-NPs (2 μM of DOX) were added to CT26 tumor cells and harvested from them right after the specific incubation time of 1 h, 9 h, 24 h, and 48 h, respectively, followed by detecting the signals of FITC and DOX fluorescence on time (Figure 2A). Specifically, green signals of FITC from the DP-NPs were increased over time in the cytoplasmic area of the CT26 tumor cells (Supplementary Materials Figure S1). In this sense, the cellular uptake of DP-NPs was shown to be successful by the incubation time. Meanwhile, there was a big difference in expression of red signals of DOX between DOX-HCl-treated and DP-NPs-treated CT26 tumor cells. From the red signal changes in the confocal images, fluorescence of DOX was largely detected in the nucleus at an incipient stage (at 1 h) in DOX-HCl-treated CT26 tumor cells, whereas it was gradually localized in the nucleus 24 h after the treatment of DP-NPs in CT26 tumor cells. Theretofore, DOX was mainly localized in the cytoplasmic area of CT26 tumor cells after the treatment of DP-NPs.

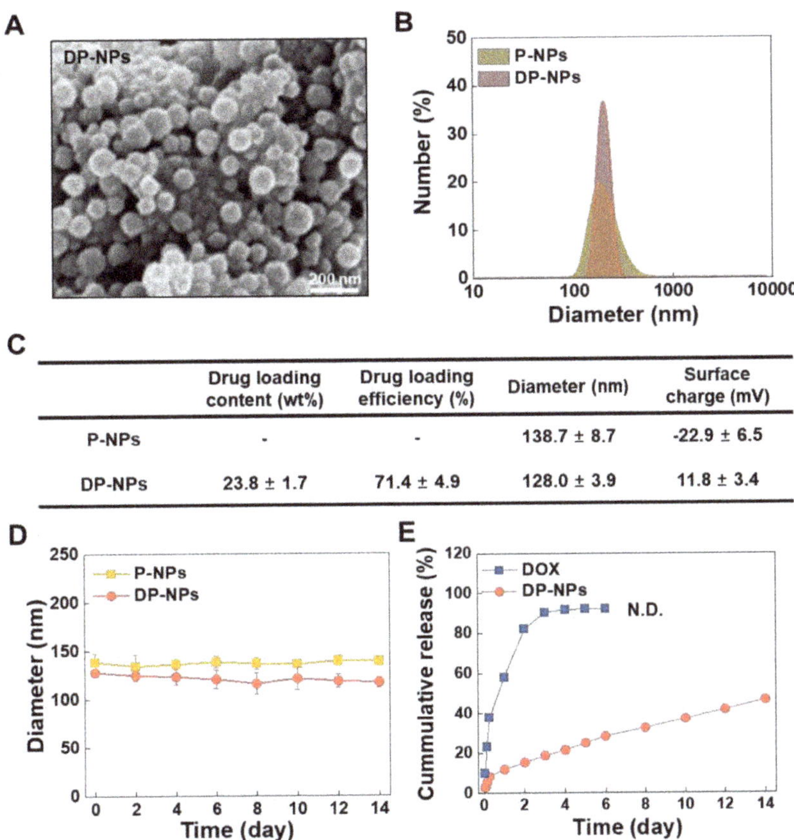

Figure 1. Physicochemical properties of the P-NPs and DP-NPs. (**A**) The morphology of the DP-NPs using FE-SEM. The scale bar indicates 200 nm. (**B**) The hydrodynamic size distribution of the P-NPs and DP-NPs using a zeta-sizer. (**C**) The DOX loading content and efficiency, particle size and surface charge of the P-NPs and DP-NPs using UV–vis spectroscopy and a zeta-sizer. (**D**) The size stability of the P-NPs and DP-NPs was monitored under 10% (v/v) FBS-containing PBS at 37 °C for 2 weeks using a zeta-sizer. (**E**) In vitro sustained release kinetics of DOX from the DP-NPs was monitored under 0.1% (v/v) tween 80-containing PBS at 37 °C for 2 weeks. The released amount of DOX was quantified using absorbance at 480 nm of UV–vis spectroscopy.

DOX-induced CT26 tumor cell cytotoxicity was determined by CCK-8 assay after the treatment of DOX-HCl and DP-NPs of which concentration ranged from 0.001 to 10 μM for 48 h (Figure 2B). The cell viability of both experimental groups was gradually decreased in proportion to the amount of the concentration of the treated DOX-HCl and DP-NPs; however, each group had somewhat different half maximal inhibitory concentration (IC$_{50}$) values. In this study, the IC$_{50}$ values obtained by CCK-8 assay against DOX-HCl- and DP-NPs-treated CT26 tumor cells were 0.227 μM and 0.641 μM, respectively. For reference, vehicle displayed no significant cytotoxicity, as if nothing was treated on the CT26 tumor cells. In comparison, the IC$_{50}$ values of DP-NPs were almost threefold higher than that of DOX-HCl in CT26 tumor cells. The primary reason of this reduced cytotoxicity of DOX from DP-NPs was its sustained-release property.

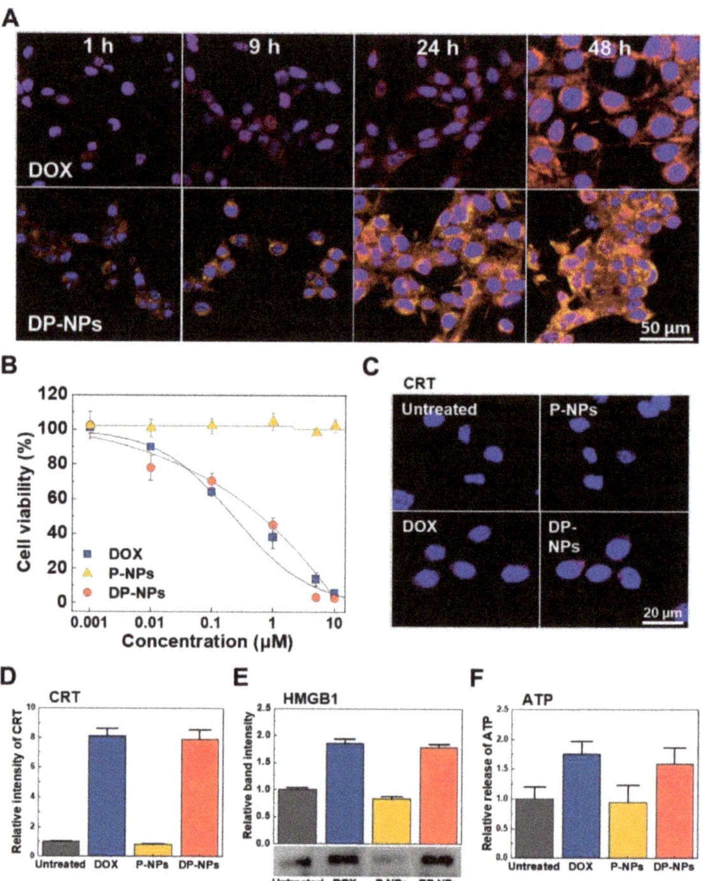

Figure 2. In vitro cellular uptake, cytotoxicity, and the subsequent ICD on the CT26 tumor cells. (**A**) Overlay confocal images of the time-dependent DOX-HCl and FITC-labeled-DP-NPs (2 μM of DOX) uptake. Blue: DAPI; Red: DOX; Green: PLGA. The scale bar indicates 50 μm. (**B**) Cell viability of DOX-HCl- and DP-NPs-treated CT26 tumor cells at increasing concentrations for 48 h using CCK-8. (**C**) Confocal images and (**D**) the quantified relative intensity of surface-exposed CRT in the DOX-HCl- and DP-NPs-treated (2 μM of DOX) CT26 tumor cells for 48 h. The scale bar indicates 20 μm. (**E**) Western blotting image and band intensity of released HMGB1 in the DOX-HCl- and DP-NPs-treated (2 μM of DOX) CT26 tumor cell culture medium for 48 h. (**F**) ATP level in the DOX-HCl- and DP-NPs-treated (2 μM of DOX) CT26 tumor cell culture medium for 48 h.

Finally, from the CT26 tumor cells treated with DOX-HCl and DP-NPs, we assessed DOX-triggered DAMPs such as CRT, HMGB1, and ATP. CRT is protein that could translocate on the cell membrane of dying cells by the mechanism of ICD. To confirm the DOX-induced ICD, we observed the CRT translocation on the DOX-HCl- and DP-NPs (2 μM of DOX)-treated CT26 tumor cells for 48 h, using a confocal microscopy (Figure 2C). It was shown the fluorescence indicating surface-exposed CRT was significantly increased on the cell membrane when the CT26 tumor cells were treated with DOX-HCl and DP-NPs, compared to that of untreated and vehicle-treated CT26 tumor cells. Additionally, the relative fluorescence intensity of CRT from DOX-HCl- and DP-NPs-treated CT26 tumor cells was 8.12- and 7.91-fold higher than that of untreated CT26 tumor cells, respectively

(Figure 2D). For reference, there were no significant differences in CRT levels between vehicle-treated and untreated CT26 tumor cells.

HMGB1 is a nuclear protein and also one of the ICD markers passively released from the dying cells. Similarly, to confirm the DOX-induced ICD, we observed the release of HMGB1 from the DOX-HCl- and DP-NPs (2 μM of DOX)-treated CT26 tumor cells for 48 h, and assessed by Western blotting. The band intensity indicating released HMGB1 was increased when the CT26 tumor cells were treated with DOX-HCl and DP-NPs, compared to that of untreated and vehicle-treated CT26 tumor cells (Figure 2E). Additionally, the relative band intensity of HMGB1 from DOX-HCl- and DP-NPs-treated CT26 tumor cells was 1.85- and 1.78-fold higher than that of untreated CT26 tumor cells, respectively. For reference, there were no significant differences in HMGB1 release between vehicle-treated and untreated CT26 tumor cells.

ATP is one of the ICD markers and a molecule emitted actively from cytoplasm and cytoplasmic organelles in dying cells. The secretion of ATP from the DOX-HCl- and DP-NPs (2 μM of DOX)-treated CT26 tumor cells for 48 h, was determined by the ATP assay to confirm the DOX-induced ICD. The measured levels indicating secreted ATP was increased when the CT26 tumor cells were treated with DOX-HCl and DP-NPs, compared to that of untreated and vehicle-treated CT26 tumor cells (Figure 2F). Additionally, the relative level of ATP from DOX-HCl- and DP-NPs-treated CT26 tumor cells was 1.75- and 1.59-fold higher than that of untreated CT26 tumor cells, respectively. For reference, there were no significant differences in ATP secretion between vehicle-treated and untreated CT26 tumor cells.

We were able to confirm that the diverse aspects of DAMPs, including CRT translocation and release of HMGB1 and ATP, were very closely correlated with the cell viability after the treatment of DOX. Based on these results, it was expected that DP-NPs could sufficiently induce DAMPs-mediating ICD to CT26 tumor cells via time-releasing DOX.

3.3. In Vivo Tumor Growth Inhibition by DP-NPs-Driven CIT

To observe in vivo release kinetics of DOX from DP-NPs, we obtained time-dependent fluorescence images by non-invasive optical imaging system, visualizing the accumulated DOX in targeted tumor tissues. In order to find the optimal drug concentration, fulfilling therapeutic window of DOX and also maximizing therapeutic efficacies, we set up several experiments with groups of various concentration of DP-NPs. In this regard, we tracked the fluorescence signals of DOX released from the DP-NPs at the targeted tumor tissues for 14 days, after the intratumoral injection of DP-NPs (5, 10, 10 × 2 mg/kg) (Figure 3A). DOX-HCl (10 mg/kg) was used as a control to compare the other groups of DP-NPs. DOX fluorescence in the DOX-HCl-treated CT26 tumor-bearing mice was extremely decreased to the level in 10 mg/kg of DP-NPs-treated group in just 2 days, and since then, it has continued to decrease. On the other hand, DP-NPs-treated groups showed mild decrease in proportion to their initial DOX concentration without a wide variation of fluorescence levels. Notably, the DOX fluorescence at the tumor site of DP-NPs 10 × 2-treated group showed gradually decreased for 6 days but recovered by second intratumoral injection, resulting in maintaining the DOX fluorescence range for 14 days. In short, there were different patterns of the change in relative DOX fluorescence intensity at the tumor that were drastically decreased, continuously decreased, or maintained a certain range of fluorescence after the initial ascending (Figure 3B). From these findings, we confirmed PK profile of DOX could be favorably controlled by sustained-released DOX from the directly injected DP-NPs. DP-NPs were confined to the tumor tissues within a given period, and DOX was successfully supplied locally, especially in the group of repeated treatment of DP-NPs. This result sufficiently gave us a sound basis for the continuous ICD and the subsequent antitumor immune responses.

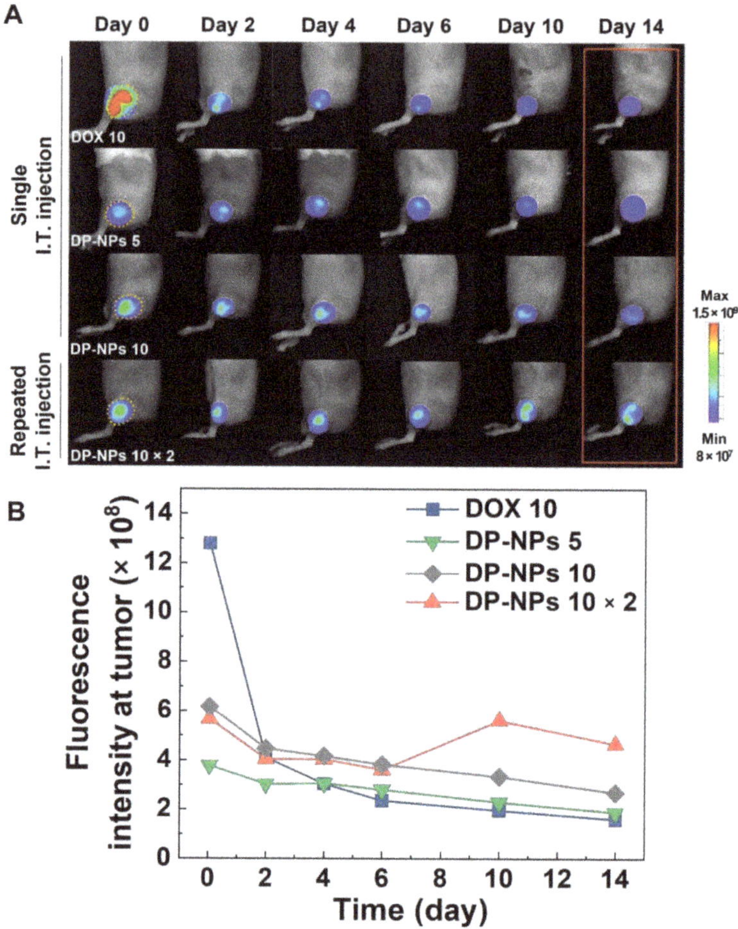

Figure 3. In vivo drug release characteristics of DP-NPs after their intratumoral injection. (**A**) Time-dependent fluorescence images of the CT26 tumor-bearing mouse intratumorally injected with DOX-HCl (10 mg/kg) and DP-NPs (5, 10, 10 × 2 mg/kg) for 2 weeks. (**B**) Quantified fluorescence intensity of DOX at the tumor tissues standing on the basis of the images from (**A**).

We confirmed the effects of DOX-based CIT and its induced TIME changes in the CT26 tumor-bearing mice model. Tumor growth of the CT26 tumor-bearing mice treated with DOX-HCl (10 mg/kg), DP-NPs (5, 10, 10 × 2 mg/kg of DOX) was measured for 4 weeks after their first intratumoral injections (Figure 4A). Injection dose of each group was the same as in the previous in vivo imaging experiment, and as described earlier, 10 × 2 mg/kg of DP-NPs was injected in twice with an interval of 10 days. For reference, PBS and P-NPs were used as controls, and we limited the maximum allowable size of subcutaneous CT26 tumors (2000 mm^3 volume in mice) in these control experimental groups. DOX-HCl (10 mg/kg; n = 7) strongly inhibited tumor growth up to 4 weeks, and the average tumor size of DOX-HCl-treated mice was 150 mm^3 on the 28th day. Meanwhile, DP-NPs (10 × 2 mg/kg; n = 6) exhibited effective tumor suppression as compared with PBS (n = 5) and P-NPs (n = 6), and also apparently it was better than DP-NPs of different doses (5, 10 mg/kg; n = 6, 5, respectively). The average tumor size of DP-NPs (10 × 2 mg/kg)-treated mice was 184 mm^3 on the same day above, and the size of all the others were over 2000 mm^3.

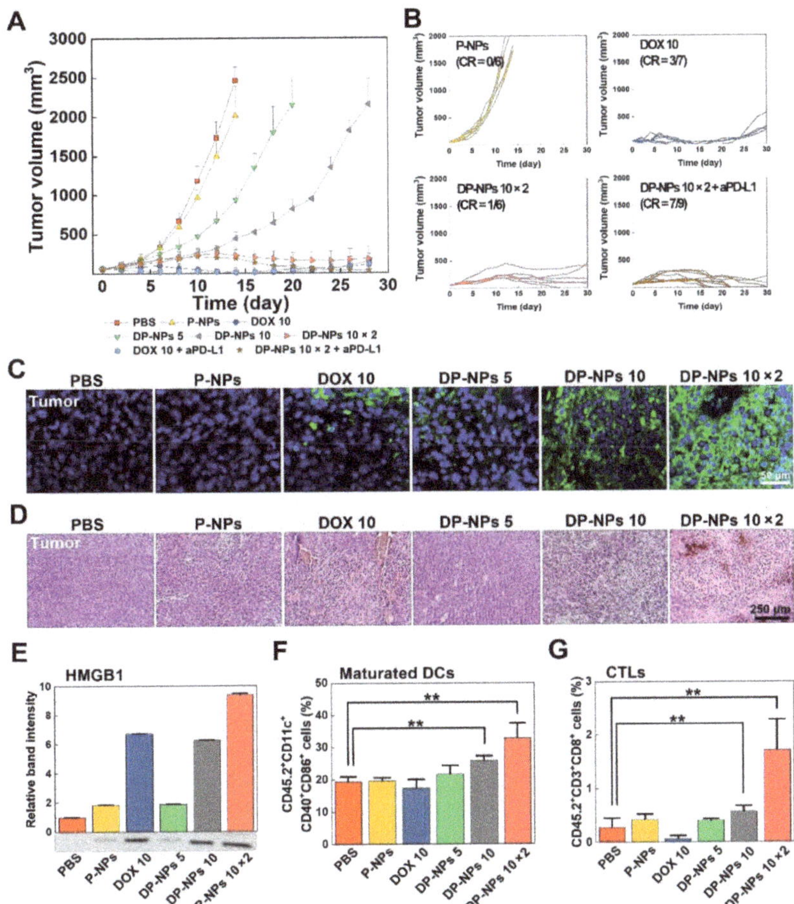

Figure 4. In vivo tumor growth inhibition and the subsequent antitumor immune responses. (**A**) Tumor growth of the CT26 tumor-bearing mice after the first intratumoral injection of PBS, P-NPs, DOX-HCl (10 mg/kg), DP-NPs (5, 10, 10 × 2 mg/kg), aPD-L1 Ab (10 × 5 mg/kg)-combined DOX-HCl (10 mg/kg), and DP-NPs (10 × 2 mg/kg) for 4 weeks. (**B**) Tumor growth for individual mice and CR rate in P-NPs-, DOX-HCl-, DP-NPs-, and aPD-L1 Ab-combined DP-NPs-treated groups of (**A**). Representative images of (**C**) TUNEL- and (**D**) H&E-stained tumor tissues at 14 days after the first treatment of PBS, P-NPs, DOX-HCl (10 mg/kg), and DP-NPs (5, 10, 10 × 2 mg/kg). TUNEL-positive cells are visible as green. The scale bars indicate 50 and 250 μm, respectively. (**E**) Western blotting image and band intensity of in vivo HMGB1 at 14 days after the first treatment of PBS, P-NPs, DOX-HCl (10 mg/kg), DP-NPs (5, 10, 10 × 2 mg/kg). Immune cell population of (**F**) maturated DCs in TDLNs and (**G**) CTLs in tumor tissues at 14 days after the first treatment of PBS, P-NPs, DOX-HCl (10 mg/kg), and DP-NPs (5, 10, 10 × 2 mg/kg), analyzed using a flow cytometer. Double asterisks (**) indicate a difference at the $p < 0.01$ significance level.

In addition to these results, we also used an additional aPD-L1 Ab for synergistic combination cancer CIT to elicit the better antitumor therapeutic effects. aPD-L1 Ab is one of the widely used ICBs in cancer immunotherapy. Engagement of PD-L1 blockade by PD-L1 expressed on tumor cells leaded to the enhancement of tumor-specific T cell-mediated adaptive immunity. Based on the ICB effect and enhancing adaptive immunity, furthermore, recent studies showed that the tumor progression was significantly inhibited by combination with PD-L1 blockade therapy, indicating the therapeutic potential for cancer CIT. Based on these hard facts with the evidence of our previous findings, PD-L1 blockade combined with DOX-HCl and DP-NPs treatment were carried out to evaluate therapeutic outcomes, respectively. Tumor growth of the CT26 tumor-bearing mice of these two groups were monitored for 4 weeks. 10 mg/kg of aPD-L1 Ab was administered to mice of all combination groups five times every 3 days via the intraperitoneal route (Supplementary Materials Figure S2). According to the previously stated results, tumor gradually regrew in DOX-HCl (10 mg/kg)-treated group, whereas the tumor in DP-NPs (10×2 mg/kg)-treated group was well-circumscribed with an average volume under 200 mm^3, approximately 2 weeks after their first treatments. In these combination study, tumor also gradually regrew in DOX-HCl (10 mg/kg) plus aPD-L1 Ab (10×5 mg/kg)-treated group 18 days after the first DOX-HCl injection, similarly in DOX-HCl-treated group. Interestingly, the extremely inhibited tumor growth was observed in DP-NPs (10×2 mg/kg) plus aPD-L1 Ab (10×5 mg/kg) group. The average tumor volume in this group gradually decreased and did not exceed 50 mm^3, 24 days after the first DP-NPs injection.

Next, we analyzed the complete response (CR) rate of each treatment based on the tumor size on the 28th day (Figure 4B and Supplementary Materials Figure S3). Generally, the term CR applies in case that all existed tumor is no longer observed after the treatment. This CR rate of DOX-HCl- and DP-NPs (10×2 mg/kg)-treated groups was 42.9% and 16.7%, respectively. It was shown the other DP-NPs (5, 10 mg/kg)-treated groups just partially responded to the treatment, and there were no tumor-suppressed mice in controls such as PBS- or P-NPs-treated groups. However, PD-L1 blockade-combined DP-NPs (10×2 mg/kg) exhibited highly increased rate of CR, compared to DOX-HCl (10 mg/kg) with PD-L1 blockade. The specific CR rate of these groups was 77.8% and 40.0%, respectively. It was shown therapeutic efficacies in ICB combination tended to correlate with reduction in tumor growth, especially in PD-L1 blockade-combined DP-NPs-treated group.

We could obtain factual basis of tumor inhibition by DOX from TUNEL assay. DOX-HCl and DP-NPs had apoptotic effects on the tumor tissues in each group, leading to the loss of tumor cell viability. The number of TUNEL-positive tumor cells were highly distributed in these groups, and it was increased as the concentration of DOX was increased (Figure 4C). The reason of significantly increased levels of TUNEL-positive cells in DP-NPs-treated tumor tissues, especially in 10×2 mg/kg of DP-NPs-treated tissues, was mainly considered as sustained drug release properties of DP-NPs. We also observed tissue morphologies and its structural abnormalities by histology from the H&E-stained tumor tissues harvested on the 14th day after the first intratumoral treatment in each group (Figure 4D). It was identified CT26 tumor tissues from PBS- or P-NPs-treated mice had few or no necrotic areas, and they displayed less or no toxicity. On the contrary, there was a large necrotic area over the tumor tissues from DOX-HCl-treated mice, indicating high toxicity was caused by DOX. There were also analogical results along the above, tumor tissues from DP-NPs-treated mice showed several similar histotoxicities, however different by their dose. It was shown higher amounts of localized DP-NPs could release a lot more DOX, actually leading to certain levels of toxicity and histological changes in TME.

Subsequently, we confirmed the immunological evidences in TME correlated with the previous results. As previously stated, ICD is generally known as an initial key process in antitumor immunity, it directly becomes a link between subsequent innate immunity and adaptive immunity. In this regard, to measure the HMGB1 levels from chemotherapeutics-treated CT26 tumor cells in vivo, we observed the release of HMGB1 in each group treated with DOX-HCl and DP-NPs with different dose levels, and assessed by Western blotting on their 14th day of our treatment. As expected from the previous in vitro results, these in vivo tests also confirmed the measurements of elevated levels of the band intensity indicating released HMGB1 in DOX-HCl- and DP-NPs-treated groups. Specifically, the relative band intensity of HMGB1 from DOX-HCl (10 mg/kg)- and DP-NPs (10, 10×2 mg/kg)-treated groups was 6.71-, 6.27-, and 9.39-fold higher than that of PBS-treated group, respectively (Figure 4E). For reference, there were no significant differences in HMGB1 level between vehicle-treated group and PBS-treated group. Especially, as shown above, exceedingly strong ICD was induced in the tumor tissues and TME of the DP-NPs (10×2 mg/kg)-treated group, in contrast with the results of controls and even DP-NPs (5 mg/kg)-treated group. This 5 mg/kg of DP-NPs was not seemed to be enough to induce and maintain the ICD continuously in vivo.

DOX-triggered ICD induce ICD-associated DAMPs in TME, thereby activating innate immunity represented by DC maturation, leading to efficient priming of CTLs, eventually resulting in effective antitumor immunity with newly changed TIME. To verify the immunological basis for this, DCs and CTLs isolated from TDLNs and tumor tissues in the CT26 tumor-bearing mice in each experimental group were examined by flow cytometry. These analyses were performed on the 14th day after the first intratumoral injection of PBS, P-NPs, DOX-HCl (10 mg/kg), and DP-NPs (5, 10, 10×2 mg/kg). According to our analytical results, populations of maturated DCs were high in DP-NPs-treated groups, and this population value was gradually increased as a dose of treated DP-NPs (Figure 4F). The value was the highest in 10×2 mg/kg of DP-NPs-treated group, wherein strongly induced ICD continuously enhanced maturation of DCs in TDLNs. On the other hand, DP-NPs-treated groups showed extremely higher CTL populations than DOX-HCl-treated group (Figure 4G). This result was rooted in whether CTLs were directly exposed to DOX or not in that CTLs were reckoned to be vulnerable to DOX-induced cytotoxicity.

Summarizing the results so far, DP-NPs successfully suppressed tumors in vivo, as shown in the results of effectively inhibited growth of CT26 tumors. It seemed we could sufficiently control the ICD being induced and maintained continuously in vivo by over a certain amount of DOX. Meanwhile, a high initial concentration of DOX in the DOX-HCl-treated group could induce severe toxic effects on major organs as well as immune cells including maturated DCs and CTLs. Furthermore, it was also identified that strongly induced ICD was not connected to adaptive immunity. Nonetheless, these results were significant in the sense that CD8-positive CTLs primed to the TIME favorably acted against the tumor due to the sustained-released DOX from the DP-NPs, even in high doses of DP-NPs (10×2 mg/kg) with reduced toxicity and much stronger ICD, leading to effective therapeutic results. Additionally, it was considered largely populated CTLs in TIME favorably killed the tumor cells better in the combination therapy group with enhanced antitumor immunity, attributable to the help of ICB.

For survival, mice with a tumor size 3000 mm^3 were counted as statistically dead. In this regard, PBS- or P-NPs-treated groups showed big changes in animal survival for 4 weeks (Figure 5A). Body weights of mice in each group were monitored with tumor volumes every 2 days. It was considered there were no severe systemic toxicities during the treatment in that there were no big differences in body weight change (Figure 5B). However, spleen weight of the DOX-HCl-treated mice was decreased by more than 20%, compared to that of PBS- or P-NPs-treated mice (Figure 5C). Furthermore, the tissue morphology of shrunken spleen from the DOX-HCl-treated mice showed a decreased size of white pulp and larger red pulp, compared to those of PBS-, P-NPs-, and DP-NPs (5, 10, 10×2 mg/kg)-treated mice. We expected that the shrinking and morphological changes of the spleen occurred because of the undesirable toxicity of DOX-HCl, resulting in underperforming normal function (Figure 5D) [44]. Additionally, intratumorally injected DP-NPs (5, 10, 10×2 mg/kg) also exhibited relatively less systemic toxicities in the spleen and other organs (liver, lung, kidney, and heart) rather than DOX-HCl, based on the histopathology data (Supplementary Materials Figure S4). From these results, exposure to therapeutic concentration levels of DOX (even in case of 10×2 mg/kg of DOX) was acceptable in some cumulative toxicity burdens by using our nanoparticulate formulation of DP-NPs.

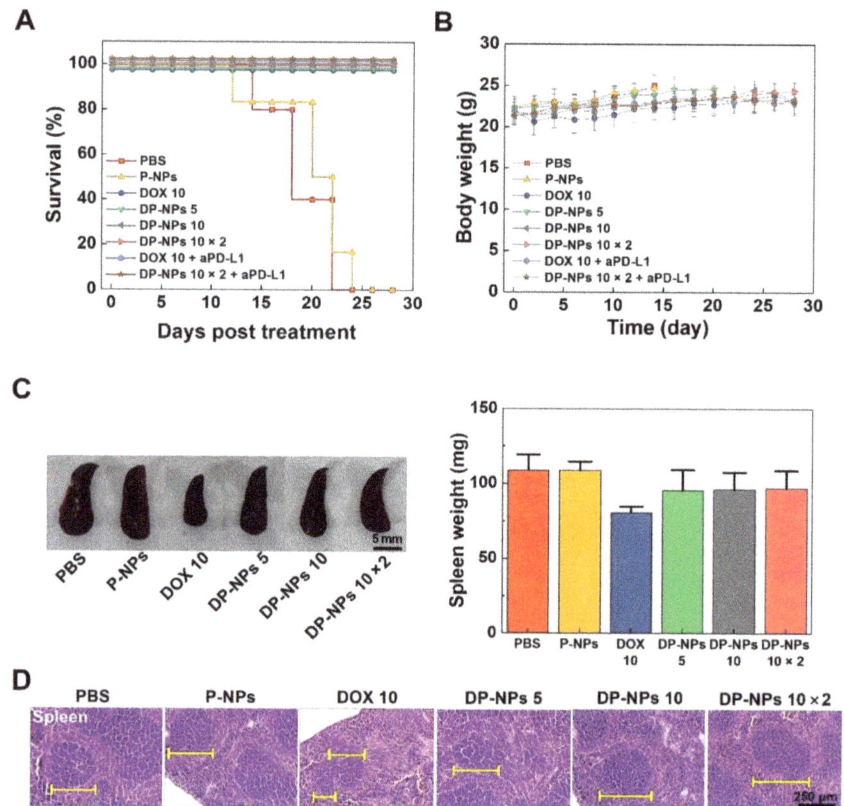

Figure 5. In vivo DOX-induced toxicity in the BALB/c mouse model. (**A**) Survival rate and (**B**) body weight change of the CT26 tumor-bearing mice after the first intratumoral injection of PBS, P-NPs, DOX-HCl (10 mg/kg), DP-NPs (5, 10, 10 × 2 mg/kg), aPD-L1 Ab (10 × 5 mg/kg)-combined DOX-HCl (10 mg/kg), and DP-NPs (10 × 2 mg/kg) for 4 weeks. (**C**) Representative photos and the weight of spleens from the CT26 tumor-bearing mice treated with PBS, P-NPs, DOX-HCl (10 mg/kg), DP-NPs (5, 10, 10 × 2 mg/kg) at 14 days after the first treatments. The scale bar indicates 5 mm. (**D**) Representative images of H&E-stained spleen tissues at 14 days after the first treatment of PBS, P-NPs, DOX-HCl (10 mg/kg), DP-NPs (5, 10, 10 × 2 mg/kg). Yellow line indicates the spleen white pulp (dark blue) in the H&E-stained histological images. The scale bar indicates 250 μm.

4. Conclusions

Chemotherapeutics are widely used and in cancer immunotherapy, and as is well known, they initiate immunogenic antitumor reactions and contribute to promoting the antitumor immunity by facilitating the infiltration of immune cells into the tumor [45]. However, uncontrolled high-dose chemotherapy can immoderately generate cytotoxic reaction and immunosuppression in TME, inevitably resulting in the negative effects on the functions of immune effectors [46]. In this sense, the sustained-release nanomedicine is necessarily required in cancer immunotherapy, especially to use the chemotherapeutics bio- and immuno-compatibly. In this study, we developed DP-NPs that could sustain the release of DOX for controlling ICD with toxicity in the TIME of the tumor tissues. Nano-sized DP-NPs released DOX for 14 days under physiological conditions in vitro. Furthermore, released DOX successfully induced ICD on the CT-26 tumor cells, resulting in significantly decreasing tumor cell viability. Intratumorally injected DP-NPs enabled prolonged retention of DOX for 14 days, compared to that of DOX-HCl. Concurrently, we confirmed the maximized therapeutic outcomes for 4 weeks created by the time-releasing DP-NPs and ICB combination effect. The continuous ICD in the tumor by DP-NPs 6.27 to 9.39-fold higher HMGB1 release than that of PBS-treated tumor tissues led to higher DC maturation and elicited a higher proportion of CTLs into tumor tissues. Furthermore, the DP-NPs-treated mice showed less toxicities in immune cells and spleen than those of DOX-HCl treated mice during CIT. Finally, we expect that combination CIT using DP-NPs and ICB will reconstitute TIME for the augmented antitumor immunity and control toxicity at the same time, can provide therapeutic strategies for the best therapeutic results.

Supplementary Materials: The following are available online at http://www.mdpi.com/xxx/s1. Figure S1: Individual images of Figure 2A. Blue: DAPI; Red: DOX; Green: PLGA. The scale bar indicates 50 μm, Figure S2: Time schedule of the whole treatment processes for CT26 tumor in BALB/c mice, Figure S3: Tumor growth for individual mice and CR rate in PBS-, DP-NPs (5, 10 mg/kg)-, and aPD-L1 Ab (10 × 5 mg/kg)-combined DOX-HCl (10 mg/kg)-treated groups of Figure 4A, Figure S4: Representative images of H&E-stained tissues (liver, lung, kidney, and heart) at 14 days after the first treatment of PBS, P-NPs, DOX-HCl (10 mg/kg), DP-NPs (5, 10, 10 × 2 mg/kg). The scale bar indicates 250 μm.

Author Contributions: Conceptualization, J.K. (Jeongrae Kim) and K.K.; methodology, J.K. (Jeongrae Kim), and Y.C.; validation, S.Y., J.L. and J.C.; formal analysis, J.K., D.-K.L. and H.Y.Y.; investigation, Y.M., J.K. (Jinseong Kim), N.S. and H.C.; resources, K.K.; data curation, Y.C. and H.Y.Y.; writing—original draft preparation, Y.C. and K.K.; visualization, M.K.S. and S.J.; supervision, K.K.; project administration, K.K.; funding acquisition, K.K. All authors have read and agreed to the published version of the manuscript.

Funding: This work was supported by grants from the National Research Foundation (NRF) of Korea, funded by the Ministry of Science (NRF-2019R1A2C3006283), the KU-KIST Graduate School of Converging Science and Technology (Korea University) and the Intramural Research Program of KIST.

Institutional Review Board Statement: The animal study protocol was approved by the Institutional Animal Care and Use Committee (IACUC) of Korea Institute of Science and Technology (KIST) (approval number: KIST-2020-070, 27 May 2020).

Informed Consent Statement: Not applicable.

Data Availability Statement: The data presented in this study are available upon request from the corresponding author.

Conflicts of Interest: The authors declare no conflict of interest.

References

1. Green, D.R.; Ferguson, T.; Zitvogel, L.; Kroemer, G. Immunogenic and tolerogenic cell death. *Nat. Rev. Immunol.* **2009**, *9*, 353–363. [CrossRef] [PubMed]
2. Kroemer, G.; Galluzzi, L.; Kepp, O.; Zitvogel, L. Immunogenic cell death in cancer therapy. *Annu. Rev. Immunol.* **2013**, *31*, 51–72. [CrossRef] [PubMed]
3. Li, Y.; Liu, X.; Zhang, X.; Pan, W.; Li, N.; Tang, B. Immunogenic cell death inducers for enhanced cancer immunotherapy. *Chem. Commun.* **2021**, *57*, 12087–12097. [CrossRef] [PubMed]
4. Krysko, D.V.; Garg, A.D.; Kaczmarek, A.; Krysko, O.; Agostinis, P.; Vandenabeele, P. Immunogenic cell death and DAMPs in cancer therapy. *Nat. Rev. Cancer* **2012**, *12*, 860–875. [CrossRef]
5. Wang, Y.J.; Fletcher, R.; Yu, J.; Zhang, L. Immunogenic effects of chemotherapy-induced tumor cell death. *Genes Dis.* **2018**, *5*, 194–203. [CrossRef]
6. Vanmeerbeek, I.; Sprooten, J.; De Ruysscher, D.; Tejpar, S.; Vandenberghe, P.; Fucikova, J.; Spisek, R.; Zitvogel, L.; Kroemer, G.; Galluzzi, L.; et al. Trial watch: Chemotherapy-induced immunogenic cell death in immuno-oncology. *Oncoimmunology* **2020**, *9*, 1703449. [CrossRef]
7. Barry, K.C.; Hsu, J.; Broz, M.L.; Cueto, F.J.; Binnewies, M.; Combes, A.J.; Nelson, A.E.; Loo, K.; Kumar, R.; Rosenblum, M.D.; et al. A natural killer-dendritic cell axis defines checkpoint therapy-responsive tumor microenvironments. *Nat. Med.* **2018**, *24*, 1178–1191. [CrossRef]
8. Chen, D.S.; Mellman, I. Oncology meets immunology: The cancer-immunity cycle. *Immunity* **2013**, *39*, 1–10. [CrossRef]
9. Serrano-Del Valle, A.; Anel, A.; Naval, J.; Marzo, I. Immunogenic Cell Death and Immunotherapy of Multiple Myeloma. *Front. Cell Dev. Biol.* **2019**, *7*, 50. [CrossRef]

10. Opzoomer, J.W.; Sosnowska, D.; Anstee, J.E.; Spicer, J.F.; Arnold, J.N. Cytotoxic Chemotherapy as an Immune Stimulus: A Molecular Perspective on Turning Up the Immunological Heat on Cancer. *Front. Immunol.* **2019**, *10*, 1654. [CrossRef]

11. Senapati, S.; Mahanta, A.K.; Kumar, S.; Maiti, P. Controlled drug delivery vehicles for cancer treatment and their performance. *Signal Transduct. Target. Ther.* **2018**, *3*, 7. [CrossRef]

12. Wei, G.; Wang, Y.; Yang, G.; Wang, Y.; Ju, R. Recent progress in nanomedicine for enhanced cancer chemotherapy. *Theranostics* **2021**, *11*, 6370–6392. [CrossRef]

13. Gonzalez-Valdivieso, J.; Girotti, A.; Schneider, J.; Arias, F.J. Advanced nanomedicine and cancer: Challenges and opportunities in clinical translation. *Int. J. Pharm.* **2021**, *599*, 120438. [CrossRef]

14. Wu, W.; Pu, Y.; Shi, J. Nanomedicine-enabled chemotherapy-based synergetic cancer treatments. *J. Nanobiotechnology* **2022**, *20*, 4. [CrossRef]

15. Meng, F.; Wang, J.; Ping, Q.; Yeo, Y. Quantitative Assessment of Nanoparticle Biodistribution by Fluorescence Imaging, Revisited. *ACS Nano* **2018**, *12*, 6458–64681. [CrossRef]

16. de Maar, J.S.; Sofias, A.M.; Porta Siegel, T.; Vreeken, R.J.; Moonen, C.; Bos, C.; Deckers, R. Spatial heterogeneity of nanomedicine investigated by multiscale imaging of the drug, the nanoparticle and the tumour environment. *Theranostics* **2020**, *10*, 1884–1909. [CrossRef]

17. Brand, W.; Noorlander, C.W.; Giannakou, C.; De Jong, W.H.; Kooi, M.W.; Park, M.V.; Vandebriel, R.J.; Bosselaers, I.E.; Scholl, J.H.; Geertsma, R.E. Nanomedicinal products: A survey on specific toxicity and side effects. *Int. J. Nanomed.* **2017**, *12*, 6107–6129. [CrossRef]

18. Giannakou, C.; Park, M.V.; de Jong, W.H.; van Loveren, H.; Vandebriel, R.J.; Geertsma, R.E. A comparison of immunotoxic effects of nanomedicinal products with regulatory immunotoxicity testing requirements. *Int. J. Nanomed.* **2016**, *11*, 2935–2952. [CrossRef]

19. Shikanov, A.; Shikanov, S.; Vaisman, B.; Golenser, J.; Domb, A.J. Cisplatin tumor biodistribution and efficacy after intratumoral injection of a biodegradable extended release implant. *Chemother Res. Pr.* **2011**, *2011*, 175054. [CrossRef]

20. Celikoglu, F.; Celikoglu, S.I.; Goldberg, E.P. Bronchoscopic intratumoral chemotherapy of lung cancer. *Lung Cancer* **2008**, *61*, 1–12. [CrossRef]

21. Goldberg, E.P.; Hadba, A.R.; Almond, B.A.; Marotta, J.S. Intratumoral cancer chemotherapy and immunotherapy: Opportunities for nonsystemic preoperative drug delivery. *J. Pharm. Pharm.* **2002**, *54*, 159–180. [CrossRef]

22. Liu, J.; Li, Z.; Zhao, D.; Feng, X.; Wang, C.; Li, D.; Ding, J. Immunogenic cell death-inducing chemotherapeutic nanoformulations potentiate combination chemoimmunotherapy. *Mater. Des.* **2021**, *202*, 109465. [CrossRef]

23. Mastria, E.M.; Cai, L.Y.; Kan, M.J.; Li, X.; Schaal, J.L.; Fiering, S.; Gunn, M.D.; Dewhirst, M.W.; Nair, S.K.; Chilkoti, A. Nanoparticle formulation improves doxorubicin efficacy by enhancing host antitumor immunity. *J. Control. Release* **2018**, *269*, 364–373. [CrossRef]

24. Gao, J.; Wang, W.-Q.; Pei, Q.; Lord, M.S.; Yu, H.-j. Engineering nanomedicines through boosting immunogenic cell death for improved cancer immunotherapy. *Acta Pharmacol. Sin.* **2020**, *41*, 986–994. [CrossRef]

25. Ahmad, S.; Idris, R.A.M.; Hanaffi, W.N.W.; Perumal, K.; Boer, J.C.; Plebanski, M.; Jaafar, J.; Lim, J.K.; Mohamud, R. Cancer Nanomedicine and Immune System—Interactions and Challenges. *Front. Nanotechnol.* **2021**, *3*, 1–9. [CrossRef]

26. Zhao, M.; Liu, M. New Avenues for Nanoparticle-Related Therapies. *Nanoscale Res. Lett.* **2018**, *13*, 136. [CrossRef]

27. Maulhardt, H.A.; Hylle, L.; Frost, M.V.; Tornio, A.; Dafoe, S.; Drummond, L.; Quinn, D.I.; Kamat, A.M.; Dizerega, G.S. Local injection of submicron particle docetaxel is associated with tumor eradication, reduced systemic toxicity and an immunologic response in uro-oncologic xenografts. *Cancers* **2019**, *11*, 577. [CrossRef]

28. Guerrouahen, B.S.; Maccalli, C.; Cugno, C.; Rutella, S.; Akporiaye, E.T. Reverting Immune Suppression to Enhance Cancer Immunotherapy. *Front. Oncol.* **2019**, *9*, 1554. [CrossRef]

29. Somarribas Patterson, L.F.; Vardhana, S.A. Metabolic regulation of the cancer-immunity cycle. *Trends Immunol.* **2021**, *42*, 975–993. [CrossRef]

30. Rivas, J.R.; Liu, Y.; Alhakeem, S.S.; Eckenrode, J.M.; Marti, F.; Collard, J.P.; Zhang, Y.; Shaaban, K.A.; Muthusamy, N.; Hildebrandt, G.C.; et al. Interleukin-10 suppression enhances T-cell antitumor immunity and responses to checkpoint blockade in chronic lymphocytic leukemia. *Leukemia* **2021**, *35*, 3188–3200. [CrossRef]

31. Petitprez, F.; Meylan, M.; de Reynies, A.; Sautes-Fridman, C.; Fridman, W.H. The Tumor Microenvironment in the Response to Immune Checkpoint Blockade Therapies. *Front. Immunol.* **2020**, *11*, 784. [CrossRef] [PubMed]

32. Zemek, R.M.; Chin, W.L.; Nowak, A.K.; Millward, M.J.; Lake, R.A.; Lesterhuis, W.J. Sensitizing the Tumor Microenvironment to Immune Checkpoint Therapy. *Front. Immunol.* **2020**, *11*, 223. [CrossRef] [PubMed]

33. Fan, Q.; Chen, Z.; Wang, C.; Liu, Z. Toward biomaterials for enhancing immune checkpoint blockade therapy. *Adv. Funct. Mater.* **2018**, *28*, 1802540. [CrossRef]

34. Jo, S.D.; Nam, G.-H.; Kwak, G.; Yang, Y.; Kwon, I.C. Harnessing designed nanoparticles: Current strategies and future perspectives in cancer immunotherapy. *Nano Today* **2017**, *17*, 23–37. [CrossRef]

35. Fan, W.; Yung, B.; Huang, P.; Chen, X. Nanotechnology for Multimodal Synergistic Cancer Therapy. *Chem. Rev.* **2017**, *117*, 13566–13638. [CrossRef]

36. Pfirschke, C.; Engblom, C.; Rickelt, S.; Cortez-Retamozo, V.; Garris, C.; Pucci, F.; Yamazaki, T.; Poirier-Colame, V.; Newton, A.; Redouane, Y.; et al. Immunogenic Chemotherapy Sensitizes Tumors to Checkpoint Blockade Therapy. *Immunity* **2016**, *44*, 343–354. [CrossRef]

37. Kepp, O.; Zitvogel, L.; Kroemer, G. Clinical evidence that immunogenic cell death sensitizes to PD-1/PD-L1 blockade. *Oncoimmunology* **2019**, *8*, 1637188. [CrossRef]
38. Swider, E.; Koshkina, O.; Tel, J.; Cruz, L.J.; de Vries, I.J.M.; Srinivas, M. Customizing poly(lactic-co-glycolic acid) particles for biomedical applications. *Acta Biomater.* **2018**, *73*, 38–51. [CrossRef]
39. Kim, K.T.; Lee, J.Y.; Kim, D.D.; Yoon, I.S.; Cho, H.J. Recent Progress in the Development of Poly(lactic-co-glycolic acid)-Based Nanostructures for Cancer Imaging and Therapy. *Pharmaceutics* **2019**, *11*, 280. [CrossRef]
40. Rezvantalab, S.; Drude, N.I.; Moraveji, M.K.; Guvener, N.; Koons, E.K.; Shi, Y.; Lammers, T.; Kiessling, F. PLGA-Based Nanoparticles in Cancer Treatment. *Front. Pharm.* **2018**, *9*, 1260. [CrossRef]
41. Engineer, C.; Parikh, J.; Raval, A. Review on hydrolytic degradation behavior of biodegradable polymers from controlled drug delivery system. *Trends Biomater. Artif. Organs* **2011**, *25*, 79–85.
42. Xu, Y.; Kim, C.S.; Saylor, D.M.; Koo, D. Polymer degradation and drug delivery in PLGA-based drug-polymer applications: A review of experiments and theories. *J. Biomed. Mater. Res. B Appl. Biomater.* **2017**, *105*, 1692–1716. [CrossRef]
43. Choi, Y.; Yoon, H.Y.; Kim, J.; Yang, S.; Lee, J.; Choi, J.W.; Moon, Y.; Kim, J.; Lim, S.; Shim, M.K.; et al. Doxorubicin-Loaded PLGA Nanoparticles for Cancer Therapy: Molecular Weight Effect of PLGA in Doxorubicin Release for Controlling Immunogenic Cell Death. *Pharmaceutics* **2020**, *12*, 1165. [CrossRef]
44. Jadapalli, J.K.; Wright, G.W.; Kain, V.; Sherwani, M.A.; Sonkar, R.; Yusuf, N.; Halade, G.V. Doxorubicin triggers splenic contraction and irreversible dysregulation of COX and LOX that alters the inflammation-resolution program in the myocardium. *Am. J. Physiol. Heart Circ. Physiol.* **2018**, *315*, 1091–1100. [CrossRef]
45. Bailly, C.; Thuru, X.; Quesnel, B. Combined cytotoxic chemotherapy and immunotherapy of cancer: Modern times. *NAR Cancer* **2020**, *2*, 1. [CrossRef]
46. Salas-Benito, D.; Perez-Gracia, J.L.; Ponz-Sarvise, M.; Rodriguez-Ruiz, M.E.; Martinez-Forero, I.; Castanon, E.; Lopez-Picazo, J.M.; Sanmamed, M.F.; Melero, I. Paradigms on Immunotherapy Combinations with Chemotherapy. *Cancer Discov.* **2021**, *11*, 1353–1367. [CrossRef]

pharmaceutics

Article

Improving Antibacterial Activity of a HtrA Protease Inhibitor JO146 against *Helicobacter pylori*: A Novel Approach Using Microfluidics-Engineered PLGA Nanoparticles

Jimin Hwang [1], Sonya Mros [2], Allan B. Gamble [1], Joel D. A. Tyndall [1] and Arlene McDowell [1,*]

1   School of Pharmacy, University of Otago, Dunedin 9054, New Zealand; hwaji220@student.otago.ac.nz (J.H.); allan.gamble@otago.ac.nz (A.B.G.); joel.tyndall@otago.ac.nz (J.D.A.T.)
2   Department of Microbiology and Immunology, University of Otago, Dunedin 9054, New Zealand; sonya.mros@otago.ac.nz
*   Correspondence: arlene.mcdowell@otago.ac.nz

Abstract: Nanoparticle drug delivery systems have emerged as a promising strategy for overcoming limitations of antimicrobial drugs such as stability, bioavailability, and insufficient exposure to the hard-to-reach bacterial drug targets. Although size is a vital colloidal feature of nanoparticles that governs biological interactions, the absence of well-defined size control technology has hampered the investigation of optimal nanoparticle size for targeting bacterial cells. Previously, we identified a lead antichlamydial compound JO146 against the high temperature requirement A (HtrA) protease, a promising antibacterial target involved in protein quality control and virulence. Here, we reveal that JO146 was active against *Helicobacter pylori* with a minimum bactericidal concentration of 18.8–75.2 µg/mL. Microfluidic technology using a design of experiments approach was utilized to formulate JO146-loaded poly(lactic-*co*-glycolic) acid nanoparticles and explore the effect of the nanoparticle size on drug delivery. JO146-loaded nanoparticles of three different sizes (90, 150, and 220 nm) were formulated with uniform particle size distribution and drug encapsulation efficiency of up to 25%. In in vitro microdilution inhibition assays, 90 nm nanoparticles improved the minimum bactericidal concentration of JO146 two-fold against *H. pylori* compared to the free drug alone, highlighting that controlled engineering of nanoparticle size is important in drug delivery optimization.

Keywords: *H. pylori*; microfluidics; design of experiments; poly(lactic-*co*-glycolic) acid; size

Citation: Hwang, J.; Mros, S.; Gamble, A.B.; Tyndall, J.D.A.; McDowell, A. Improving Antibacterial Activity of a HtrA Protease Inhibitor JO146 against *Helicobacter pylori*: A Novel Approach Using Microfluidics-Engineered PLGA Nanoparticles. *Pharmaceutics* **2022**, *14*, 348. https://doi.org/10.3390/pharmaceutics14020348

Academic Editors: Christian Celia and Oya Tagit

Received: 7 December 2021
Accepted: 29 January 2022
Published: 1 February 2022

1. Introduction

The escalating threat of antibiotic resistance warrants development of antibacterials with a novel mechanism of action and an improved mode of delivery. The majority of antibacterial agents in clinical development share common existing mechanisms of action, targeting bacterial cell wall biosynthesis or deterring protein synthesis on ribosomes, to which multiple resistance mechanisms are already established [1]. There is consensus in the drug design and discovery field that new lead compounds against these existing targets may have already been exhausted and are already exposed to the risk of antibacterial resistance; thus, identification of novel bacterial targets (e.g., bacterial proteases) has been acknowledged as a key approach to mitigate antibacterial insusceptibility [2–5].

Previously, JO146 (Figure 1) has been identified as a small molecule peptide-based inhibitor of the *Chlamydia trachomatis* high temperature requirement A (HtrA) protease [6]. HtrA is an emerging novel antibacterial target essential in protein quality control and survival under stress conditions [7]. Currently, there are no protease inhibitors used to treat bacterial infections in clinical settings. HtrA is commonly located in the periplasm of Gram-negative bacteria; however, it was demonstrated that *Helicobacter pylori* (*H. pylori*) HtrA is secreted extracellularly and directly contributes to the pathogenesis of the infection [8]. The secreted fraction of *H. pylori* HtrA cleaves E-cadherin, the adherens junction protein in gastric epithelial cells, resulting in disruption of the gastric epithelium and promoting bacterial invasion and paracellular transmigration across host tissues [9]. Although HtrA proteins are evolutionarily well-preserved proteases in bacteria, differences in substrate specificity and mechanistic and/or functional roles in different bacteria are speculated to ascribe JO146 to be a narrow-spectrum antibacterial. JO146 resulted in complete loss of infectious progeny in *Chlamydia* spp. in in vitro and in vivo models, while lacking inhibition against *Staphylococcus aureus*, *Pseudomonas aeruginosa*, and *Escherichia coli* in vitro [10]. In addition, JO146 does not cause toxicity in human, koala, and mice cell lines [6,11]. In the current study, we reveal that JO146 elicits antibacterial activity against *H. pylori*, suggesting that it could be used as an anti-*H. pylori* agent with a novel mechanism of action. Identification of an inhibitor with species-specific activity has become an important strategy of developing antibiotics, which could generate fewer off-target effects on the microbiome and decrease the pressure of antibiotic resistance [5].

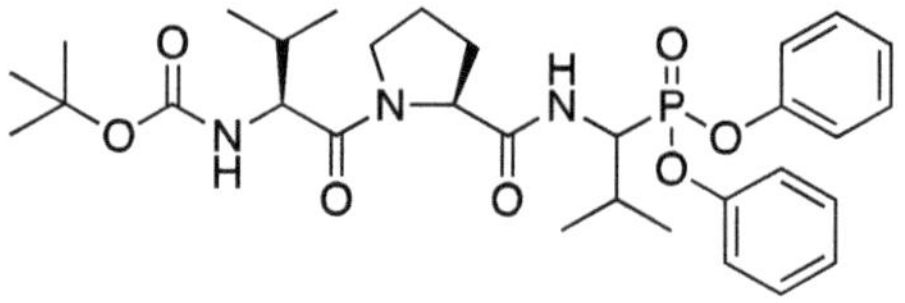

Figure 1. Structure of JO146 initially identified as a lead compound of HtrA protease in *Chlamydia* by high-through screening [6]. JO146 was loaded into PLGA NPs to study the drug delivery to *H. pylori*.

Nanoformulations have recently become a way to combat antibacterial resistance and challenges associated with delivery of antibacterial agents, such as low bioavailability, sub-therapeutic drug accumulation in microbial reservoirs, drug-related toxicities, and frequent drug dosing regimens [12]. Nanoparticles (NPs) enable a controlled release of antimicrobials, as well as improving accessibility of the drugs to pathogens and drug targets that are often hard-to-reach due to intracellular localization within bacteria cell membrane barriers containing multiple efflux pumps [13,14]. *H. pylori* infection remains challenging to treat as it mainly colonizes beneath the deep gastric mucosa and adheres to epithelial cells of the stomach [15]. Although *H. pylori* is traditionally viewed as an extracellular Gram-negative bacteria, approximately 1–5% of epithelial cells surrounded by *H. pylori* contain an intracellular population of bacteria, which is implicated in the bacteria's ability to cause life-long persistence and resistance to antibiotics, particularly those that exhibit poor intracellular accumulation [16,17]. Encapsulation of antimicrobial drugs into NPs can improve muco-adhesion and penetration of cell membranes, as well as intracellular delivery of antibiotics to combat various bacteria (e.g., *Staphylococcus aureus* [18–20], *Escherichia coli* [21], *Mycobacterium abscessus* [19], *Klebsiella pneumonia* [22], *Chlamydia trachomatis* [23] and *Mycobacterium tuberculosis* [24,25]), including those that are classified as extracellular pathogens but also show intracellular residence in host cells. Despite its promise as an alternative avenue for combating bacteria, understanding the interaction between NP and cells and management of nanomaterial toxicity often associated with inorganic nanomaterials remain as an important consideration for nanomaterial-based therapeutics [26].

H. pylori is one of the most prevalent bacterial pathogens that infects over half the global population and is linked to gastroduodenal diseases, including gastric cancer and peptic ulcers, posing a substantial healthcare burden [27]. *H. pylori* is currently listed as a high priority pathogen by the World Health Organization for research and development of new antibiotics [28]. Currently, triple therapy based on a combination of two antibiotics (clarithromycin plus either amoxicillin or metronidazole) with a proton pump inhibitor is the recommended first-line treatment for *H. pylori* infection [29]. However, *H. pylori* has developed resistance to clarithromycin, amoxicillin, metronidazole, and other antibiotics including levofloxacin, which has resulted in a high rate of treatment failures [30–32].

Therefore, new and effective anti-*H. pylori* treatments with a multipronged approach of combining chemical biology and nanoformulation would bring a valuable contribution to the global effort in addressing resistance development.

Based on this initial activity as a novel anti-*H. pylori* agent, we formulated JO146 in poly(lactic-*co*-glycolic) acid (PLGA) NPs to investigate whether the antibacterial efficacy could be enhanced. PLGA NPs are among the most extensively researched drug delivery vehicles due to their outstanding biocompatibility, tuneable degradation features, and a long history of clinical use [33,34]. PLGA NPs have also been utilized to improve drug pharmacokinetics, including low solubility and high sensitivity to chemical and enzymatic degradation [23–25]. Hence, these properties may be useful clinically for delivering lipophilic peptide agents such as JO146 (cLogP = 5.60).

The size of NPs is an important colloidal feature that directly influences cellular uptake, biodistribution, and hence therapeutic efficacy of the particles, determining the in vivo fate of the therapeutic cargo [35–37]. A myriad of studies has investigated the influence of particle size on the intracellular uptake in human cells (either normal or cancerous) [36,38–41]. Although the impact of particle size on the antibacterial activity of metallic NPs has been explored [42–44], little is known about their correlation with polymeric NPs.

Even though size is a critical physicochemical property of NPs that governs bacterial interaction, the lack of well-defined size control technology has limited the exploration of appropriate NP sizing for targeting bacterial cells. The uniformity of particle characteristics, such as size, is significantly affected by the manufacturing processes [45]. Conventional production techniques based on bulk emulsification methods lack the precision, reproducibility, and uniformity over particle size [46]. In recent years, microfluidics technology has emerged as a highly promising technique to produce particles in a highly controlled manner [47,48]. Microfluidics involves mixing of nanolitre scale fluids at high speeds and allows for automation of formulation parameters for production of particles with specific sizes [49,50]. The continuous nature of the microfluidic formulation process is inherently scalable by increasing the amounts of solvents pumped through the system or parallelizing multiple channels and mixers, allowing optimization with minimum requirement of materials that might be scarce or costly [51].

In this study, we report on a combined microfluidics with design of experiments (DoE) approach to produce size-tuneable PLGA particles, which enabled precise assessment of different formulation parameters on physicochemical characteristics of NPs. The DoE approach offers a strategy to make systematic and simultaneous variations in microfluidics settings and rapidly optimize various NP formulations to draw statistical interpretations with a minimum number of experimental runs [52–55]. This offers a major advantage over traditional "trial and error" approach of changing one variable at a time (OVAT), which is time-consuming and uneconomical. We explored the feasibility of generating different sized nanoparticles with low polydispersity index (PDI) by varying process and formulation parameters such as total flow rate and flow rate ratio of organic and aqueous phases in a DoE approach. We tuned the particle size at sub-100 nm, ~150 nm, and ~220 nm diameter, which represent biologically relevant sizes that influence intracellular particle uptake mechanisms, particle biodistribution, and clearance in vivo [23,35]. Encapsulation efficiency, release profile, and antibacterial efficacy of JO146-loaded PLGA NPs were also assessed.

2. Materials and Methods

2.1. Materials

PLGA 50:50, ester terminated, MW 15,980 Da, was purchased from Durect Lactel (Cupertino, CA, USA). Acetonitrile (ACN; HPLC grade) was supplied by Merck (Darmstadt, Germany). Mowiol® 4–88 (polyvinyl alcohol, PVA, MW 31,000), phosphate buffered saline (PBS), sucrose (≥99.5%), and tetracycline were purchased from Sigma–Aldrich (St. Louis, MO, USA). Trifluoroacetic acid (TFA, HPLC grade, purity > 99%) was purchased from

AK Scientific, Inc. (Union City, CA, USA). Distilled, ultrapure water was obtained from a Milli-Q® Water Millipore Purification System (Billerica, MA, USA). Mueller-Hinton agar (BD Difco, Franklin Lakes, NJ, USA), sheep blood (Fort Richard Laboratories, Auckland, New Zealand), Brucella broth (BD BBL, Franklin Lakes, NJ, USA), foetal bovine serum (FBS; Sigma, Auckland, New Zealand), and tryptic soy agar plates with 5% Sheep blood (Fort Richard Laboratories, Auckland, New Zealand) were utilized to maintain *H. pylori* culture. DMSO (Analytical Reagent Grade) was supplied by Fisher Chemical (Waltham, MA, USA) and MTT by Invitrogen Thermo Fisher Scientific (Waltham, MA, USA). AnaeroPack oxygen absorber-CO_2 generator was purchased for microaerophilic cultivation from Mitsubishi Gas Chemical (Tokyo, Japan).

2.2. Microfluidic Preparation of JO146-PLGA NPs with a Design of Experiments Approach

JO146 was prepared by in-house synthesis according to our previously reported methods [56]. Two independent sets of D-optimal DoE were conducted using the MODDE GO 12 software (Version 12.0, Umetrics, Umeå, Sweden).

The first DoE aimed at investigating the effects of formulation parameters on NP size (Z-average size) and polydispersity index (PDI). The second DoE was carried out with a goal of determining the lower limit of the PLGA NP size that could be achieved using the microfluidic method. In DoE studies, four input parameters—PLGA concentration (mg/mL), drug concentration (mg/mL), total flow rate (TFR; mL/min), and flow rate ratio (FRR; aq:org v/v)—were investigated at multilevel settings (low, medium, and high levels, i.e., −1, 0, +1), as described in Tables 1 and 2. The DoE models were analysed by PLS (partial least squares) regression [57] and one-way ANOVA to evaluate the statistical significance ($p < 0.05$) of the interactions between the input parameters and average particle size and PDI (Tables S1 and S2) [58]. Statistical testing was further conducted for each DoE regression model to obtain value of significance (p) and lack of fitness to estimate the error of the model and evaluate the quality of the model by R^2 and Q^2 (Table S3) [57,59].

Table 1. Input parameters of DoE-1 for the formulation of JO146-PLGA NPs with sizes in the range of 110–270 nm. Total flow rate = 5, 10, 15 mL/min and flow rate ratio = 1, 2, 5.

Parameters	−1	0	+1
PLGA concentration (mg/mL)	5	10	15
TFR (mL/min)	5	10	15
FRR (aqueous:organic)	1 0.5	2 1	5 2

Table 2. Input parameters of DoE-2 for the formulation of JO146-PLGA NPs with size targeted below 100 nm. Total flow rate = 10, 15 mL/min and flow rate ratio = 3, 5, 7.

Parameters	−1	0	+1
PLGA concentration (mg/mL)	2	4	6
TFR (mL/min)	10	-	15
FRR (aqueous:organic)	3 0.5	5 -	7 1

PLGA NPs were prepared using a microfluidic staggered herringbone mixer (SHM) with a NanoAssemblr® Benchtop instrument (Precision NanoSystems Inc., Vancouver, BC, Canada). The organic phase containing PLGA and JO146 in ACN and the 2% PVA aqueous phase (w/v) were injected into the two inlets of the microfluidic cartridge by a syringe pump. The organic and aqueous phases were mixed through the SHM by chaotic advection, leading to nanoprecipitation of JO146-loaded NPs. The formulation parameters (FRR and TFR) of the mixing conditions were programmed into the microfluidic setting according to the DoEs. An initial volume of 0.15–0.25 mL and final volume of 0.05 mL were discarded to collect the NP formulation from the outlet of the cartridge (Figure 2).

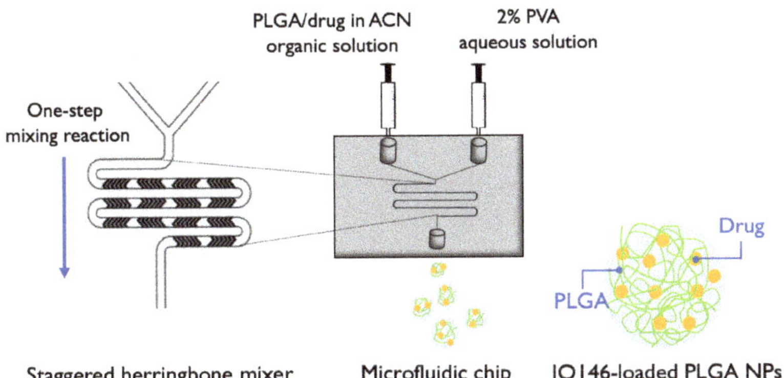

Figure 2. Schematic of microfluidic-assisted NP formulation via nanoprecipitation method using a Y-junction staggered herringbone mixer (modified from [60]). PLGA: poly(lactic-*co*-glycolic acid); ACN: acetonitrile; PVA: polyvinyl alcohol; NP: nanoparticle. Four formulation parameters—TFR, FRR, and polymer and drug concentrations—were investigated in the settings for NP optimization.

2.3. Characterization of NPs

JO146-PLGA NPs were characterized for their size (Z-Average size) and PDI by dynamic light scattering using a Malvern® Zetasizer (Malvern® Nano ZS, Model Zen 3600, Malvern Instruments, Malvern, UK), equipped with a 633 nm laser and 173° detection optics. The zeta potential (ZP) of the NPs were measured by laser Doppler electrophoresis using the same machine but in Malvern® folded capillary zeta cells (Malvern, UK). The NP samples were prepared by diluting 50 μL of the formulations collected from the NanoAssemblr® in 800 μL of ultrapure water and 10 mM NaCl and analysed for characterization in triplicate at 25 °C. Each sample measurement generated an NP count greater than 200 kcps and the derived attenuation factor ranging between 6 and 10.

2.4. Measuring Encapsulation of JO146 in NPs

Drug encapsulation efficiency (EE) and drug loading (DL) of three selected formulations of different NP sizes (90 nm, 150 nm, and 220 nm) were measured directly from the amount of JO146 entrapped in the NPs using Equations (1) and (2):

$$Encapsulation\ efficiency\ (\%) = (amount\ of\ the\ drug\ in\ the\ pellet \times 100) / (total\ amount\ of\ the\ drug\ added) \tag{1}$$

$$Drug\ loading\ (\%) = (amount\ of\ the\ drug\ in\ the\ pellet \times 100) / (total\ amount\ of\ the\ drug\ and\ PLGA\ added) \tag{2}$$

The JO146-NP suspensions from the microfluidic preparation were separated in an ultracentrifuge twice, at 45,000 rpm (~138,656× *g*) for 90 nm, 25,000 rpm (~42,795× *g*) for 150 nm, and 15,000 rpm (~15,406× *g*) for 220 nm, at 4 °C for 15 min. Centrifugation speed was optimized for each NP size formulation to facilitate quick re-dispersibility of the pellet, and sufficient centrifugal force was provided to ensure less than 2% loss of NPs in the supernatant. The pellet was dissolved in 2:1:1 DMSO:ACN:H$_2$O and sonicated to release the drug from the lysed NPs. The remaining suspension was separated using a centrifuge at 14,500 rpm (14,100× *g*) for 30 min, and the JO146 concentration of the supernatant was analysed by gradient elution using analytical RP-HPLC:deionized water with 0.5% TFA for phase A and ACN with 0.5% TFA for phase B at a flow rate of 1 mL/min (Table 3).

Table 3. Gradient method for the analytical RP-HPLC used for JO146 quantification.

Time (min)	Phase A%	Phase B%
1.00	40	60
10.00	15	85
11.00	40	60
15.00	40	60

2.5. Stability of JO146-PLGA in Simulated Physiological Conditions

A 2 mL aliquot of each JO146-PLGA NP formulation was prepared as described in the DoE (Tables 1 and 2). After ultracentrifugation as described in Section 2.4., the pellet from each formulation was re-suspended in 9 mL of PBS (pH 7.4) and divided into three equal aliquots, each of which was transferred to a 7 mL scintillation vial with a magnetic stirrer. The formulations were stirred in closed vials at 100 rpm at 37 °C on a hot plate for 96 h. An aliquot of a formulation sample in each vial was withdrawn, and the NPs were characterized for their size, PDI, and ZP at 0, 12, 24, 48, and 96 h during the stability testing.

2.6. Measuring JO146 Solubility by Turbidity and RP-HPLC

The solubility of JO146 was determined at 37 °C in PBS and PBS with 0.5% Tween 80 (v/v) used as a surfactant. A turbidity assay was used as an initial screening test to determine the range of JO146 solubility. A stock solution of JO146 in DMSO (20 mg/mL) was introduced into the first well containing the aqueous media, leading to drug precipitation. The suspension in the first well was serially diluted from 750 to 100 µg/mL in PBS (pH 7.4) containing 0.5% Tween 80 (v/v) and 100 to 21.0 µg/mL in PBS alone (pH 7.4). The turbidity of the wells with different JO146 concentrations was measured by a POLARstar® Omega microplate reader (BMG LABTECH, Ortenberg, Germany) in the endpoint mode at the wavelength of 600 nm, at which solubilised JO146 could not be detected. The absorbance of each well was determined by the average of five different measurement points with a 5 s shaking prior to the first measurement. Each concentration level was tested in triplicate. The absorbance readings were normalized by the average reading of the triplicate media controls in the absence of the drug. The solubility range of the drug was defined here as between the concentration that yielded statistically significant increase in the absorbance compared to the media control and the concentration tested one level below.

2.7. In Vitro JO146 Release Kinetics

After preparations of 90-nm, 150-nm, and 220-nm NPs, the pellets were re-suspended in 1.5% (g/mL) sucrose in ultrapure water as a cryoprotectant for freeze-drying of NPs. Lyophilisation of NPs was processed by a Labconco® freeze-drier (Kansas City, MO, USA). The NPs were freeze-dried around 0.04 mbar, −30–40°C overnight, and then progressively warmed to room temperature over the next 1–2 days. The lyophilized NPs were suspended in the release media (PBS, pH 7.4, with 0.5% Tween 80, 37 °C) at approximately 1 mg/mL and stirred in closed scintillation vials in triplicate at 100 rpm, 37 °C on a hot plate. For a control group, free JO146 was used. At 0, 0.5, 1, 2, 24, 48, and 96 h time points, 150 µL was sampled and replaced by fresh media to maintain sink conditions. Samples were centrifuged at 14,500 rpm (14,100× g) for 15 min at ambient temperature then the supernatants were kept in a −80 °C freezer until analysis using the analytical HPLC to quantify JO146 concentrations. At the end of the release experiment, the remaining NP suspensions were separated using centrifugation and the pellets were dissolved in 2:1:1 DMSO:ACN:H_2O to determine the concentration of JO146 that was not released. The results of the amount of drug released at pre-determined time points were reported as average values and standard deviations of three independent sets of experiments.

2.8. MIC and MBC of Free JO146 against H. pylori

The minimum inhibitory concentration (MIC) of free JO146 against *H. pylori* (ATCC® 43504) was investigated by a broth microdilution method. A stock solution of 25 mg/mL JO146 in DMSO was diluted in the assay media (Brucella broth with 3% FBS) to give resulting concentrations ranging from 500 μM (300.84 μg/mL) to 1.56 μM (0.94 μg/mL). The highest JO146 concentration tested for the MBC formulation assays was 200 μM (equivalent to 120.34 μg/mL, containing 0.5% DMSO), which was serially diluted by 2-fold to 1.56 μM (0.94 μg/mL) in a 96-well plate (resulting in 100 μL of inhibitor solution in each well). Likewise, a tetracycline control was prepared by adding 100 μL of the drug in the assay media at concentrations ranging from 5.0 to 0.039 μg/mL to the wells. Growth controls without the inhibitor were included in the assays. A 10 μL sample of *H. pylori* broth culture (OD 0.100; final concentration of ~5.0×10^5–10.0×10^5 CFU/mL) was then added to each well, except for the sterility control wells, prior to incubation of the plate for 4–6 days at 37 °C in a microaerophilic environment. To measure MIC, the turbidity of each well was measured by UV spectroscopy at 600 nm. The MIC was defined by the lowest concentration of the inhibitor that resulted in the same average optical density as the negative controls. The results were cross-checked by an MTT colorimetric assay, in which addition of 10 μL of 5 mg/mL MTT dye (3-(4,5-dimethylthiazol-2-yl)-2,5-diphenyltetrazolium bromide) to each well caused a colour change from yellow to pink or light purple in the presence of viable bacteria. Assays were conducted in triplicate and repeated on three independent days with fresh batches of *H. pylori* culture.

The minimum bactericidal concentration (MBC) of JO146 was measured by taking 10 μL from each well and inoculating onto fresh tryptic soy agar plates with 5% sheep blood. The agar plates were further incubated in a microaerophilic environment at 37 °C for 4–6 days to detect growth of any remaining bacteria. The MBC of each formulation was determined as the minimum concentration required for complete inhibition (as indicated qualitatively by no bacterial growth on the agar plate) from 10 μL of undiluted culture and compared to that of free JO146.

2.9. Antibacterial Activity of JO146 and JO146-PLGA NPs against H. pylori

In each of the three independent sets of MBC experiments, a new batch of JO146-PLGA NP formulations was prepared, and each was divided into two aliquots, one of which was used in the antibacterial assays and the other for characterizing the NP size and drug EE%. Tetracycline, free JO146, and empty PLGA NPs were used as controls; a positive growth control (*H. pylori* in assay media without JO146) and negative growth control (assay media only; Brucella broth with 3% FBS and 0.5% DMSO) were included. Freeze-dried JO146-PLGA NPs were re-suspended in the assay media so that the resulting concentration of JO146 entrapped in the NPs was 200 μM. The average concentration of PLGA in the NP resuspensions was 2.8 mg/mL, thus the NP control was also re-suspended at 2.8 mg/mL. Immediately after the re-suspension, 200 μL of each formulation was added to the plate wells in triplicate and serially diluted two-fold down to the eighth concentration (each well containing 100 μL of the NP suspension). Next, 10 μL of *H. pylori* broth culture (OD 0.100 measured at 600 nm UV wavelength; final concentration of ~5.0×10^5–10.0×10^5 CFU/mL) was added to each well, except for the negative growth control wells. The assay plates were incubated for 96 h at 37 °C in microaerophilic environment and shaken at 100 rpm. After the incubation, the assays were inoculated onto fresh tryptic soy agar plates with 5% sheep blood and further incubated to detect growth of any remaining bacteria.

3. Results and Discussion

3.1. NP Formulation by the Microfluidic-DoE Method and Characterization

A commercially available NanoAssemblr™ cartridge was used in a microfluidic device to optimize NPs with three discrete size categories in a range from sub-100 to 250 nm, which was proposed to be appropriate for endocytic uptake by mammalian host cells, which is required to reach the invasive, intracellular population of *H. pylori*.

Using the SHM, a set of preliminary experiments was carried out by a traditional OVAT approach, varying one factor at a time for TFR (5, 10, 15 mL/min) and FRR (aq:org 1:1, 2:1, 4:1, 5:1, 6:1 *v/v*) at constant concentrations of the polymer and drug to define the approximate range of process parameter settings for the DoE. There was a decreasing trend in the NP size with increasing TFR (Table S4). The NPs also reduced in size as the FRR increased from 1:1 to 5:1, with a small increase in NP size with a change in FRR from 5:1 to 6:1 (Table S4). Taking these preliminary data into consideration, three levels of FRR—1:1, 2:1, and 5:1 (aq:org)—were selected to be investigated for DoE-1.

To systematically assess the SHM-assisted nanoprecipitation method feasibility, a D-optimal design was used to identify the best subset of experiments that provides flexibility in the input parameters [58,61,62]. The first DoE (DoE-1) was used to assess four formulation parameters (concentration of PLGA polymer and drug, FRR, and TFR) in a microfluidic process and investigate the weighted influence of each parameter on the NP size and PDI. Based on the results obtained from DoE-1, the second DoE (DoE-2) was conducted with the aim to investigate the formulation conditions that enabled finding the lower limit of NP size attainable by the PLGA polymer using the NanoAssemblr® microfluidic system. The use of DoE also allowed rational optimization of the experimental conditions to formulate the NP in the target range of sizes with a narrow mean distribution (PDI < 0.3) and high inter-batch reproducibility.

JO146-PLGA NPs ranging from 70 nm to 349 nm in size were obtained through precise control of the formulation parameters (Tables S5 and S6). All the experiments yielded PDI with <0.3 except experiments no. 14 and 16 in DoE-2, which produced the smallest NP sizes of 70 ± 8.5 and 77 ± 4.3 nm (PDIs of 0.343 and 0.316, respectively; Table S6). The smallest particle size was related to experiments conducted with a high level of TFR (15 mL/min) and FRR (aq:org = 7:1) and a low concentration of PLGA (2 mg/mL). The increase in the PDI with the smallest particle sizes (Figure 3) suggests that when NP size was below 90 nm, the NPs were less stable, hence more prone to aggregation [63,64]. This was supported by the observation that NPs below or close to 80 nm often increased in size by approximately two-fold after ultracentrifugation (data not included). For the purpose of investigating the effect of PLGA NP size on the efficacy of antibacterial delivery, three sizes (90 nm, 150 nm, and 220 nm) were selected (Table 4).

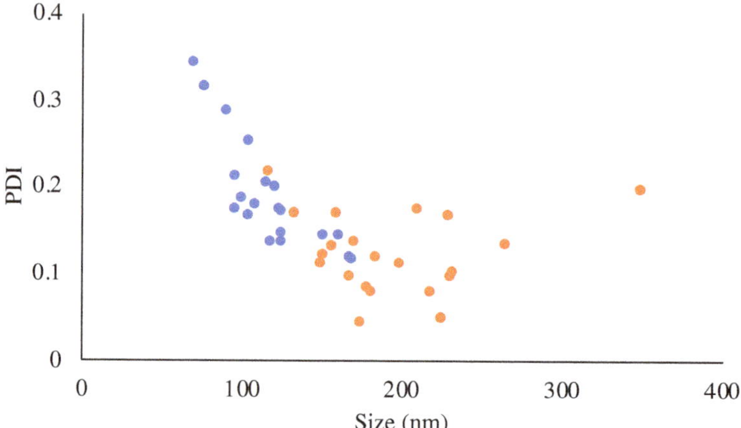

Figure 3. Relationship between the NP size and PDI obtained from DoE-1 (orange) and DoE-2 (blue). Data points are individual measurements.

Table 4. Three selected sizes of JO146-PLGA NPs from the design of experiments analysis.

Size (nm)	[PLGA] (mg/mL)	FRR (aq:org)	TFR (mL/min)	[JO146] (mg/mL)	PDI	ZP (mv)	EE%	DL%
90	10	1	10	1	0.096	−9.6 ± 2.0	5	1.3
150	5	5	10	2	0.170	−8.0 ± 2.0	16	1.6
220	2	3	15	0.5	0.173	−13.8 ± 0.4	25	3.9

The mean ZP of the NP formulations in DoE-1 and DoE-2 (Tables S5 and S6) was −13 ± 3.78 mV in deionized water and −0.6 ± 0.39 mV in 10 mM NaCl [65]. The ZP of resulting NPs was preserved regardless of the variations in the polymer and drug concentrations, TFR, or FRR. This indicated that chemical properties of the NP surface are maintained despite varying the microfluidic settings.

The significance of each input parameter on the NP size was illustrated by a contour relationship diagram (Figure 4). Particle size is largely influenced by the solvent diffusion coefficient (D), mixing time (t_{mix}), as well as the width of the focused stream (w_f) that is determined by the architectural design of the microfluidic cartridge mixer according to Equation (3) [66]:

$$t_{mix} \approx w_f^2/4D \tag{3}$$

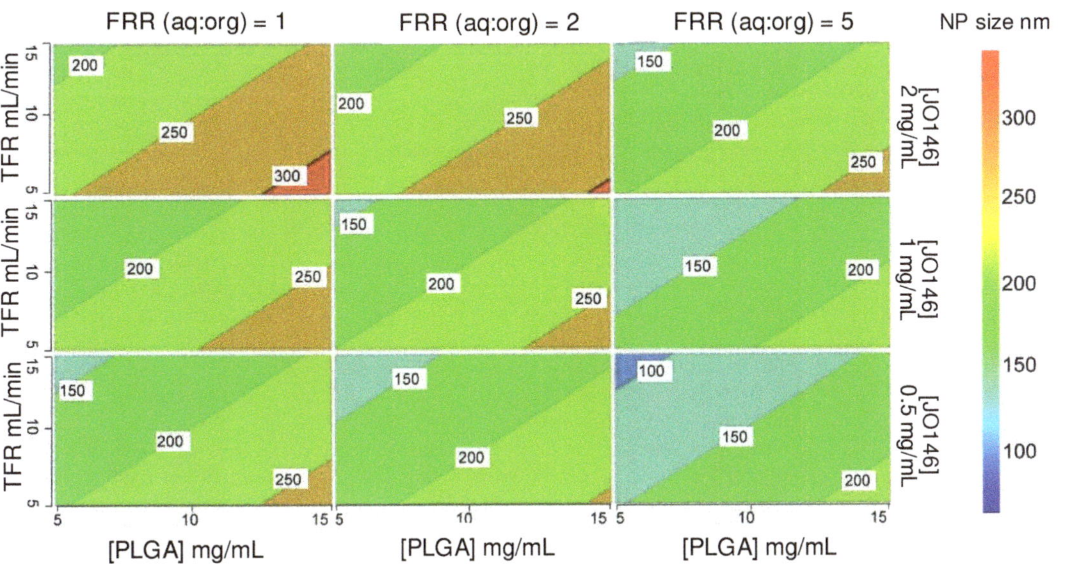

Figure 4. Contour representation of the relationships between the four input parameters and the particle size from DoE-1. TFR: total flow rate; FRR: flow rate ratio.

Increasing the polymer concentration in the organic solvent phase led to an increase in the NP size (Figure 4). This is ascribed to the higher viscosity of the organic phase with an increased polymer concentration, which resulted in slower diffusion of the organic phase to the aqueous phase, thereby increasing the mixing time [67]. Furthermore, increasing the amount of aqueous phase relative to the organic phase (i.e., higher FRR) accelerated the mixing time, thus decreasing the NP size [68]. In addition, a nucleation mechanism can account for NP formation and growth. Increasing the FRR causes supersaturation of the drug during the mixing process, which leads to an increase in the number of nuclei formed, ultimately decreasing the size of the NPs produced [69]. Additionally, a higher TFR decreased the size of the NPs produced. This is because increasing the speed of solvents pumped into the mixer decreases the size of the droplet formed during the breakup of the organic phase in the aqueous phase, thus the number of nuclei that are formed within the droplet. The decreased number of nuclei within each droplet results in the decrease in the final NP size [69].

Mathematical modelling identified PLGA concentration and FRR as the key variables in the microfluidic process with the highest impact on the NP size in DoE-1 (Table S1). The order of the most influential parameters on the NP size in DoE-1 based on the normalized regression coefficients was found to be PLGA concentration (0.53) > FRR (-0.50) > TFR (-0.42) > drug concentration (0.31; Table S1). In comparison, the influence of each parameter in DoE-2 had a similar but more distinctive ranking: PLGA concentration (0.73) >> TFR (-0.58) >> FRR (-0.26, n.s., $p > 0.05$) >> drug concentration (0.04, n.s., $p > 0.05$; Table S2). The fact that FRR showed a statistically not significant (n.s.) influence on size in DoE-2 could be interpreted that once the volume difference between the aqueous and organic phases exceeds a certain threshold (close to aq:org 5:1 as observed in the preliminary study using the OVAT approach), the correlation between the size and FRR becomes less conspicuous. This could be due to an excessive pressure difference between the aqueous and organic phases flowing into the microchannel, resulting in a backflow of the solvents into the organic inlet. Only FRR conferred significant change in PDI in both DoEs with increasing FRR (aq:org) giving a lower PDI, hence a narrower distribution of size (Tables S1 and S2).

The calculated values of R^2 and Q^2 in the DoEs were all above the recommended values from the literature ($R^2 > 0.75$, $Q^2 > 0.60$) [70], indicating that the model generated in each DoE showed high accuracy in fitting the existing data and predicting future data, respectively (Table S3). In addition, the difference between R^2 and Q^2 was below 0.3, suggesting that a high R^2 value was not due to over-fitting of the data between the parameters and responses [71,72]. These results reinforce the ability of the microfluidic technology to precisely control the input parameters and therefore produce NPs with corresponding characteristics with low variability and high reproducibility.

3.2. JO146-PLGA Stability

The physical stability of PLGA NPs was assessed in vitro in PBS (pH 7.4) at 37 °C for 96 h. The evaluation of NP stability was essential to ensure that the in vitro drug release kinetics were not influenced by drug leakage or dumping due to NP instability or degradation. The 90-nm NPs increased in size by 20% over 96 h. The increase in size of 90-nm NPs was statistically significant after 12 h (98 ± 2.4 nm), 48 h (100 ± 1.4 nm) and 96 h (109 ± 5.5 nm) compared to the initial size at 0 h (Figure 5). The size of the 150-nm NPs increased only slightly from 148 to 153 nm after 96 h ($p \leq 0.05$), and the 220-nm NPs remained the same size over the whole incubation period (one-way ANOVA analysis; Dunnett's multiple comparisons test). This size-dependent stability of NPs could be explained by the swelling mechanism of the PLGA polymeric matrix as the water penetrates inside the NPs during the bulk diffusion process [73,74]. Small NPs with a high surface to volume ratio could be more susceptible to hydrolytic degradation and physical instability as suggested by higher PDI and aggregation during ultracentrifugation as discussed above. This could be further supported by the complementary increase in PDI as the 90-nm NP size increases throughout the incubation time. Despite this, given that the PDI remained below 0.3 after 96 h, 90-nm NPs were regarded as monodisperse and relatively stable, suitable to deliver JO146 during the incubation of the drug treatment with *H. pylori* in in vitro antibacterial assays [75]. In addition, the ZP of the NPs remained constant regardless of their size at all sampling time points. The stability results of JO146-PLGA NPs showed that low ZP of the NPs does not necessarily reflect low stability of the NPs.

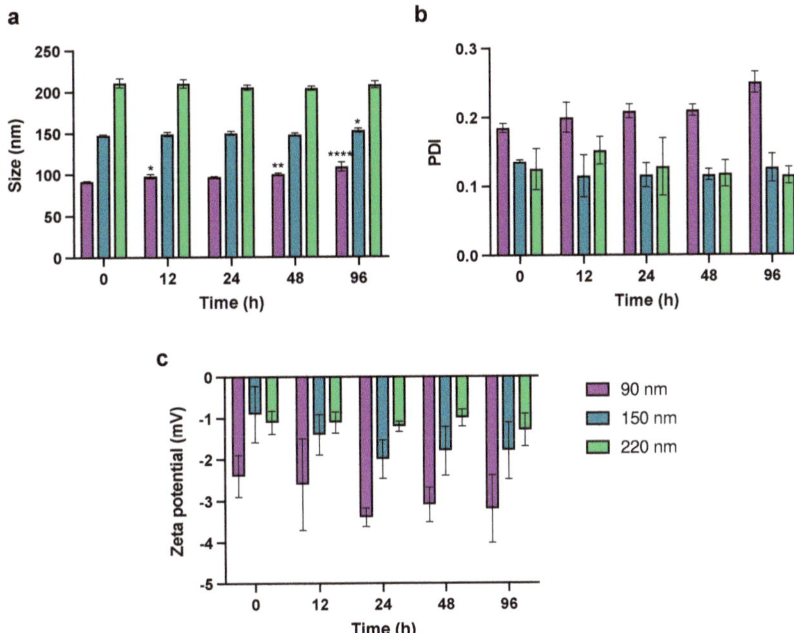

Figure 5. Monitoring of change in NP (**a**) size, (**b**) PDI, and (**c**) zeta potential in PBS at pH 7.4, 37 °C over 96 h. Values represent the mean ± SD (* $p \leq 0.05$, ** $p \leq 0.01$, **** $p \leq 0.001$; one-way ANOVA; Dunnett's multiple comparisons test).

3.3. In Vitro Release Kinetics

Release kinetics of JO146 from three different sizes of PLGA NPs were investigated using in vitro studies in PBS (pH 7.4) containing 0.5% Tween 80. Due to the poor aqueous solubility of JO146, 0.5% Tween 80 was utilized as a surfactant to allow JO146 to remain in solution for subsequent quantification using analytical HPLC. The solubility range of JO146 in PBS alone and PBS containing 0.5% (v/v) Tween 80 was initially screened by turbidity assays [76]. The solubility of JO146 in PBS alone and PBS containing 0.5% Tween 80 (v/v) were determined to lie in the range of 64–80 µg/mL and 237–316 µg/mL, respectively (Figure S2). This showed that adding 0.5%. Tween 80 (v/v) increased the aqueous solubility of JO146 in PBS, which aids the investigation of the drug release kinetics.

The release of the drug from NPs was biphasic with a burst where the majority of drug release occurs over the first 2 h followed by on onset of slower drug release up to 96 h (Figure 6). The 90-nm and 150-nm NPs demonstrated very similar first-order drug release kinetics with a release of approximately 80% of the entrapped drug within the first hour ($R^2 = 0.851$ and 0.852 for 90 and 150 nm, respectively). In comparison, 220-nm NPs released less than 50% during the first half an hour ($R^2 = 0.876$), which was about 1.5-fold less than the smaller NPs. This result aligns well with the theory that smaller NPs have a higher surface to volume ratio, hence facilitating greater release of drug in the initial burst release phase [77]. The second phase of release was zero order, unlike 90-nm 150-nm NPs, which released up to 97% and 95% of the drug after 96 h ($R^2 = 0.803$ and 0.914, for 90 and 150 nm, respectively); 220-nm NPs exhibited released up to 84% by the end of the experiment ($R^2 = 0.954$) (Figure 6). The average concentration of the free JO146 in the control group was constant throughout the 96-h experiment, confirming that the quantified JO146 concentrations accurately reflect the amount released from the NPs at each time point because JO146 did not degrade under the release conditions.

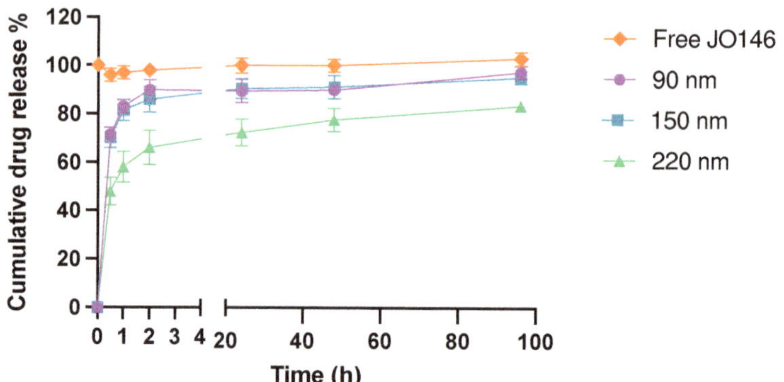

Figure 6. In vitro drug release kinetics of JO146 in PBS (pH 7.4, 37 °C) with 0.5% Tween 80 (*n* = 3). Error bars represent standard error of the mean (SEM).

3.4. Antimicrobial Activities of Free and Formulated JO146 against H. pylori

HtrA is crucial for the growth of *H. pylori* and recognized as an important virulence factor attributed to its ability to degrade E-cadherin-based adherens junctions, which allows efficient bacterial transmission during infections [78]. Previously, small-molecule inhibitors have been identified by a series of virtual screening and fragment-based, *de novo* design strategies to target the *H. pylori* HtrA serine protease [79], corroborating that HpHtrA protease can be targeted by small molecules. However, no peptide-based or substrate-based protease inhibitors have yet been reported against HpHtrA.

Three independent MIC assays were conducted with tetracycline as a positive control, which inhibited *H. pylori* with MIC and MBC values of 0.63–1.25 µg/mL and 0.625–2.5 µg/mL, respectively (Table S7). The variability in the inhibitory concentrations was consistent with values reported in the literature for *H. pylori* [80,81]. JO146 exhibited moderately strong antibacterial activity against *H. pylori* with a MIC range of 50–62.5 µM (30–37.6 µg/mL), which is a promising level of inhibition, comparable to the antibacterial activity against *C. trachomatis* [6].

Despite a lack of antibacterial activity against *E. coli*, *P. aeruginosa*, and *S. aureus* [6,10], the ability of JO146 to inhibit the growth of *H. pylori* could indicate that *H. pylori* HtrA has more similar substrate specificity to CtHtrA than those in other bacteria [82].

The MBC of JO146 was determined to be 31.3–125 µM (18.8–75.2 µg/mL), which was within the range of the MIC reported above, indicating that the drug likely acts via a bactericidal mechanism rather than being bacteriostatic [83]. This was not unexpected given that JO146 is a covalent transition state analogue that irreversibly binds to the biological target. Haemolysis of the sheep blood cells was not observed with any concentrations of JO146 tested. To further validate the bactericidal activity of JO146, it would be necessary to perform bacterial-killing kinetics assays in subsequent studies.

The 150-nm and 220-nm NPs did not affect the MBC of JO146 (50 µM) compared to that of free JO146 (Figure 7). However, the MBC of JO146 delivered by 90-nm NPs improved two-fold (25 µM). This was despite the finding that 90-nm and 150-nm NPs showed similar drug release kinetics, indicating that the difference in the efficacy of JO146 delivered by the two NP formulations could be attributed to the NP size. In comparison to mammalian cells that can endocytose NPs up to several hundred nanometres, bacteria can have very limited uptake of NPs due to their cell wall forming a barrier against simple diffusion and the lack of facilitated uptake such as endocytosis [84,85]. However, NPs may interact with bacteria by different mechanisms: (1) fusion with the microbial cell wall or cell membrane, transporting the antibacterial agent within the membrane barrier [86], (2) adsorption to the cell wall surface and continuous release of the antibacterial agent [86], and (3) alteration of the bacterial membrane potential to enhance membrane permeability [87]. Although 220-nm NPs displayed more sustained release (releasing ~80% over 96 h) compared to

90-nm and 150-nm NPs, the rate of drug release did not seem to affect the overall viability of *H. pylori* in vitro after the 96-h incubation. However, NP size had greater influence on the bacterial growth, suggesting that the superior surface-to-volume ratio of 90-nm NPs allowed greater interaction with the bacterial cell wall [37,88], thereby increased drug exposure to the periplasmic target site where HtrA proteins are commonly localized in Gram-negative bacteria [7]. Future research against an intracellular population of *H. pylori* and in vivo testing of the formulations would provide insight on the clinical benefits of utilizing a highly defined distribution of NP sizes for the treatment of *H. pylori* infections that are resistant to traditional antibiotics.

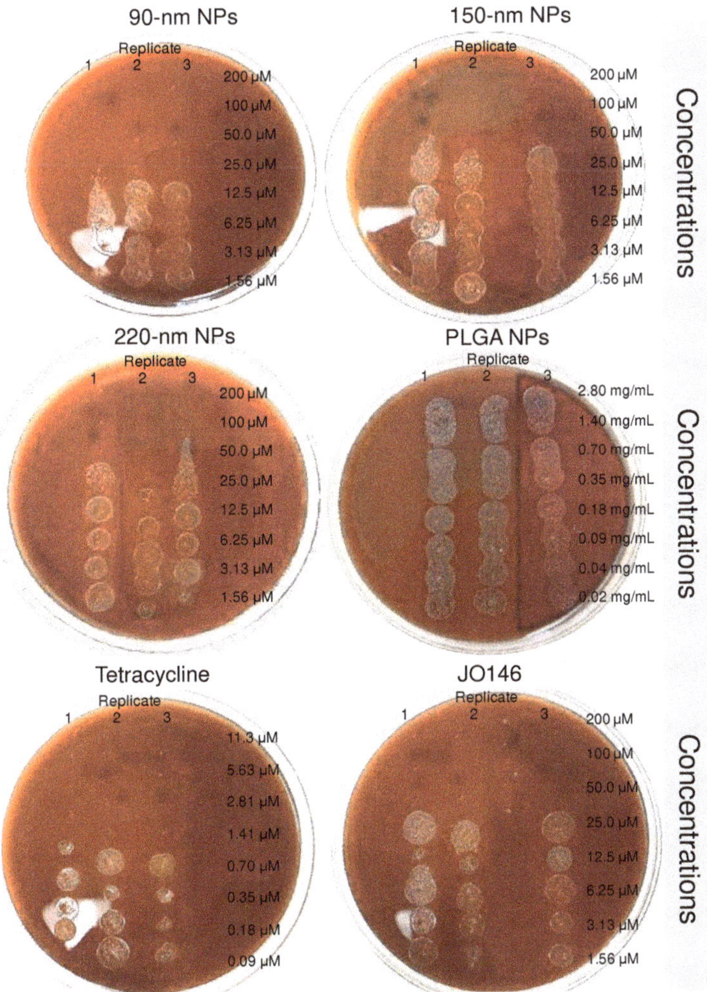

Figure 7. Minimum bactericidal concentration assays conducted to determine viability of *H. pylori* treated with JO146-PLGA NPs, drug-free PLGA NPs, tetracycline, and JO146. The efficacy of JO146 formulated in PLGA NPs was determined as MBC required to kill *H. pylori* (i.e., no visible growth of bacterial colonies). MBC assays were conducted in triplicate on tryptic soy agar plates with 5% sheep blood.

4. Conclusions

Although PLGA NPs have been extensively studied with a wide range of therapeutics, the optimal NP size for targeting bacteria is unclear, and the optimization of the NPs using the traditional bulk mixing methodology remains challenging.

In this paper, a combined microfluidics with DoE approach enabled assessment of the four key formulation parameters (drug and polymer concentrations, TFF, FRR) on physicochemical characteristics of NPs. Three different sized JO146-PLGA NPs (90-, 150-, and 220-nm) were formulated with a narrow distribution of size to investigate the optimal NP size to facilitate improved delivery of JO146 to *H. pylori*. These results showed the NanoAssemblr® microfluidic platform enabled precise control of the parameters to fine-tune physicochemical properties of NPs that might be required for specific clinical applications. NP size significantly influenced the particle stability and drug release kinetics, and 90-nm NPs improved MBC against *H. pylori* two-fold compared to 150-nm and 220-nm NPs and free JO146 in in vitro antibacterial assays, indicating that NP size plays a critical role in drug delivery and enhancing the antibacterial potency of the encapsulated drug. In vivo study of the antibacterial activity of the NP formulations compared to JO146 is required to investigate the biological and clinical significance of the size-controlled NPs.

Supplementary Materials: The following are available online at https://www.mdpi.com/article/10.3390/pharmaceutics14020348/s1. Table S1: Normalized regression coefficients of DoE-1, Table S2: Normalized regression coefficients of DoE-2, Table S3: Diagnostic parameters of DoE model validity, Table S4: OVAT investigation of the effects of TFR and FRR on NP size, Table S5: Experimental conditions of DoE-1 and JO146-PLGA nanoparticle characterization, Table S6: Experimental conditions of DoE-2 and JO146-PLGA nanoparticle characterization, Table S7: Quantitative and qualitative estimation of MIC of free JO146 by MTT colorimetric assays. Figure S1: DoE-2 contour plot of the influence of the input parameters on NP size, Figure S2: JO146 solubility determination by turbidity assays.

Author Contributions: Conceptualization, J.H., J.D.A.T., A.B.G. and A.M.; methodology, J.H., S.M. and A.M.; validation, S.M. and A.M.; formal analysis, J.H.; investigation, J.H.; resources, J.D.A.T., A.B.G., A.M. and S.M.; data curation, J.H.; writing—original draft preparation, J.H.; writing—review and editing, J.H., A.B.G., S.M., J.D.A.T. and A.M.; supervision, A.B.G., J.D.A.T. and A.M.; project administration, J.D.A.T. All authors have read and agreed to the published version of the manuscript.

Funding: This research was funded by University of Otago Doctoral Scholarship (J.H.) and MBIE (Ministry of Business, Innovation and Employment), Endeavour Research Programme (S.M.).

Institutional Review Board Statement: Not applicable.

Informed Consent Statement: Not applicable.

Data Availability Statement: Data available on request.

Acknowledgments: The authors would like to thank Sarah Streck for technical training to use the microfluidics and Gregory M. Cook for the provision of resources used for conducting the *H. pylori* antibacterial assays in his lab. We also thank Scott Ferguson for his helpful input at the review stage and the reviewers for their comments to improve the manuscript.

Conflicts of Interest: The authors declare no conflict of interest.

References

1. Walsh, C.T.; Wencewicz, A.T. Prospects for new antibiotics: A molecule-centered perspective. *J. Antibiot.* **2014**, *67*, 7–22. [CrossRef] [PubMed]
2. Culp, E.; Wright, G.D. Bacterial proteases, untapped antimicrobial drug targets. *J. Antibiot.* **2017**, *70*, 366–377. [CrossRef] [PubMed]
3. Fischbach, M.A.; Walsh, C.T. Antibiotics for Emerging Pathogens. *Science* **2009**, *325*, 1089–1093. [CrossRef] [PubMed]
4. Lewis, K. Platforms for antibiotic discovery. *Nat. Rev. Drug Discov.* **2013**, *12*, 371–387. [CrossRef]
5. Brown, E.D.; Wright, G.D. Antibacterial drug discovery in the resistance era. *Nature* **2016**, *529*, 336–343. [CrossRef]
6. Gloeckl, S.; Ong, V.A.; Patel, P.; Tyndall, J.; Timms, P.; Beagley, K.; Allan, J.A.; Armitage, C.; Turnbull, L.; Whitchurch, C.; et al. Identification of a serine protease inhibitor which causes inclusion vacuole reduction and is lethal to *Chlamydia trachomatis*. *Mol. Microbiol.* **2013**, *89*, 676–689. [CrossRef]

7. Skorko-Glonek, J.; Figaj, D.; Zarzecka, U.; Przepióra, T.; Renke, J.; Lipinska, B. The Extracellular Bacterial HtrA Proteins as Potential Therapeutic Targets and Vaccine Candidates. *Curr. Med. Chem.* **2017**, *24*, 2174–2204. [CrossRef]

8. Zarzecka, U.; Modrak-Wójcik, A.; Figaj, D.; Apanowicz, M.; Lesner, A.; Bzowska, A.; Lipinska, B.; Pawlik, A.; Backert, S.; Skorko-Glonek, J. Properties of the HtrA Protease from Bacterium *Helicobacter pylori* Whose Activity Is Indispensable for Growth Under Stress Conditions. *Front. Microbiol.* **2019**, *10*, 961. [CrossRef]

9. Zawilak-Pawlik, A.; Zarzecka, U.; Żyła-Uklejewicz, D.; Lach, J.; Strapagiel, D.; Tegtmeyer, N.; Böhm, M.; Backert, S.; Skorko-Glonek, J. Establishment of serine protease htrA mutants in *Helicobacter pylori* is associated with secA mutations. *Sci. Rep.* **2019**, *9*, 11794. [CrossRef]

10. Hwang, J.; Strange, N.; Phillips, M.J.; Krause, A.L.; Heywood, A.; Gamble, A.B.; Huston, W.M.; Tyndall, J.D. Optimization of peptide-based inhibitors targeting the HtrA serine protease in *Chlamydia*: Design, synthesis and biological evaluation of pyridone-based and N-Capping group-modified analogues. *Eur. J. Med. Chem.* **2021**, *224*, 113692. [CrossRef]

11. Lawrence, A.; Fraser, T.; Gillett, A.; Tyndall, J.; Timms, P.; Polkinghorne, A.; Huston, W.M. Chlamydia Serine Protease Inhibitor, targeting HtrA, as a New Treatment for Koala *Chlamydia* infection. *Sci. Rep.* **2016**, *6*, 31466. [CrossRef] [PubMed]

12. Kirtane, A.R.; Verma, M.; Karandikar, P.; Furin, J.; Langer, R.; Traverso, G. Nanotechnology approaches for global infectious diseases. *Nat. Nanotechnol.* **2021**, *16*, 369–384. [CrossRef] [PubMed]

13. Abed, N.; Couvreur, P. Nanocarriers for antibiotics: A promising solution to treat intracellular bacterial infections. *Int. J. Antimicrob. Agents* **2014**, *43*, 485–496. [CrossRef]

14. Fulaz, S.; Vitale, S.; Quinn, L.; Casey, E. Nanoparticle–Biofilm Interactions: The Role of the EPS Matrix. *Trends Microbiol.* **2019**, *27*, 915–926. [CrossRef] [PubMed]

15. Jain, S.K.; Haider, T.; Kumar, A.; Jain, A. Lectin-Conjugated Clarithromycin and Acetohydroxamic Acid-Loaded PLGA Nanoparticles: A Novel Approach for Effective Treatment of *H. pylori*. *AAPS PharmSciTech* **2015**, *17*, 1131–1140. [CrossRef] [PubMed]

16. Dubois, A.; Borén, T. *Helicobacter pylori* is invasive and it may be a facultative intracellular organism. *Cell. Microbiol.* **2007**, *9*, 1108–1116. [CrossRef]

17. Amieva, M.R.; Salama, N.R.; Tompkins, L.S.; Falkow, S. *Helicobacter pylori* enter and survive within multivesicular vacuoles of epithelial cells. *Cell. Microbiol.* **2002**, *4*, 677–690. [CrossRef]

18. Lacoma, A.; Uson, L.; Mendoza, G.; Sebastián, V.; Garcia-Garcia, E.; Muriel-Moreno, B.; Domínguez, J.; Arruebo, M.; Prat, C. Novel intracellular antibiotic delivery system against *Staphylococcus aureus*: Cloxacillin-loaded poly(d,l-lactide-co-glycolide) acid nanoparticles. *Nanomedicine* **2020**, *15*, 1189–1203. [CrossRef]

19. Dimer, F.A.; De Souza Carvalho-Wodarz, C.; Goes, A.; Cirnski, K.; Herrmann, J.; Schmitt, V.; Pätzold, L.; Abed, N.; De Rossi, C.; Bischoff, M.; et al. PLGA nanocapsules improve the delivery of clarithromycin to kill intracellular *Staphylococcus aureus* and *Mycobacterium abscessus*. *Nanomed. Nanotechnol. Biol. Med.* **2019**, *24*, 102125. [CrossRef]

20. Pillai, R.R.; Somayaji, S.N.; Rabinovich, M.; Hudson, M.C.; Gonsalves, E.K. Nafcillin-loaded PLGA nanoparticles for treatment of osteomyelitis. *Biomed. Mater.* **2008**, *3*, 034114. [CrossRef]

21. Agel, M.R.; Baghdan, E.; Pinnapireddy, S.R.; Lehmann, J.; Schäfer, J.; Bakowsky, U. Curcumin loaded nanoparticles as efficient photoactive formulations against gram-positive and gram-negative bacteria. *Colloids Surf. B Biointerfaces* **2019**, *178*, 460–468. [CrossRef] [PubMed]

22. Jiang, L.; Greene, M.K.; Insua, J.L.; Pessoa, J.S.; Small, D.M.; Smyth, P.; McCann, A.P.; Cogo, F.; Bengoechea, J.A.; Taggart, C.C.; et al. Clearance of intracellular *Klebsiella pneumoniae* infection using gentamicin-loaded nanoparticles. *J. Control. Release* **2018**, *279*, 316–325. [CrossRef] [PubMed]

23. Toti, U.S.; Guru, B.; Hali, M.; McPharlin, C.M.; Wykes, S.M.; Panyam, J.; Whittum-Hudson, J.A. Targeted delivery of antibiotics to intracellular chlamydial infections using PLGA nanoparticles. *Biomaterials* **2011**, *32*, 6606–6613. [CrossRef] [PubMed]

24. De Faria, T.J.; Roman, M.; De Souza, N.M.; De Vecchi, R.; De Assis, J.V.; Dos Santos, A.L.G.; Bechtold, I.; Winter, N.; Soares, M.; Silva, L.P.; et al. An Isoniazid Analogue Promotes Mycobacterium tuberculosis-Nanoparticle Interactions and Enhances Bacterial Killing by Macrophages. *Antimicrob. Agents Chemother.* **2012**, *56*, 2259–2267. [CrossRef] [PubMed]

25. Pandey, R.; Sharma, A.; Zahoor, A.; Sharma, S.; Khuller, G.K.; Prasad, B. Poly (DL-lactide-co-glycolide) nanoparticle-based inhalable sustained drug delivery system for experimental tuberculosis. *J. Antimicrob. Chemother.* **2003**, *52*, 981–986. [CrossRef] [PubMed]

26. Naskar, A.; Kim, K.-S. Nanomaterials as Delivery Vehicles and Components of New Strategies to Combat Bacterial Infections: Advantages and Limitations. *Microorganisms* **2019**, *7*, 356. [CrossRef] [PubMed]

27. O'Connor, A.; O'Morain, A.O.C.A.; Ford, A. Population screening and treatment of *Helicobacter pylori* infection. *Nat. Rev. Gastroenterol. Hepatol.* **2017**, *14*, 230–240. [CrossRef]

28. World Health Organisation. Global Priority List of Antibiotic-Resistant Bacteria to Guide Research, Discovery, and Development of New Antibiotics. 2017. Available online: https://www.who.int/medicines/publications/WHO-PPL-Short_Summary_25Feb-ET_NM_WHO.pdf (accessed on 4 March 2019).

29. Safavi, M.; Sabourian, R.; Foroumadi, A. Treatment of *Helicobacter pylori* infection: Current and future insights. *World J. Clin. Cases* **2016**, *4*, 5–19. [CrossRef] [PubMed]

30. Savoldi, A.; Carrara, E.; Graham, D.Y.; Conti, M.; Tacconelli, E. Prevalence of Antibiotic Resistance in *Helicobacter pylori*: A Systematic Review and Meta-analysis in World Health Organization Regions. *Gastroenterology* **2018**, *155*, 1372–1382.e17. [CrossRef]

31. Thung, I.; Aramin, H.; Vavinskaya, V.; Gupta, S.; Park, J.Y.; Crowe, S.E.; Valasek, M.A. Review article: The global emergence of *Helicobacter pylori* antibiotic resistance. *Aliment. Pharmacol. Ther.* **2016**, *43*, 514–533. [CrossRef]
32. Tshibangu-Kabamba, E.; Yamaoka, Y. *Helicobacter pylori* infection and antibiotic resistance—from biology to clinical implications. *Nat. Rev. Gastroenterol. Hepatol.* **2021**, *18*, 613–629. [CrossRef] [PubMed]
33. Astete, C.E.; Sabliov, C.M. Synthesis and characterization of PLGA nanoparticles. *J. Biomater. Sci. Polym. Ed.* **2006**, *17*, 247–289. [CrossRef] [PubMed]
34. Gentile, P.; Chiono, V.; Carmagnola, I.; Hatton, P.V. An Overview of Poly(lactic-co-glycolic) Acid (PLGA)-Based Biomaterials for Bone Tissue Engineering. *Int. J. Mol. Sci.* **2014**, *15*, 3640–3659. [CrossRef] [PubMed]
35. Hoshyar, N.; Gray, S.; Han, H.; Bao, G. The effect of nanoparticle size on in vivo pharmacokinetics and cellular interaction. *Nanomedicine* **2016**, *11*, 673–692. [CrossRef] [PubMed]
36. Kulkarni, S.A.; Feng, S.-S. Effects of Particle Size and Surface Modification on Cellular Uptake and Biodistribution of Polymeric Nanoparticles for Drug Delivery. *Pharm. Res.* **2013**, *30*, 2512–2522. [CrossRef] [PubMed]
37. Yeh, Y.-C.; Huang, T.-H.; Yang, S.-C.; Chen, C.-C.; Fang, J.-Y. Nano-Based Drug Delivery or Targeting to Eradicate Bacteria for Infection Mitigation: A Review of Recent Advances. *Front. Chem.* **2020**, *8*, 286. [CrossRef] [PubMed]
38. Rosenblum, D.; Joshi, N.; Tao, W.; Karp, J.M.; Peer, D. Progress and challenges towards targeted delivery of cancer therapeutics. *Nat. Commun.* **2018**, *9*, 1410. [CrossRef]
39. Foroozandeh, P.; Aziz, A.A. Insight into Cellular Uptake and Intracellular Trafficking of Nanoparticles. *Nanoscale Res. Lett.* **2018**, *13*, 339. [CrossRef] [PubMed]
40. Sabourian, P.; Yazdani, G.; Ashraf, S.; Frounchi, M.; Mashayekhan, S.; Kiani, S.; Kakkar, A. Effect of Physico-Chemical Properties of Nanoparticles on Their Intracellular Uptake. *Int. J. Mol. Sci.* **2020**, *21*, 8019. [CrossRef]
41. Xia, Q.; Huang, J.; Feng, Q.; Chen, X.; Liu, X.; Li, X.; Zhang, T.; Xiao, S.; Li, H.; Zhong, Z.; et al. Size- and cell type-dependent cellular uptake, cytotoxicity and in vivo distribution of gold nanoparticles. *Int. J. Nanomed.* **2019**, *14*, 6957–6970. [CrossRef]
42. Wu, M.; Guo, H.; Liu, L.; Liu, Y.; Xie, L. Size-dependent cellular uptake and localization profiles of silver nanoparticles. *Int. J. Nanomed.* **2019**, *14*, 4247–4259. [CrossRef]
43. Applerot, G.; Lellouche, J.; Lipovsky, A.; Nitzan, Y.; Lubart, R.; Gedanken, A.; Banin, E. Understanding the Antibacterial Mechanism of CuO Nanoparticles: Revealing the Route of Induced Oxidative Stress. *Small* **2012**, *8*, 3326–3337. [CrossRef] [PubMed]
44. Pareek, V.; Gupta, R.; Panwar, J. Do physico-chemical properties of silver nanoparticles decide their interaction with biological media and bactericidal action? A review. *Mater. Sci. Eng. C* **2018**, *90*, 739–749. [CrossRef] [PubMed]
45. Sah, E.; Sah, H. Recent Trends in Preparation of Poly(lactide- co -glycolide) Nanoparticles by Mixing Polymeric Organic Solution with Antisolvent. *J. Nanomater.* **2015**, *2015*, 794601. [CrossRef]
46. Operti, M.C.; Fecher, D.; van Dinther, E.A.; Grimm, S.; Jaber, R.; Figdor, C.G.; Tagit, O. A comparative assessment of continuous production techniques to generate sub-micron size PLGA particles. *Int. J. Pharm.* **2018**, *550*, 140–148. [CrossRef] [PubMed]
47. Ding, S.; Anton, N.; Vandamme, T.F.; Serra, C.A. Microfluidic nanoprecipitation systems for preparing pure drug or polymeric drug loaded nanoparticles: An overview. *Expert Opin. Drug Deliv.* **2016**, *13*, 1447–1460. [CrossRef]
48. Karnik, R.; Gu, F.; Basto, P.; Cannizzaro, C.; Dean, L.; Kyei-Manu, W.; Langer, R.; Farokhzad, O.C. Microfluidic Platform for Controlled Synthesis of Polymeric Nanoparticles. *Nano Lett.* **2008**, *8*, 2906–2912. [CrossRef] [PubMed]
49. Garg, S.; Heuck, G.; Ip, S.; Ramsay, E. Microfluidics: A transformational tool for nanomedicine development and production. *J. Drug Target.* **2016**, *24*, 821–835. [CrossRef] [PubMed]
50. Roces, C.B.; Lou, G.; Jain, N.; Abraham, S.; Thomas, A.; Halbert, G.W.; Perrie, Y. Manufacturing Considerations for the Development of Lipid Nanoparticles Using Microfluidics. *Pharmaceutics* **2020**, *12*, 1095. [CrossRef]
51. Gdowski, A.; Johnson, K.; Shah, S.; Gryczynski, I.; Vishwanatha, J.; Ranjan, A. Optimization and scale up of microfluidic nanolipomer production method for preclinical and potential clinical trials. *J. Nanobiotechnology* **2018**, *16*, 12. [CrossRef]
52. Mandlik, S.K.; Ranpise, N.S. Implementation of experimental design methodology in preparation and characterization of zolmitriptan loaded chitosan nanoparticles. *Int. Curr. Pharm. J.* **2017**, *6*, 16–22. [CrossRef]
53. Streck, S.; Neumann, H.; Nielsen, H.M.; Rades, T.; McDowell, A. Comparison of bulk and microfluidics methods for the formulation of poly-lactic-co-glycolic acid (PLGA) nanoparticles modified with cell-penetrating peptides of different architectures. *Int. J. Pharm. X* **2019**, *1*, 100030. [CrossRef] [PubMed]
54. Vu, H.T.; Streck, S.; Hook, S.M.; McDowell, A. Utilization of Microfluidics for the Preparation of Polymeric Nanoparticles for the Antioxidant Rutin: A Comparison with Bulk Production. *Pharm. Nanotechnol.* **2019**, *7*, 469–483. [CrossRef] [PubMed]
55. Streck, S.; Hong, L.; Boyd, B.; McDowell, A. Microfluidics for the Production of Nanomedicines: Considerations for Polymer and Lipid-based Systems. *Pharm. Nanotechnol.* **2019**, *7*, 423–443. [CrossRef] [PubMed]
56. Agbowuro, A.A.; Hwang, J.; Peel, E.; Mazraani, R.; Springwald, A.; Marsh, J.W.; McCaughey, L.; Gamble, A.B.; Huston, W.M.; Tyndall, J.D. Structure-activity analysis of peptidic *Chlamydia* HtrA inhibitors. *Bioorganic Med. Chem.* **2019**, *27*, 4185–4199. [CrossRef]
57. Kettaneh-Wold, N. Analysis of mixture data with partial least squares. *Chemom. Intell. Lab. Syst.* **1992**, *14*, 57–69. [CrossRef]
58. Eriksson, L.; Johansson, E.; Kettaneh-Wold, N.; WikstrÄom, C.; Wold, S. Design of Experiments, Principles and Applications. *J Chemom.* **2001**, *15*, 495–496. [CrossRef]

59. Sprunk, A.; Strachan, C.J.; Graf, A. Rational formulation development and in vitro assessment of SMEDDS for oral delivery of poorly water soluble drugs. *Eur. J. Pharm. Sci.* **2012**, *46*, 508–515. [CrossRef]

60. Belliveau, N.M.; Huft, J.; Lin, P.J.C.; Chen, S.; Leung, A.K.K.; Leaver, T.J.; Wild, A.W.; Lee, J.B.; Taylor, R.J.; Tam, Y.K.; et al. Microfluidic synthesis of highly potent limit-size lipid nanoparticles for in vivo delivery of siRNA. *Mol. Ther. Nucleic Acids* **2012**, *1*, e37. [CrossRef]

61. Lewis, G.A.; Mathieu, D.; Phan-Tan-Luu, R. *Pharmaceutical Experimental Design*; Marcel Dekker: New York, NY, USA, 1999.

62. Lewis, G.A. Non Classical Experimental Designs in Pharmaceutical Formulation. *Drug Dev. Ind. Pharm.* **1991**, *17*, 1551–1570. [CrossRef]

63. Menon, J.U.; Kona, S.; Wadajkar, A.S.; Desai, F.; Vadla, A.; Nguyen, K.T. Effects of surfactants on the properties of PLGA nanoparticles. *J. Biomed. Mater. Res. Part A* **2012**, *100A*, 1998–2005. [CrossRef]

64. Sharma, S.; Parmar, A.; Kori, S.; Sandhir, R. PLGA-based nanoparticles: A new paradigm in biomedical applications. *TrAC Trends Anal. Chem.* **2016**, *80*, 30–40. [CrossRef]

65. Caputo, F. Measuring Zeta Potential. 2015. Available online: https://www.researchgate.net/profile/J-Tarafdar/post/What-buffer-to-use-in-Zetasiser-Nano-analysis/attachment/5f4e296c6a5a0300017c19ab/AS%3A930911189229574%4015989579319 83/download/EUNCL-PCC-002.pdf (accessed on 10 August 2020).

66. Chiesa, E.; Dorati, R.; Pisani, S.; Conti, B.; Bergamini, G.; Modena, T.; Genta, I. The Microfluidic Technique and the Manufacturing of Polysaccharide Nanoparticles. *Pharmaceutics* **2018**, *10*, 267. [CrossRef] [PubMed]

67. Halayqa, M.; Domańska, U. PLGA Biodegradable Nanoparticles Containing Perphenazine or Chlorpromazine Hydrochloride: Effect of Formulation and Release. *Int. J. Mol. Sci.* **2014**, *15*, 23909–23923. [CrossRef]

68. Chiesa, E.; Dorati, R.; Modena, T.; Conti, B.; Genta, I. Multivariate analysis for the optimization of microfluidics-assisted nanoprecipitation method intended for the loading of small hydrophilic drugs into PLGA nanoparticles. *Int. J. Pharm.* **2018**, *536*, 165–177. [CrossRef] [PubMed]

69. van Ballegooie, C.; Man, A.; Andreu, I.; Gates, B.D.; Yapp, D. Using a Microfluidics System to Reproducibly Synthesize Protein Nanoparticles: Factors Contributing to Size, Homogeneity, and Stability. *Processes* **2019**, *7*, 290. [CrossRef]

70. Mandenius, C.-F.; Brundin, A. Bioprocess optimization using design-of-experiments methodology. *Biotechnol. Prog.* **2008**, *24*, 1191–1203. [CrossRef]

71. Veerasamy, R.; Rajak, H.; Jain, A.; Sivadasan, S.; Varghese, C.P.; Agrawal, R.K. Validation of QSAR models-strategies and importance. *Int. J. Drug Des. Discov.* **2011**, *2*, 511–519.

72. Stiolica, A.T.; Popescu, M.; Bubulica, M.V.; Oancea, C.N.; Nicolicescu, C.; Manda, C.V.; Neamtu, J.; Croitoru, O. Optimization of Gold Nanoparticles Synthesis using Design of Experiments Technique. *Rev. de Chim.* **2017**, *68*, 1423–1518. [CrossRef]

73. Dinarvand, R.; Sepehri, N.; Manouchehri; Rouhani, H.; Atyabi, F. Polylactide-co-glycolide nanoparticles for controlled delivery of anticancer agents. *Int. J. Nanomed.* **2011**, *6*, 877–895. [CrossRef]

74. Makadia, H.K.; Siegel, S.J. Poly lactic-co-glycolic acid (PLGA) As biodegradable controlled drug delivery carrier. *Polymers* **2011**, *3*, 1377–1397. [CrossRef] [PubMed]

75. Andersen, L.P.; Wadström, T. *Helicobacter pylori: Physiology and Genetics. Chapter 4 Basic Bacteriology and Culture*; ASM Press: Washington, DC, USA, 2001.

76. Fürer, R.; Geiger, M. A simple method of determining the aqueous solubility of organic substances. *Pestic. Sci.* **1977**, *8*, 337–344. [CrossRef]

77. Esmaeili, F.; Ghahremani, M.H.; Esmaeili, B.; Khoshayand, M.R.; Atyabi, F.; Dinarvand, R. PLGA nanoparticles of different surface properties: Preparation and evaluation of their body distribution. *Int. J. Pharm.* **2008**, *349*, 249–255. [CrossRef] [PubMed]

78. Hoy, B.; Löwer, M.; Weydig, C.; Carra, G.; Tegtmeyer, N.; Geppert, T.; Schröder, P.; Sewald, N.; Backert, S.; Schneider, G.; et al. *Helicobacter pylori* HtrA is a new secreted virulence factor that cleaves E-cadherin to disrupt intercellular adhesion. *EMBO Rep.* **2010**, *11*, 798–804. [CrossRef] [PubMed]

79. Perna, A.M.; Rodrigues, T.; Schmidt, T.P.; Böhm, M.; Stutz, K.; Reker, D.; Pfeiffer, B.; Altmann, K.; Backert, S.; Wessler, S.; et al. Fragment-Based De Novo Design Reveals a Small-Molecule Inhibitor of *Helicobacter pylori* HtrA. *Angew. Chem. Int. Ed.* **2015**, *54*, 10244–10248. [CrossRef]

80. An, B.; Moon, B.S.; Kim, H.; Lim, H.C.; Lee, Y.C.; Lee, G.; Kim, S.-H.; Park, M.; Kim, J.B. Antibiotic Resistance in *Helicobacter pylori* Strains and its Effect on *H. pylori* Eradication Rates in a Single Center in Korea. *Ann. Lab. Med.* **2013**, *33*, 415–419. [CrossRef]

81. Lee, H.-K.; Song, H.E.; Lee, H.-B.; Kim, C.-S.; Koketsu, M.; Ngan, L.T.M.; Ahn, Y.-J. Growth Inhibitory, Bactericidal, and Morphostructural Effects of Dehydrocostus Lactone from *Magnolia sieboldii* Leaves on Antibiotic-Susceptible and -Resistant Strains of *Helicobacter pylori*. *PLoS ONE* **2014**, *9*, e95530. [CrossRef]

82. Tegtmeyer, N.; Moodley, Y.; Yamaoka, Y.; Pernitzsch, S.R.; Schmidt, V.; Traverso, F.R.; Schmidt, T.P.; Rad, R.; Yeoh, K.G.; Bow, H.; et al. Characterisation of worldwide *Helicobacter pylori* strains reveals genetic conservation and essentiality of serine protease HtrA. *Mol. Microbiol.* **2015**, *99*, 925–944. [CrossRef]

83. Laboratories, Q. Minimum Inhibitory (MIC) and Minimum Bactericidal Concentration (MBC) Evaluations as R&D Tools. Available online: https://www.qlaboratories.com/minimum-inhibitory-mic-and-minimum-bactericidal-concentration-mbc-evaluations-as-rd-tools/ (accessed on 16 July 2021).

84. Doak, S.; Manshian, B.; Jenkins, G.; Singh, N. In vitro genotoxicity testing strategy for nanomaterials and the adaptation of current OECD guidelines. *Mutat. Res. Toxicol. Environ. Mutagen.* **2012**, *745*, 104–111. [CrossRef]

85. Landsiedel, R.; Kapp, M.D.; Schulz, M.; Wiench, K.; Oesch, F. Genotoxicity investigations on nanomaterials: Methods, preparation and characterization of test material, potential artifacts and limitations—Many questions, some answers. *Mutat. Res. Mutat. Res.* **2009**, *681*, 241–258. [CrossRef]

86. Lotfipour, F.; Valizadeh, H.; Milani, M.; Bahrami, N.; Ghotaslou, R. Study of Antimicrobial Effects of Clarithromycin Loaded PLGA Nanoparticles against Clinical Strains of *Helicobacter pylori*. *Drug Res.* **2015**, *66*, 41–45. [CrossRef] [PubMed]

87. Vazquez-Muñoz, R.; Meza-Villezcas, A.; Fournier, P.; Soria-Castro, E.; Juarez-Moreno, K.; Gallego-Hernandez, A.; Bogdanchikova, N.; Vazquez-Duhalt, R.; Huerta-Saquero, A. Enhancement of antibiotics antimicrobial activity due to the silver nanoparticles impact on the cell membrane. *PLoS ONE* **2019**, *14*, e0224904. [CrossRef] [PubMed]

88. Aruguete, D.M.; Hochella, M.F. Bacteria—Nanoparticle interactions and their environmental implications. *Environ. Chem.* **2010**, *7*, 3–9. [CrossRef]

pharmaceutics

Industrial Scale Manufacturing and Downstream Processing of PLGA-Based Nanomedicines Suitable for Fully Continuous Operation

Maria Camilla Operti [1,2], Alexander Bernhardt [2], Vladimir Sincari [3], Eliezer Jager [3], Silko Grimm [2], Andrea Engel [4], Martin Hruby [3], Carl Gustav Figdor [1] and Oya Tagit [1,*]

[1]   Department of Tumor Immunology, Radboud Institute for Molecular Life Sciences, Radboud University Medical Center, 6500 HB Nijmegen, The Netherlands; maria-camilla.operti@evonik.com (M.C.O.); carl.Figdor@radboudumc.nl (C.G.F.)
[2]   Evonik Operations GmbH, Research Development & Innovation, 64293 Darmstadt, Germany; alexander.bernhardt@evonik.com (A.B.); silko.grimm@evonik.com (S.G.)
[3]   Institute of Macromolecular Chemistry CAS, Heyrovsky Square 2, 162 06 Prague, Czech Republic; sincarivl@gmail.com (V.S.); jagereliezer@gmail.com (E.J.); mhruby@centrum.cz (M.H.)
[4]   Evonik Corporation, Birmingham Laboratories, Birmingham, AL 35211, USA; andrea.engel@evonik.com
*   Correspondence: oyatagit@gmail.com

Citation: Operti, M.C.; Bernhardt, A.; Sincari, V.; Jager, E.; Grimm, S.; Engel, A.; Hruby, M.; Figdor, C.G.; Tagit, O. Industrial Scale Manufacturing and Downstream Processing of PLGA-Based Nanomedicines Suitable for Fully Continuous Operation. *Pharmaceutics* 2022, 14, 276. https://doi.org/10.3390/pharmaceutics14020276

Academic Editor: Giovanni Tosi

Received: 22 December 2021
Accepted: 21 January 2022
Published: 25 January 2022

Publisher's Note: MDPI stays neutral with regard to jurisdictional claims in published maps and institutional affiliations.

Abstract: Despite the efficacy and potential therapeutic benefits that poly(lactic-co-glycolic acid) (PLGA) nanomedicine formulations can offer, challenges related to large-scale processing hamper their clinical and commercial development. Major hurdles for the launch of a polymeric nanocarrier product on the market are batch-to-batch variations and lack of product consistency in scale-up manufacturing. Therefore, a scalable and robust manufacturing technique that allows for the transfer of nanomedicine production from the benchtop to an industrial scale is highly desirable. Downstream processes for purification, concentration, and storage of the nanomedicine formulations are equally indispensable. Here, we develop an inline sonication process for the production of polymeric PLGA nanomedicines at the industrial scale. The process and formulation parameters are optimized to obtain PLGA nanoparticles with a mean diameter of 150 ± 50 nm and a small polydispersity index (PDI < 0.2). Downstream processes based on tangential flow filtration (TFF) technology and lyophilization for the washing, concentration, and storage of formulations are also established and discussed. Using the developed manufacturing and downstream processing technologies, production of two PLGA nanoformulations encasing ritonavir and celecoxib was achieved at 84 g/h rate. As a measure of actual drug content, encapsulation efficiencies of 49.5 ± 3.2% and 80.3 ± 0.9% were achieved for ritonavir and celecoxib, respectively. When operated in-series, inline sonication and TFF can be adapted for fully continuous, industrial-scale processing of PLGA-based nanomedicines.

Keywords: PLGA; poly(lactic-co-glycolic acid); nanomedicine; nanoparticles; scale-up manufacturing; clinical translation; inline sonication; tangential flow filtration; lyophilization; downstream processing

1. Introduction

Polymeric nanoparticles composed of biodegradable and biocompatible polymers such as poly(lactic-co-glycolic acid) (PLGA) display a promising future for various biomedical and pharmaceutical applications. However, the development of parenteral nanoparticle formulations for clinical and commercial use often faces the challenges of process scaling in a sterile or aseptic environment that complies with Good Manufacturing Practices (GMP) [1]. The main requirements for clinical and commercial development of nanoparticle formulations are high therapeutic efficacy and safety as well as production scalability and process robustness, which are closely connected with the material characteristics and applied production technology [2–4]. Therefore, the ability to establish large-scale processes for GMP-compliant production is indispensable for translating nanoparticle formulations from the bench to the bedside [1,5].

Emulsion-based methods are among the most exploited approaches for the preparation of PLGA nanoparticles, in which organic solvents are removed by evaporation or extraction. Equipment used currently used for the crucial homogenization step comprise probe sonicators, high-shear mixers, high-pressure homogenizers, and microfluidic systems [1,6]. In particular, emulsification by direct sonication with a transducer probe immersed in the processed medium is a highly common laboratory-scale approach due to the ease of operation that allows for rapid formulation screening [6,7]. However, direct contact of the drug product with the metal probe can lead to cross-contamination (e.g., with heavy metals), and the harsh homogenization shear and high temperature created during the cavitation phenomena can affect sensitive APIs and even result in the degradation of the polymeric carrier chains [1,7–9]. Besides, increased throughput can lead to changes in the formulation properties, making the technology only suitable for small batch preparations [1,6,7]. Sonication can also be applied indirectly via distributing the energy to the sample container itself, preventing the risk of sample contamination. The process can be entirely constructed using disposable materials (e.g., tubing and vessels) and operated inline in fully enclosed containers, which makes the process scalable and suitable for aseptic manufacturing [7,10].

During scale-up manufacturing, downstream processes are also necessary to isolate materials in the desired form and purity. Based on the preparation method, various impurities that can be toxic are present in the final product [11]. These impurities may include organic processing solvents such as dichloromethane (DCM) and dimethyl sulfoxide (DMSO), emulsifiers and stabilizers, monomeric residues, free unbound payloads, and salts. Such substances can lead to potential biological intolerance and alter the physicochemical characteristics of the nanoformulations. Therefore, an effectual cleaning strategy of nanoformulations is essential for controlling the quality and characteristics of nanomedicine products [11]. Lab-scale purification of nanoparticles is usually achieved by centrifugation, extraction, or filtration-based techniques. Most of these traditional techniques are time-consuming and often lack reproducibility, which can make these processes relatively inefficient [1,11]. Moreover, the requirement for manual handling can lead to more laborious processes when scaled up and is difficult to perform in an aseptic environment. Tangential flow filtration (TFF) is an alternative purification method that involves fluid flow along the surface of a membrane rather than passing through a filter. Scale-up of TFF processes is possible and is commonly considered straightforward since membrane cartridges are linearly scalable. Moreover, TFF can be operated in two modes within downstream unit operations: batch and single pass. The former involves recirculation of the processed solution over the membrane surface multiple times, while the latter operates in a single step, enabling a completely continuous operation [12]. As a continuous and gentle process that can significantly reduce membrane clogging compared to other traditional filtration methods, TFF is suitable for downstream processing of nanoparticles at the industrial level [1].

In this study, we developed PLGA nanoparticle manufacturing and downstream processes suitable for industrial scale-up. A scalable inline indirect sonication technique similar to that previously reported by Freitas et al. [10] was herein adapted, developed, and enhanced for the preparation of PLGA nanoparticles of approximately $150 \pm 50\,\text{nm}$ diameter with a polydispersity index (PDI) < 0.2. The particle size, PDI, zeta potential, and PLGA polymer chain integrity of different formulations were studied. TFF was tested and evaluated for different diafiltration volumes to determine the amount of impurities in the final drug product before storage. The need to use a cryoprotective agent during freeze-drying of the drug product was demonstrated. The results were evaluated following a statistical analysis to determine the significance of the difference among the compared groups. Ultimately, the inline sonication technique was scaled up for the production of nanomedicines containing two small drug molecules: ritonavir and celecoxib. The two model drugs used in this work have already been employed in the literature for the production of PLGA-based nanoparticles to overcome problems such as low patient therapy adherence [13] and important side effects [14]. The main reasons for the choice of these substances for large-scale experiments are the low toxicity of the compounds that are to be used in large quantities and their common availability.

2. Experimental Section

2.1. Materials

RESOMER RG 502 H (PLGA) (lactide–glycolide mole ratio of 50:50, inherent viscosity 0.16–0.24 dL/g) is an in-house product of Evonik Nutrition & Care GmbH (Darmstadt, Germany). Dichloromethane (DCM) ≥ 99.5% and ethyl acetate (EtOAc) ≥ 99.5% were purchased from Avantor Performance Materials (Gliwice, Poland) and used without further purification. Dimethyl sulfoxide (DMSO) 99.9%, USP grade, was acquired from WAK-Chemie Medical GmbH (Steinbach, Germany), and polyvinyl alcohol (PVA) was procured from Sigma-Aldrich (St. Louis, MO, USA). Trehalose dihydrate, boric acid, and Lugol's iodine solution were purchased from Carl Roth GmbH + Co. KG (Karlsruhe, Germany). Ritonavir ≥ 98% was acquired from Angene International Limited (Hong Kong, China) and celecoxib 99.92% was procured from Aarti Drugs Ltd. (Mumbai, IN, USA).

2.2. Nanoparticle Production

2.2.1. Placebo Formulation Development

Probe Direct Sonication

To obtain placebo particles of approx. 150 ± 50 nm and a PDI < 0.2, 4.06 mL of an organic solution composed of 3.9 wt% PLGA, 74.4 wt% DCM, and 21.7 wt% DMSO (dispersed phase (DP)) and 12.18 mL of an aqueous phase containing 2 wt% PVA (continuous phase (CP)) were emulsified together using a UP200St Ultrasonic Lab Homogenizer (Hielscher Ultrasonic GmbH, Teltow, Germany) equipped with an S26d2D needle probe. The process parameters were set at 100% amplitude and 100% phase. The treatment duration was 2 min. During the process, the sample container was kept immersed in an ice bath to prevent the degradation of sensitive material due to the high temperature generated by ultrasound. Ultimately, 290 mL of MilliQ water was added to the dispersion. Process parameters are summarized in Table 1.

Table 1. Process parameters for the classic probe sonication method.

		Probe Sonication					
Exp. No.	DP Solution	DP Volume (mL)	CP Solution	CP Volume (mL)	Total Sonication Time (min)	EP Solution	EP Volume (mL)
1	3.9 wt% PLGA, 74.4 wt% DCM, 21.7 wt% DMSO	4.06	2 wt% PVA	12.18	2	MilliQ water	290

Inline Indirect Sonication

To produce particles in an inline fashion, a GDmini2 Ultrasonic Inline Micro-Reactor (Hielscher Ultrasonic GmbH, Teltow, Germany) was used. This inline indirect sonotrode, consisting of a resonating stainless-steel jacket around a borosilicate glass tube, was connected to a pressurized coolant that transmitted the vibrations to the glass cannula during the process. Briefly, the same DP and CP solutions described above were prepared and filled inside two separate 60 mL BD Plastipak Luer-Lock single-use syringes and assembled on two Nexus 6000 syringe pumps (Chemyx Inc., Stafford, TX, USA). By connecting the two syringes to a T-junction, the DP and the CP were passed through the GDmini2 Ultrasonic Inline Micro-Reactor (Hielscher Ultrasonic GmbH, Teltow, Germany). DP:CP flow rate ratio was kept 1:3 and the total flow rate (TFR) was set at 2 mL/min. The pressurized coolant surrounding the glass cannula was maintained at a temperature of 10 °C through a cooling circulator (Ministat cc, Peter Huber Kältemaschinenbau AG., Offenburg, Germany) to avoid damage of sensitive raw materials during ultrasound treatment. Both the amplitude and phase of the indirect sonication process were kept at 100%. Prior to collection, another T-junction was coupled to the sonicator outlet, and MilliQ water (extraction phase (EP)) was pumped in through an ISCO pump (Teledyne ISCO, Lincoln, NE, USA) at 36 mL/min. Table 2 summarizes the process parameters applied. A schematic illustration of the process is depicted in Figure 1.

Table 2. Process parameters for the inline sonication method.

		Inline Sonication					
Exp. No.	DP Solution	DP Flowrate (mL/min)	CP Solution	CP Flowrate (mL/min)	EP Solution	EP Flowrate (mL/min)	Throughput g/h
2	3.9 wt% PLGA, 74.4 wt% DCM, 21.7 wt% DMSO	0.5	2 wt% PVA	1.5	MilliQ water	36	1.6

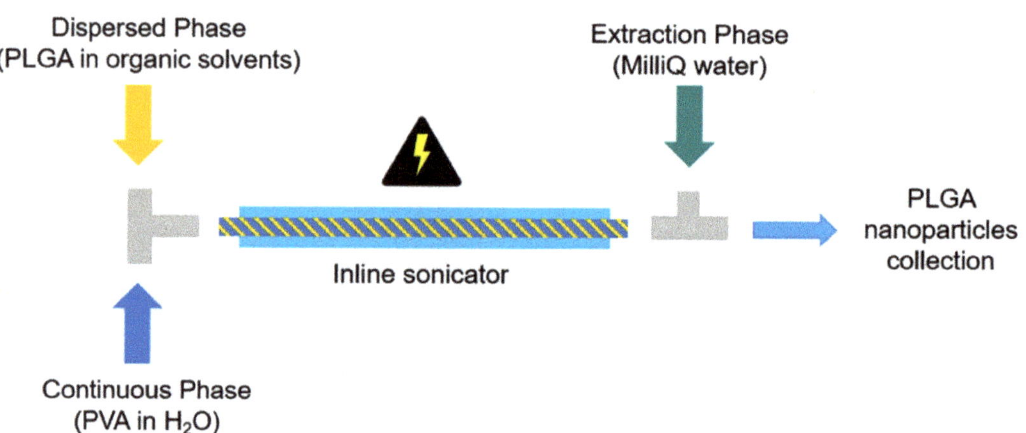

Figure 1. Scheme of the indirect inline sonication process setup for the manufacturing of PLGA nanoparticles. PLGA: poly(lactic-co-glycolic acid); PVA: polyvinyl alcohol.

2.2.2. Scale-Up of Inline Sonication

For the dispersed phase, 5, 10, and 20 wt% solutions of PLGA in EtOAc were prepared, mixed with DMSO, and processed at 8, 16, and 32 mL/min of TFR (Table 3, exp. 3–7). After having established the parameters to apply for the scale-up, 4 g of a 5 wt% solution of ritonavir or celecoxib in DMSO was added to 9 g of the 20 wt% solution of PLGA in EtOAc. The applied TFR of the DP/CP joint stream was set at 32 mL/min, and the resulting suspension was diluted with MilliQ water pumped at a rate of 240 mL/min by the ISCO pump (Table 3, exp. 8–9). After 1 h of stirring, the samples were purified and lyophilized. The final theoretical yield was calculated based on the composition and flow rate of the DP as 84 g/h. The obtained total volume containing the polymer and the API was divided by the flow rate of the DP to find the time necessary to process the materials. Finally, the grams of PLGA and APIs together were divided by the time to obtain the quantity of material processed per unit of time.

Table 3. Process parameters evaluated for the inline sonication method scale-up.

					Inline Sonication			
Exp. No.	**API**	**DP Solution**	**DP Flowrate (mL/min)**	**CP Solution**	**CP Flowrate (mL/min)**	**EP Solution**	**EP Flowrate (mL/min)**	**Throughput (g/h)**
3	-	4.5 wt% PLGA, 85.9 wt% EtOAc, 9.6 wt% DMSO	2	2 wt% PVA	6	MilliQ water	60	5
4	-	8.3 wt% PLGA, 74.3 wt% EtOAc, 5.7 wt% DMSO	2	2 wt% PVA	6	MilliQ water	60	10
5	-	14.1 wt% PLGA, 56.2 wt% EtOAc, 29.7 wt% DMSO	2	2 wt% PVA	6	MilliQ water	60	19
6	-	14.1 wt% PLGA, 56.2 wt% EtOAc, 29.7 wt% DMSO	4	2 wt% PVA	12	MilliQ water	120	38
7	Placebo	14.1 wt% PLGA, 56.2 wt% EtOAc, 29.7 wt% DMSO	8	2 wt% PVA	24	MilliQ water	240	76
8	Ritonavir	13.9 wt% PLGA, 55.4 wt% EtOAc, 1.5 wt% ritonavir, 29.2 wt% DMSO	8	2 wt% PVA	24	MilliQ water	240	84
9	Celecoxib	13.9 wt% PLGA, 55.4 wt% EtOAc, 1.5 wt% ritonavir, 29.2 wt% DMSO	8	2 wt% PVA	24	MilliQ water	240	84

The scalability of the probe batch technique was also investigated (Table 4, exp. 10). Twelve milliliters of CP was added to 4 mL of the DP solution containing 20 wt% of PLGA in EtOAc mixed with DMSO, and the sample was sonicated for 30 s at 100% amplitude using an S26d2D needle probe. The final suspension was diluted with 81 mL of MilliQ water and stirred for 1 h prior to particle size characterization.

Table 4. Process parameters evaluated for the classic probe sonication method scale-up.

					Probe Sonication			
Exp. No.	**API**	**DP Solution**	**DP Volume (mL)**	**CP Solution**	**CP Volume (mL)**	**Total Sonication Time (min)**	**EP Solution**	**EP Volume (mL)**
10	Placebo	14.1 wt% PLGA, 56.2 wt% EtOAc, 29.7 wt% DMSO	4	2 wt% PVA	12	0.5	MilliQ water	81.23

2.3. Analysis of Particle Size, Polydispersity Index, and Zeta Potential

The mean particle size diameter (Z-Average) and the polydispersity index (PDI) were determined via dynamic light scattering using a Zetasizer Nano ZS (Malvern Panalytical, Malvern, UK). Three measurements at 25 °C with a 173° scattering angle were conducted on each sample, which was previously diluted in sterile filtrated MilliQ water (0.2 μm). The surface charge of the nanoparticles was investigated by zeta potential measurement at 25 °C with the same instrument using the Smoluchowski equation. The two measurements together call for approx. 8 min of analysis time.

2.4. Downstream Processes

2.4.1. Purification

The formulations were purified via tangential flow filtration (TFF) technique employing a KrosFlo KR2i TFF System (Repligen). A Spectrum hollow fiber filter of the MidiKros module family based on modified polyethersulfone (mPES) material was chosen with a molecular weight cut-off (MWCO) of 750 kDa, fiber ID of 0.5 mm, and 20 cm effective length (D02-E750-05-N). Various diafiltration volumes (DVs) (3, 5, and 8) were investigated, and the ability to remove the excess of stabilizer and residual organic solvents was evaluated for each DV.

2.4.2. Analysis of Impurities

Proton Nuclear Magnetic Resonance (^{1}H-NMR)

^{1}H-NMR spectroscopy method was chosen as the analytical method for DMSO and DCM quantification. Despite gas chromatography–mass spectrometry (GC-MS) being a more sensitive, accurate, and suitable method for regulatory purposes, it is time-consuming and expensive. Conversely, the NMR method is advantageous for routine analysis because it minimizes sample preparation and has a fast measurement interval, saving time, money, and environmentally unfriendly organic solvents [15]. Although a disadvantage of NMR could be the higher detection limit compared to GC-MS, the amount of DCM and DMSO detected in our studies resulted satisfactory for the specific purpose.

Briefly, 200 µL of deuterated water (D2O) was added to 400 µL of each sample. Spectra were recorded using a Bruker 600 MHz spectrometer. Calibration was obtained based on the peak integrals related to a known amount of pure solvents. This method was used before, showing reproducible results [16]. The quantification limits (LOQs) for 1H-NMR are in a range of 1–15 µg/mL according to the literature [17–19].

Colorimetric Assay

The amount of residual PVA was determined by a colorimetric method based on the formation of a colored complex between two adjacent hydroxyl groups of PVA and an iodine molecule [20,21]. Briefly, 500 µL of each sample suspension was treated with 200 µL of 0.5 M sodium hydroxide solution for 15 min in a bath sonicator (BANDELIN electronic GmbH & Co. KG., Berlin, Germany) at 60 °C. Afterward, samples were neutralized with 90 µL of 1 M hydrochloric acid. To each sample, 300 µL of a 0.65 M solution of boric acid and 50 µL of Lugol's iodine solution (I2/KI) were added. Finally, after 15 min of incubation at 700 rpm, the absorbance of the samples was measured by setting the absorbance of an Infinite M200 PRO (Tecan, Männedorf, Switzerland) at a wavelength of 689 nm. A standard plot of PVA was prepared using known PVA concentrations under identical conditions.

2.4.3. Lyophilization

The purified suspensions were lyophilized using an Epsilon 2–6D LSCPlus freeze-dryer (Martin Christ Gefriertrocknungsanlagen GmbH, Osterode am Harz, Germany) in the presence of a cryoprotectant. A 40 wt% trehalose dihydrate stock solution was prepared and added to the colloidal suspensions to reach a 3 wt% final concentration. Figure 2 summarizes the experimental details of the freeze-drying protocol.

2.5. Molecular Weight Characterization

The molecular weight distribution (Mw, Mn, Mw/Mn) and inherent viscosity (η_{inh}) of the processed polymer were determined by size exclusion chromatography (SEC) and viscometry methods. Experiments 1 and 2 were reproduced with the exception that the CP phase consisted of pure MilliQ water to avoid the interference of PVA with the PLGA chain measurements. Samples were lyophilized in the absence of cryoprotectant before the analysis.

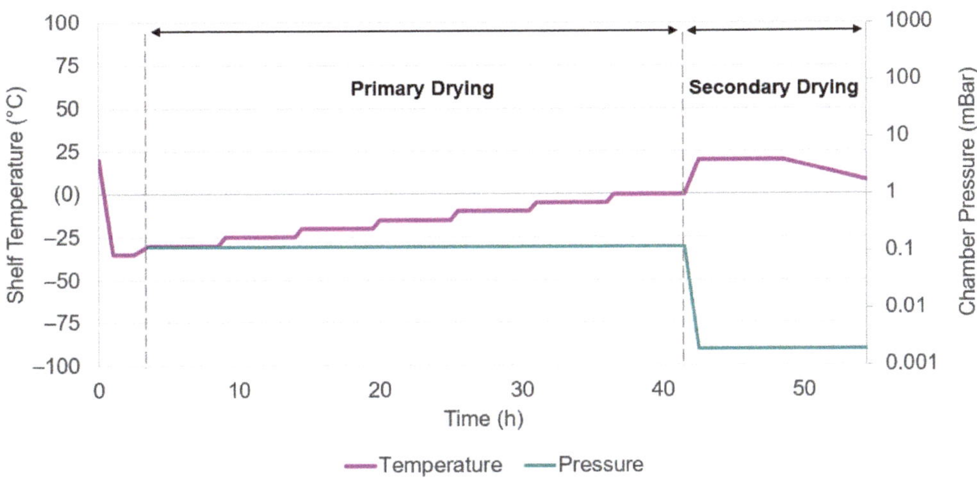

Figure 2. Freeze-drying protocol of the PLGA-based nanoparticles.

2.5.1. Gel Permeation Chromatography (GPC)

GPC analysis was performed on a setup equipped with VD400 Viscometer detector, RI200 RI detector, A5250 autosampler (Watrex Praha s.r.o., Prague, Czech Republic), and S 3210 UV/VIS detector (SYKAM GmbH, Eresing, Germany), using PLgel 5 μm 100 Å (Agilent Technologies, SC, USA) and DeltaGel Mixed-B (Watrex Praha s.r.o., Prague, Czech Republic) columns. Briefly, 5 mg of each sample was dissolved in 1 mL of a mixture consisting of chloroform:trimethylamine:isopropanol in 94:4:2 vol% composition. The same mixture was also employed as the mobile phase with a flow rate of 1 mL/min. The weight-average molecular weight (Mw), number-average molecular weight (Mn), and polydispersity (Đ) values were calculated using poly(methylmethacrylate) as standards (PSS Polymer Standards Service GmbH, Mainz, Germany).

2.5.2. Inherent Viscosity (η_{inh})

One hundred milligrams of PLGA was dissolved in 100 mL chloroform and stirred for a minimum of 6 h to obtain a polymer solution. Before starting the measurements, an Ubbelohde viscometer was filled and equilibrated to 25 °C for 5 min. The flux time of the polymer solution between two marks was recorded, and the flux time of pure chloroform was measured as a reference. Two measurements were performed for each sample, and the η_{inh} was calculated using Equation (1):

$$\eta_{inh} = \ln \eta_{rel}/c \tag{1}$$

where η_{rel} is the relative viscosity, which is the flux time ratio of the polymer solution to pure chloroform, and c is the polymer concentration having the unit of grams per deciliter.

2.6. API Content Analysis

For the API content analysis, high-performance liquid chromatography (HPLC) composed of a DIONEX UltiMate 3000 Pump and Diode Array Detector (UV-Vis) (Thermo Fisher Scientific, Waltham, MA, USA) was employed.

Drug encapsulation efficiency (EE) was calculated using Equation (2):

$$\text{EE (\%)} = \frac{\text{Weight of drug found in the nanoparticles}}{\text{Weight of drug initially used}} \times 100 \tag{2}$$

2.6.1. Ritonavir

Ritonavir quantification was conducted using a Nucleosil 100-7 C18 (125 × 4.6 mm; 7 µm) column. Standard calibration curves were performed at a fixed wavelength of 215 nm. The column temperature was maintained at 22 °C. The mobile phase comprised 30 mM potassium dihydrogen phosphate with pH adjusted to 4.0 with orthophosphoric acid and acetonitrile (45:55 vol%) with a flow rate of 1.5 mL/min. Injection volume was 10 µL. Retention time was 4.2 min. DMSO was employed to prepare the standard calibration solutions and the nanoparticle samples.

2.6.2. Celecoxib

Celecoxib was detected using a Nucleosil 100-7 C18 (125 × 4.6 mm; 7 µm) column. Eluents were monitored at a wavelength of 254 nm using a mixture (50:50:0.15 vol%) of acetonitrile, water, and triethylamine (TEA) with pH adjusted to 3.0 with orthophosphoric acid. Flow rate was kept at 1.8 mL/min and the column temperature was maintained at 40 °C. Injection volume was 5 µL and retention time was recorded at 4.1 min. Standard calibration and samples were formulated in DMSO.

2.7. Statistical Analysis

Student's *t*-test (two-tailed distribution, homoscedastic) ($n = 3$) was used in the polymer chain length analysis to determine the significance of the difference ($p < 0.05$) in Mw and η_{inh} among the compared groups with regards to the unprocessed PLGA powder. It was also exploited to verify the significance of the difference in mean particle size values for downstream processing impact analysis.

3. Results and Discussion

3.1. Assessment of Process and Formulation Parameters for Lab-Scale Production

PLGA nanoformulations were developed at a laboratory scale using both probe sonication and inline sonication with the same dispersed phase (DP) composition (3.9 wt% PLGA, 74.4 wt% DCM, 21.7 wt% DMSO; Tables 1 and 2). Chlorinated DPs for the emulsion-based preparation of nanosized PLGA formulations are commonly found in the scientific literature [1,22]. DMSO was added to the organic phase prior to sonication in order to mimic the exact conditions commonly used for the encapsulation of (bio)therapeutic payloads [23]. DMSO is a polar aprotic solvent with the ability to dissolve both polar and nonpolar compounds and is miscible with a wide range of organic solvents [24]. Addition of DMSO to the limitedly water-miscible DCM usually results in the formation of smaller emulsion droplets for a given set of process and formulation parameters due to lowering of the interfacial tension by DMSO [6,25,26], which is unlimitedly miscible with water [27,28]. Following the sonication process (probe sonication), an excess amount of MilliQ water was mixed with the crude emulsion to extract the organic solvents from the suspension to harden the particles, thereby accelerating the downstream processes. In view of a possible commercial and GMP-compliant manufacturing process, it is desirable to have a fast particle surface hardening process to decrease the waiting time of evaporation of a volatile solvent such as DCM. Depending on the water solubility of the organic solvent and its affinity for PLGA [29], the hardening process may take a variable amount of time. Based on the solubility of DCM in water (1.3 g/100 mL at 25 °C [30]), approx. 77 mL of water was added for each gram of DCM used for an efficient extraction.

The starting formulation tested for the probe sonication method was kept unchanged and processed at the inline sonicator. In the inline indirect sonication method, the sample flows continuously inside of a glass tube. The glass cannula is surrounded by a pressurized water coolant, which transmits the ultrasonic waves generated by the sonotrode by means of mechanical vibrations, allowing the formation of a fine emulsion. In this process, the glass tube is the only material in contact with the sample, rendering the procedure free from cross-contamination (e.g., metal shards from the probe, impurities derived from the surrounding environment) and potentially suitable for sterile processing [7]. The glass tube can be for single use or sterilized before usage for GMP-compliant processes.

During the process, flow stability was maintained through the use of syringe pumps, which can provide a constant fluid flow. Although syringe pumps are limited by the loading capacity of the syringes, they can be easily replaced by HPLC pumps to process larger volumes. In either case, pipes must conform to certain specifications with regard to compatibility with harsh solvents. In addition, it is important to note that an inline technology consisting of multiple tubes can generate a dead volume and some of the material can therefore be lost. In the setup described in this work, approximately 15 mL of dead volume was generated within the inline apparatus, which can be reduced by, for example, decreasing the length or diameter of the tubes in the system.

The placebo nanoparticle characteristics obtained for both sonication methods are shown in Figure 3. Figure 3A shows that a slightly smaller particle size compared to the classical probe method was obtained with the inline sonication method and highly monodisperse particles (PDI < 0.1) were generated for both preparations. The zeta potential value of PLGA nanoparticles obtained by the probe and inline sonication methods was around −40 mV (Figure 3B) due to the utilization of the same formulation parameters in terms of PLGA and PVA concentrations.

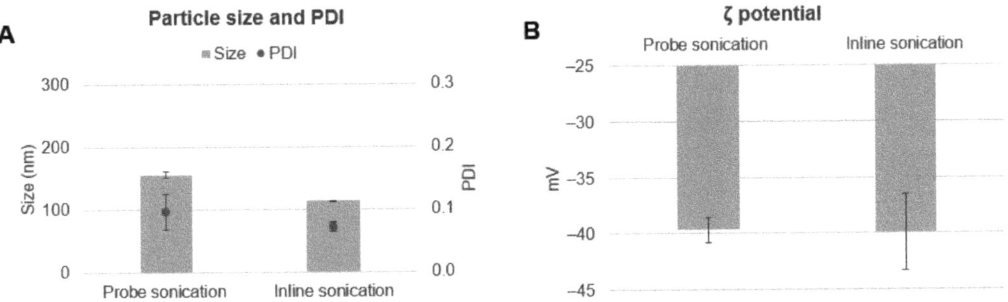

Figure 3. (**A**) Particle size and PDI and (**B**) zeta potential of placebo PLGA nanoparticles generated via probe direct and inline indirect sonication.

3.2. Downstream Process Development for Washing, Concentration, and Storage of PLGA Nanoparticles

The obtained diluted placebo suspension was first subjected to a concentration treatment corresponding to 6 times the initial volume, and then to three, five, and eight washing cycles (called diafiltration volumes (DVs)) using the TFF. The reason for choosing such DVs was dictated by the fact that typically, when washing samples by centrifugation, at least three wash cycles are used. In this work it was decided to start from a similar concept, i.e., to use three DVs and then scale up to five and eight. A single DV corresponds to the sample volume that is processed when the diafiltration starts. The washing liquid is added at the same rate as the discarded filtrate liquid is generated. When the volume of filtrate collected is equal to the initial sample volume, one DV is processed. The organic solvent content as well as the amount of stabilizer present in the suspensions were analyzed before and after each passage. The particle size and polydispersity were monitored during the process to ensure that particle properties were not affected by the TFF process. Furthermore, zeta potential was checked after each tested DV given its shielding effect on PLGA nanoparticle surface charge [21]. Figure 4 demonstrates the impact of diafiltration volumes on the PVA (Figure 3A), DMSO (Figure 4B), and DCM (Figure 4C) levels. While three diafiltration volumes were enough to completely eliminate DCM (Figure 4C), a significant drop in DMSO and PVA content after three diafiltration volumes was registered, and a gradual reduction after five and eight volumes was observed. Table 5 summarizes the reduction in DCM, DMSO, and PVA using TFF, showing that no traces of DCM were found in suspension, and that approx. 80% of DMSO and more than 60% of total PVA amount were removed already after three diafiltration volumes. The zeta potential measurements revealed negative values (Figure 4D), which are typically observed for uncapped end carboxylic groups of PLGA or hydrolyzed PLGA. However, these values were observed to be less negative for the nanoformulation not yet washed, thus containing a higher quantity of PVA, compared to those where PVA was substantially removed. In fact, as already observed by other groups [21,31], PVA presence is directly connected to the resulting zeta potential. Partially hydrolyzed PVA is an amphiphilic "multiblock" copolymer possessing poly(vinyl acetate) and poly(vinyl alcohol) blocks. This polymer structure is preferred because the hydrolysis rate of vinyl acetate monomeric units is strongly accelerated if the unit in the neighborhood is already hydrolyzed. Thus, once a unit in the poly(vinyl acetate) is hydrolyzed, it speeds up the hydrolysis of its neighbors. The hydrophobic part composed of poly(vinyl acetate) may be found associated with the polymer matrix of PLGA, while the hydrophilic poly(vinyl alcohol) can protrude from the surface of the nanoparticle [21,31]. Thereby, most of the PVA located close to the particle surface can shield the negative charges generated by PLGA [32]. Figure 4D highlights that three DVs are sufficient to remove the excess of PVA which greatly shielded the surface charge of the particles, leaving only the necessary surfactant that is attached to the surface and contributes to the overall stability of the nanoparticles. Finally, Figure 5 shows that the particle size and PDI remained unaffected by the flow filtration process, meaning that the process is well tolerated by the colloidal suspensions regardless of the number of diafiltrations performed. Based on these results, three DVs were chosen for washing the API-containing PLGA particles since the concentration of DCM and DMSO was reduced by more than approx. 80%, excess of PVA was removed revealing a more negative zeta potential of the particles, and size and PDI were not altered.

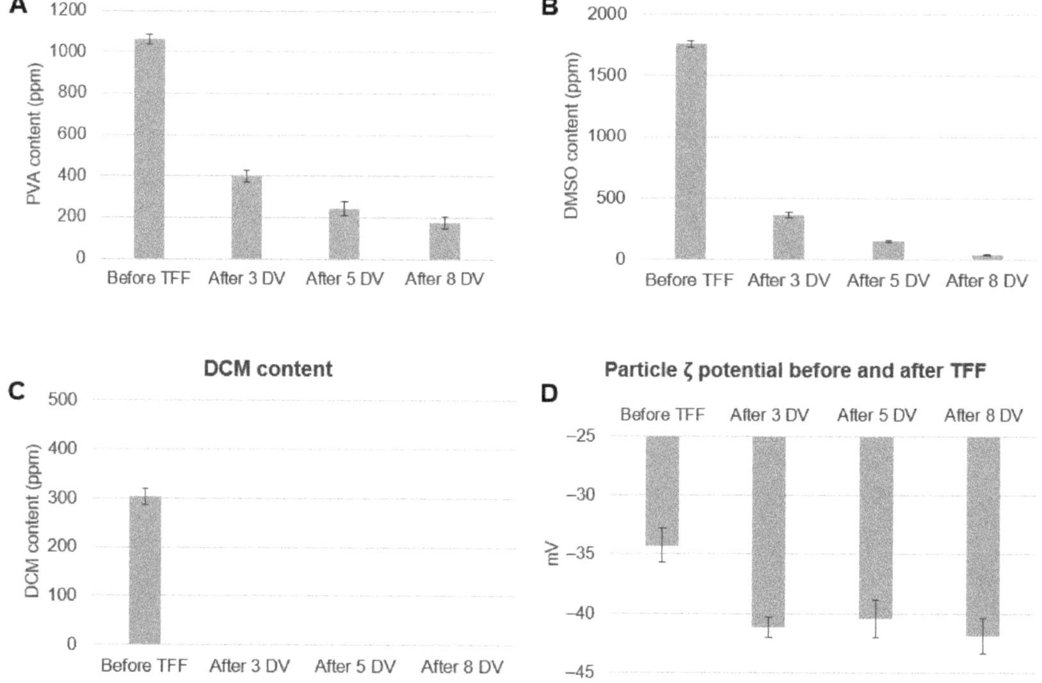

Figure 4. Downstream process evaluation. Reduction in (**A**) PVA, (**B**) DMSO, and (**C**) DCM content and (**D**) particle zeta potential tested before and after 3, 5, and 8 DVs.

Table 5. Impact of different TFF diafiltration volumes in reducing PVA, DMSO, and DCM content.

Diafiltration Volumes	PVA Reduction (%)	DMSO Reduction (%)	DCM Reduction (%)
3	62.6 ± 3.3	79.4 ± 1.4	100.0
5	77.1 ± 3.7	91.4 ± 0.7	100.0
8	83.3 ± 2.4	97.5 ± 0.2	100.0

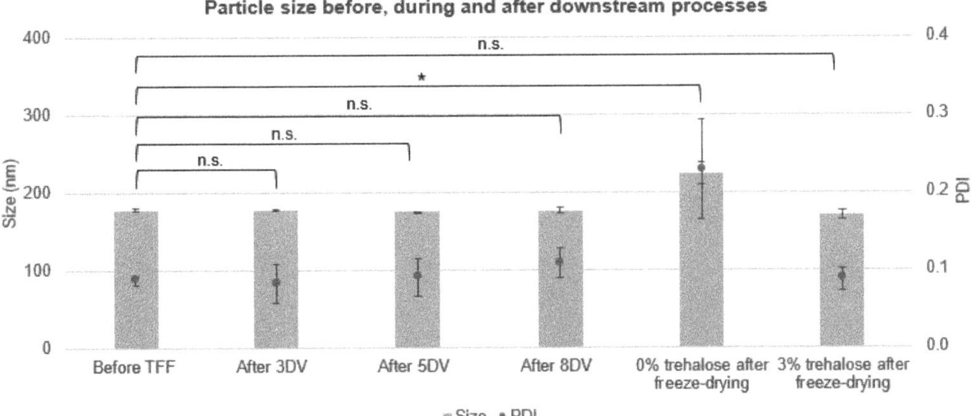

Figure 5. Size and size polydispersity of the nanoformulations before, during, and after the downstream processes. Significant, $p < 0.05$ (*); nonsignificant, $p > 0.05$ (n.s.).

In order to prevent the nanoparticle agglomeration, flocculation, and premature release of APIs from the polymeric carrier, a lyophilization process was developed for nanoparticle storage. For these trials, the efficacy of a cryoprotectant (i.e., trehalose) to preserve the particle size and polydispersity of the final formulations was investigated. Trehalose is a nonreducing disaccharide consisting of two D-glucose monomers linked by an α,α–1,1 glycosidic bond [33]. It is found in bacteria, fungi, plants, and animals (e.g., tardigrades) that can survive extreme temperatures and withstand dehydration [33]. In the dried state, carbohydrates such as trehalose exert their protective effect by acting as a water substitute [34]. Besides, trehalose is an excipient already used in approved injectable products such as Herceptin (trastuzumab), Advate (octocog alfa), and Avastin (bevacizumab). As shown in Figure 5, when the lyophilization was performed without the addition of the cryoprotectant, the mean particle size increased above 200 nm and the PDI doubled in value. Conversely, the addition of trehalose in the suspension (in 3 wt% final concentration) prevented particles from agglomerating and preserved nanoparticle size and PDI.

3.3. Characterization of Polymer Chain Integrity after Sonication Processes

Having obtained similar particles by both techniques, the methods were compared regarding their ability to keep the PLGA polymer chains intact as high shear forces exerted during sonication step can cause polymer chain rupture and degradation [35]. This is highly undesirable since it can both modify the structure of the carrier, which could result in altered phenomena (e.g., encapsulation efficiency and drug release), and generate reactive structures that could react with other species in suspension (e.g., payloads, stabilizers, etc.) [8]. Therefore, the PLGA chain lengths were analyzed using gel permeation chromatography (GPC) and viscometry before and after the ultrasound treatment in comparison to unprocessed PLGA powder. The normalized GPC spectra are shown in Figure 6, and the Tables 6 and 7 summarize the results.

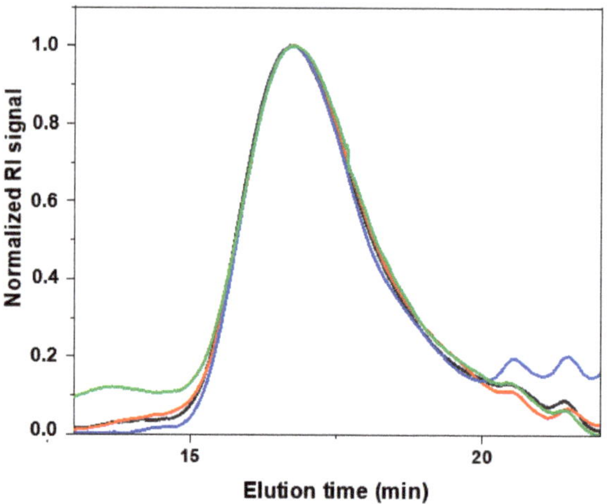

Figure 6. GPC spectra of the processed and untreated polymer under ultrasound. Data obtained for PLGA pure powder untreated (black), PLGA dissolved in the organic solvent (green), and PLGA treated under probe (red) and inline (blue) ultrasound technologies are presented.

Table 6. Weight (Mw) and number (Mn) averaged molecular weight and PDI (Mw/Mn) of PLGA polymer treated and untreated via ultrasound. Significance of the difference ($p < 0.05$) among the compared groups was determined with regards to the unprocessed PLGA powder.

Gel Permeation Chromatography				
Sample	M_w **(kDa)** $\pm$ **SD**	M_n **(kDa)** $\pm$ **SD**	**PDI**	p **Value (M$_w$)**
Untreated PLGA	18.77 ± 0.51	11.67 ± 0.51	1.59	
PLGA dissolved in DCM, untreated	19.52 ± 0.45	14.23 ± 0.99	1.35	0.13
PLGA treated with probe sonication	18.72 ± 0.56	12.41 ± 1.31	1.50	0.92
PLGA treated with inline sonication	18.80 ± 1.00	12.57 ± 1.43	1.50	0.96

Table 7. η_{inh} of the PLGA polymer treated and untreated via ultrasound. Significance of the difference ($p < 0.05$) among the compared groups was determined with regards to the unprocessed PLGA powder.

Viscometry		
Sample	η_{inh} **(dl/g)** $\pm$ **SD**	p **Value**
Untreated PLGA powder	0.219 ± 0.008	
PLGA dissolved in DCM, untreated	0.213 ± 0.001	0.24
PLGA treated with probe sonication	0.216 ± 0.004	0.57
PLGA treated with inline sonication	0.206 ± 0.001	0.05

As sonication may cause a drop in molecular weight of polymers due to shear-induced chain breaks [8,9], the molecular weight of PLGA before and after processing was analyzed and compared. GPC analysis showed that the type of PLGA employed (RG 502 H) has a weight-average molecular weight Mw of approx. 19 kDa and an inherent viscosity of approx. 0.22 dL/g, which is in line with the product specifications reported in the certificate of analysis of RESOMER RG 502 H (0.16–0.24 dL/g). Although the drop in inherent viscosity results of PLGA treated with the inline sonication method showed a statistical significance at $p \leq 0.05$, the difference is still only 6% of the original value, meaning that neither of the techniques affected the polymer chains during the manufacturing of particles. Negligibility of this difference and preservation of the original polymer properties is underlined by the fact that the original PLGA has a relatively wide distribution of molecular weights (PDI 1.59), so differences in molecular weight of ca. 4% are meaningless (also, within the experimental error of GPC).

3.4. Scale-Up of Formulation

The inline indirect sonication method was further investigated for PLGA-based nanomedicine scale-up manufacturing. First, the organic solvent used to prepare the polymer solution was switched from DCM to EtOAc due to the lower toxicity and higher water solubility of EtOAc compared to DCM. In fact, in accordance with the ICH guidelines [36], DCM is a class 2 solvent with a permitted daily exposure (PDE) equivalent to 6.0 mg/day. Conversely, EtOAc lies in class 3, meaning that the PDE corresponds to 50 mg/day [36]. Therefore, working with large amount of a less toxic solvent increases the safety of employees at work while reducing the complexity of safety protocols that must be prepared and heeded. Also, the higher degree of water solubility of EtOAc (8 g/100 mL versus 1.3 g/100 mL for DCM at 25 °C [30,37]) would require smaller volumes of water as the extraction phase and render the process easier to handle. Although therefore many substances should be reduced or directly replaced with greener ones, it is not always possible to choose the desired solvents because not all substances have the same solubility in various solvents. For example, super critical CO_2 offers the advantages of being non-toxic and leaving no residue. On the other hand, the limited choice of soluble materials and the compatibility of organic solvents with the technical apparatus hinder its industrial scalability [1,38]. Therefore, in this case, during scale-up experiments, EtOAc was chosen over DCM because of its lower toxicity and unaltered ability to solubilize PLGA, while DMSO was selected for its ability to dissolve a wide variety of active ingredients. Thus, the DMSO/EtOAc mix was identified as optimal for most formulations to be manufactured. Nonetheless, it is important to note that, although notoriously toxic, the use of DCM in particle production is still widespread and has been reported for the production of formulations used in phase I clinical trials [39]. Keeping in mind the required specification of 150 ± 50 nm and PDI < 0.2, a set of particles was generated via indirect inline sonication technique with the scope of scaling-up the technology. The experiment parameters are reported in Table 3, exp. 3–7. The influence of polymer concentration and the TFR on the particle size and PDI were studied systematically. Figure 7A shows the variations of particle size as a function of PLGA concentration (5 wt%, 10 wt% and 20 wt%) obtained at a TFR of 8 mL/min. Given the greater water solubility of EtOAc compared to DCM, an extraction phase ratio of 3 (3 folds the amount of water required to solubilize EtOAc) was adopted, bringing the EP flow rate at 60 mL/min. This was done in order to dilute the particle suspension enough to be processed on the same day at the TFF, avoiding potential damage of the hollow fiber filter caused by the solvent. As expected, the use of EtOAc resulted in smaller particles than DCM due to the lower interfacial tension of the solvent with water, as previously observed in the case of DMSO addition. Also, as already experienced in one of our previous works [6], a gradual increase in particle size and PDI was noticed for higher concentrations of PLGA. While 5 wt% polymer resulted in particles with a mean diameter of approx. 85 nm, 125 nm particles were obtained with 20 wt% of polymer concentration with a PDI of approx. 0.1 for all formulations. As 10 wt% and 20 wt% PLGA were in the specification range, the higher PLGA concentration in EtOAc was chosen as the polymer concentration for the subsequent experiments due to the higher throughput of 19 g/h obtained with this polymer concentration. The influence of the TFR on the particle size and PDI is shown in Figure 7B. Accordingly, an increase in particle size was observed at higher TFR. This is due to shorter sonication time for the faster-flowing samples, leading to less homogenizing treatment time. 32 mL/min was found to result in particles still in the target size, therefore, it was chosen for the production of the loaded particles given the high throughput obtained from the high continuous flow (approx. 76 g/h).

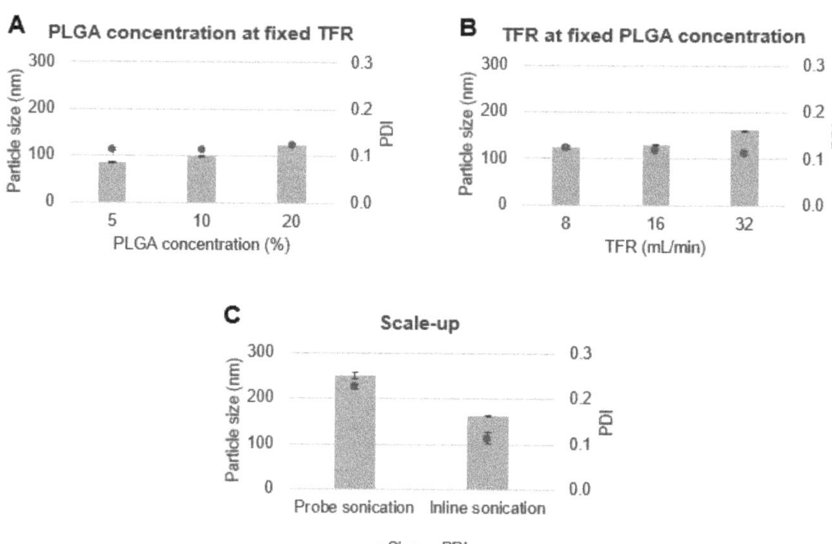

Figure 7. Scale-up of nanoformulations. (**A**) Size and PDI of the particles obtained at varied PLGA concentrations and fixed total flow rate of 8 mL/min. (**B**) Particle size and size distribution of placebo particles obtained applying different TFRs by using an initial PLGA concentration of 20 wt% in EtOAc. (**C**) Comparison of the particles obtained with the scaled-up indirect inline technology versus particles achieved at a higher processed volume and lower total sonication time of direct probe batch method.

As a comparison, the batch production method was also implemented in large scale (Table 4, exp. 10). Briefly, 20 wt% *w/w* PLGA in EtOAc was mixed with DMSO and sonicated for 0.5 min with PVA 2 wt%. Subsequently, the homogenized suspension was transferred to 81 mL of water used as the extraction phase. Figure 7C shows the particle size produced by this method. As it can be noted, the particle size as well as the PDI are out of the required specifications, and the values are considerably higher than those obtained with the inline method. This shows that inline sonication is easier to scale up and that, to allow for increased batch scale, a new process with a different probe would have to be entirely reevaluated.

Ritonavir and Celecoxib Nanoparticles

To further confirm the usefulness of the inline production method, ritonavir and celecoxib were chosen as model drugs and were encapsulated within PLGA nanoparticles. Ritonavir is an antiretroviral protease inhibitor API that is widely used in combination with other medications for the treatment of human immunodeficiency virus (HIV) infection, which causes the acquired immunodeficiency syndrome (AIDS) [13]. Celecoxib is a cyclo-oxygenase-2 (COX-2) selective inhibitor used in the treatment of pain and inflammation [14]. It is one of the most commonly prescribed COX-2 specific inhibitors since its use effectively reduces clinical gastrointestinal events in comparison to other nonsteroidal anti-inflammatory drugs (NSAIDs). Both of these APIs have already been investigated regarding their ability to be entrapped into PLGA nanoparticles in order to overcome problems such as low patient therapy adherence [13] and important side effects [14].

Since ritonavir and celecoxib are hydrophobic compounds, PLGA and API are dissolved together in the organic phase and then emulsified with the aqueous solution containing the surfactant. Otherwise, encapsulation of a hydrophilic API would require an initial formation of a water-in-oil emulsion, in which the API is in the aqueous phase and the polymer is in the organic phase. Next, the water-in-oil emulsion is mixed with a second aqueous solution containing the surfactant, creating a water-in-oil-in-water system [1]. The experimental conditions for large-scale manufacturing of ritonavir and celecoxib nanoparticles are shown in Table 3 (exp. 8–9). Both the nanoparticle types were produced achieving a total yield of 84 g/h, which corresponds to approx. 2 kg/day when run continuously. The characteristics of the nanoparticles are shown in Figure 8. The particle size achieved with this method was below 200 nm for both formulations. Ritonavir encapsulation led to particles of 188.5 ± 10.2 nm and PDI of 0.19 ± 0.09. Zeta potential was registered at −37.4 ± 3.0 mV, meaning that the particles were overall stable. Celecoxib resulted in particles of similar size and PDI of 184.7 ± 1.2 nm and 0.17 ± 0.01. Zeta potential was slightly more negative, being −42.8 ± 2.5 mV. Encapsulation efficiency of both the APIs was determined with HPLC. As shown in Table 8, the efficiency of ritonavir and celecoxib encapsulation was approx. 50% and 80%, respectively. Although ritonavir and celecoxib may possess some similar characteristics such as no formal charge and very poor water solubility, their functional molecular groups as well as molecular weights are substantially different [40,41]. Therefore, it is reasonable that each difference may decree a variation in the interaction of the API with the polymer matrix, resulting in a unique encapsulation capacity. Table 9 summarizes some of the most prominent physicochemical characteristics of the APIs.

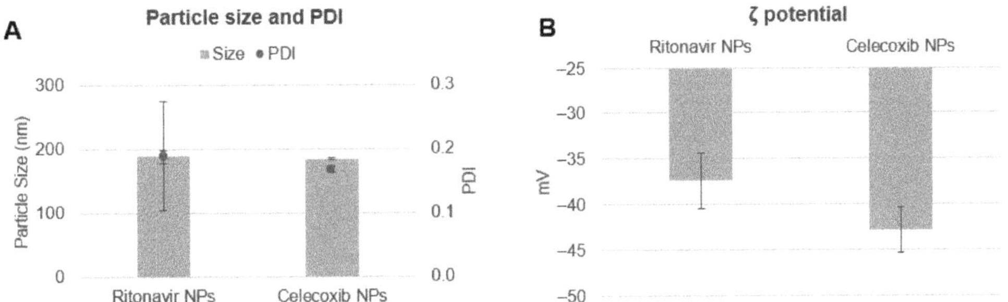

Figure 8. (**A**) Particle size and PDI and (**B**) zeta potential of ritonavir and celecoxib nanoparticles obtained with the inline scaled-up sonication method.

Table 8. Ritonavir and celecoxib nanoparticle encapsulation efficiency and relative drug load obtained via the scaled-up indirect inline continuous method.

Drug Content	Ritonavir PLGA Nanoparticles	Celecoxib PLGA Nanoparticles
EE (%)	49.5 ± 3.2	80.3 ± 0.9
Drug load (mg/g)	4.95 ± 0.32	8.03 ± 0.09

Table 9. Ritonavir and celecoxib physicochemical characteristics. Data obtained consulting PubChem and DrugBank databases.

Physicochemical Characteristics	Ritonavir	Celecoxib
API Type	Small molecule	Small molecule
Mw	720.9	381.4
Log P	3.9	3.53
Hydrogen Bond Donors	4	1
Hydrogen Bond Acceptors	9	7
Formal Charge	0	0
Water Solubility (mg/L, 25 °C)	1.1×10^{-4}	4.3

4. Conclusions

Technologies that rely on inline continuous manufacturing processes allow the scaled-up production of nanomedicines without changing formulation specifications. Although inline processes may suffer from material loss due to dead volume in the tubes that necessarily increases the cost of upstream materials, this reliance makes such technologies attractive from a commercial and clinical development standpoint. In the present study, the indirect inline sonication method was found to be comparable to the direct probe method in terms of particle size, PDI, and zeta potential. Furthermore, it was confirmed to be safe as it does not alter the length of the polymer chains of PLGA during the sonication process. The inline sonication technology proved to be scalable, achieving a production yield of 84 g/h (approximately 2 kg/day) for PLGA nanoparticles containing ritonavir and celecoxib as modal APIs. Downstream processes have been developed and shown to be suitable to purify colloidal suspensions from impurities and residual organic solvents. Finally, due to the application of fully enclosed pipes and containers that can be easily sterilized or replaced, the inline indirect sonication method together with the TFF technique can potentially be considered for GMP and aseptic manufacturing processes. Overall, the developed manufacturing process proved to be suitable for the production of PLGA-based nanomedicine products on a clinical and commercial scale.

Author Contributions: Conceptualization and visualization: M.C.O., A.B., C.G.F. and O.T.; investigation and data curation: M.C.O., V.S. and E.J.; writing—original draft preparation: M.C.O.; writing—review and editing: A.B., V.S., E.J., S.G., A.E., M.H., C.G.F. and O.T.; supervision: A.B., S.G., A.E., M.H., C.G.F. and O.T.; project administration and funding acquisition: S.G., M.H. and C.G.F. All authors have read and agreed to the published version of the manuscript.

Funding: This research was funded by EU grant PRECIOUS (686089).

Institutional Review Board Statement: Not applicable.

Informed Consent Statement: Not applicable.

Data Availability Statement: Not applicable.

Acknowledgments: C.F. received the NWO Spinoza grant, ERC Advanced Grant Pathfinder (269019), and Dutch cancer society award 2009-4402.

Conflicts of Interest: M.C.O., S.G. and A.E. have a patent pending for the presented inline technology. The remaining authors have no conflict of interest to declare.

References

1. Operti, M.C.; Bernhardt, A.; Grimm, S.; Engel, A.; Figdor, C.G.; Tagit, O. PLGA-based nanomedicines manufacturing: Technologies overview and challenges in industrial scale-up. *Int. J. Pharm.* **2021**, *605*, 120807. [CrossRef] [PubMed]
2. Agrahari, V.; Agrahari, V. Facilitating the translation of nanomedicines to a clinical product: Challenges and opportunities. *Drug Discov. Today* **2018**, *23*, 974–991. [CrossRef] [PubMed]
3. Agrahari, V.; Hiremath, P. Challenges associated and approaches for successful translation of nanomedicines into commercial products. *Nanomedicine* **2017**, *12*, 819–823. [CrossRef] [PubMed]
4. Paliwal, R.; Babu, R.J.; Palakurthi, S. Nanomedicine scale-up technologies: Feasibilities and challenges. *AAPS PharmSciTech* **2014**, *15*, 1527–1534. [CrossRef]
5. Sun, Q.; Radosz, M.; Shen, Y. Challenges in design of translational nanocarriers. *J. Control. Release* **2012**, *164*, 156–169. [CrossRef]
6. Operti, M.C.; Fecher, D.; van Dinther, E.A.W.; Grimm, S.; Jaber, R.; Figdor, C.G.; Tagit, O. A comparative assessment of continuous production techniques to generate sub-micron size PLGA particles. *Int. J. Pharm.* **2018**, *550*, 140–148. [CrossRef]
7. Schiller, S.; Hanefeld, A.; Schneider, M.; Lehr, C.-M. Focused Ultrasound as a Scalable and Contact-Free Method to Manufacture Protein-Loaded PLGA Nanoparticles. *Pharm. Res.* **2015**, *32*, 2995–3006. [CrossRef]
8. Reich, G. Ultrasound-induced degradation of PLA and PLGA during microsphere processing: Influence of formulation variables. *Eur. J. Pharm. Biopharm.* **1998**, *45*, 165–171. [CrossRef]
9. Mohod, A.V.; Gogate, P.R. Ultrasonic degradation of polymers: Effect of operating parameters and intensification using additives for carboxymethyl cellulose (CMC) and polyvinyl alcohol (PVA). *Ultrason. Sonochem.* **2011**, *18*, 727–734. [CrossRef]
10. Freitas, S.; Hielscher, G.; Merkle, H.P.; Gander, B. Continuous contact- and contamination-free ultrasonic emulsification—A useful tool for pharmaceutical development and production. *Ultrason. Sonochem.* **2006**, *13*, 76–85. [CrossRef]
11. Dalwadi, G.; Benson, H.A.; Chen, Y. Comparison of diafiltration and tangential flow filtration for purification of nanoparticle suspensions. *Pharm. Res.* **2005**, *22*, 2152–2162. [CrossRef] [PubMed]

12. Clutterbuck, A.; Beckett, P.; Lorenzi, R.; Sengler, F.; Bisschop, T.; Haas, J. Single-Pass Tangential Flow Filtration (SPTFF) in Continuous Biomanufacturing. In *Continuous Biomanufacturing—Innovative Technologies and Methods*; John Wiley & Sons, Inc.: Hoboken, NJ, USA, 2017; pp. 423–456.
13. Destache, C.J.; Belgum, T.; Christensen, K.; Shibata, A.; Sharma, A.; Dash, A. Combination antiretroviral drugs in PLGA nanoparticle for HIV-1. *BMC Infect. Dis.* **2009**, *9*, 198. [CrossRef] [PubMed]
14. Cooper, D.L.; Harirforoosh, S. Effect of Formulation Variables on Preparation of Celecoxib Loaded Polylactide-Co-Glycolide Nanoparticles. *PLoS ONE* **2014**, *9*, e113558.
15. Grosch, S.; Monakhova, Y.B.; Kuballa, T.; Ruge, W.; Kimmich, R.; Lachenmeier, D.W. Comparison of GC/MS and NMR for quantification of methyleugenol in food. *Eur. Food Res. Technol.* **2013**, *236*, 267–275. [CrossRef]
16. Kram, T.C.; Turczan, J.W. Determination of DMSO in solutions and ointments by NMR. *J. Pharm. Sci.* **1968**, *57*, 651–652. [CrossRef]
17. Khatun, R.; Hunter, H.; Magcalas, W.; Sheng, Y.; Carpick, B.; Kirkitadze, M. Nuclear Magnetic Resonance (NMR) Study for the Detection and Quantitation of Cholesterol in HSV529 Therapeutic Vaccine Candidate. *Comput. Struct. Biotechnol. J.* **2017**, *15*, 14–20. [CrossRef]
18. Maes, P.; Monakhova, Y.B.; Kuballa, T.; Reusch, H.; Lachenmeier, D.W. Qualitative and Quantitative Control of Carbonated Cola Beverages Using 1H NMR Spectroscopy. *J. Agric. Food Chem.* **2012**, *60*, 2778–2784. [CrossRef]
19. Sun, S.; Jin, M.; Zhou, X.; Ni, J.; Jin, X.; Liu, H.; Wang, Y. The Application of Quantitative 1H-NMR for the Determination of Orlistat in Tablets. *Molecules* **2017**, *22*, 1517. [CrossRef]
20. Joshi, D.P.; Lan-Chun-Fung, Y.L.; Pritchard, J.G. Determination of poly(vinyl alcohol) via its complex with boric acid and iodine. *Anal. Chim. Acta* **1979**, *104*, 153–160. [CrossRef]
21. Sahoo, S.K.; Panyam, J.; Prabha, S.; Labhasetwar, V. Residual polyvinyl alcohol associated with poly (d,l-lactide-co-glycolide) nanoparticles affects their physical properties and cellular uptake. *J. Control. Release* **2002**, *82*, 105–114. [CrossRef]
22. Sahana, D.K.; Mittal, G.; Bhardwaj, V.; Kumar, M.N.V.R. PLGA Nanoparticles for Oral Delivery of Hydrophobic Drugs: Influence of Organic Solvent on Nanoparticle Formation and Release Behavior In Vitro and In Vivo Using Estradiol as a Model Drug. *J. Pharm. Sci.* **2008**, *97*, 1530–1542. [CrossRef] [PubMed]
23. Dölen, Y.; Valente, M.; Tagit, O.; Jäger, E.; Van Dinther, E.A.W.; van Riessen, N.K.; Hruby, M.; Gileadi, U.; Cerundolo, V.; Figdor, C.G. Nanovaccine administration route is critical to obtain pertinent iNKt cell help for robust anti-tumor T and B cell responses. *OncoImmunology* **2020**, *9*, 1738813. [CrossRef] [PubMed]
24. Galvao, J.; Davis, B.; Tilley, M.; Normando, E.; Duchen, M.R.; Cordeiro, M.F. Unexpected low-dose toxicity of the universal solvent DMSO. *FASEB J.* **2014**, *28*, 1317–1330. [CrossRef] [PubMed]
25. Lagreca, E.; Onesto, V.; Di Natale, C.; La Manna, S.; Netti, P.A.; Vecchione, R. Recent advances in the formulation of PLGA microparticles for controlled drug delivery. *Prog. Biomater.* **2020**, *9*, 153–174. [CrossRef] [PubMed]
26. Song, K.C.; Lee, H.S.; Choung, I.Y.; Cho, K.I.; Ahn, Y.; Choi, E.J. The effect of type of organic phase solvents on the particle size of poly(d,l-lactide-co-glycolide) nanoparticles. *Colloids Surf. A Physicochem. Eng. Asp.* **2006**, *276*, 162–167. [CrossRef]
27. PubChem. Dimethyl Sulfoxide Solubility. Available online: https://pubchem.ncbi.nlm.nih.gov/compound/679#section=Solubility (accessed on 18 July 2021).
28. Wong, D.B.; Sokolowsky, K.P.; El-Barghouthi, M.I.; Fenn, E.E.; Giammanco, C.H.; Sturlaugson, A.L.; Fayer, M.D. Water Dynamics in Water/DMSO Binary Mixtures. *J. Phys. Chem. B* **2012**, *116*, 5479–5490. [CrossRef]
29. Murakami, H.; Kobayashi, M.; Takeuchi, H.; Kawashima, Y. Preparation of poly(dl-lactide-co-glycolide) nanoparticles by modified spontaneous emulsification solvent diffusion method. *Int. J. Pharm.* **1999**, *187*, 143–152. [CrossRef]
30. PubChem. Dichloromethane Solubility. Available online: https://pubchem.ncbi.nlm.nih.gov/compound/Dichloromethane#section=Solubility (accessed on 18 July 2021).
31. Srihaphon, K.; Lamlertthon, S.; Pitaksuteepong, T. Influence of stabilizers and cryoprotectants on the characteristics of freeze-dried PLGA nanoparticles containing *Morus alba* stem extract. *Songklanakarin J. Sci. Technol.* **2021**, *43*, 72–79.
32. Operti, M.C.; Dölen, Y.; Keulen, J.; van Dinther, E.A.; Figdor, C.G.; Tagit, O. Microfluidics-Assisted Size Tuning and Biological Evaluation of PLGA Particles. *Pharmaceutics* **2019**, *11*, 590. [CrossRef]
33. Teramoto, N.; Sachinvala, N.D.; Shibata, M. Trehalose and trehalose-based polymers for environmentally benign, biocompatible and bioactive materials. *Molecules* **2008**, *13*, 1773–1816. [CrossRef]
34. Nema, S.; Brendel, R.J. Excipients and their role in approved injectable products: Current usage and future directions. *PDA J. Pharm. Sci. Technol.* **2011**, *65*, 287–332. [CrossRef] [PubMed]
35. Paulusse, J.; Sijbesma, R. Ultrasound in polymer chemistry: Revival of an established technique. *J. Polym. Sci. Part A Polym. Chem.* **2006**, *44*, 5445–5453. [CrossRef]
36. International Council for Harmonisation of Technical Requirements for Pharmaceuticals for Human Use (ICH). Impurities: Guideline for Residual Solvents. Available online: https://database.ich.org/sites/default/files/Q3C-R6_Guideline_ErrorCorrection_2019_0410_0.pdf (accessed on 27 August 2021).
37. PubChem. Ethyl Acetate Solubility. Available online: https://pubchem.ncbi.nlm.nih.gov/compound/8857#section=Solubility (accessed on 27 August 2021).
38. Gangapurwala, G.; Vollrath, A.; De San Luis, A.; Schubert, U.S. PLA/PLGA-Based Drug Delivery Systems Produced with Supercritical CO(2)-A Green Future for Particle Formulation? *Pharmaceutics* **2020**, *12*, 1118. [CrossRef] [PubMed]

39. Dölen, Y.; Gileadi, U.; Chen, J.-L.; Valente, M.; Creemers, J.H.A.; Van Dinther, E.A.W.; van Riessen, N.K.; Jäger, E.; Hruby, M.; Cerundolo, V.; et al. PLGA Nanoparticles Co-encapsulating NY-ESO-1 Peptides and IMM60 Induce Robust CD8 and CD4 T Cell and B Cell Responses. *Front. Immunol.* **2021**, *12*, 641703. [CrossRef]
40. PubChem. Ritonavir. Available online: https://pubchem.ncbi.nlm.nih.gov/compound/392622 (accessed on 1 October 2021).
41. PubChem. Celecoxib. Available online: https://pubchem.ncbi.nlm.nih.gov/compound/2662#section=Computed-Properties (accessed on 1 October 2021).

pharmaceutics

Article

Targeting Cancer Cell Tight Junctions Enhances PLGA-Based Photothermal Sensitizers' Performance In Vitro and In Vivo

Victoria O. Shipunova [1,2,3,*], Vera L. Kovalenko [1], Polina A. Kotelnikova [2], Anna S. Sogomonyan [2], Olga N. Shilova [2], Elena N. Komedchikova [1], Andrei V. Zvyagin [2], Maxim P. Nikitin [1,3] and Sergey M. Deyev [2]

[1] Moscow Institute of Physics and Technology, 9 Institutskiy per., 141701 Dolgoprudny, Russia; kovalenko.vl@phystech.edu (V.L.K.); lena-kom08@rambler.ru (E.N.K.); max.nikitin@phystech.edu (M.P.N.)

[2] Shemyakin-Ovchinnikov Institute of Bioorganic Chemistry, Russian Academy of Sciences, 16/10 Miklukho-Maklaya St., 117997 Moscow, Russia; kotelnikova@phystech.edu (P.A.K.); annasogomonyan2012@mail.ru (A.S.S.); olchernykh@yandex.ru (O.N.S.); andrei.zvyagin@mq.edu.au (A.V.Z.); biomem@mail.ru (S.M.D.)

[3] Department of Nanobiomedicine, Sirius University of Science and Technology, 1 Olympic Ave., 354340 Sochi, Russia

* Correspondence: viktoriya.shipunova@phystech.edu

Citation: Shipunova, V.O.; Kovalenko, V.L.; Kotelnikova, P.A.; Sogomonyan, A.S.; Shilova, O.N.; Komedchikova, E.N.; Zvyagin, A.V.; Nikitin, M.P.; Deyev, S.M. Targeting Cancer Cell Tight Junctions Enhances PLGA-Based Photothermal Sensitizers' Performance In Vitro and In Vivo. *Pharmaceutics* **2022**, *14*, 43. https://doi.org/10.3390/pharmaceutics14010043

Academic Editor: Oya Tagit

Received: 23 November 2021
Accepted: 22 December 2021
Published: 26 December 2021

Publisher's Note: MDPI stays neutral with regard to jurisdictional claims in published maps and institutional affiliations.

Abstract: The development of non-invasive photothermal therapy (PTT) methods utilizing nanoparticles as sensitizers is one of the most promising directions in modern oncology. Nanoparticles loaded with photothermal dyes are capable of delivering a sufficient amount of a therapeutic substance and releasing it with the desired kinetics in vivo. However, the effectiveness of oncotherapy methods, including PTT, is often limited due to poor penetration of sensitizers into the tumor, especially into solid tumors of epithelial origin characterized by tight cellular junctions. In this work, we synthesized 200 nm nanoparticles from the biocompatible copolymer of lactic and glycolic acid, PLGA, loaded with magnesium phthalocyanine, PLGA/Pht-Mg. The PLGA/Pht-Mg particles under the irradiation with NIR light (808 nm), heat the surrounding solution by 40 °C. The effectiveness of using such particles for cancer cells elimination was demonstrated in 2D culture in vitro and in our original 3D model with multicellular spheroids possessing tight cell contacts. It was shown that the mean inhibitory concentration of such nanoparticles upon light irradiation for 15 min worsens by more than an order of magnitude: IC50 increases from 3 µg/mL for 2D culture vs. 117 µg/mL for 3D culture. However, when using the JO-4 intercellular junction opener protein, which causes a short epithelial–mesenchymal transition and transiently opens intercellular junctions in epithelial cells, the efficiency of nanoparticles in 3D culture was comparable or even outperforming that for 2D (IC50 = 1.9 µg/mL with JO-4). Synergy in the co-administration of PTT nanosensitizers and JO-4 protein was found to retain in vivo using orthotopic tumors of BALB/c mice: we demonstrated that the efficiency in the delivery of such nanoparticles to the tumor is 2.5 times increased when PLGA/Pht-Mg nanoparticles are administered together with JO-4. Thus the targeting the tumor cell junctions can significantly increase the performance of PTT nanosensitizers.

Keywords: photothermal therapy; phthalocyanine; SKOVip-kat; Katushka; TurboFP635; JO-4; PLGA; orthotopic tumors; 3D culture; spheroids

1. Introduction

The methods of photodynamic (PDT) and photothermal (PTT) cancer therapy are based on the use of photosensitizers that are accumulated in the tumor and lead to tumor elimination after light irradiation. Upon absorption of light of a certain wavelength, the photosensitizer switches to an excited state, from which it can come back to the ground state either radiatively with fluorescence emission or non-radiatively with the release of thermal energy. Photosensitizers can also react with cell components via electron transfer, which leads to the formation of free radicals or transfer energy to oxygen with the formation of highly reactive singlet oxygen [1]. Thus, photosensitizers can lead to oxidative stress on a cancer cell or perform local hyperthermia. The increased sensitivity of cancer cells to heating up to 41–47 °C underlies the effectiveness of photothermal therapy [2].

The main advantages of PDT and PTT are non-invasiveness and spatial selectivity. The photosensitizer should have minimal toxicity in the dark, efficiently penetrate the tumor and accumulate inside cancer and stromal cells. Thus PDT and PTT therapies can significantly reduce side effects and improve the effectiveness and specificity of cancer treatment. However, photosensitizers used in the clinic today can accumulate in the organism, significantly increasing the photosensitivity of the skin [3]. Many photosensitizers suffer from poor water solubility, low bioavailability, and instability in physiological conditions. Due to these negative effects, the use of PDT is limited in clinical practice, but these difficulties can be overcome by chemical modification, PEGylation, or by photosensitizer encapsulation in nanocarriers of various nature [4–6].

However, as well as the above-mentioned limitations, there are additional problems that arise in the development of phototherapy and other cancer treatment methods. In particular, solid tumors of the epithelial origin are characterized by tight intercellular contacts limiting the penetration of active substances deeper than 3–4 layers of cells [7]. Preservation of epithelial tissue intercellular contacts is typical for cancer cells and makes traditional chemotherapies as well as targeted therapies with monoclonal antibodies and supramolecular agents ineffective.

To effectively penetrate the tumor through anatomical barriers and actively diffuse within solid tumors, therapeutic agents must bypass the intercellular contacts that seal the boundaries of normal endothelial cells and intercellular spaces within the tumor. To date, the most promising agents that open up cell contacts are the junction opener proteins (JO) obtained from human adenovirus serotype 3 [8–11]. Protein JO-1 and its improved variant JO-4 bind to desmoglein 2 on the cell surface, which leads to the activation of MAP-kinases that activate the metalloprotease ADAM17, which breaks down desmoglein of cell contacts [10]. Activation of MAP-kinases leads to transient transdifferentiation of epithelial cells, including a decrease in the expression of adhesion and blocking cell contact proteins, thus solving the problem of the diffusion of drugs within the tumor [9]. Thus, JO-1 and JO-4 induce a partial epithelial–mesenchymal transition (EMT), increasing the permeability of the tumor to high molecular weight compounds and protein molecules, including antibodies [12]. The improvement of drug delivery to tumors using JO proteins has been demonstrated thoroughly for antibodies and chemotherapy drugs, but the effect of JO on nanostructures delivery is poorly understood. It was shown that JO significantly increases bulk tumor accumulation of 35 nm but not 120 nm gold nanoparticles [13] and significantly enhances the efficacy of liposomes loaded with doxorubicin in vivo [8].

Here we describe the synthesis and characterization of biocompatible polymer nanocontainers loaded with magnesium phthalocyanine (Pht-Mg) as effective photothermal sensitizers. We showed their effectiveness when exposed to near-IR light in terms of selective destruction of cancer cells in 2D culture. However, during the transition from 2D tests to 3D cell culture, the effectiveness of such agents decreased by more than 10 times. Nevertheless, the use of a junction opener protein JO-4 led to the efficiency of nanostructures in 3D culture, comparable or outperforming that for 2D culture, and significantly increased the efficiency of accumulation of nanoparticles in orthotopic mouse tumors in vivo.

2. Materials and Methods

2.1. PLGA Nanoparticle Synthesis

Poly(D, L-lactide-co-glycolide) (RG 858 S, Poly(D,L-lactide-co-glycolide) ester terminated, lactide:glycolide 85:15, Mw 190,000–240,000 Da, Sigma, Darmstadt, Germany) was used as a matrix for the nanoparticle synthesis with the "oil-in-water" microemulsion method. The emulsion was formed by dropping the solution of 40 g/L PLGA and 1 g/L magnesium phthalocyanine (Sigma #402737-1G, Pht-Mg) in chloroform into the 3 mL of 3% PVA (Mowiol® 4-88, Sigma, Darmstadt, Germany) in Milli-Q water with 1 g/L of chitosan oligosaccharide lactate (5 kDa, Sigma, Darmstadt, Germany). Alternatively, manganese(II) phthalocyanine (Sigma 379557-1G), copper(II) phthalocyanine (Sigma #546682-200MG) or 29H,31H-phthalocyanine (Sigma #253103-1G) at 1 g/L were used for nanoparticle synthesis instead of magnesium phthalocyanine. The emulsion was continuously sonicated with the Bandelin Sonopuls HD 2200 (Bandelin, Germany) for 1 min with 25% amplitude of 200 W sonicator. The resulting nanoparticles were washed by triple centrifugation with PBS and finally resuspended in 300 µL of PBS. The final concentration of the particles was determined with air drying at 90 °C.

2.2. Scanning Electron Microscopy

Scanning electron microscopy images of as-synthesized PLGA/Pht-Mg nanoparticles were obtained with a MAIA3 Tescan (Tescan, Brno-Kohoutovice, Czech Republic) microscope at an accelerating voltage of 7 kV. Samples of PLGA/Pht-Mg particles in water at 10 µg/mL were air-dried on a silicon wafer on carbon film and analyzed immediately. SEM images were evaluated using ImageJ software to get a particle size distribution.

2.3. Photothermal Properties Study in Cell-Free System

A colloidal solution of nanoparticles in phosphate-buffered saline was irradiated using a 1200 mW 808 nm laser in a 2 mL test tube. The temperature change was recorded using a FLIR One Pro (Teledyne FLIR, Santa Barbara, CA, USA) thermal imaging camera. Temperature changes are shown as a function of time with substracted the initial room temperature (usually 23–24 °C).

2.4. Fluorescence Spectroscopy

The excitation and emission spectra of PLGA/Pht-Mg nanoparticles at 1 g/L were recorded using an Infinite M1000 Pro microplate reader (Tecan, Grödig, Austria) in 50%: 50% DMSO: H_2O solution. The excitation spectrum was recorded using the emission wavelength of 750 nm within the 280–740 nm range. The emission spectrum was recorded using the excitation wavelength of 350 nm within the 360–850 nm range.

2.5. Cell Culture

SKOVip-kat and EMT6/P cells were cultured in DMEM medium (HyClone, Logan, UT, USA) supplemented with 10% fetal bovine serum (HyClone, Logan, UT, USA) and 2 mM L-glutamine (PanEko, Moscow, Russia). Cells were incubated under a humidified atmosphere with 5% CO_2 at 37 °C. For the cytotoxicity tests cells SKOVip-ka stably expressing red fluorescent protein Katushka were used. These cells were developed by us earlier [14].

For the cell toxicity studies previously obtained human ovarian cancer cells stably expressing fluorescent protein Katushka (far-red fluorescent protein TurboFP635), SKOVip-kat was used [15]. These cells stably express both Katushka protein in the cytoplasm and transmembrane oncomarker HER2 on the cell surface, thus being an ideal model for the study of therapeutic targeting compounds in vivo. These cells have excitation and emission wavelengths, 588 nm and 635 nm, and are suitable for most in vitro and in vivo imaging devices.

2.6. Fluorescence Microscopy

For fluorescent microscopy analysis, cells were seeded on a 96-well plate at $5 \cdot 10^3$ cells per well in 100 µL of DMEM medium supplemented with 10% FBS and cultured for 10 h. Next, Pht-Mg-loaded PLGA nanoparticles were added to get a final concentration of 10 µg/mL. Cells were incubated with nanoparticles for 1 h or 4 h at 37 °C, washed from non-bound particles, and analyzed with epifluorescent Zeiss microscope at the following conditions: for Katushka protein imaging: excitation filter −560/40 nm, emission filter −630/75 nm; for PLGA/Pht-Mg nanoparticles imaging: excitation filter −595–645 nm, emission filter −670–725 nm.

2.7. Cytotoxicity Assay

Cytotoxicity of PLGA/Pht-Mg nanoparticles in 2D cell culture was determined using a resazurin-based toxicity assay. SKOVip-kat cells were incubated with nanoparticles at different concentrations in 200 µL of full phenol red-free DMEM medium in a 2-mL Eppendorf tube for 4 h at 37 °C with 5% CO_2 and then light irradiated with 808 nm laser for different time intervals. Next, cells were diluted with full medium and seeded on a 96-well plate at 2×10^3 cells per well in 100 µL of DMEM medium supplemented with 10% FBS and cultured for 72 h. Next, wells were washed from non-bound particles, and 100 µL of resazurin solution (13 mg/L in phosphate-buffered saline) was added to each well. Samples were incubated for 3 h at 37 °C in a humidified atmosphere with 5% CO_2. The fluorescence of each well was measured using the CLARIOstar microplate reader (BMG Labtech, Ortenberg, Germany) at wavelengths of $\lambda ex = 570$ nm, $\lambda em = 600$ nm. Data are presented as percent from non-treated cells.

2.8. 3D Culture

A Form3 3D printer (FormLabs, Somerville, CA, USA) with FormLabs Gray Resin 1 L photopolymer resin (FormLabs, Somerville, CA, USA) was used to create master molds for pouring agarose gel to create microwells for cell growth. One percent agarose in phenol red-free growth DMEM medium (Gibco, Carlsbad, CA, USA) without FBS was used. Spheroids were prepared by dropping SKOVip-kat cell suspension into agarose wells with subsequent cultivation in a 12-well plate (Nunc, Roskilde, Denmark) with DMEM growth medium with 10% FBS grown for 3 days with 5% CO_2 at 37 °C. The formed spheroids were incubated with nanoparticles with or without subsequent irradiation and analyzed with the Axiovert 200 (Carl Zeiss, Göttingen, Germany) fluorescence microscope. The images of fluorescent spheroids were analyzed using ImageJ software to calculate the total spheroid fluorescence using an "integrated density" parameter.

2.9. Proteins Purification

JO-4 and DARP-LoPE were purified as described by us previously [15].

2.10. Confocal Microscopy

Three dimensional spheroids were incubated with JO-4 at final concentration of 10 µg/mL for 2 h followed with 4-h incubation of 50 µg/mL of PLGA/Pht-Mg nanoparticles in the full cell culture media. Spheroids were imaged with confocal laser scanning miscroscopy. Confocal microscopy images of spheroids were obtained with a FV3000 laser-scanning confocal microscope (Olympus Optical Co Ltd, Tokyo, Japan) using LUCPLFLN $20\times$ objective ($20\times$ magnification, 0.45 numerical aperture) with 640 nm laser and GaAsP detector (500 V) (650–750 nm).

2.11. Chemical Conjugation

JO-4 labeled with FITC (JO-4-FITC) was prepared by rapid mixing of 100 µL of JO-4 at 1 g/L in phosphate buffer, pH 7.4 with 10 µL of FITC at 0.5 g/L in DMSO. JO-4 labeled with Cy5.5 (JO-4-Cy5.5) was prepared by rapid mixing of 100 µL of JO-4 at 1 g/L in phosphate buffer, pH 7.4 with 10 µL of Cy5.5-NHS ester at 1.0 g/L in DMSO. The solutions of labeled proteins were incubated for 8 h, and the excess of unreacted FITC and Cy5.5-NHS molecules was removed with Zeba Spin Desalting Columns (7 kDa MWCO) according to the manufacturer's recommendations.

2.12. Flow Cytometry

Spheroids were taken out from agarose molds, disaggregated with trypsin/EDTA solution (0.25%) for 15 min at 5% CO_2 at 37 °C, washed twice with PBS. For the cell viability assays, cells were stained with propidium iodide at 2.5 µg/mL final concentration 5 min before the analysis. The cell populations were analyzed using BD Accuri C6 (BD) flow cytometer using the excitation laser 488 nm and the emission filter 615/20 nm. For nanoparticle binding assay, cells were analyzed with Novocyte 2000R flow cytometer (Acea Biosciences, San Diego, CA, USA) with an excitation laser of 640 nm and emission filter 675/30 nm.

2.13. Photothermal Properties Study In Vitro

A total of 2×10^6 cells were incubated with PLGA/Pht-Mg nanoparticles to get a final concentration of nanoparticles equal to 0.5 g/L in 200 µL in 2-mL eppendorf tube. The samples were incubated for 4 h at 5% CO_2 at 37 °C. Next, the tube with nanoparticles and cells were incubated at 5% CO_2 at 37 °C for 4 h and irradiated with 808-nm laser. The thermal imaging was performed using a FLIR One Pro (Teledyne FLIR, Santa Barbara, CA, USA) thermal imaging camera. Temperature changes were recorded using a smartphone and are presented as thermal images of the tubes with cells with or without nanoparticles in dependence on irradiation time.

2.14. Tumor-Bearing Mice

Female BALB/c mice of 18–22 g weight were used in the experiments. All procedures with animals were approved by the IBCh RAS Institutional Animal Care and Use Committee. The animals were anesthetized with a mixture of Zoletil (Virbac, Carros, France) and Rometar (Bioveta, Ivanovice na Hané, Czech Republic) at a dose of 25/5 mg/kg. Mice were injected with $4 \cdot 10^6$ EMT6/P cells in the mammary fat pad to create orthotopic tumors. The tumor size was measured with a caliper using the formula $V = width^2 \times length/2$.

For bioimaging experiments, when tumor size reached 90–100 mm^3, mice were i.v. injected with 400 µg of PLGA nanoparticles with or without pre-injection of JO-4 (1 h before nanoparticles at 4 mg/kg dose in 0.15 M NaCl).

2.15. Ex Vivo Study

For ex vivo tumor study, mice were i.v. injected with JO-4-Cy5.5 (4 mg/kg dose in 0.15 M NaCl). Next, mice were sacrificed with cervical dislocation and cryosections of tumor tissue were obtained using FSE cryostat (Thermo). Confocal microscopy images of spheroids were obtained with a FV3000 laser-scanning confocal microscope (Olympus Optical Co. Ltd., Tokyo, Japan) using UPLSAPO 40×2 objective (40× magnification, 0.95 numerical aperture) with 640 nm laser And GaAsP detector (500 V) (650–750 nm).

2.16. In Vivo Imaging

In vivo imaging was performed with a LumoTrace FLUO bioimaging system (Abisense, Sochi, Russia) as follows: mice were anesthetized 3 h after injection of nanoparticles and imaged with fluorescence excitation at $\lambda_{ex} = 630$ nm and 655/40 nm filter.

2.17. In Vivo Therapy

For in vivo therapy study, when tumor size reached 90–100 mm^3, mice were i.v. injected with 400 µg of PLGA nanoparticles with or without pre-injection of JO-4 (1 h before nanoparticles at 4 mg/kg dose in 0.15 M NaCl). Next, the tumor area was irradiated with 808 nm laser (1200 mW) as follows: 5 min irradiation, 15 min pause, 5 min irradiation. The tumor size was measured with a caliper using the formula $V = width^2 \times length/2$ every one or two days.

3. Results

The following experimental scheme was designed to assess the effectiveness of PLGA-based nanocarriers. Biocompatible nanoparticles loaded with magnesium phthalocyanine were synthesized by the microemulsion method and used as nanoagents for photothermal therapy and diagnostics in vitro and in vivo (Figure 1a). The resulting nanoparticles were used both for visualization of cancer cells and for their destruction when exposed to light in the transparency window of biological tissue. For the cell toxicity studies previously obtained, human ovarian cancer cells stably expressing fluorescent protein Katushka (far-red fluorescent protein TurboFP635), SKOVip-kat were used [14]. These cells have excitation and emission wavelengths, 588 nm and 635 nm, and are suitable for most in vitro and in vivo imaging devices.

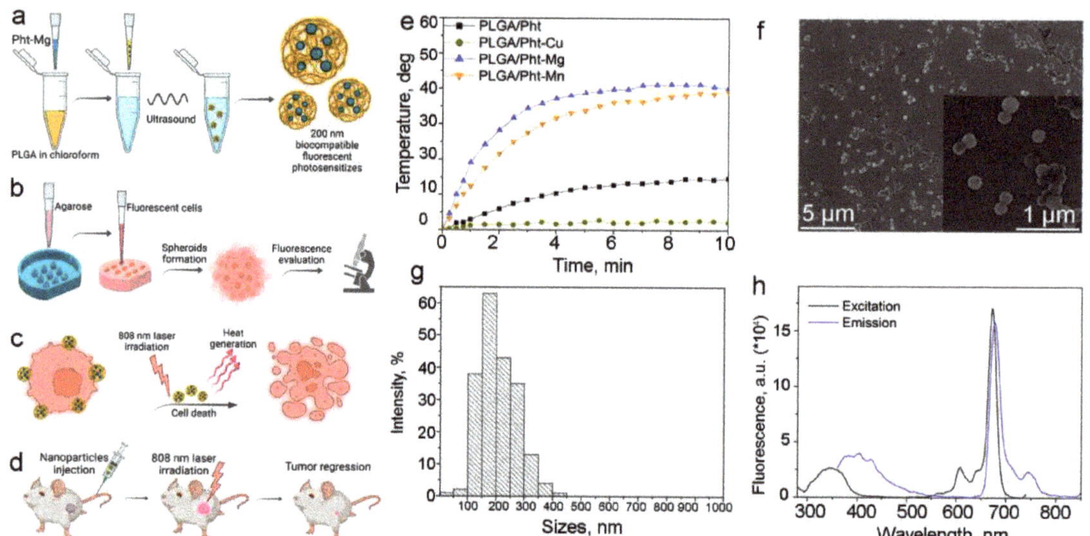

Figure 1. Synthesis and characterization of PLGA-based nanocarriers for photothermal therapy. (**a**) Scheme of nanoparticle synthesis. The mixture of PLGA and Pht-Mg in chloroform is poured into the PVA with chitosan oligosaccharide lactate solution in water and subjected to intensive sonication. (**b**) 3D spheroids formation in agarose gel. Agarose solution was added to the plastic molds and was allowed to solidify at room temperature. Next, a suspension of cells is added to the agarose microwells in order to obtain spheroids on 3–6 days of cultivation. The fluorescence of obtained 3D cell culture is further evaluated using epifluorescent microscopy. (**c**) As obtained 3D spheroids were incubated with PLGA/Pht-Mg nanoparticles followed by 808 nm laser irradiation that leads to the heating of nanoparticles and cell death. (**d**) As-synthesized nanoparticles were used for imaging and treatment of orthotopic tumors in vivo. (**e**) Photothermal properties of PLGA, PLGA/Pht-Mg, PLGA/Pht-Mn, and PLGA/Pht-Cu nanoparticles under the 808 nm light irradiation. (**f**) Scanning electron microscopy images of PLGA nanoparticles. (**g**) Size distribution of PLGA/Pht-Mg nanoparticles, mean size is 208 ± 66 nm. (**h**) Fluorescence excitation and emission spectra of PLGA/Pht-Mg nanoparticles.

The effect of synthesized particles on 3D multicell spheroids, which were obtained by culturing a cell suspension in agarose microwells, was investigated (Figure 1b). Nanoparticles at various concentrations were added to the formed spheroids and exposed to light with a wavelength of 808 nm (Figure 1c). After the selection the optimal conditions for affecting the spheroids, nanoparticles were i.v. injected into mice for visualization and treatment of orthotopic tumors in vivo.

3.1. Synthesis and Characterization of PLGA-Based Photothermal Sensitizers

Nanoparticles of poly-lactide-co-glycolide (PLGA) loaded with magnesium phthalocyanine were synthesized by the microemulsion "oil-in-water" method. The use of PLGA as a drug carrier, attracts the attention of researchers in modern biomedicine, since it provides a stable and controlled release of the compound, thereby reducing side effects [16–20]. PLGA degrades by hydrolysis of its ester linkages in the presence of water and the by-products, lactic acid and glycolic acid, are non-toxic, biocompatible, and rapidly metabolized in the human body.

The photothermal properties of the synthesized nanoparticles were studied under irradiation with an 808 nm laser for 7 min. We tested different NIR dyes to find the optimal for the best performance as a photothermal sensitizer. Among the various NIR dyes tested in this work, including phthalocyanines, magnesium phthalocyanine proved to be the most effective agent for photothermal therapy. Namely, we synthesized PLGA nanoparticles loaded with phthalocyanine (PLGA/Pht), magnesium phthalocyanine (PLGA/Pht-Mg), copper (II) phthalocyanine (PLGA/Pht-Cu), and manganese (II) phthalocyanine (PLGA/Pht-Mn) and tested their photothermal properties (Figure 1d). We showed that the most effective in terms of energy transfer to heat under the 808 nm light irradiation is PLGA nanoparticles loaded with magnesium phthalocyanine (PLGA/Pht-Mg). It was found that the solution of nanoparticles is capable of effectively generating heat energy (Figure 1g). Namely, we have shown that the solution with nanoparticles at a concentration of 1 g/L heats up to $\Delta T = 40\ ^\circ C$ for 3 min and reaches a plateau. Moreover, we have shown that repeated heating of nanoparticles up to three times leads to the same heating of the solution, which indicates that nanoparticles do not degrade and can be used for repeated irradiation to achieve the desired effect.

As obtained nanoparticles PLGA/Pht-Mg were characterized by scanning electron microscopy (Figure 1f). According to the results of SEM image processing, the particles possess a spherical shape and the size of PLGA/Pht-Mg particles is 208 ± 66 nm (Figure 1g).

The fluorescence properties of the synthesized nanoparticles were studied using the fluorescence spectroscopy method. Data presented in Figure 1h confirm that PLGA/Pht-Mg nanoparticles are suitable for fluorescence detection in NIR region, e.g., under the excitation with 640 nm laser as the most commonly used in various visualization devices.

3.2. PLGA/Pht-Mg Nanoparticles Interaction with Cells in 2D Cell Culture

The fluorescence microscopy study was carried out to assess the effectiveness of synthesized nanoparticles for cell labeling. SKOVip-kat cells were seeded on 96-well plate, incubated with nanoparticles for 1 h or 4 h, and analyzed with an epifluorescent microscope. We showed that under the excitation with 595–645 nm filter and with emission filter of 670–725 nm particles can be effectively used for imaging purposes (Figure 2a). Data presented in Figure 2a demonstrate that after 1 h of incubation the particles are preferably presented outside the cells; however, after 4 h of incubation at 37 °C particles are intensively accumulated in cells.

Moreover, the surface properties of nanoparticles can be effectively used for fine-tuning the particle penetration into cells. To enhance the saturation of cell surface with PLGA/Pht-Mg and PLGA/Pht-Mg nanoparticles accumulation in cells, the surface of particles was modified with chitosan oligosaccharide lactate during the synthesis process (Figure 1a). We showed that using equal experimental conditions, the accumulation of chitosan-coated nanoparticles is more effective in comparison with pristine PLGA (Figure 2a).

3.3. Light-Induced Phototoxicity of PLGA Nanoparticles In Vitro

According to the results of fluorescence microscopy, 4 h is enough for the penetration of the main portion of particles into cells. For light-induced cytotoxicity tests, SKOVip-kat cells were incubated with PLGA/Pht-Mg nanoparticles for 4 h and irradiated with 808 nm laser for 1, 2, 3, 4, 5, 10, and 15 min. The cytotoxicity was assessed 72 h later using the resazurin-based cytotoxicity test.

We have shown that the cytotoxicity of nanoparticles was of a concentration-dependent manner and significantly increased when the nanoparticles were heated by a laser. At the same time, the effect of the laser on cells without particles did not have a cytotoxic effect. We showed that IC50 of PLGA/Pht-Mg without irradiation is equal to 98 μg/mL (Figure 2c). However, when cells with particles were irradiated with light for, e.g., 5 and 15 min, the IC50 value significantly decreases—up to 32 and 3 μg/mL, respectively.

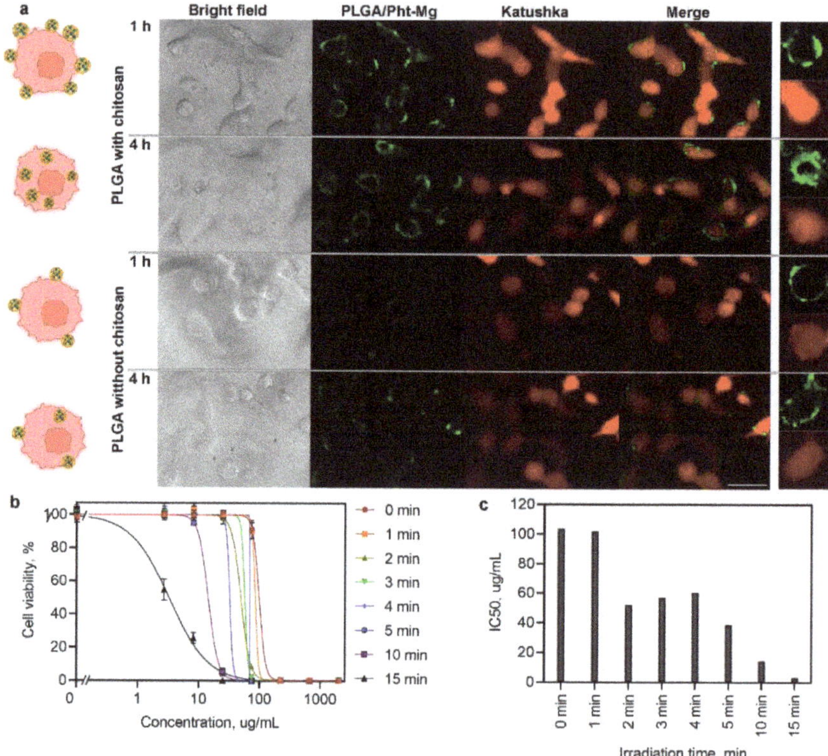

Figure 2. PLGA nanoparticles loaded with Pht-Mg and coated with chitosan oligosaccharide lactate effectively penetrate the cell membrane and possess cytotoxic properties under the 808 nm light irradiation. (**a**) Fluorescence microscopy study of PLGA/Pht-Mg nanoparticles interaction with cells. SKOVip-Kat cells were incubated with PLGA/Pht-Mg particles loaded with Pht-Mg with or without chitosan coating during the synthesis. Cells were imaged 1 h and 4 h after the incubation. Images were taken in a bright field, in the fluorescence channel corresponding to the PLGA/Pht-Mg fluorescence (excitation filter −560/40 nm, emission filter −630/75 nm) and Katushka fluorescence (excitation filter −560/40 nm, emission filter −630/75 nm). Scale bar, 20 μm. (**b**) Light-induced photothermal toxicity of PLGA/Pht-Mg particles for SKOVip-kat cells. Cells were incubated with PLGA/Pht-Mg and irradiated with an 808 nm laser. Cell viability was evaluated 72 h after the irradiation in dependence on particle concentration and irradiation time. (**c**) IC50 dependence on irradiation time for 72 h viability assay.

3.4. 3D Culture

Since the efficacy of both low molecular weight compounds and supramolecular structures differs significantly within in vitro and in vivo models, 3D spheroids are often used as an intermediate step as a more relevant tool for studying drug efficacy [21–25]. In particular, in spheroids, cells form tight intercellular contacts, which are often the main reason for the poor penetration of therapeutic substances into the tumor.

To test the PLGA/Pht-Mg nanoparticles' efficacy in 3D culture, we developed the 3D cultivation method based on cells culturing in agarose microwells. The method is based on pouring agarose into microwells of a master-mold and subsequent cell cultivation in the obtained agarose microwells.

The master mold that was designed using 3D-printing technology schematically illustrated in Figure 1b. The .stl file for 3D-printing can be uploaded as Supporting File 2 and used for the printing of the plastic master-mold for agarose pouring.

Agarose solution was poured onto the master mold and then left to solidify at room temperature under sterile conditions resulting in the formation of nine 2.3 × 3.3 mm U-shaped microchambers for the cell culture. The suspension of fluorescent SKOVip-kat cells was then added to the wells leading to the formation of spheroids with tight junctions at 3–6 days of cultivation.

The developed method presents a universal tool for 3D culturing of mammalian cancer cells based on reusable molds for agarose, which allows the reproducible formation of multicellular spheroids with tight contacts. The formation of dense cell spheroids in the wells of an agarose gel allows both real-time bright-field and fluorescent visualization, since agarose is an optically transparent material, and the analysis of spheroids separately when they are removed from the molds with a 200 μL pipette tip.

Data presented in Figure 3a demonstrate the formation of reproducible spheroids from SKOVip-kat cells. Moreover, the spheroids retain their integrity after the removal from agarose microchambers thus confirming that as-obtained 3D cell structures do possess tight junctions. This fact was also confirmed using the 2 mM EDTA solution for the spheroid disaggregation. We noted that 2 mM EDTA which is usually used for the detachment of the SKOVip-kat cells from the plastic surface is absolutely ineffective even after 3 h of incubation for spheroid disaggregation: the spheroids retain their integrity after the incubation. The spheroid can only be disaggregated using concentrated trypsin (0.25%) solution for 15–20 min.

Thus obtained spheroids can be used for further research, for example, for the analysis by flow cytometry after disaggregation, or further cultivation in more complex systems, for example, including cells of the stromal tumor microenvironment.

We showed that the fluorescence intensity calculation using image processing is an adequate tool for the study of cytotoxicity of different compounds in 3D culture in the real-time mode without affecting cell viability and the fluorescence intensity calculation was used in further tests (Figure S1).

3.5. Cytotoxicity of PLGA-Based Nanocarriers in 3D Culture

As obtained spheroids were used for evaluating the effectiveness of PLGA carriers in 3D culture mimicking solid tumors.

For the experiment, concentrations within the 74–2000 μg/mL and irradiation time of 0, 5, and 15 min were selected as most clearly reflecting the cell response on incubation with nanoparticles and light irradiation according to 2D cytotoxicity test. Spheroids were overnight incubated with nanoparticles followed by light irradiation and the fluorescence intensity analysis reflecting the cells' viability was performed on 6 day of cultivation.

Data presented in Figure 3f,g demonstrate that although the cytotoxicity of PLGA/Pht-Mg nanoparticles is quite similar to 2D cytotoxicity tests, much higher doses are required to achieve the desired cell death. For instance, the IC50 of pristine PLGA/Pht-Mg increased to 3.5 times in 3D culture (98 μg/mL for 2D vs. 347 μg/mL for 3D) and to 8.4 times (32 μg/mL for 2D vs. 269 μg/mL for 3D) for 5 min-irradiation and 39 times (3 μg/mL for 2D vs. 117 μg/mL for 3D) for 15 min-irradiation.

To enhance the efficacy of PLGA/Pht-Mg photothermal sensitizers, we used JO-4 protein that induces the short EMT thus allowing nanoparticles to pass through tight cells junctions. JO-4 was added to cells 1 h before the nanoparticle addition and the cytotoxicity test was performed in the same way as for without JO-4. We confirmed that JO-4 conjugated with FITC does interact with the tight junctions of SKOVip-Kat cells when 2D cell culture reaches confluent monolayer (Figure 3b).

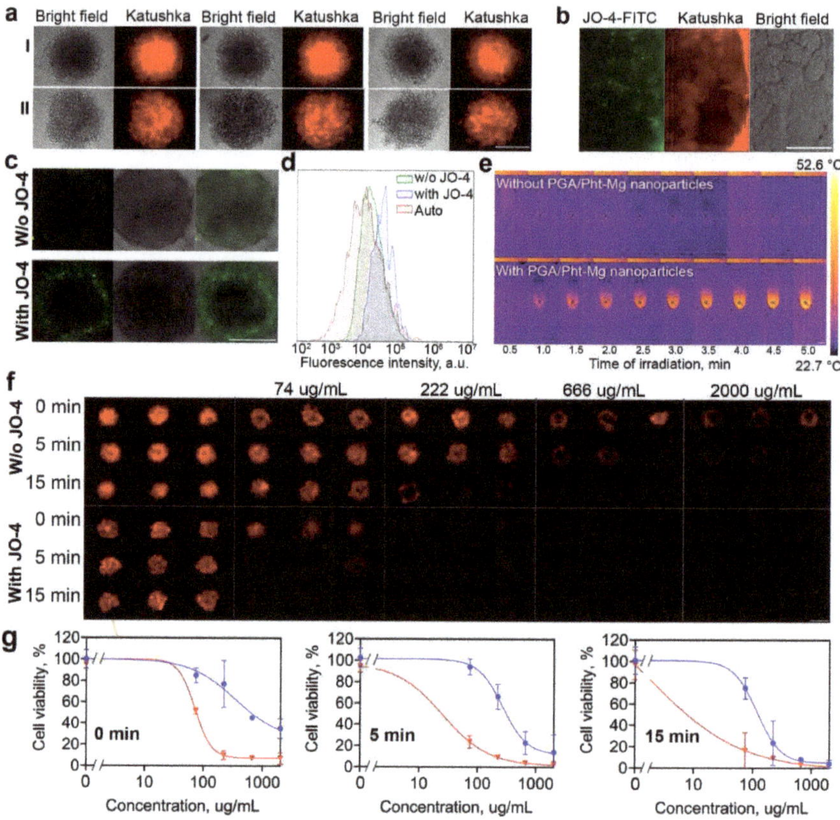

Figure 3. JO-4 enhances the therapeutic and diagnostic capabilities of polymer PLGA-based photothermal sensitizers in vitro. (**a**) Bright-field and fluorescent images of 3D cell spheroids based on SKOVip-Kat cells cultured in agarose molds for 9 days before (I) and after (II) the removal from agarose microwells. Scale bar, 250 μm. (**b**) JO-4 conjugated with FITC interacts with the tight junctions of SKOVip-Kat cells. Scale bar, 20 μm. (**c**) Confocal microscopy images of 3D spheroids labeled with PLGA/Pht-Mg nanoparticles with or without preincubation with JO-4 protein. Images are presented in the fluorescence channel corresponding to the fluorescence of Pht-Mg (excitation 640 nm, emission 660–760 nm, the image is pseudocolored with green color), in the bright field and as a merged image of bright field and fluorescence. Scale bar, 250 μm. (**d**) Flow cytometry assay on evaluation of PLGA/Pht-Mg labeling of cells in spheroids. Spheroids were labeled with PLGA/Pht-Mg with or without preincubation with JO-4 protein, disaggregated with trypsin solution and analyzed with flow cytometry (excitation laser 640 nm, emission filter 675/30 nm). (**e**) Thermal imaging of cell suspension with or without PLGA/Pht-Mg nanoparticles under the irradiation with 808 nm laser with FLIR One camera. Scale bar, 1 cm. (**f**) Epifluorescent images of 3D spheroids of SKOVip-Kat cells incubated with PLGA/Pht-Mg nanoparticles loaded with Pht-Mg and irradiated with NIR light for 0, 5, and 15 min. Scale bar, 250 μm. (**g**) Fluorescence-based cytotoxicity curves for SKOVip-kat spheroids incubated with PLGA/Pht-Mg and irradiated for 0, 5, or 15 min with or without pre-incubation with JO-4 protein.

Using the laser confocal scanning microscopy, we confirmed that JO-4 indeed enhances the permeability of fluorescent PLGA/Pht-Mg nanoparticles inside the 3D spheroids (Figure 3c). These data are accompanied by flow cytometry assay data (Figure 3d) performed on cells from disaggregated spheroids incubated with PLGA/Pht-Mg nanoparticles with or without pre-incubation with JO-4 protein. Data presented in Figure 3d confirm that

JO-4 enhances the particle accumulation in the total cell population within the spheroid. Thus, the cell tight junctions targeting with JO-4 protein and cancer cell heating with PLGA/Pht-Mg sensitizers were used for selective destruction of cancer cells in spheroids. Before that, the retaining of hyperthermic properties of PLGA/Pht-Mg particles after cell binding was confirmed with photothermal imaging using FLIR One pro camera (Figure 3e).

Data shown in Figure 3f,g confirm that the approach is very effective leading to nanoparticle performance increase in 3D culture as it is in 2D culture. Indeed, the IC50 value calculated for viability curve of PLGA/Pht-Mg particles without irradiation was found to be 72 μg/mL (comparable to 98 μg/mL for 2D), 25 μg/mL for 5 min irradiation (comparable to 32 μg/mL for 2D) and 1.9 μg/mL for 15 min irradiation (comparable to 3 μg/mL for 2D).

3.6. JO-4 Enhances Nanoparticle Accumulation in Tumor In Vivo

As-obtained nanoparticle PLGA/Pht-Mg performance was assessed in vivo using orthotopic mouse mammary tumors. BALB/c mice were injected with EMT6/P cells to get tumors in the mammary fat pad.

Mice were i.v. injected with JO-4-Cy5.5 and cryosections of tumor tissues were analyzed with confocal microscopy thus confirming the accumulation of fluorescently labeled JO-4-Cy5.5 protein in the upper layers of the tumor (Figure 4a).

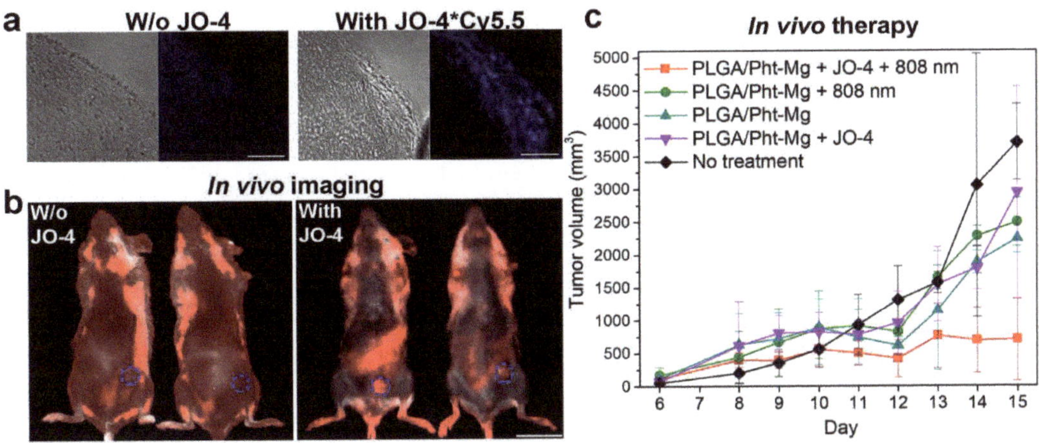

Figure 4. Targeting cancer cell junctions with JO-4 enhances the therapeutic and diagnostic capabilities of polymer PLGA-based photothermal sensitizers in vivo. (**a**) Cryosections of tumor tissue demonstrate the Cy5.5-labeled JO-4 accumulation in the tumor. Confocal laser scanning microscopy images are presented in the fluorescence channel corresponding to the Cy5.5 fluorescence (excitation laser 640 nm, emission 660–760 nm) and in the bright field. Scale bars, 100 μm. (**b**) In vivo living imaging of BALB/c mice with orthotopic tumors i.v. injected with PLGA/Pht-Mg particles with or without pre-injection of JO-4 in the fluorescence channel corresponding to the nanoparticle's fluorescence. Scale bar, 2 cm. (**c**) Tumor growth dynamics under the treatment with PLGA/Pht-Mg particles with or without JO-4 pre-treatment and with or without 808 nm laser irradiation. Data are presented as mean ± s.d.

For bioimaging tests, mice were inoculated with 400 μg of PLGA/Pht-Mg nanoparticles with or without pre-injection of JO-4 and 4 h later visualized with LumoTrace imaging system (Figure 4b) using the excitation diodes of 630 nm and 655 nm emission filters. Data shown in Figure 4b confirm the effective PLGA/Pht-Mg particles accumulation in the tumor site (blue circled area) only in the case when JO-4 was pre-injected. The fluorescence intensity calculations showed that using the pre-injection of JO-4 prior to nanoparticles leads to the tumor fluorescence intensity increase up to 2.5 times (2.49×10^6 vs. 9.86×10^5 a.u.).

3.7. JO-4 Enhances the Therapeutic Capabilities of PLGA-Based Nanosensitizers In Vivo

Finally, we evaluated the therapeutic capabilities of PLGA/Pht-Mg particles in vivo using their photothermal properties. Mice were i.v. injected with 400 µg of PLGA/Pht-Mg nanoparticles with or without pre-injection of JO-4 and irradiated with 808 nm laser by two repeated cycles (5 min each). The dynamics of tumors' growth recorded for two weeks are presented in Figure 4c. Data in Figure 4c confirm that the affecting the tumor with PLGA/Pht-Mg is effective only in the case of pre-injecting with JO-4 and 808 nm light irradiation; the tumor growth inhibition at 15 day calculated as %TGI = (1 − {Tt/T0/Ct/C0}/1 − {C0/Ct}) × 100 where Tt = median tumor volume of treated at time t, T0 = median tumor volume of treated at time 0, Ct = median tumor volume of control at time t and C0 = median tumor volume of control at time 0 was found to be equal to TGI = 92%.

4. Discussion

A large number of photosensitizers require excitation with visible light, and the penetration into deep tissues, especially into solid tumors, is limited. The use of near-infrared wavelengths allows deeper penetration into biological tissues and has less effect on endogenous chromophores such as hemoglobin. Optimal irradiation is provided in the so-called biological "transparency window" of 650–950 nm [26]. Longer wavelength ranges lying beyond 1000 nm (the maximum absorption of water) are also reported [27,28]. It should be noted that light in the visible and near-IR ranges is relatively safe for surrounding tissues in a much wider range of irradiation energy used, compared to, e.g., radiation therapy of tumors.

Phthalocyanine-based compounds weakly absorb light at 400–600 nm and practically do not lead to photosensitivity of the skin [29]. However, their ability to absorb light and fluorescence at 650–800 nm and generate singlet oxygen can be used to destroy both skin and deep tumors [30–33]. The photodynamic and photothermal properties of phthalocyanines depend on their structure. The efficiency of heating or generation of ROS depends on the ion at the base of the complex (for example, heavy metal ions can increase the generation of ROS), and the chemical modification can increase the solubility of phthalocyanines or change the excitation wavelength of their photosensitizing properties [34,35]. Due to the high stability of the phthalocyanine molecules [35], they can be used for repeated cycles of irradiation without any additional injections, in contrast to, e.g., well-studied indocyanine green photothermal sensitizer.

A number of drugs based on phthalocyanine and its derivatives are at various stages of clinical trials and even approved in the clinic thus confirming their efficiency for PDT and PTT. The phototherapy drugs used include hydroxyaluminium trisulfophthalocyanine (Photosens), which has been tested on a variety of tumor types, including eye and eyelid cancer [36], bladder cancer and cervical cancer [37–39]. A number of drugs based on phthalocyanine derivatives have undergone different stages of clinical trials and include zinc phthalocyanine derivatives (CGP55847 [40] and Photocyanine [41]), silicon (Pc 4 [42]). Currently, a large number of phthalocyanine-based drugs are being developed for photodynamic [43] (Li et al. 2018) and photothermal therapy [44,45] cancer.

In this work, nanoparticles based on magnesium phthalocyanines were obtained, which are effectively heated when irradiated with light of a wavelength of 808 nm. We have shown that such particles, when stabilized by chitosan oligosaccharide lactate, are selectively internalized into cells and can be effectively used for laser-induced cell death in 2D in vitro culture.

However, when 3D cellular spheroids were used, the efficiency of these particles decreased significantly more than an order of magnitude in terms of IC50. To meet the challenge and increase the nanoparticle performance, especially their penetration into tight multicellular structures, we used the combination of PLGA-based photothermal sensitizers and the protein JO-4 that weakened intercellular contacts by activating metalloprotease ADAM17, which breaks down desmoglein of cell contacts and increased tissue permeability. After the binding to desmoglein 2, these proteins cause temporary transdifferentiation of cells, a partial epithelial–mesenchymal transition (EMT). JO-4 was obtained by screening mutant variants and has a higher affinity for desmoglein-2 than JO-1. JO proteins are able to induce the opening of cell contacts in the normal stratified epithelium; however, JO-1 has been shown to preferentially act on intercellular contacts of tumors and does not cause serious side effects at therapeutic doses [9,11,12]. This was confirmed for JO-1 in xenograft models [9,12], the safety of JO-4 has also been shown in macaques [8]. Thus, JO proteins can be used to overcome anatomical barriers and improve the diffusion of therapeutic agents within a tumor. At the same time, the change in the cellular phenotype of the tumor is temporary, and long-term experiments have shown that the use of JO in vivo does not increase the likelihood of metastases [11].

Of course, the application of photothermal therapy with photothermal nanosensitizers and NIR light irradiation has serious limitations regarding light penetration into tissues. However, light sources in the 800–900 nm wavelength range are the most effective for PTT from UV-Vis-NIR range since the light penetration depth is about 4–5 mm in this range in comparison with, e.g., 0.5 mm for blue light. The described method of PTT can be significantly improved with the use of invasive light sources or in the combination with chemotherapy or radiotherapy that does not require external light sources. Namely, the PLGA/Pht-Mg sensitizers can be additionally loaded with chemotherapeutic drugs— doxorubicin or paclitaxel and used in co-administration with junction opener proteins to perform better for oncotherapy.

Previously the effectiveness of the JO-4 protein in relation to nanoparticles was studied in relation to PEGylated liposomes and gold nanoparticles only. Thus, the effect of this protein on enhancing the effectiveness of liposomes loaded with doxorubicin in vivo [8] was demonstrated, and it was shown that JO protein significantly increases bulk tumor accumulation of 35 nm but not 120 nm gold nanoparticles [13]. Here we show that targeting tight junctions is effective for quite large 200 nm polymer nanoparticles acting as sensitizers of PTT thus opening up new possibilities in cancer treatment using polymer-based nano-cargo. However, the systemic toxicity of such treatment still remains an open question and should be thoroughly investigated after this proof-of-concept study. Nevertheless, based on previous studies [8], we believe that this method is safe for use in oncology in combination with the use of nanoparticles. Previously, when studying the combination of Doxil and JO-4 in healthy non-human primates the non-significant effects regarding the systemic toxicity were observed: no changes in animal health/behavior were observed; histological analysis revealed mild gastro-enterocolitis, the hematological analysis revealed mild lymphopenia on day 1, and thrombocytopenia at days 1, 2, and 3 after the injection. A mild transaminitis was observed on day 1 after JO-4 injection which could be related to residual bacterial endotoxin in the JO-4 preparation. No significant damage of liver, kidney and spleen was observed. When combined with Doxil, analysis of clinical symptoms and blood parameters did not show remarkable signs of toxicity. Overall, the study (Richter et al., 2015) did not show critical JO-4-related toxicity or an increase of Doxil-related side effects, thus proving the clinical safety of the proposed method for the use in combination with nanomedications [8].

5. Conclusions

Despite the significant progress achieved in the treatment of oncological diseases in recent decades, the treatment of solid tumors of epithelial origin possessing intercellular contacts is still a serious problem. Considerable efforts are being made to enhance the efficacy of nanoagents of different origins, e.g., aimed at extending their bloodstream circulation [46,47] or changing their biodistribution [48]. Nevertheless, tight intercellular contacts typical for cancer cells make both traditional chemotherapies and therapy with monoclonal antibodies and supramolecular agents ineffective. Therefore, the active targeting of tumor cell contacts in the development of new and improvement of existing approaches to the therapy of solid tumors is one of the most urgent areas of modern biomedicine.

Here we show that the co-administration of a junction opener protein JO-4 and quite large 200 nm PLGA nanoparticles loaded with magnesium phthalocyanine is an effective method both for therapeutic purposes and for the diagnosis of solid tumors. The developed technology can also be used for other types of polymer nanoparticles acting as PTT or PDT sensitizers.

Since several PLGA-based non-targeted drugs formulations are already entered the clinic, e.g., Lupron Depot (Abbvie Endocrine Inc., North Chicago, IL, USA) and Trelstar (Allergan Sales Inc., Madison, NJ, USA) for prostate cancer treatment, the described technique can probably be used for already approved micro- and nanoparticles for improving their efficiency. We believe that the co-administration of agents targeting tight tumor contacts together with therapeutic nanoparticles can significantly improve both the quality of diagnosis and therapy of solid, including metastatic, tumors.

Supplementary Materials: The following are available online at https://www.mdpi.com/article/10.3390/pharmaceutics14010043/s1, Supplementary Note 1: Cytotoxicity tests in 3D in vitro culture, Figure S1: Comparison of cytotoxicity tests for assessing the efficacy of anticancer drugs in 3D culture with multicellular spheroids possessing tight contacts.

Author Contributions: Conceptualization, V.O.S.; Data curation, V.O.S.; Formal analysis, V.O.S.; Funding acquisition, A.V.Z., M.P.N. and S.M.D.; Investigation, V.O.S. and O.N.S.; Methodology, V.O.S., V.L.K., P.A.K., A.S.S., and E.N.K.; Project administration, V.O.S.; Resources, A.V.Z., M.P.N., and S.M.D.; Software, V.O.S., V.L.K., P.A.K., and O.N.S.; Supervision, V.O.S.; Validation, V.O.S.; Visualization, V.O.S. and V.L.K.; Writing—original draft, V.O.S., P.A.K., and E.N.K.; Writing—review and editing, V.O.S. All authors have read and agreed to the published version of the manuscript.

Funding: Different aspects and parts of this multidisciplinary research was funded by Russian Science Foundation grant No. 21-74-30016 (cell culture), Russian Foundation for Basic Research No. 19-29-04012 (in vivo study), and by Sirius University (nanoparticle characterization).

Institutional Review Board Statement: All experimental animal procedures were approved by the Animal Care and Use Committee of the Shemyakin-Ovchinnikov Institute of Bioorganic Chemistry Russian Academy of Science, protocol No. 298/2020, dated 29 May 2020.

Informed Consent Statement: Not applicable.

Data Availability Statement: All data are presented within the manuscript and supporting information.

Acknowledgments: Schemes in table of content and Figure 1 were created with BioRender.com. This study was made possible through the support of the Applied Genetics Resource Facility of MIPT.

Conflicts of Interest: Maxim P. Nikitin is the founder of Abisense company. The company had no role in the design of the study; in the collection, analyses, or interpretation of data; in the writing of the manuscript, and in the decision to publish the results.

References

1. Chen, J.; Fan, T.; Xie, Z.; Zeng, Q.; Xue, P.; Zheng, T.; Chen, Y.; Luo, X.; Zhang, H. Advances in nanomaterials for photodynamic therapy applications: Status and challenges. *Biomaterials* **2020**, *237*, 119827. [CrossRef]
2. Mallory, M.; Gogineni, E.; Jones, G.C.; Greer, L.; Simone, C.B. Therapeutic hyperthermia: The old, the new, and the upcoming. *Crit. Rev. Oncol. Hematol.* **2016**, *97*, 56–64. [CrossRef]
3. Baskaran, R.; Lee, J.; Yang, S.-G. Clinical development of photodynamic agents and therapeutic applications. *Biomater. Res.* **2018**, *22*, 25. [CrossRef] [PubMed]
4. Jung, H.S.; Verwilst, P.; Sharma, A.; Shin, J.; Sessler, J.L.; Kim, J.S. Organic molecule-based photothermal agents: An expanding photothermal therapy universe. *Chem. Soc. Rev.* **2018**, *47*, 2280–2297. [CrossRef]
5. Fernandes, N.; Rodrigues, C.F.; Moreira, A.F.; Correia, I.J. Overview of the application of inorganic nanomaterials in cancer photothermal therapy. *Biomater. Sci.* **2020**, *8*, 2990–3020. [CrossRef] [PubMed]
6. Hu, J.-J.; Cheng, Y.-J.; Zhang, X.-Z. Recent advances in nanomaterials for enhanced photothermal therapy of tumors. *Nanoscale* **2018**, *10*, 22657–22672. [CrossRef]

7.   Lammers, T.; Kiessling, F.; Hennink, W.E.; Storm, G. Drug targeting to tumors: Principles, pitfalls and (pre-) clinical progress. *J. Control. Release* **2012**, *161*, 175–187. [CrossRef] [PubMed]

8.   Richter, M.; Yumul, R.; Wang, H.; Saydaminova, K.; Ho, M.; May, D.; Baldessari, A.; Gough, M.; Drescher, C.; Urban, N.; et al. Preclinical safety and efficacy studies with an affinity-enhanced epithelial junction opener and PEGylated liposomal doxorubicin. *Mol. Ther. Methods Clin. Dev.* **2015**, *2*, 15005. [CrossRef]

9.   Wang, H.; Li, Z.-Y.; Liu, Y.; Persson, J.; Beyer, I.; Möller, T.; Koyuncu, D.; Drescher, M.R.; Strauss, R.; Zhang, X.-B.; et al. Desmoglein 2 is a receptor for adenovirus serotypes 3, 7, 11 and 14. *Nat. Med.* **2011**, *17*, 96–104. [CrossRef]

10.   Wang, H.; Ducournau, C.; Saydaminova, K.; Richter, M.; Yumul, R.; Ho, M.; Carter, D.; Zubieta, C.; Fender, P.; Lieber, A. Intracellular Signaling and Desmoglein 2 Shedding Triggered by Human Adenoviruses Ad3, Ad14, and Ad14P1. *J. Virol.* **2015**, *89*, 10841–10859. [CrossRef]

11.   Beyer, I.; Cao, H.; Persson, J.; Song, H.; Richter, M.; Feng, Q.; Yumul, R.; van Rensburg, R.; Li, Z.; Berenson, R.; et al. Coadministration of epithelial junction opener JO-1 improves the efficacy and safety of chemotherapeutic drugs. *Clin. Cancer Res.* **2012**, *18*, 3340–3351. [CrossRef]

12.   Beyer, I.; van Rensburg, R.; Strauss, R.; Li, Z.; Wang, H.; Persson, J.; Yumul, R.; Feng, Q.; Song, H.; Bartek, J.; et al. Epithelial junction opener JO-1 improves monoclonal antibody therapy of cancer. *Cancer Res.* **2011**, *71*, 7080–7090. [CrossRef]

13.   Wang, C.E.; Yumul, R.C.; Lin, J.; Cheng, Y.; Lieber, A.; Pun, S.H. Junction opener protein increases nanoparticle accumulation in solid tumors. *J. Control. Release* **2018**, *272*, 9–16. [CrossRef]

14.   Sokolova, E.A.; Shilova, O.N.; Kiseleva, D.V.; Schulga, A.A.; Balalaeva, I.V.; Deyev, S.M. HER2-Specific Targeted Toxin DARPin-LoPE: Immunogenicity and Antitumor Effect on Intraperitoneal Ovarian Cancer Xenograft Model. *Int. J. Mol. Sci.* **2019**, *20*, 2399. [CrossRef]

15.   Shipunova, V.O.; Komedchikova, E.N.; Kotelnikova, P.A.; Zelepukin, I.V.; Schulga, A.A.; Proshkina, G.M.; Shramova, E.I.; Kutscher, H.L.; Telegin, G.B.; Kabashin, A.V.; et al. Dual Regioselective Targeting the Same Receptor in Nanoparticle-Mediated Combination Immuno/Chemotherapy for Enhanced Image-Guided Cancer Treatment. *ACS Nano* **2020**, *14*, 12781–12795. [CrossRef]

16.   Cunha, A.; Gaubert, A.; Latxague, L.; Dehay, B. PLGA-Based Nanoparticles for Neuroprotective Drug Delivery in Neurodegenerative Diseases. *Pharmaceutics* **2021**, *13*, 1042. [CrossRef]

17.   Varani, M.; Campagna, G.; Bentivoglio, V.; Serafinelli, M.; Martini, M.L.; Galli, F.; Signore, A. Synthesis and Biodistribution of 99mTc-Labeled PLGA Nanoparticles by Microfluidic Technique. *Pharmaceutics* **2021**, *13*, 1769. [CrossRef] [PubMed]

18.   Miele, D.; Xia, X.; Catenacci, L.; Sorrenti, M.; Rossi, S.; Sandri, G.; Ferrari, F.; Rossi, J.J.; Bonferoni, M.C. Chitosan Oleate Coated PLGA Nanoparticles as siRNA Drug Delivery System. *Pharmaceutics* **2021**, *13*, 1716. [CrossRef]

19.   Abdelkader, D.H.; Abosalha, A.K.; Khattab, M.A.; Aldosari, B.N.; Almurshedi, A.S. A Novel Sustained Anti-Inflammatory Effect of Atorvastatin-Calcium PLGA Nanoparticles: In Vitro Optimization and In Vivo Evaluation. *Pharmaceutics* **2021**, *13*, 1658. [CrossRef] [PubMed]

20.   Shipunova, V.O.; Sogomonyan, A.S.; Zelepukin, I.V.; Nikitin, M.P.; Deyev, S.M. PLGA Nanoparticles Decorated with Anti-HER2 Affibody for Targeted Delivery and Photoinduced Cell Death. *Molecules* **2021**, *26*, 3955. [CrossRef] [PubMed]

21.   Sokolova, E.; Kutova, O.; Grishina, A.; Pospelov, A.; Guryev, E.; Schulga, A.; Deyev, S.; Balalaeva, I. Penetration Efficiency of Antitumor Agents in Ovarian Cancer Spheroids: The Case of Recombinant Targeted Toxin DARPin-LoPE and the Chemotherapy Drug, Doxorubicin. *Pharmaceutics* **2019**, *11*, 219. [CrossRef] [PubMed]

22.   Quarta, A.; Gallo, N.; Vergara, D.; Salvatore, L.; Nobile, C.; Ragusa, A.; Gaballo, A. Investigation on the Composition of Agarose-Collagen I Blended Hydrogels as Matrices for the Growth of Spheroids from Breast Cancer Cell Lines. *Pharmaceutics* **2021**, *13*, 963. [CrossRef]

23.   Cavaco, M.; Fraga, P.; Valle, J.; Andreu, D.; Castanho, M.A.R.B.; Neves, V. Development of Breast Cancer Spheroids to Evaluate Cytotoxic Response to an Anticancer Peptide. *Pharmaceutics* **2021**, *13*, 1863. [CrossRef]

24.   Zhao, Y.; Zheng, Y.; Zhu, Y.; Zhang, Y.; Zhu, H.; Liu, T. M1 Macrophage-Derived Exosomes Loaded with Gemcitabine and Deferasirox against Chemoresistant Pancreatic Cancer. *Pharmaceutics* **2021**, *13*, 1493. [CrossRef]

25.   Winter, S.J.; Miller, H.A.; Steinbach-Rankins, J.M. Multicellular Ovarian Cancer Model for Evaluation of Nanovector Delivery in Ascites and Metastatic Environments. *Pharmaceutics* **2021**, *13*, 1891. [CrossRef]

26.   Altinoğlu, E.I.; Adair, J.H. Near infrared imaging with nanoparticles. *Wiley Interdiscip. Rev. Nanomed. Nanobiotechnol.* **2010**, *2*, 461–477. [CrossRef]

27.   Smith, A.M.; Mancini, M.C.; Nie, S. Bioimaging: Second window for in vivo imaging. *Nat. Nanotechnol.* **2009**, *4*, 710–711. [CrossRef] [PubMed]

28.   Duan, Y.; Liu, B. Recent Advances of Optical Imaging in the Second Near-Infrared Window. *Adv. Mater.* **2018**, *30*, 1802394. [CrossRef]

29.   Hamblin, M.R. *Advances in Photodynamic Therapy: Basic, Translational, and Clinical*; Artech House: Norwood, MA, USA, 2008.

30.   Li, X.; Zheng, B.-D.; Peng, X.-H.; Li, S.-Z.; Ying, J.-W.; Zhao, Y.; Huang, J.-D.; Yoon, J. Phthalocyanines as medicinal photosensitizers: Developments in the last five years. *Coord. Chem. Rev.* **2019**, *379*, 147–160. [CrossRef]

31.   Zheng, B.-D.; He, Q.-X.; Li, X.; Yoon, J.; Huang, J.-D. Phthalocyanines as contrast agents for photothermal therapy. *Coord. Chem. Rev.* **2021**, *426*, 213548. [CrossRef]

32.   Josefsen, L.B.; Boyle, R.W. Unique diagnostic and therapeutic roles of porphyrins and phthalocyanines in photodynamic therapy, imaging and theranostics. *Theranostics* **2012**, *2*, 916–966. [CrossRef]

33. Machacek, M.; Cidlina, A.; Novakova, V.; Svec, J.; Rudolf, E.; Miletin, M.; Kučera, R.; Simunek, T.; Zimcik, P. Far-red-absorbing cationic phthalocyanine photosensitizers: Synthesis and evaluation of the photodynamic anticancer activity and the mode of cell death induction. *J. Med. Chem.* **2015**, *58*, 1736–1749. [CrossRef] [PubMed]
34. Lo, P.-C.; Rodríguez-Morgade, M.S.; Pandey, R.K.; Ng, D.K.P.; Torres, T.; Dumoulin, F. The unique features and promises of phthalocyanines as advanced photosensitisers for photodynamic therapy of cancer. *Chem. Soc. Rev.* **2020**, *49*, 1041–1056. [CrossRef] [PubMed]
35. Wong, R.C.H.; Lo, P.-C.; Ng, D.K.P. Stimuli responsive phthalocyanine-based fluorescent probes and photosensitizers. *Coord. Chem. Rev.* **2019**, *379*, 30–46. [CrossRef]
36. Budzinskaia. Experimental assessment of the capacities of use of photosense. Communication 2. Photodynamic therapy for epibulbar and choroid tumors. *Vestn. Oftalmol.* **2005**, *121*, 17.
37. Apolikhin. Adjuvant photodynamic therapy (PDT) with photosensitizer photosens for superficial bladder cancer. Experimental investigations to treat prostate cancer by PDT with photosens. *Soc. Photo-Opt. Instrum.* **2007**, *8*, 45.
38. Trushina, O.I.; Novikova, E.G.; Sokolov, V.V.; Filonenko, E.V.; Chissov, V.I.; Vorozhtsov, G.N. Photodynamic therapy of virus-associated precancer and early stages cancer of cervix uteri. *Photodiagnosis Photodyn. Ther.* **2008**, *5*, 256–259. [CrossRef]
39. Sekkat, N.; van den Bergh, H.; Nyokong, T.; Lange, N. Like a bolt from the blue: Phthalocyanines in biomedical optics. *Molecules* **2011**, *17*, 98–144. [CrossRef]
40. Love, W.G.; Duk, S.; Biolo, R.; Jori, G.; Taylor, P.W. Liposome-mediated delivery of photosensitizers: Localization of zinc (II)-phthalocyanine within implanted tumors after intravenous administration. *Photochem. Photobiol.* **1996**, *63*, 656–661. [CrossRef]
41. Shao, J.; Xue, J.; Dai, Y.; Liu, H.; Chen, N.; Jia, L.; Huang, J. Inhibition of human hepatocellular carcinoma HepG2 by phthalocyanine photosensitiser PHOTOCYANINE: ROS production, apoptosis, cell cycle arrest. *Eur. J. Cancer* **2012**, *48*, 2086–2096. [CrossRef]
42. Miller, J.D.; Baron, E.D.; Scull, H.; Hsia, A.; Berlin, J.C.; McCormick, T.; Colussi, V.; Kenney, M.E.; Cooper, K.D.; Oleinick, N.L. Photodynamic therapy with the phthalocyanine photosensitizer Pc 4: The case experience with preclinical mechanistic and early clinical-translational studies. *Toxicol. Appl. Pharmacol.* **2007**, *224*, 290–299. [CrossRef] [PubMed]
43. Li, X.; Peng, X.-H.; Zheng, B.-D.; Tang, J.; Zhao, Y.; Zheng, B.-Y.; Ke, M.-R.; Huang, J.-D. New application of phthalocyanine molecules: From photodynamic therapy to photothermal therapy by means of structural regulation rather than formation of aggregates. *Chem. Sci.* **2018**, *9*, 2098–2104. [CrossRef] [PubMed]
44. Li, X.; Kim, C.-Y.; Lee, S.; Lee, D.; Chung, H.-M.; Kim, G.; Heo, S.-H.; Kim, C.; Hong, K.-S.; Yoon, J. Nanostructured Phthalocyanine Assemblies with Protein-Driven Switchable Photoactivities for Biophotonic Imaging and Therapy. *J. Am. Chem. Soc.* **2017**, *139*, 10880–10886. [CrossRef]
45. Du, L.; Qin, H.; Ma, T.; Zhang, T.; Xing, D. In Vivo Imaging-Guided Photothermal/Photoacoustic Synergistic Therapy with Bioorthogonal Metabolic Glycoengineering-Activated Tumor Targeting Nanoparticles. *ACS Nano* **2017**, *11*, 8930–8943. [CrossRef] [PubMed]
46. Zelepukin, I.V.; Yaremenko, A.V.; Shipunova, V.O.; Babenyshev, A.V.; Balalaeva, I.V.; Nikitin, P.I.; Deyev, S.M.; Nikitin, M.P. Nanoparticle-based drug delivery via RBC-hitchhiking for the inhibition of lung metastases growth. *Nanoscale* **2019**, *11*, 1636–1646. [CrossRef] [PubMed]
47. Nikitin, M.P.; Zelepukin, I.V.; Shipunova, V.O.; Sokolov, I.L.; Deyev, S.M.; Nikitin, P.I. Enhancement of the blood-circulation time and performance of nanomedicines via the forced clearance of erythrocytes. *Nat. Biomed. Eng.* **2020**, *4*, 717–731. [CrossRef]
48. Zelepukin, I.V.; Mashkovich, E.A.; Lipey, N.A.; Popov, A.A.; Shipunova, V.O.; Griaznova, O.Y.; Deryabin, M.S.; Kurin, V.V.; Nikitin, P.I.; Kabashin, A.V.; et al. Direct photoacoustic measurement of silicon nanoparticle degradation promoted by a polymer coating. *Chem. Eng. J.* **2022**, *430*, 132860. [CrossRef]

Article

Synthesis and Biodistribution of ^{99m}Tc-Labeled PLGA Nanoparticles by Microfluidic Technique

Michela Varani *, Giuseppe Campagna, Valeria Bentivoglio, Matteo Serafinelli, Maria Luisa Martini, Filippo Galli and Alberto Signore

Nuclear Medicine Unit, Department of Medical-Surgical Sciences and of Translational Medicine, Faculty of Medicine and Psychology, "Sapienza" University of Rome, 00161 Rome, Italy; gius.campagna@gmail.com (G.C.); valeria.benti@gmail.com (V.B.); matteo.serafinelli@gmail.com (M.S.); luisetta84m@libero.it (M.L.M.); filippo.galli@uniroma1.it (F.G.); alberto.signore@uniroma1.it (A.S.)
* Correspondence: varanimichela@gmail.com; Tel.: +39-6-3377-5037

Abstract: The aim of present study was to develop radiolabeled NPs to overcome the limitations of fluorescence with theranostic potential. Synthesis of PLGA-NPs loaded with technetium-99m was based on a Dean-Vortex-Bifurcation Mixer (DVBM) using an innovative microfluidic technique with high batch-to-batch reproducibility and tailored-made size of NPs. Eighteen different formulations were tested and characterized for particle size, zeta potential, polydispersity index, labeling efficiency, and in vitro stability. Overall, physical characterization by dynamic light scattering (DLS) showed an increase in particle size after radiolabeling probably due to the incorporation of the isotope into the PLGA-NPs shell. NPs of 60 nm (obtained by 5:1 PVA:PLGA ratio and 15 mL/min TFR with ^{99m}Tc included in PVA) had high labeling efficiency (94.20 ± 5.83%) and >80% stability after 24 h and showed optimal biodistribution in BALB/c mice. In conclusion, we confirmed the possibility of radiolabeling NPs with ^{99m}Tc using the microfluidics and provide best formulation for tumor targeting studies.

Keywords: radiolabeled nanoparticles; poly (lactic-co-glycolic acid) (PLGA); nuclear medicine; microfluidics

Citation: Varani, M.; Campagna, G.; Bentivoglio, V.; Serafinelli, M.; Martini, M.L.; Galli, F.; Signore, A. Synthesis and Biodistribution of ^{99m}Tc-Labeled PLGA Nanoparticles by Microfluidic Technique. *Pharmaceutics* **2021**, *13*, 1769. https://doi.org/10.3390/pharmaceutics13111769

Academic Editor: Oya Tagit

Received: 1 October 2021
Accepted: 18 October 2021
Published: 22 October 2021

Publisher's Note: MDPI stays neutral with regard to jurisdictional claims in published maps and institutional affiliations.

1. Introduction

The design of targeted drug delivery systems using nanomaterials has rapidly spread in medicine, due to the possibility of improving the targeting and release of drugs from these nano-systems, avoiding their premature biodegradation, off-target and systemic toxicity [1]. The ideal delivery system should satisfy several requirements like biodegradability, biocompatibility, non-immunogenicity and non-toxicity in a biological system. Additionally, they should have the capability of a high load of drugs and administer them to the target with controlled release and distribution, with minimal losses and a prolonged release in the desired site [2,3]. Nanomaterials match these characteristics as they can be used to fabricate several types of nanoparticles (NPs), formulated with organic, inorganic, or hybrid core. The physico-chemical parameters of NPs such as size, shape, surface charge, and materials of the structure, are fundamental for their biodistribution, excretion, pharmacokinetics, targeting, and therefore therapeutic efficacy [4,5]. Polymeric NPs are composed of biodegradable and biocompatible polymers, are easily synthetized and can encapsulate or absorb surface insoluble molecules, therefore they have been extensively studied as vehicles for controlled drug delivery [6,7]. The polylactide-co-glycolic acid (PLGA) are polymeric NPs already approved by Food and Drug Administration (FDA) and European Medicine Agency (EMA) to deliver therapeutic agents parentally administered [8]. PLGA-NPs are naturally degraded in the body by hydrolysis of ester bonds between the lactic (PLA) and glycolic acid (PGA) monomers, metabolized by the Krebs cycle, and excreted by the lungs as carbon dioxide and water or by urine in a non-toxic way [9,10]. The retention of PLGA-NPs at the target site and also the effect of the entrapped drug depends on the composition of polymers. The PLA:PGA ratio influences the degradation kinetics of NPs since a high PLA monomer ratio causes hydrophobicity of PLGA-NPs and their degradation occurs more slowly [11]. PLGA with a ratio of 50:50 is the most used copolymer in nanomedicine, due to its fastest degradation rate and therefore faster drug release rate than other formulations [12]. Drug release can be achieved by passive diffusion from polymer barrier or by erosion of the PLGA-NPs structure [13]. When PLGA-NPs are synthetized, it is important to consider the physical and chemical properties of polymers, and also the complexity of the process to achieve the optimal drug delivery system [14]. In recent years, an innovative synthesis method has been explored, called 'microfluidic technique', that overcomes some limitations given by the bulk nanoprecipitation method [15,16]. In this method, an aqueous solution has been mixed with a solvent phase (as acetonitrile, acetone, or ethanol) and this causes the precipitation and nucleation of polymers, assembling in NPs [17]. The bulk-synthesis technique has been extensively studied for NPs production, but suffers from poor batch-to-batch reproducibility, long preparation times, and several steps of manipulation that can lead to NPs aggregation [18,19]. The microfluidic method offers the potential of a controlled system in which the organic and the aqueous phases are mixed in a microfluidic chip with precision settings, such as the total flow ratio (TFR) and the flow rate ratio (FRR) [20]. The TFR is the total speed in mL/min at which both the fluids are mixed in the microfluidic platform, and the FRR is the volumetric ratio of the mixed organic and aqueous phases. This leads to precise NPs size control and a high degree of particle uniformity (polydispersity index below 0.2), allowing batch-to-batch reproducibility. These are crucial factors for GMP production of radiolabeled PLGA-NPs to be used in clinical practice. In our study, we explored the possibility to produce radiolabeled PLGA-NPs with a novel microfluidic system based on a Dean-Vortex-Bifurcation Mixer (DVBM) [21]. Despite the wide use of PLGA-NPs as delivery drug system, the knowledge of a reproducible incorporation of imaging agents into PLGA-NPs is yet to be explored. In our previous study, we demonstrated how fluorescent PLGA-NPs were able to passively accumulate in a syngeneic tumor model of sarcoma, with maximum uptake at 72 h [22]. Then, the possibility to combine diagnostic agents with therapeutics molecules in NPs, makes them a promising theranostic tool. The aim of this study was to investigate a novel microfluidic process for manufacturing PLGA-NPs loaded with diagnostic isotope, as a necessary step for future theranostic application. The radioactive isotope allows a signal detection from deeper tissues, overcoming the limitation of fluorescence and enabling a translational potential of NPs as imaging probe. Therefore, formulation parameters (TFR and FRR) were set to achieve the best encapsulation efficiency of technetium-99m (^{99m}Tc).

2. Materials and Methods

2.1. Materials

Poly (lactic-co-glycolic acid) (lactide to glycolide ratio 50:50, acid endcap, Mw: 35.000–45.000 Da) was purchased by Akina, Inc. (West Lafayette, IN, USA). Polyvinyl alcohol (PVA Mw: 31,000 Da). Acetonitrile (ACN) (HPLC grade) and ultrapure water (HPLC grade) were supplied by Merck (Darmstadt, Germany). Technetium-99m was obtained by elution from ^{99}Mo/^{99m}Tc generator.

2.2. Synthesis of PLGA-NPs

An innovative microfluidic mixer (NanoAssemblr® Benchtop, Precision NanoSystems Inc., Vancouver, BC, Canada) was used to manufacture PLGA-NPs. The platform uses compatible syringes to inject solutions in two different channels on a chip incorporated in a disposable cartridge inserted into a microfluidic device controlled by a laptop. The chip is composed by a plurality of mixers, defined as Dean-Vortex-Bifurcation Mixer (DVBM). Each DVBM is characterized by a bifurcation with four toroidal elements arranged in series to mix the liquid provided [23]. Each DVBM in turn contains a plurality of toroidal mixing elements. The fluids under a constant pressure are pumped into the mixer and flow through curved channels. The increased speed of the flow in the center of the channel causes it to deflect outward due to the higher centripetal force. This flow is opposed to another inward flow at the top and the bottom of the channel, creating a pair of counter-rotating vortices, known as Dean vortex. DVBM has a fully planar geometry which reduces the possibility of material clogging or adsorption and also allows for higher flow rates on larger devices. The non-turbulent process condition ensures reproducible results with high NPs quality.

To synthetize PLGA-NPs, the polymeric materials (PLGA 50:50) were dissolved in acetonitrile (ACN) at a concentration of 5 mg/mL. PVA (2%, *w/v*), dissolved in ultrapure water, was used as aqueous phase. The solutions were loaded in two different and disposable syringes and injected into a separate inlet point on the mixer.

In order to evaluate the effect of the different instrument parameters, TFR 8, 12, and 15 mL/min and FRR 1:1, 3:1, and 5:1, were tested. Nine parallel batches were synthetized and analyzed (Table 1).

Table 1. Selected parameters for the formulation of PLGA-NPs using NanoAssemblr®. FRR = flow rate ratio; TFR = total flow rate.

Batch	PLGA (mg/mL)	FRR (Aqueous:Organic)	TFR (mL/min)
#1	5	1:1	8
#2	5	3:1	8
#3	5	5:1	8
#4	5	1:1	12
#5	5	3:1	12
#6	5	5:1	12
#7	5	1:1	15
#8	5	3:1	15
#9	5	5:1	15

2.3. Synthesis of ^{99m}Tc-PLGA-NPs

In order to find the best radiolabeling approach, two different approaches were tested for the microfluidic syntheses of ^{99m}Tc-radiolabeled PLGA-NPs (Figure S1).

In the first one, 100 μL of ^{99m}TcO$_4$$^-$ (NaCl) were added dropwise to 1 mL of PLGA polymers (5 mg/mL ACN) and injected into the organic inlet of the microfluidic mixer. PVA (2%, *w/v*) dissolved in ultrapure water was injected through the other inlet of the microfluidic mixer. We called these batches ^{99m}Tc-polymer-(PLGA-NPs).

In the second one, 100 μL of ^{99m}TcO$_4$$^-$ (NaCl) were added dropwise to 2 mL of the aqueous phase (PVA, 2%) and injected into the appropriate inlet of the microfluidic mixer. PLGA polymers (5 mg/mL ACN) were injected into the other inlet. We called these formulations ^{99m}Tc-PVA-(PLGA-NPs).

For both the radiolabeling approaches, the different instrument parameters described in Table 1 were tested, thus obtaining a total of 18 different formulations.

To radiolabel each batch with the same activity of ^{99m}TcO$_4$$^-$ (5 mCi), we used different starting concentrations of isotope, considering the different FFR, as shown in Table 2.

Table 2. Effective incorporation of ^{99m}TcO$_4$$^-$ based on the amount of aqueous and solvent phase used for the synthesis of ^{99m}Tc-PLGA-NPs.

FRR (Aqueous:Organic)	PVA (mL)	PLGA Polymers (mL)	^{99m}TcO$_4$$^-$ in PVA (mCi/mL)	^{99m}TcO$_4$$^-$ in PLGA Polymers (mCi/mL)
1:1	1	1	5	-
3:1	1.5	0.5	3.3	-
5:1	1.67	0.33	2.99	-
1:1	1	1	-	5
3:1	1.5	0.5	-	10
5:1	1.67	0.33	-	15.15

2.4. Quality Controls

2.4.1. Particle Size Distribution and Zeta Potential (ζ) Measurements

Volume-average diameter (nm) and polydispersity index (PDI) of NPs were analyzed by dynamic light scattering (DLS) using a NanoZetaSizer analyzer (Malvern Instruments Ltd., Malvern, UK). After the batch synthesis with NanoAssemblr®, native PLGA-NPs, ^{99m}Tc-polymer-(PLGA-NPs), and ^{99m}Tc-PVA-(PLGA-NPs) were loaded into the instrument and analyzed. All the batches were analyzed pre- and post-PD-10 purification step.

Briefly, 10 µL of each sample were suspended with 90 µL of ultrapure water and loaded in Sarstedt polystyrol/polystyrene cuvettes (10 × 10 × 45 mm) for the measurements performed at 25 °C. To study the surface charge of NPs, the zeta potential analysis was performed with the same instrument. In total, 20 µL of NPs were suspended with 980 µL of ultrapure water, sonicated, and loaded in Malvern folded capillary cells for measurements. All measurements were performed in triplicate and mean values ± standard deviation (SD) are reported.

2.4.2. Labeling Efficiency and Yield Calculation of ^{99m}Tc-PLGA-NPs

The ITLC was used to evaluate the presence of unlabeled ^{99m}Tc in the preparation (free pertechnetate ^{99m}TcO$_4{}^-$) with sodium chloride (NaCl 0.9%) as mobile phase. The free pertechnetate present in the solution migrates to the front of the strip (Retention Factor = 0.9), while the radiolabeled compound remained at the bottom (Retention Factor = 0.1). ITLC strips were spotted with 2–3 µL of the solution at 1 cm above the bottom, and immediately put into a falcon vial with the mobile phase. After migration of the mobile phase up to 1 cm from the top, the strip was dried and analyzed in a Radio-TLC Imaging Scanner (Bioscan, Inc, Poway, CA, USA). Labeling efficiency was calculated as percentage of NPs activity over total.

After the synthesis, a PD-10 desalting column containing Sephadex G-25 resin (GE Healthcare, Uppsala, Sweden) was used to purify the radiolabeled NPs from the unlabeled technetium present in solution. Briefly, 1 mL of ^{99m}Tc-PLGA-NPs was added to column, then 1.5 mL of PBS were added to completely fill the column volume and the sample eluted was discarded. In total, 10 mL of PBS were eluted and collected in 20 fractions (500 µL each). The acetonitrile and the other small contaminants in solution were removed in the same purification step.

Each fraction was counted in a single-well NaI γ-counter (AtomLab, 500-Biodex) and the yield was calculated as follows:

$$\text{YIELD (\%)} = 100 \times [\text{mCi}_{(^{99m}\text{Tc-PLGA-NPs})}/\text{mCi}_{(\text{tot})}]$$

where $\text{mCi}_{(\text{tot})}$ is the starting activity used (5 mCi), while $\text{mCi}_{(^{99m}\text{Tc-PLGA-NPs})}$ is the amount of mCi from the first 5 fractions in which the labeled NPs were eluted.

2.4.3. "In Vitro" Release Study

^{99m}Tc-polymers (PLGA-NPs) and ^{99m}Tc-PVA (PLGA-NPs) synthetized with the formulation parameters of batch #9 were selected for further experimentations.

In vitro release of technetium-99 from selected radiolabeled NPs was evaluated in NaCl 0.9%. The radiolabeled compound was incubated at room temperature for 1, 3, 6, and 24 h. At the end of each time point, each batch was analyzed by ITLC as previously described.

Size and PDI of labeled NPs were also measured by DLS over time, being important indicators of NPs aggregation.

2.5. In Vivo Biodistribution Studies

All animal experiments were carried out in compliance with the local ethics committee and in agreement with the National rules and the EU regulation (Study 204/2018-PR). Biodistribution of ^{99m}Tc-PLGA-NPs was studied in 18 normal 8-week-old female BALB/c purchased from Envigo. Mice were divided in two groups of nine mice each. Group 1 was injected ^{99m}Tc-polymers (PLGA-NPs) batch #9 formulation, group 2 was injected ^{99m}Tc-PVA (PLGA-NPs) batch #9 formulation. The 100 µCi (100 µL) of radiolabeled nanoparticles were injected into the lateral tail vein of each animal. After each time point (1, 6, 24 h) three mice per group were sacrificed; blood samples and major organs (small and large bowel, kidneys, spleen, stomach, liver, muscle, bone, lungs, heart) were collected and weighted. The radioactivity from each vial was counted in a single-well γ-counter (PerkinElmer, Waltham, MA, USA). The percentage of injected dose per organ (%ID) and percentage of injected dose per gram (%ID/g) were calculated. ^{99m}Tc standards were prepared by dilution method and appropriate decay corrections were applied to all the samples.

3. Statistical Analysis

Continuous variables are shown as mean ± SD (standard deviation). The differences between native vs. ^{99m}Tc-polymers (PLGA-NPs) vs. ^{99m}Tc-PVA (PLGA-NPs) of the continuous variables were tested by GLM (General Linear Model) when the normality of the residuals was verified or otherwise Kruskal–Wallis. The normality of the residuals was evaluated by Shapiro–Wilk test. Post-hoc analysis was performed by Tukey method when the homoscedasticity (homogeneity of the variance) was verified or Games–Howell test in presence of heteroscedasticity. Homoscedasticity was verified by check of the box-plots relatively to three groups and also analyzing the studentized residuals.

Analysis of variables pre and post-purification of each group ((native, ^{99m}Tc-polymers (PLGA-NPs) and ^{99m}Tc-PVA (PLGA-NPs)) was performed by paired t test and the normality of the differences pre–post was evaluated by Shapiro–Wilk test. Wilcoxon test was used when the normality of the differences failed.

The stability over time (baseline vs. 1 h vs. 3 h vs. 6 h vs. 24 h) was analyzed by GLIMMIX (Generalized Linear Mixed Model) with Gaussian distribution and identity link for repeated measures.

The analysis of interaction between the groups (^{99m}Tc-PVA (PLGA-NPs) and ^{99m}Tc-polymers (PLGA-NPs)) and time (group*time) relative to organs was tested by GLIMMIX (Generalized Linear Mixed Model) with Gaussian distribution and identity link.

The multiple comparisons were corrected by Benjamini–Hochberg (FDR) method. All analyses were performed by SAS v. 9.4 and JMP PRO v. 16 (SAS Institute Inc., Cary, NC, USA). A p-value < 0.05 was considered detectable.

4. Results

4.1. Particle Size Distribution and Zeta Potential (ζ) Measurements

To assess the effects of different microfluidic parameters (batch #1 to #9) on NPs size, charge and PDI, we first synthetized and analyzed native PLGA-NPs. DLS results confirmed that particle size is affected by both TFR and FRR as already published by others [24]. Indeed, by increasing TFR of the two phases (from 8 to 15 mL/min) and by varying FRR from 1:1 to 5:1 (aqueous:organic phase), we detected a reduction in size, especially for radiolabeled NPs (Figure 1, Table S1). The uniformity on NPs size distribution was confirmed by PDI values, ranging from 0.176 to 0.273 (Table S2). All batches had a negative surface charge (ZP) ranging from −11.62 to −24.14 mV (Table S3). A purification step, using PD-10 desalting column, was also performed for native PLGA-NPs to remove the ACN present in solution after synthesis. After PD10 purification we observed a considerable difference in size for all tested batches, except for batch #9 (Table S4).

In particular, the formulation with an FRR of 1:1 and 3:1 at different TFRs (8, 12, or 15 mL/min) showed the greatest increase in size. By contrast, all formulations with FRR of 5:1 and TFR of 8, 12, or 15 mL/min, had a small increase in size. (Figure 1).

PDI values after purification showed no detectable differences (Table S5).

Zeta potential values of native NPs showed a detectable decrease after purification only for batch #1 (Table S6).

The same formulations of native NPs were used to synthetize radiolabeled NPs.

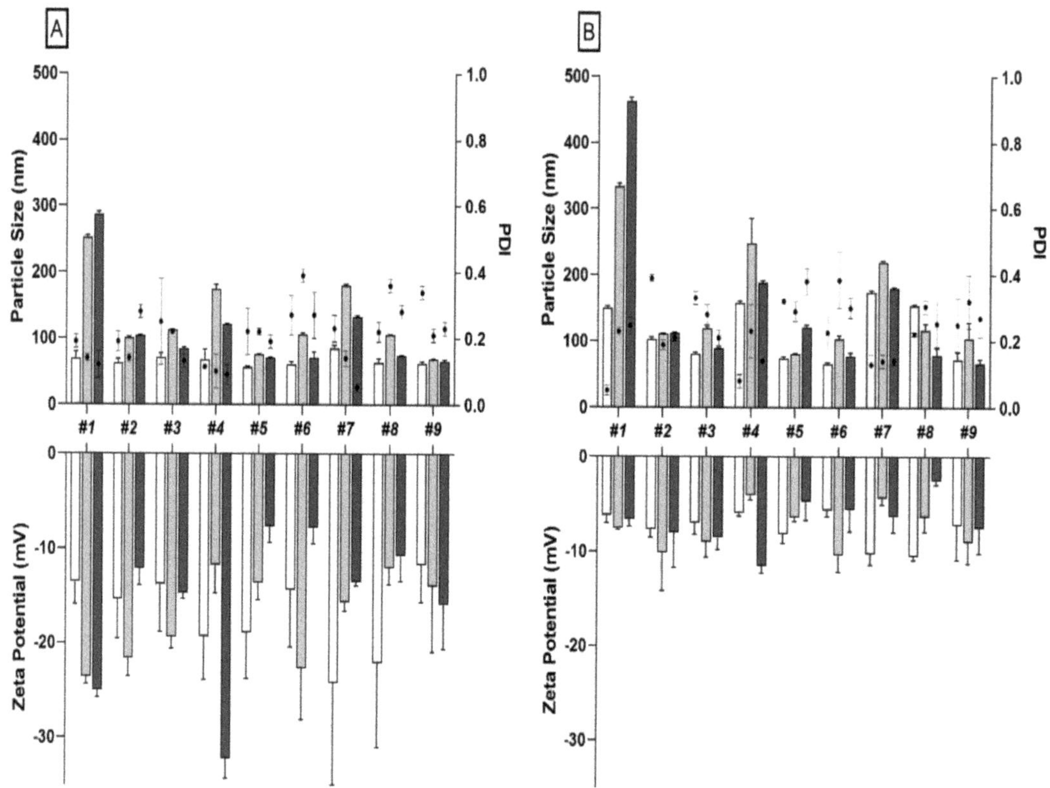

Figure 1. (**A**) Particle size distribution and zeta potential measurements (bars) and PDI (dot) for each formulation of native (white bars), ^{99m}Tc-PVA (PLGA-NPs) (grey bars), and ^{99m}Tc-polymers (PLGA-NPs) (black bars) pre-PD10 purification. (**B**) Particle size distribution and zeta potential measurements (bars) and PDI (dot) for each formulation of native (white bars), ^{99m}Tc-PVA (PLGA-NPs) (grey bars) and ^{99m}Tc-polymers (PLGA-NPs) (black bars) post-PD10 purification.

Independently from the radiolabeling method (^{99m}Tc-PVA- or ^{99m}Tc-polymers- PLGA-NPs), all formulations showed detectable increase in size after labeling as compared to the respective native formulation (Figure 1, Table S1). ^{99m}Tc-PVA-PLGA-NPs of batch #9 did not show a relivable increase in size after radiolabeling, retaining small size: p 63.11 ± 2.53 nm vs. 65.80 ± 2.28 nm, p = 0.18 (Table S1).

Similarly to native NPs, labeled NPs showed an increase in size after PD10 purification. In particular, relivable difference was shown as follows: for batches #1, #2, #4, #5, #7, and #9 of ^{99m}Tc-polymers (PLGA-NPs) with p = 0.009, p = 0.006, p = 0.03, p = 0.006, p = 0.003, p = 0.01, respectively; for batches #1, #2, #3, #4, #5, #6, and #7 of ^{99m}Tc-PVA (PLGA-NPs) with p = 0.0007, p = 0.008, p = 0.02, p = 0.0009, p = 0.001, p = 0.02, p = 0.0006, respectively (Figure 1, Table S4).

The smallest NPs size was obtained by labeling batches #5, #6, #8, and #9 with ^{99m}Tc dissolved in aqueous phase (Table S1).

Zeta potential decreased after purification, with values relivable only for batches #1, #3, #4, #5, and #7 of ^{99m}Tc-polymers (PLGA-NPs) (Table S6).

PDI did not show detectable modifications after PD10 purification for batches #3, #6, #7, and #8 of ^{99m}Tc-polymers (PLGA-NPs) and for batches #2, #4, #6, #7, #8, and #9 of ^{99m}Tc-PVA (PLGA-NPs) (Table S5).

4.2. "In Vitro" Studies

ITLC results showed a high LE for all batches ranging from 73 to 99% for ^{99m}Tc-polymers-(PLGA-NPs) and from 85 to 99% for ^{99m}Tc-PVA-(PLGA-NPs), as shown in Table 3. Each batch was then purified with a PD-10 desalting columns to remove unlabeled technetium and other small contaminants, obtaining a final LE of 100%. Radiolabeled yield after purification ranged from 11 to 34% for ^{99m}Tc-polymers (PLGA-NPs), and from 19 to 41% for ^{99m}Tc-PVA (PLGA-NPs).

Table 3. Encapsulation efficiency (%) and yield (%) of ^{99m}Tc-PVA (PLGA-NPs) and ^{99m}Tc-polymers (PLGA-NPs). Results represent triplicate measurements.

Batch	^{99m}Tc-Polymers (PLGA-NPs)		^{99m}Tc-PVA (PLGA-NPs)	
N°	LE Mean ± SD	YIELD Mean ± SD	LE Mean ± SD	YIELD Mean ± SD
#1	96.87 ± 1.74	26.70 ± 0.14	98.25 ± 2.47	19.06 ± 0.72
#2	99.25 ± 1.06	34.05 ± 0.19	99.20 ± 1.13	27.50 ± 0.58
#3	98.83 ± 0.40	22.54 ± 0.26	99.45 ± 0.78	31.60 ± 0.36
#4	95.50 ± 1.20	22.12 ± 0.17	99.61 ± 0.56	32.93 ± 0.32
#5	85.09 ± 10.96	11.60 ± 0.88	85.28 ± 17.23	29.82 ± 0.18
#6	73.15 ± 26.90	14.78 ± 0.17	92.57 ± 10.51	41.33 ± 0.07
#7	95.70 ± 1.48	19.10 ± 0.21	95.10 ± 2.06	31.73 ± 0.21
#8	95.11 ± 1.92	13.8 ± 0.10	85.83 ± 5.25	20.71 ± 0.27
#9	88.53 ± 1.43	10.92 ± 0.15	94.20 ± 5.83	33.69 ± 0.22

ITLC results showed a high LE of both radiolabeled PLGA-NPs up to 6 h, with a slight decrease at 24 h (Table 4).

Table 4. Labeling efficiency calculation (%) at different time points (1, 3, 6, and 24 h) to evaluate the technetium-99m release in 0.9% NaCl from ^{99m}Tc-PVA (PLGA-NPs) and ^{99m}Tc-polymers (PLGA-NPs) batch #9. Results represent triplicate measurements.

^{99m}Tc PLGA-NPs	1 h Mean ± SD	3 h Mean ± SD	6 h Mean ± SD	24 h Mean ± SD
99mTc-polymers	98.34 ± 1.79	97.97 ± 1.82	97.92 ± 1.86	84.34 ± 7.57
99mTc-PVA	97.17 ± 3.66	97.92 ± 2.38	97.13 ± 2.65	84.21 ± 8.22

DLS results showed no significant differences in size and zeta potential over time for both ^{99m}Tc-PVA (PLGA-NPs) and ^{99m}Tc-polymers (PLGA-NPs) (Figure 2).

4.3. In Vivo Biodistribution Studies

Due to their small size, high LE%, and radiochemical stability, ^{99m}Tc-PVA (PLGA-NPs) and ^{99m}Tc-polymers (PLGA-NPs) from batch #9 were selected for biodistribution studies in BALB/c mice.

As shown in Figure 3, single organ counting at 1 h post-injection (p.i.) indicated a rapid clearance of radiolabeled PLGA-NPs from the bloodstream and their accumulation in liver and kidneys. In particular, at 1 h p.i. the highest uptake was detected in the liver with values of 2.07 ± 0.03 and 1.87 ± 0.43%ID per organ, for ^{99m}Tc-PVA (PLGA-NPs) and ^{99m}Tc-polymers (PLGA-NPs), respectively. The kidneys showed the highest dose per gram of tissue at 1 h p.i.: 3.13 ± 0.65 and 3.64 ± 0.32%ID/g, for ^{99m}Tc-PVA (PLGA-NPs) and ^{99m}Tc-polymers (PLGA-NPs), respectively. Nevertheless, kidney kinetics was fast and radioactivity almost disappeared after 24 h for both formulations ((1 h vs. 24 h), $p < 0.0001$ for ^{99m}Tc-polymers (PLGA-NPs), and (1 h vs. 24 h) $p = 0.0005$ for ^{99m}Tc-PVA (PLGA-NPs)).

Overall, ^{99m}Tc-polymers (PLGA-NPs) showed higher uptake in the liver than ^{99m}Tc-PVA (PLGA-NPs) with a relevant difference at 6 h ((^{99m}Tc-polymers (PLGA-NPs) vs.

^{99m}Tc-PVA (PLGA-NPs), %ID/g $p = 0.006$ and %ID per organ $p = 0.01$)) and with significant organ retention up to 24 h (Figure 3).

Tissues or organs such as muscle, bones, large bowel, and heart showed low or negligible uptake of both radiolabeled NPs.

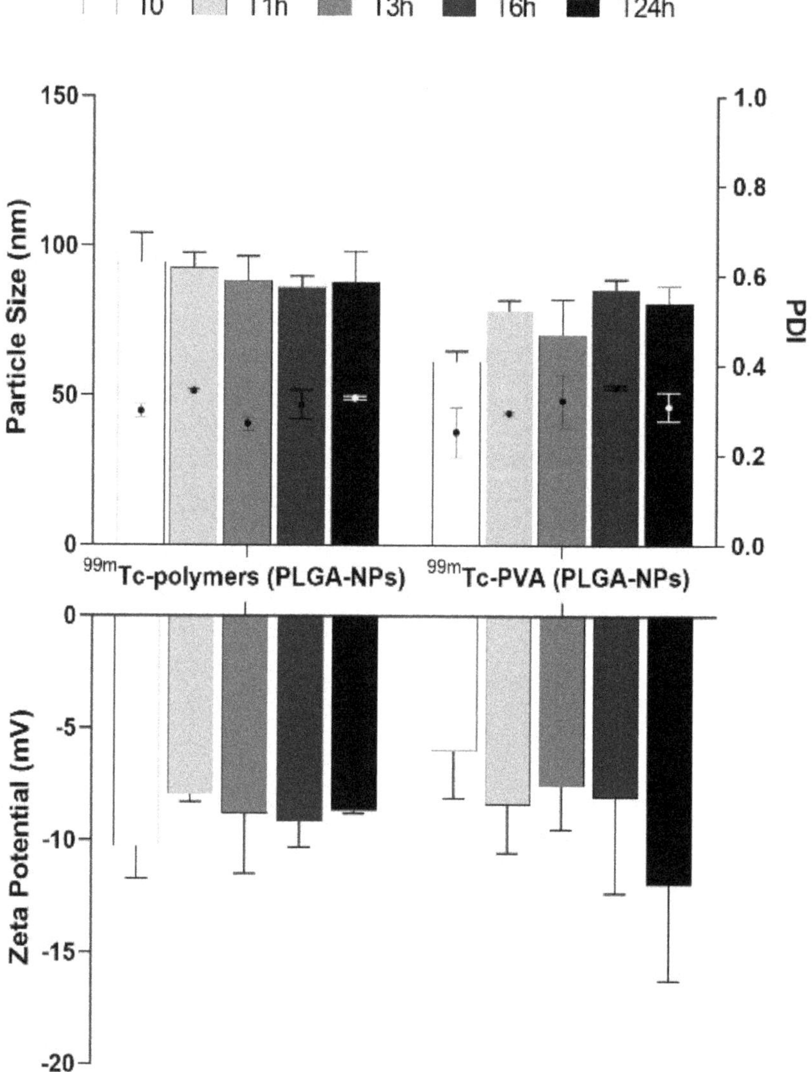

Figure 2. Particle size distribution and zeta potential measurements (bars) and PDI (dot) of batch #9 for both ^{99m}Tc-PVA (PLGA-NPs) and ^{99m}Tc-polymers (PLGA-NPs) at different time points (1, 3, 6, 24 h).

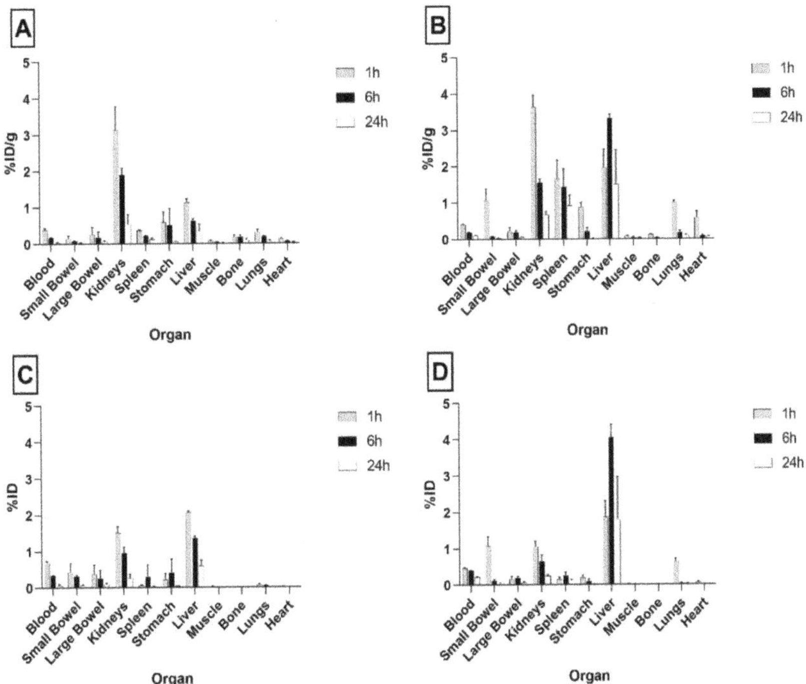

Figure 3. Biodistribution of ^{99m}Tc-PVA (PLGA-NPs) (**A,C**) and ^{99m}Tc-polymers (PLGA-NPs) (**B,D**) batch #9 in BALB/c mice. Data are ex vivo counts of each organ at different time points (1, 6, 24 h). The measured activity is expressed as %ID/organ (mean ± SD) (**A,B**) and %ID (mean ± SD) (**C,D**) detected from three different mice per time point.

5. Discussion

Physicochemical characteristics of NPs such as size and PDI have an important impact for their in vivo distribution and, therefore, for their application. In the process of NPs synthesis, obtaining reproducible results is a major challenge [25]. The innovative microfluidic technique allows to set different parameters to have dimensional uniformity and low polydispersity of samples [26]. PDI value is important as an indicator of different particles size in solution, that could have different pharmacokinetics profiles in vivo. Values below 0.2 in PDI indicate a high monodisperse solution [27].

Zeta potential depends on the composition of the polymer and indicates the stability of the solution. Values close to zero indicate a low repulsion between the NPs which can lead to the formation of colloids [28]. Stolnik et al. reported as zeta potential of PLGA-NPs, in neutral buffer without the presence of PVA in the aqueous phase, a value of about −45 mV, due to the uncapped end carboxyl groups in the particle surface [29]. The use of PVA 2% in aqueous phase, during the synthesis process, reduces the zeta potential of PLGA-NPs. Furthermore, by increasing PVA concentration in solution, the zeta potential becomes less negative, with possible increase of sample instability and NP aggregation over time [30,31]. However, in our experiments, we did not observe NP aggregation as zeta potential remained between −7 and −12 mV. Indeed, PVA being a non-ionic surfactant may promote emulsification of solutions, avoiding aggregation of PLGA-NPs.

Several studies have demonstrated that the innovative microfluidic technique allows to encapsulate drugs or small-RNA during the synthesis of PLGA-NPs with reproducible results [32]. Chiesa et al. showed that a small hydrophilic drug can be easily encapsulated in PLGA-NPs with the novel microfluidic-based device. They demonstrated that settings of 15 mL/min TFR and a 1:5 FRR, provided the best encapsulation efficiency over 65% [33]. Furthermore, they showed that using the bulk mixing nanoprecipitation method there was a high drug release from NPs (90%) in the first 60 min, compared to 20% of release with microfluidic method. This is important to explain how the microfluidic technique allows a drug or a radioisotope to be encapsulated inside the NPs, avoiding the rapid drug release that occurs when they are absorbed on the surface of NPs. A similar approach was used to encapsulate a hydrophobic drug on PLGA-NPs by Garg et al. They tested different TFR parameters (2–12 mL/min) to reach the best drug encapsulation efficiency. A TFR of 12 mL/min and FRR of 1:1 showed the maximum encapsulation of 75% [34].

In our study, we found the formulation parameters of batch #9 (TFR 15 mL/min and FRR 5:1) as the best formulation in terms of LE, labeling yield and size, before and after PD10 purification.

To perform the encapsulation, the drug can be dissolved in the organic or in aqueous phase. For this reason, in our study we explored two different approaches of encapsulation: dissolving the $^{99m}TcO_4^-$ in ACN or in PVA aqueous phase. The advantage of incorporating a radioisotope into NPs is to avoid its direct exposition to biological molecules in vivo, that could lead to radiochemical instability. Furthermore, the encapsulation of $^{99m}TcO_4^-$ allows to avoid the use of reducing agents (as $SnCl_2$), and consequently the possible formation of radiocolloids [35]. The low release of $^{99m}TcO_4^-$ from NPs up to 6 h also confirmed the encapsulation of the radioisotope on NPs, avoiding the instability that occurs when it is absorbed on NPs surface.

It has been previously demonstrated that, in vivo, increased NPs size correlates with increased recognition by the mononuclear phagocytic system (MPS), like Kupffer cells in the liver, red-pulp macrophages in the spleen and alveolar macrophages in lungs [36].

Indeed, both formulations analyzed in our in vivo study showed some uptake in liver, spleen, and lungs, as primary organs of the reticuloendothelial system (Figure 3). ^{99m}Tc-polymers (PLGA-NPs) with a colloidal size greater than ^{99m}Tc-PVA-(PLGA-NPs), showed much higher retention in these organs, particularly in liver (Figure 3A,B).

Of particular note is the kidney uptake of radiolabeled NPs.

Considering that the effective threshold of the glomerulus filtration apparatus for circulating particles is approximately 10 nm, the kidney uptake can be explained by the size of NPs [37]. Indeed, the glomerulus filtration apparatus is composed by three layers with different size threshold. The first barrier is composed by the fenestrated endothelial cells with a size cut-off of ~100 nm. Then, NPs with size below 100 nm are able to access the endothelial fenestration, reaching the mesangium [38]. Choi et al. demonstrated that NPs with a size between 20 and 100 nm passing through the endothelial pores accumulate in the mesangium in different percentages depending on their size. In particular, it has been shown that NPs with size ~50 nm (hydrodynamic size in water ~70 nm) have the highest accumulation in the mesangium due to phagocytosis of mesangial cells [39]. After digestion and degradation $^{99m}TcO_4^-$ is released in the urine with sharp reduction of renal activity.

In our study, both ^{99m}Tc-PVA (PLGA-NPs) and ^{99m}Tc-polymers (PLGA-NPs) showed some kidney uptake (Figure 3C,D). Nevertheless, ^{99m}Tc-polymers (PLGA-NPs) did accumulate much more in other RES tissues such as the liver (Figure 3D).

The low accumulation of radioisotope in the stomach indicated a low release of isotope from NPs, in that free $^{99m}TcO_4^-$ normally accumulated in this organ [40].

More experiments are now being constructed using several formulations of NPs and different tumor models to investigate the role of NP size in tumor targeting.

6. Conclusions

In this study, we selected an efficient radiolabeling procedure for PLGA-NPs using an innovative microfluidic method. The ability to load the isotope into PLGA-NPs in a single step, with high yield, allows PLGA-NPs to be considered an attractive candidate for molecular imaging and therapy. Further studies are necessary to explore the tumor binding and retention capacity of these NPs.

Supplementary Materials: The following are available online at https://www.mdpi.com/article/10.3390/pharmaceutics13111769/s1, Figure S1: Schematic illustration of two different radiolabeling approaches during ^{99m}Tc-NPs synthesis via microfluidic technique method using a Dean-Vortex-Bifurcation Mixer (DVBM) cartridge, Table S1: Particle size distribution (nm) of native-, ^{99m}Tc-PVA-, and ^{99m}Tc-polymers -(PLGA-NPs) pre-PD10 purification, Table S2: PDI values of native-, ^{99m}Tc-PVA-, and ^{99m}Tc-polymers -(PLGA-NPs) pre-PD10 purification, Table S3: Zeta potential measurements (mV) of native-, ^{99m}Tc-PVA-, and ^{99m}Tc-polymers -(PLGA-NPs) pre-PD10 purification, Table S4: Particle size distribution (nm) of native-, ^{99m}Tc-PVA-, and ^{99m}Tc-polymers -(PLGA-NPs) post-PD10 purification, Table S5: PDI values of native-, ^{99m}Tc-PVA-, and ^{99m}Tc-polymers -(PLGA-NPs) post-PD10 purification, Table S6: Zeta potential measurements (mV) of native-, ^{99m}Tc-PVA-, and ^{99m}Tc-polymers (PLGA-NPs) post-PD10 purification.

Author Contributions: Conceptualization, M.V. and A.S.; methodology, F.G., M.V. and A.S.; software, M.L.M.; validation, A.S.; investigation, M.V., M.S. and V.B.; resources, A.S. and F.G.; data curation and data analyzing, G.C.; writing—original draft preparation, M.V.; writing—review and editing, A.S. and F.G., funding acquisition, A.S. All authors have read and agreed to the published version of the manuscript.

Funding: This research was funded by "Associazione Italiana Ricerca Cancro" (AIRC), grant number IG-Grant 20411 and "Sapienza" University of Rome.

Institutional Review Board Statement: All animal experiments were carried out in compliance with the local ethics committee and in agreement with the National rules and the EU regulation (Study 204/2018-PR).

Informed Consent Statement: Not applicable.

Data Availability Statement: The data presented in this study are available within this article. Any other data is available on request from the corresponding author.

Acknowledgments: We would like to acknowledge Nuclear Medicine Discovery (Nu.Me.D.) and the Animal Facility of University Tor Vergata of Rome for providing support during animal studies.

Conflicts of Interest: The authors declare no conflict of interest.

References

1. Yohan, D.; Chithrani, B.D. Applications of nanoparticles in nanomedicine. *J. Biomed. Nanotechnol.* **2014**, *10*, 2371–2392. [CrossRef]
2. Bi, Y.; Hao, F.; Yan, G.; Teng, L.; Lee, R.J.; Xie, J. Actively targeted nanoparticles for drug delivery to tumor. *Curr. Drug Metab.* **2016**, *17*, 763–782. [CrossRef] [PubMed]
3. Yetisgin, A.A.; Cetinel, S.; Zuvin, M.; Kosar, A.; Kutlu, O. Therapeutic nanoparticles and their targeted delivery applications. *Molecules* **2020**, *25*, 2193. [CrossRef]
4. Ernsting, M.J.; Murakami, M.; Roy, A.; Li, S.D. Factors controlling the pharmacokinetics, biodistribution and intratumoral penetration of nanoparticles. *J. Control. Release* **2013**, *172*, 782–794. [CrossRef] [PubMed]
5. Longmire, M.; Choyke, P.L.; Kobayashi, H. Clearance properties of nano-sized particles and molecules as imaging agents: Considerations and caveats. *Nanomedicine* **2008**, *3*, 703–717. [CrossRef] [PubMed]
6. Dinarvand, R.; Sepehri, N.; Manoochehri, S.; Rouhani, H.; Atyabi, F. Polylactide-co-glycolide nanoparticles for controlled delivery of anticancer agents. *Int. J. Nanomed.* **2011**, *6*, 877–895. [CrossRef] [PubMed]
7. Prabhu, R.H.; Patravale, V.B.; Joshi, M.D. Polymeric nanoparticles for targeted treatment in oncology: Current insights. *Int. J. Nanomed.* **2015**, *2*, 1001–1018. [CrossRef]
8. Bobo, D.; Robinson, K.; Islam, J.; Thurecht, J.K.; Corrie, S.R. Nanoparticle-based medicines: A review of FDA-approved materials and clinical trials to date. *Pharm. Res.* **2016**, *33*, 2373–2387. [CrossRef]
9. Rezvantalab, S.; Drude, N.I.; Moraveji, M.K.; Güvener, N.; Koons, E.K.; Shi, Y.; Lammers, T.; Kiessling, F. PLGA-Based Nanoparticles in Cancer Treatment. *Front. Pharmacol.* **2018**, *9*, 1260. [CrossRef]
10. Semete, B.; Booysen, L.; Lemmer, Y.; Kalombo, L.; Katata, L.; Verschoor, J.; Swai, H.S. In vivo evaluation of the biodistribution and safety of PLGA nanoparticles as drug delivery systems. *Nanomedicine* **2010**, *6*, 662–671. [CrossRef] [PubMed]
11. Keles, H.; Naylor, A.; Clegg, F.; Sammon, C. Investigation of factors influencing the hydrolytic degradation of single PLGA microparticles. *Polym. Degrad. Stab.* **2015**, *119*, 228–241. [CrossRef]
12. Swider, E.; Koshkina, O.; Tel, J.; Cruz, L.J.; de Vries, I.J.M.; Srinivas, M. Customizing poly(lactic-co-glycolic acid) particles for biomedical applications. *Acta Biomater.* **2018**, *73*, 38–51. [CrossRef] [PubMed]

13. Houchin, M.L.; Topp, E.M.J. Review Chemical degradation of peptides and proteins in PLGA: A review of reactions and mechanisms. *J. Pharm. Sci.* **2008**, *97*, 2395–2404. [CrossRef]

14. Morikawa, Y.; Tagami, T.; Hoshikawa, A.; Ozeki, T. The use of an efficient microfluidic mixing system for generating stabilized polymeric nanoparticles for controlled drug release. *Biol. Pharm. Bull.* **2018**, *41*, 899–907. [CrossRef] [PubMed]

15. Bally, F.; Garg, D.K.; Serra, C.A.; Hoarau, Y.; Anton, N.; Brochon, C.; Parida, D.; Vandamme, T.; Hadziioannou, G. Improved size-tunable preparation of polymeric nanoparticles by microfluidic nanoprecipitation. *Polymer* **2012**, *53*, 5045–5051. [CrossRef]

16. Capretto, L.; Carugo, D.; Mazzitelli, S.; Nastruzzi, C.; Zhang, X. Microfluidic and lab-on-a-chip preparation routes for organic nanoparticles and vesicular systems for nanomedicine applications. *Adv. Drug. Deliv. Rev.* **2013**, *65*, 1496–1532. [CrossRef]

17. Schubert, S.; Delaney, J.T.; Schubert, U.S. Nanoprecipitation and nanoformulation of polymers: From history to powerful possibilities beyond poly (lactic acid). *Soft Matter* **2011**, *7*, 1581–1588. [CrossRef]

18. Donno, R.; Gennari, A.; Lallana, E.; De La Rosa, J.M.R.; d'Arcy, R.; Treacher, K.; Hill, K.; Ashford, M.; Tirelli, N. Nanomanufacturing through microfluidic-assisted nanoprecipitation: Advanced analytics and structure-activity relationships. *Int. J. Pharm.* **2017**, *534*, 97–107. [CrossRef]

19. Streck, S.; Neumann, H.; Nielsen, H.M.; Rades, T.; McDowell, A. Comparison of bulk and microfluidics methods for the formulation of poly-lactic-co-glycolic acid (PLGA) nanoparticles modified with cell-penetrating peptides of different architectures. *Int. J. Pharm. X* **2019**, *1*, 100030. [CrossRef]

20. Duncanson, W.J.; Lin, T.; Abate, A.R.; Seiffert, S.; Shah, R.K.; Weitz, D.A. Microfluidic synthesis of advanced microparticles for encapsulation and controlled release. *Lab. Chip.* **2012**, *12*, 2135–2145. [CrossRef]

21. Howell, P.B., Jr.; Mott, D.R.; Golden, J.P.; Ligler, F.S. Design and evaluation of a Dean vortex-based micromixer. *Lab. Chip.* **2004**, *4*, 663–669. [CrossRef]

22. Varani, M.; Galli, F.; Capriotti, G.; Mattei, M.; Cicconi, R.; Campagna, G.; Panzuto, F.; Signore, A. Theranostic Designed Near-Infrared Fluorescent Poly (Lactic-co-Glycolic Acid) Nanoparticles and Preliminary Studies with Functionalized VEGF-Nanoparticles. *J. Clin. Med.* **2020**, *9*, 1750. [CrossRef] [PubMed]

23. Chen, J.J.; Chen, C.H.; Shie, S.R. Optimal designs of staggered dean vortex micromixers. *Int. J. Mol. Sci.* **2011**, *12*, 3500–3524. [CrossRef] [PubMed]

24. Chiesa, E.; Greco, A.; Dorati, R.; Conti, B.; Bruni, G.; Lamprou, D.; Genta, I. Microfluidic-assisted synthesis of multifunctional iodinated contrast agent polymeric nanoplatforms. *Int. J. Pharm.* **2021**, *599*, 120447. [CrossRef]

25. Astete, C.E.; Sabliov, C.M. Synthesis and characterization of PLGA nanoparticles. *J. Biomater. Sci. Polym. Ed.* **2006**, *17*, 247–289. [CrossRef]

26. Gañán-Calvo, A.M.; Gordillo, J.M. Perfectly monodisperse microbubbling by capillary flow focusing. *Phys. Rev. Lett.* **2001**, *87*, 274501. [CrossRef] [PubMed]

27. Danaei, M.; Dehghankhold, M.; Ataei, S.; Davarani, F.H.; Javanmard, R.; Dokhani, A.; Khorasani, S.; Mozafari, M.R. Impact of Particle Size and Polydispersity Index on the Clinical Applications of Lipidic Nanocarrier Systems. *Pharmaceutics* **2018**, *10*, 57. [CrossRef]

28. Jain, A.K.; Thareja, S. In vitro and in vivo characterization of pharmaceutical nanocarriers used for drug delivery. *Artif. Cells Nanomed. Biotechnol.* **2019**, *47*, 524–539. [CrossRef]

29. Stolnik, M.C.; Garnett, M.C.; Davies, M.C.; Illum, L.; Bousta, M.; Vert, M.; Davis, S.S. The colloidal properties of surfactant-free biodegradable nanospheres from poly (β-malic acid-co-benzyl malate)s and poly (lactic acid-co-glycolide). *Coll. Surf.* **1995**, *97*, 235–245. [CrossRef]

30. Roces, C.B.; Christensen, D.; Perrie, Y. Translating the fabrication of protein-loaded poly(lactic-co-glycolic acid) nanoparticles from bench to scale-independent production using microfluidics. *Drug Deliv. Transl. Res.* **2020**, *10*, 582–593. [CrossRef]

31. Sahoo, S.K.; Panyam, J.; Prabha, S.; Labhasetwar, V. Residual polyvinyl alcohol associated with poly (D,L-lactide-co-glycolide) nanoparticles affects their physical properties and cellular uptake. *J. Control. Release* **2002**, *82*, 105–114. [CrossRef]

32. Damiati, S.; Kompella, U.B.; Damiati, S.A.; Kodzius, R. Microfluidic Devices for Drug Delivery Systems and Drug Screening. *Genes* **2018**, *9*, 103. [CrossRef] [PubMed]

33. Chiesa, E.; Dorati, R.; Modena, T.; Conti, B.; Genta, I. Multivariate analysis for the optimization of microfluidics-assisted nanoprecipitation method intended for the loading of small hydrophilic drugs into PLGA nanoparticles. *Int. J. Pharm.* **2018**, *536*, 165–177. [CrossRef]

34. Garg, S.M.; Thomas, A.; Heuck, G.; Armstead, A.; Singh, J.; Leaver, T.J.; Wild, A.W.; IP, S.; Taylor, R.J.; Ramsay, E.C. *PLGA Nanoparticles Production and In Situ Drug Loading Using the NanoAssemblr™ Benchtop Instrument and the Impact of Solvent Removal Methods*; Precision NanoSystems Inc.: Vancouver, BC, Canada, 2017. Available online: https://www.precisionnanosystems.com/docs/default-source/pni-files/app-notes/plga---drug-loading-and-particle-purification-(pni-app-bt-009-ext)-e.pdf?sfvrsn=88e91c0e_2 (accessed on 10 April 2020).

35. Psimadas, D.; Bouziotis, P.; Georgoulias, P.; Valotassiou, V.; Tsotakos, T.; Loudos, G. Radiolabeling approaches of nanoparticles with (99m) Tc. *Contrast Media Mol. Imaging* **2013**, *8*, 333–339. [CrossRef] [PubMed]

36. Owens, D.E.; Peppas, N.A. Opsonization, biodistribution, and pharmacokinetics of polymeric nanoparticles. *Int. J. Pharm.* **2006**, *307*, 93–102. [CrossRef] [PubMed]

37. Choi, H.S.; Liu, W.; Misra, P.; Tanaka, E.; Zimmer, J.P.; Ipe, B.I.; Bawendi, M.G.; Frangioni, J.V. Renal clearance of quantum dots. *Nat. Biotechnol.* **2007**, *25*, 1165–1170. [CrossRef]

38. Zuckerman, J.E.; Choi, C.H.J.; Han, H.; Davis, M.E. Polycation-siRNA nanoparticles can disassemble at the kidney glomerular basement membrane. *Proc. Natl. Acad. Sci. USA* **2012**, *109*, 3137–3142. [CrossRef]
39. Choi, C.H.; Zuckerman, J.E.; Webster, P.; Davis, M.E. Targeting kidney mesangium by nanoparticles of defined size. *Proc. Natl. Acad. Sci. USA* **2011**, *108*, 6656–6661. [CrossRef]
40. Franken, P.R.; Guglielmi, J.; Vanhove, C.; Koulibaly, M.; Defrise, M.; Darcourt, J.; Pourcher, T. Distribution and dynamics of (^{99m})Tc-pertechnetate uptake in the thyroid and other organs assessed by single-photon emission computed tomography in living mice. *Thyroid* **2010**, *20*, 519–526. [CrossRef]

Article

A Novel Sustained Anti-Inflammatory Effect of Atorvastatin—Calcium PLGA Nanoparticles: In Vitro Optimization and In Vivo Evaluation

Dalia H. Abdelkader [1,*,†], Ahmed Kh. Abosalha [1], Mohamed A. Khattab [2], Basmah N. Aldosari [3] and Alanood S. Almurshedi [3,*,†]

1   Pharmaceutical Technology Department, Faculty of Pharmacy, Tanta University, Tanta 31111, Egypt; ahmed.khaled@pharm.tanta.edu.eg
2   Department of Cytology and Histology, Faculty of Veterinary Medicine, Cairo University, Giza 12211, Egypt; mabdelrazik@cu.edu.eg
3   Department of Pharmaceutics, College of Pharmacy, King Saud University, Riyadh 11451, Saudi Arabia; baldosari@ksu.edu.sa
*   Correspondence: dalia.abdelkader@pharm.tanta.edu.eg or mhdalia86@gmail.com (D.H.A.); marshady@ksu.edu.sa (A.S.A.)
†   These authors contributed equally to this work.

Abstract: Atorvastatin Calcium (At-Ca) has pleiotropic effect as anti-inflammatory drug beside its main antihyperlipidemic action. Our study was conducted to modulate the anti-inflammatory effect of At-Ca to be efficiently sustained for longer time. Single oil-water emulsion solvent evaporation technique was used to fabricate At-Ca into polymeric nanoparticles (NPs). In vitro optimization survey was performed on Poly(lactide-co-glycolide) (PLGA) loaded with At-Ca regrading to particle size, polydispersity index (PDI), zeta potential, percent entrapment efficiency (% EE), surface morphology and in vitro release pattern. In vitro drug-polymers interactions were fully scanned using Fourier-Transform Infrared Spectroscopy (FTIR) and Differential Scanning calorimetry (DSC) proving that the method of fabrication is an optimal strategy maintaining the drug structure with no interaction with polymeric matrix. The optimized formula with particle size (248.2 $\pm$ 15.13 nm), PDI (0.126 $\pm$ 0.048), zeta potential (-12.41 ± 4.80 mV), % EE (87.63 $\pm$ 3.21%), initial burst (39.78 $\pm$ 6.74%) and percent cumulative release (83.63 $\pm$ 3.71%) was orally administered in Male Sprague–Dawley rats to study the sustained anti-inflammatory effect of At-Ca PLGA NPs after carrageenan induced inflammation. In vivo results demonstrate that AT-Ca NPs has a sustained effect extending for approximately three days. Additionally, the histological examination revealed that the epidermal/dermal layers restore their typical normal cellular alignment with healthy architecture.

Keywords: atorvastatin calcium; poly(lactide-co-glycolide); polymeric nanoparticles; carrageenan induced inflammation; anti-inflammatory; sustained release

Citation: Abdelkader, D.H.; Abosalha, A.K.; Khattab, M.A.; Aldosari, B.N.; Almurshedi, A.S. A Novel Sustained Anti-Inflammatory Effect of Atorvastatin—Calcium PLGA Nanoparticles: In Vitro Optimization and In Vivo Evaluation. *Pharmaceutics* **2021**, *13*, 1658. https://doi.org/10.3390/pharmaceutics13101658

Academic Editor: Oya Tagit

Received: 18 August 2021
Accepted: 6 October 2021
Published: 11 October 2021

Publisher's Note: MDPI stays neutral with regard to jurisdictional claims in published maps and institutional affiliations.

1. Introduction

Statins are classified according to their solubility into lipophilic such as atorvastatin, lovastatin and simvastatin or hydrophilic-like pravastatin and fluvastatin. In addition, they are divided into two generations depending on their originality: naturally occurring (simvastatin and pravastatin) or chemically synthetized (e.g., atorvastatin and rosuvastatin) agents. Generally, statins have various pharmacokinetics and pharmacodynamic properties affecting on their ADME (Absorption, distribution, Metabolism and Elimination) behavior, lipid reducing actions and side effects upon administration [1].

Basically, the antihyperlipidemic effect of statins includes inhibition of HMG-CoA (β-Hydroxy β-methylglutaryl-CoA) reductase enzyme which converts 3-hydroxy-3-methylglutaryl-CoA into L-mevalonate (the main precursor for cholesterol formation).

L-mevalonate pathway has several biofunctions beside cholesterol production, e.g., enhancing cellular proliferation and migration, controlling formation of myelin sheath and dynamicity of filamentous network of cellular proteins. This ensures that statins have some auxiliary pharmacological effects known as "pleiotropic effects" [2].

Recently, several studies investigated the anti-inflammatory action of statins independently on their lipid lowering effect [1–4]. The pattern recognition receptors (PRRs) of different immune cells such as macrophages, neutrophils, dendritic cells and etc. are located on cell membrane such as C-type lectin receptors (CTLs) and Toll-like receptors (TLRs) or embedded in the cytoplasmic microenvironments such as RIG-I-like receptors (RLRs) and NOD-like receptors (NLRs) [1]. Specially, NLRs and TLRs are two crucial types acting as a link to induce or inhibit the inflammation process as summarized in Figure 1. Statins can efficiently interrupt the inflammation cycle via inhibiting and stimulating the action of NLRs and TLRs, respectively [1]. Rosuvastatin and atorvastatin could suppress the activation of T cells through oxidized LDL (low-density lipoprotein). T cell activation via the influence of dendritic cells plays a major role in flaring up the inflammation [5]. Statins could disrupt the miRNA sequencing which effectively regulate NF-kB levels in dendritic cells leading to NF-kB deactivation as illustrated in Frostegård et al., study [5]. Other supposed anti-inflammatory mechanism is that statin could modulate cellular level of Rho proteins, one of small guanine triphosphate (GTP) binding proteins, which is highly expressed in pro-inflammatory cytokines [6].

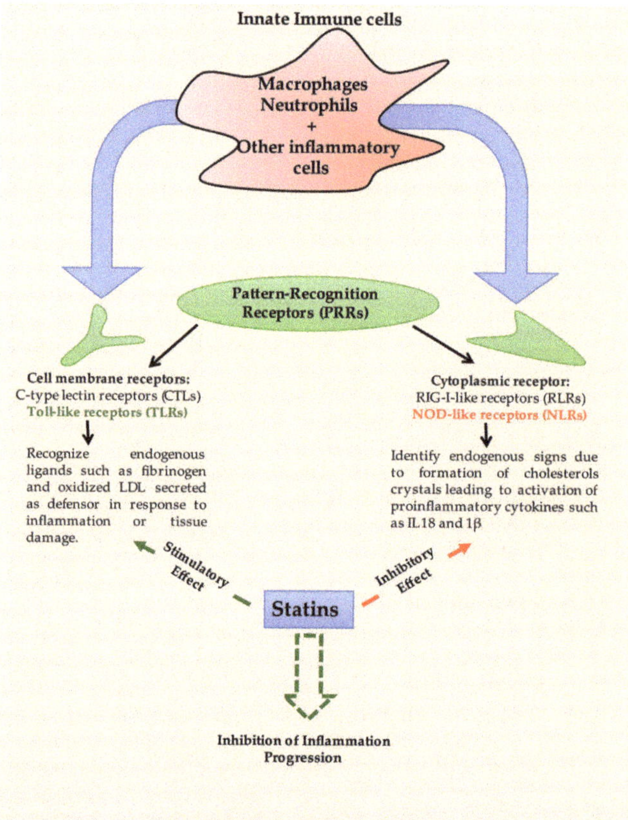

Figure 1. Pharmacological pathways of statins to inhibit inflammation progression via drug-receptors interactions of innate immune cells [1].

Our model drug atorvastatin calcium (At-Ca) has various formulation obstacles preventing implementation of desired release pattern with high bioavailability [7]. At-Ca has poor aqueous solubility with low oral bioavailability approximately equal to 14% due to extensive first pass hepatic metabolism [8]. Different formulation strategies have been utilized to overcome these obstacles such as solid dispersion [9], co-crystallization [10], incorporation of self-emulsifying vehicle [11], etc.

Our approach is to fabricate At-Ca into polymeric PLGA nanoparticles (NPs). PLGA is a biodegradable and biocompatible polymer [12]. It is safely employed in oral delivery systems [13]. Encapsulation powered drug into nanosized polymeric matrix will enhance aqueous drug solubility. Drug particles would be amorphously formulated achieving higher dispersibility with better wetting properties [14]. NPs could efficiently escape enzymatic degradation occurring in liver and gastrointestinal tract. NPs reach directly to systemic circulation via lymphatic pathway through M-cells [7]. Polymeric NPs is ultimately considered the optimum delivery system for drug candidate sensitive to first pass metabolism with poor aqueous solubility.

At-Ca has been previously formulated into polymeric nanoparticles. Zhenbao Li et al., used probe ultrasonication and evaporation method to prepare At-Ca loaded PLGA NPs. The oral bioavailability of At-Ca has been enhanced to approximately 4 folds comparing with free At-Ca suspension [15]. Additionally, At-Ca chitosan nanocrystals fabricated into PLGA polymeric background showed minimum drug degradation, optimized pharmacokinetic parameters and efficient hypolipidemic action [16]. Furthermore, brain delivery of At-Ca loaded amphiphilic PLGA-b-PEG NPs was investigated by Soner Şimşek et al., demonstrating that the concentration of polyethylene glycol (PEG) and polysorbate 80 have a determining effect on the particle size, drug loading efficiency, release behavior concurrently with the cellular uptake to endothelial cells. They concluded that At-Ca loaded amphiphilic PLGA-b-PEG NPs, able to penetrate the blood brain barrier achieving significant targeting 1 h post injection, are considered promising nanocarriers for treatment of several neurological disorders [17]. In this manuscript, we are concerning with the anti-inflammatory effect of atorvastatin calcium (At-Ca) after carrageenan induced inflammation in Male Sprague–Dawley rats. We study the effect of encapsulating At-Ca into PLGA NPs and compare it with free At-Ca and placebo NPs (contain no medication).

2. Materials and Methods

2.1. Materials

Poly (D,L-lactide-co-glycolide, acid terminated, lactide:glycolide 50:50, MW 24,000–38,000), Poly (vinyl alcohol) (MW = 31,000–50,000, 87–89% hydrolysed), poly (ethylene glycol) (PEG) flakes (MW 2000 Da), Carrageenan, suitable for gel preparation, were all purchased from Sigma Aldrich, Gillingham, UK. Atorvastatin calcium was gifted from El Obour Modern Pharmaceutical Industries-Egypt. Cal-Ex™ II Fixative/Decalcifier reagent was purchased from Fisher Scientific, Pittsburgh, PA, USA). Phosphate Buffered Saline Tablets, Dulbecco A (PBS, Oxoid™, ThermoFisher Scientific, Waltham, MA, USA). Dichloromethane (DCM), dimethylformamide (DMF) and sodium chloride (NaCL) were all purchased from Al—Gomhoria Co. for medicines and medical supplies, Egypt. All other used solvents, reagents and chemical are of appropriate standards with no further purification needed. Millipore® deionized water was used throughout all the experiments.

2.2. Atorvastatin Calcium (At-Ca) PLGA NPs Fabrication

At-Ca loaded NPs were prepared using single oil-water emulsion solvent evaporation technique [18] with slight modifications for proper NPs fabrication. Firstly, PEG (5% w/w) comixed with different concentrations of PLGA (% w/v) were dissolved in DCM (Table 1). The weight ratio between PLGA and PEG was kept to being 19:1 in all NPs formulae (Table 1). At-Ca (10 mg) was dissolved in DMF. Then, the organic phase with volume ratio of DMF (1): DCM (9) was well vortexed for 2 min. O/W emulsion was formed by probe homogenization (Cole-Parmer Model 50 Cp T 4710 Series Ultrasonic Homogenizer,

Chicago, IL, USA) at an operating frequency 20 KHz for 3 min of the organic phase into the aqueous phase containing diverse concentrations of PVA (% w/v) as shown by the formulae codes in Table 1. The organic/aqueous phase volume ratio was fixed at 1:100 throughout the study The prepared formulae were stayed overnight on magnetic stirring to allow the evaporation of organic solvents and ensure solidification of NPs. The nanoparticles were then collected by centrifugation (Hettich Microliter centrifuge MIKRO 220, Tuttlingen, Germany) at $10,000\times g$ for 20–25 min at 4 °C and washed three times with deionized water to remove excess PVA and free unentrapped At-Ca in supernatants [7]. Three individually prepared batches of At-Ca PLGA NPs were employed for each data point's analysis for in vitro characterizations mentioned in Sections 2.3, 2.4 and 2.7. The mean ±Standard Deviation (SD) was calculated.

Table 1. Formulae coding and composition of At-Ca PLGA NPs.

Formula Code	PLGA Concentration (% w/v)	PEG Content in Primary Emulsion (% w/w)	PVA Concentration in Aqueous External Phase (% w/v)	Drug Loading (mg)	Organic/Aqueous Phase Volume Ratio
F1	2.50				
F2	5.00	5.00	0.50	10	1:100
F3	7.50				
F4	2.50				
F5	5.00	5.00	1.00	10	1:100
F6	7.50				
F7	2.50				
F8	5.00	5.00	1.50	10	1:100
F9	7.50				

2.3. Particle Size and Zeta Potential Measurements

Particle size (diameter, nm), polydispersity index (PDI) and zeta potential of At-Ca PLGA NPs were determined using laser diffraction technique (NanoBrook 90Plus, Brookhaven Instruments Corporation, New York, NY, USA). The prepared At-Ca PLGA NPs samples were appropriately diluted to tenfold of its volume with deionized water for proper analysis. The measurement parameters were adjusted (wavelength (658 nm), viscosity (0.798 cP) and refractive index (1.33)) using Broakhaven software.

2.4. Determination of Percent Entrapment Efficiency (EE)

The amount of At-Ca encapsulated into polymeric matrix composed from PLGA and PEG (19:1) was determined indirectly by assaying the free unentrapped drug dispersed in the supernatant obtained after centrifugation step mentioned in Section 2.2 [19]. The amount of At-Ca was measured using UV-visible spectrophotometry (cuvette type) at detection wavelength of 282 nm (Evolution 300 spectrophotometer, Thermo Scientific, Waltham, MA, USA) [7]. The indirect % EE was calculated as follow:

$$\text{Indirect \%EE} = \frac{\text{Total mass of drug used (mg)} - \text{mass of drug in supernatant (mg)}}{\text{Total mass of drug used (mg)}} \times 100 \tag{1}$$

2.5. Morphological Imaging

At-Ca PLGA NPs were loaded on carbon coated Cu grids (200 mesh) and kept for appropriate time till fixation. At-Ca PLGA NPs were visualized using transmission electron microscope (TEM) at 200 kV (TEM, HRTEM, JEM 2100, JEOL, Tokyo, Japan).

2.6. Analysis of Drug—Polymer Interaction

2.6.1. Fourier-Transform Infrared Spectroscopy (FTIR)

The FTIR spectrophotometer (Bruker Tensor 27, Borken, Germany) was employed for detecting the FTIR spectral bands of PLGA 50:50, PEG2000, pure At-Ca and the prepared At-Ca NPs (F5). The samples in solid state were physically mixed with potassium bromide and compressed in spherical disks. FTIR was run in the range of 4000 to 400 cm^{-1} for each sample.

2.6.2. Differential Scanning Colorimetry (DSC)

Thermal scanning of PLGA 50:50, PEG 2000, free At-Ca and At-Ca NPs (F5) were investigated using differential scanning calorimeter (PerkinElmer, Inc. STA 6000, Waltham, MA, USA). Before starting, nitrogen gas should be purged at a rate equal to 40 mL/min through DSC cell. Heating rate was fixed at 10 °C/min to achieve equality of thermal scanning. the endothermic peaks were recorded in the range of 20–200 °C.

2.7. In Vitro Drug Release

In vitro release platform of different At-Ca PLGA NPs formulae was fully investigated using dialysis membrane method. Bags of dialysis membrane (Molecular weight cut off 12–14 kDa Fisher Scientific, Pittsburgh, PA, USA) were soaked overnight in release media (Phosphate Buffer Saline, PBS, 7.4) prior to in vitro release experiment. Using beaker 100 mL, the prepared nano-formulations containing At-Ca equivalent to 10 mg were placed in dialysis bags that should be tightly sealed to avoid leakage. Sink condition was maintained by adjusting the volume of release medium to be three times larger that is required to prepare a saturated solution of At-Ca [20]. Dialysis bag was immersed in 50 mL PBS and stirred at 37 °C. At predetermined time intervals (1, 2, 4, 6, 8, 24, 48 and 72 h), two mL samples were withdrawn for analysis. Same volume of fresh medium should be replaced to keep the volume constant at 50 mL [7]. The concentrations of At-Ca in collected samples were assayed using UV-visible spectroscopy as mentioned before in Section 2.4. A release of free At-Ca was performed as a control to evaluate the effect of PLGA NPs on the sustained release of nano-formulations. Each experiment was carried out in triplicate and the percent cumulative release was expressed as a mean ± SD.

2.8. In Vivo Study

All animal experiments including handling, study design and euthanization were carried out in accordance with the rules approved by Faculty of Pharmacy and the University of Tanta's Animal Ethics Committee guidelines.

2.8.1. Induction of Carrageenan Induced Inflammation

Carrageenan induced inflammation was initiated by subcutaneous (S.Q) injection of carrageenan solution freshly prepared in normal saline [21]. The experimental animal model was kept unanesthetized during S.Q injection of 0.2 mL carrageenan solution (1% w/v) in subplanter right hind paw. The left hind paw acts as control [22].

2.8.2. Experimental Groups

Five groups (n = 6, at each time interval) of male Sprague–Dawley rats weighing 200–250 g were employed throughout this study. The first group serves as a standard non-inflamed model, not injected with carrageenan. Half an hour before carrageenan injection, normal saline (untreated control group), placebo NPs (contain no drug), free At-Ca and At-Ca PLGA NPs were orally administered to the other four remaining groups. All groups were fasted for approximately eight hours before receiving the medication. The amount/mass of AT-Ca PLGA NPs was calculated to adjust At-Ca equivalent to 10 mg/kg of rat's weight [7]. At different three-time intervals of 6, 24 and 72 h, all right and lefts paws should be precisely excised at the same place and weighed within maximally 0.5 h to avoid water loss and hence weight variation [22]. The percent decrease in paw edema's weight was calculated using this equation:

$$\% \text{ Decrease in paw edema's weight} = \frac{Wt\ left - Wt\ right}{Wt\ left} \times 100 \qquad (2)$$

where, *Wt left* and *Wt right* are weights of left (control, not injected) and right (injected with carrageenan) paws, respectively. Only numerical results were considered regardless of the negative sign.

2.8.3. Histological Examination

Paws samples were fixed in 10% neutral buffered formalin for 48 h followed by decalcification using Cal-X II for 14 days, then mid sagittal section was made to the whole paws using a surgical blade, followed by samples processing using serial grades of ethanol, cleared in xylene followed by infiltration and embedding in paraplast tissue embedding media. Five microns 2–3 step serial tissue sections were made by rotatory microtome from each sample and mounted on glass slides for hematoxylin and eosin (H and E) staining for general histological examination of tissue samples and grading of recorded cutaneous lesions. All guidelines for samples fixation, processing and staining were carried out according to Culling techniques [23]. Samples were examined and micrographs were obtained by Leica Full High Definition (HD) microscopic imaging system (Leica Microsystems GmbH, Wentzler, Germany).

2.9. Statistical Analysis

All experiments were performed for at least in triplicates. Results are shown as the mean ±SD. Prism 5 (Graph-Pad Software) was used for statistical analysis. A one-way ANOVA, followed by a pairwise comparison post-hoc test, were utilized wherever appropriate. ($p < 0.05$) is considered statistically significant.

3. Results

In this study, two formulation variables (PLGA and PVA concentrations) with three levels of each were conducted as demonstrated in Table 1. The influences of these two independent factors were screened on the mean particle size, PDI, zeta potential, entrapment efficiency and In vitro release pattern of At-Ca NPs. TEM was employed to examine NPs morphology and any interaction occurred between the polymeric content and the encapsulated drug was investigated using FTIR and DSC. In vivo study was further performed to compare the anti-inflammatory effect of placebo NPs (contain no drug), free At-Ca and the optimized At-Ca NPs that showed a well-controlled in vitro characterization.

3.1. Effect of Formulation Variables

3.1.1. PLGA Concentration

PLGA constitutes the main ingredient of the polymeric matrix encapsulating At-Ca. We utilized three different concentrations of PLGA at 2.5, 5 and 7.5% *w/v*. Increasing PLGA concentration significantly ($p < 0.05$) results in NPs with greater mean particle size (nm). As shown in Table 2 and Figure 2, the particle size of F7, F8 and F9 were equal to 238.1 ± 22.58, 289.5 ± 17.89 and 326.9 ± 16.74 nm for PLGA concentrations of 2.5, 5 and 7.5% *w/v*, respectively. In addition, variations in size distribution significantly increase with increasing PLGA concentration. PDI values for F1, F2 and F3 are equal to 0.096 ± 0.011, 0.134 ± 0.025 and 0.145 ± 0.012, respectively (Table 2), for the same previously mentioned different PLGA concentrations. Zeta potential of At-Ca NPs are highly controlled by PLGA concentration (Table 2 and Figure 2). Increasing PLGA concentration could significantly ($p < 0.05$) increase the net negative surface charge of the resulting NPs, for example F7, F8 and F9 have values of zeta potential equal to −11.71 ± 0.78, −14.55 ± 2.95 and −19.28 ± 0.77 mV, respectively. % EE significantly increases with increasing of PLGA concentration (Figure 3). Table 2 shows that % EE of F1, F2 and F3 is increasing by approximately 10% when PLGA concentration increases by 2.5% increments from 2.5% to 7.5% *w/v*.

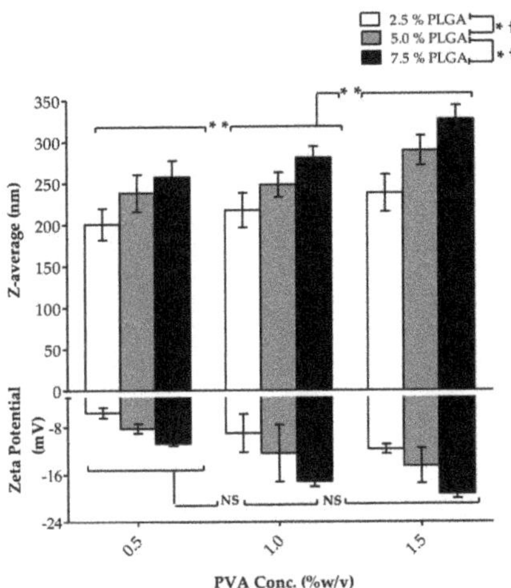

Figure 2. Effect of PLGA and PVA concentrations on the particle size and zeta potential of At-Ca NPs. Results show mean ±SD (n = 3). (* $p < 0.05$) and (** $p < 0.05$), the significant effect of increasing PLGA and PVA concentrations, respectively, on the particle size of At-Ca NPs. († $p < 0.05$) and NS, the significant and non-significant effects of increasing PLGA and PVA concentrations, respectively, on zeta potential value of At-Ca NPs.

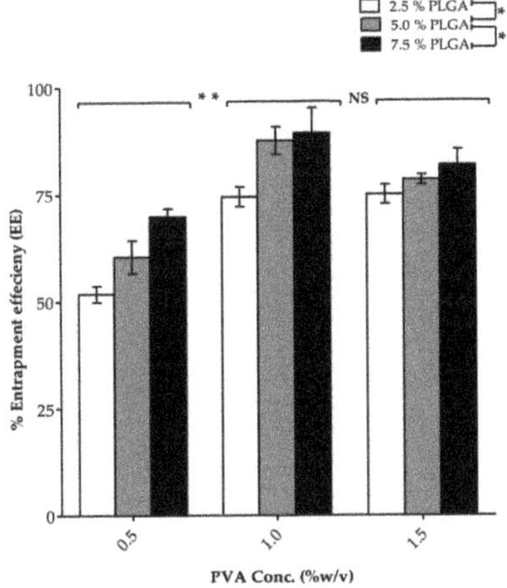

Figure 3. Effect of PLGA and PVA concentrations on the percent entrapment efficiency (% EE) of At-Ca NPs. Results show mean ±SD (n = 3). (* $p < 0.05$), the significant effect for increasing PLGA concentration. (** $p < 0.05$) and NS, the significant and non-significant effects for increasing PVA concentration to 1 and 1.5% w/v, respectively.

Table 2. Physicochemical properties of At-Ca PLGA NPs formulations (F1–F9). Data represent mean values ± SD of at least three measurements.

Formula Code	Z-Average (nm)	Polydispersity Index (PDI)	Zeta Potential (mV)	% EE
F1	200.7 ± 18.85	0.096 ± 0.011	−5.57 ± 0.88	51.78 ± 1.89
F2	238.3 ± 22.43	0.134 ± 0.025	−8.23 ± 0.82	60.42 ± 3.84
F3	257.9 ± 19.53	0.145 ± 0.012	−10.77 ± 0.36	69.81 ± 1.84
F4	217.5 ± 20.87	0.107 ± 0.066	−9.02 ± 3.22	74.52 ± 2.35
F5	248.2 ± 15.13	0.126 ± 0.048	−12.41 ± 4.80	87.63 ± 3.21
F6	280.8 ± 13.55	0.151 ± 0.039	−17.18 ± 0.87	89.36 ± 5.89
F7	238.1 ± 22.58	0.076 ± 0.047	−11.71 ± 0.78	75.15 ± 2.24
F8	289.5 ± 17.89	0.102 ± 0.029	−14.55 ± 2.95	78.95 ± 1.17
F9	326.9 ± 16.74	0.110 ± 0.012	−19.28 ± 0.77	82.02 ± 3.59

3.1.2. PVA Concentration

PVA is a common polymer widely used in multiple drug delivery systems due to its feasibility and biodegradability [24,25]. PVA was chosen as a surfactant added to the aqueous phase with varieties of concentrations (% w/v) as demonstrated in Table 1. Increasing PVA concentration from 0.5 to 1.5% w/v results in a significant ($p < 0.05$) increase in particle size (nm) (Table 2 and Figure 2). For example, the particle size of F6 and F9 are equal to 280.8 ± 13.55 and 326.9 ± 16.74 nm for PVA concentration of 1 and 1.5% w/v, respectively, at the same PLGA concentration equal to 7.5% w/v. NPs fabrication at higher PVA concentration have lower PDI with homogenous size distribution. As shown in Table 2, F2, F5 and F8 have PDI values equal to 0.134 ± 0.025, 0.126 ± 0.048 and 0.102 ± 0.029, respectively. The effect of PVA concentration on zeta potential of NPs are well illustrated in Table 2 and Figure 2 showing non-significant effect of PVA concentration on NPs net surface charge. PVA concentration has a diverse effect on % EE of NPs. A significant increase in % EE when PVA concentration increases from 0.5 to 1% w/v, whereas a non-significant effect observed on the values of % EE between PVA concentration of 1% compared to 1.5% w/v (Table 2, Figure 3). For example, % EE of F1, F4 and F7 were equal to 51.78 ± 1.89, 74.52 ± 2.35 and 75.15 ± 2.24 for PVA concentrations of 0.5, 1 and 1.5% w/v, respectively, with a significant difference ($p < 0.05$) between F1 and F4 and non-significant variation between F4 and F7.

3.2. Morphological Imaging

Transmission electron microscopy (TEM) is utilized to visualize the geometry and shape of At-Ca NPs (Figure 4). NPs displayed spherical nanostructure with a clearly observed increase in particle size of F1, F4 and F7 equal to 180, 216.6 and 276.6 nm (the particle size is determined by calculating the mean of the individual particles labelled in Figure 4). These results are closely matched with the numerical values measured by laser diffraction technique (Table 2).

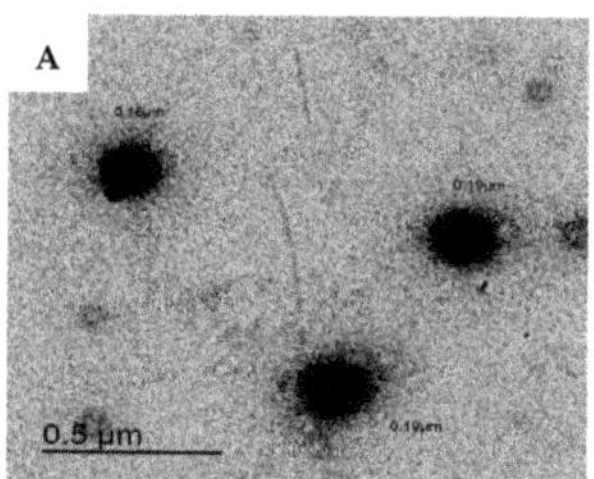
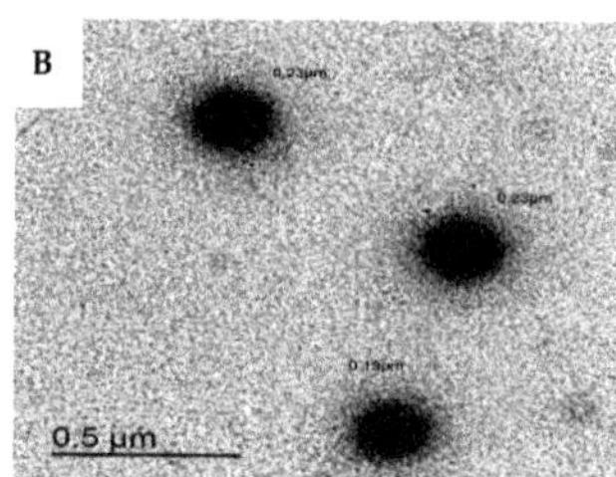
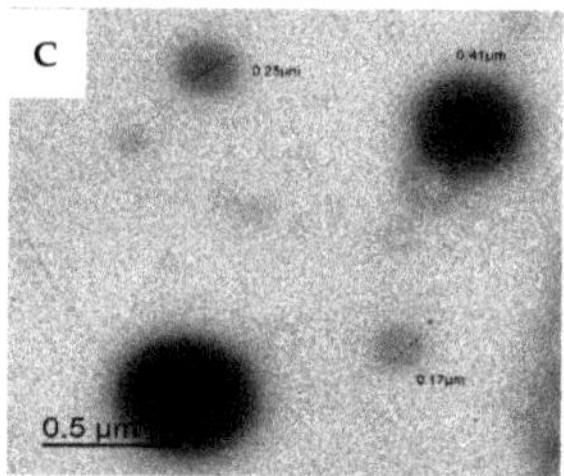

Figure 4. Transmission electron micrographs of At-Ca NPs. A, B and C for F1, F4 and F7, respectively. Formulae code is shown in Table 1.

3.3. In Vitro Drug Stability

3.3.1. Fourier-Transform Infrared Spectroscopy (FTIR)

After fabrication of PLGA NPs encapsulating At-Ca, there is no major shifting in any of the characteristic peaks for all the components of PLGA, PEG and At-Ca. FTIR spectrum of At-Ca PLGA NPs (Figure 5) shows peaks at 3456 cm^{-1} (O-H stretching), 3001-2955-2850 cm^{-1} (C-H stretching), 1759 cm^{-1} (C=O stretching), 1457-1386-1275 cm^{-1} (C-H bending), 1173 cm^{-1} (=C-O asymmetric stretching) and 1092 cm^{-1}(C-O-C stretching) that are closely matched with the positions of their corresponding peaks of PLGA spectrum (Figure 5, Table 3).

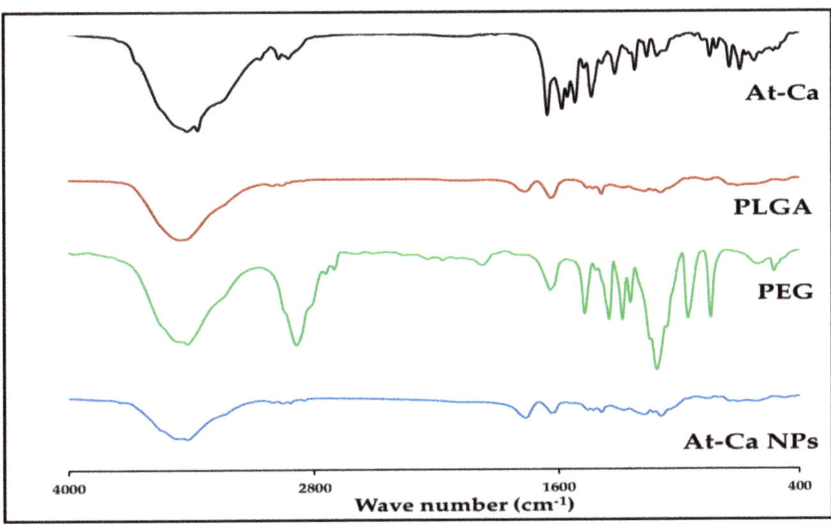

Figure 5. FTIR spectrum of free At-Ca, PLGA 50:50, PEG 2000 and At-Ca NPs (F5).

Table 3. Characteristic IR peaks of PLGA 50:50, PEG 2000 and At-Ca.

	Assignment	Approximate Frequency (cm^{-1})	Description
	O-H (very broad)	3452.21	Stretching
	C-O-C	1093.27	Stretching
	=C-O	1169.23	Asymmetric Stretching
PLGA 50:50 [26–28]	C-H	3001.64 (CH3) 2955.70 (CH2) 2851.21 (CH2)	Stretching
		1455.76 (CH3) 1385.24 (CH3) 1276.49 (CH2)	Bending
	C=O	1758.52	Stretching
		1427.20 1385.24	Bending
PEG [29–31]	O-H (very broad)	3414.43	Stretching
	C-H	2886.97	
	C-O	1112.8	
	C-C	842.13	

Table 3. *Cont.*

	Assignment	Approximate Frequency (cm^{-1})	Description
At-Ca [7,32,33]	O-H (very broad)	3416.25	Stretching
	N-H	3366.44	
	C-H	3059.59 2969.41 29922.27	
	C=O	1649.89 1579.24	
	C-C	1515.04 1433.1	
	CH3 and -CH2	1316.51	Deformation Bending
	C-N	1217.72	Stretching
	C-O	1157.55	
	aromatic C-H	843.52 747.00	Out plane Bending
	C-H	695.58 623.99	Deformation Bending

Furthermore, peaks at 3365 cm^{-1} (N-H stretching), 1634 cm^{-1} (C=O stretching), 1428 cm^{-1} (C-C stretching), 748 cm^{-1} (aromatic C-H out plane bending) and 627 cm^{-1} (C-H deformation bending) are presented on FTIR spectrum of At-Ca NPs (Figure 5) resembling the same characteristic peaks of free At-Ca (Table 3). It was demonstrated that peak at 3418 cm^{-1} become flatter and wider in At-Ca NPs spectrum indicating that hydrogen bond might be enhanced between O-H stretching peaks of PEG (3414 cm^{-1}) and At-Ca drug (3416 cm^{-1}) [34].

3.3.2. Differential Scanning Colorimetry (DSC)

DSC study is mainly preformed to investigate the change in the physical state of At-Ca after encapsulation into polymeric nanocarriers. In our study free At-Ca, PLGA 50:50 and PEG 2000 show their endothermic peaks at 157.75 °C [7], 47.75 °C [35] and 55.1 °C [36,37], respectively (Figure 6). DSC thermogram of At-Ca NPs demonstrates an endothermic peak at 51.27 °C in a position similar to the individual peaks of PLGA and PEG with no peak in the range of 155 to 158 °C (Figure 6) indicating the disappearance of the characteristics At-Ca peak.

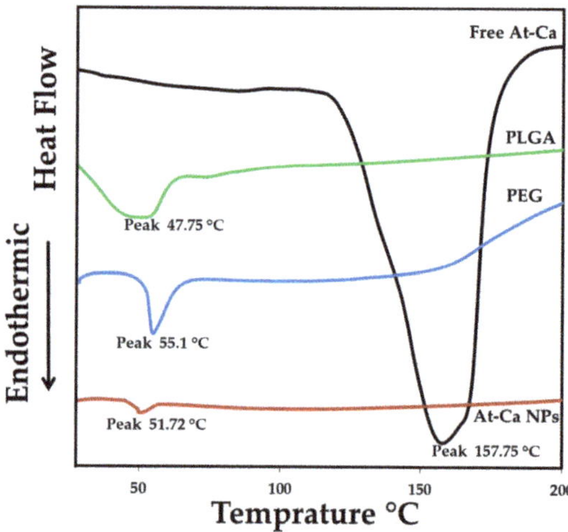

Figure 6. DSC thermograms for At-Ca, PLGA 50:50, PEG 2000 and At-Ca NPs (F5).

All At-Ca loaded NPs showed biphasic release platform with an initial burst at 6 h followed by sustained release profile extended to 72 h (Figure 7A–C). It is clearly observed that % cumulative release of free drug reached to approximately 100% at 6 h ensuring that dialysis membrane method offers no sustained effect for in vitro release study. All NPs formulations have significant ($p < 0.05$) more sustained release behavior compared with free At-Ca (control). At-Ca NPs with smaller diameter exhibits a significant higher initial burst, for example F7 (238.1 ± 22.58 nm), F8 (289.5 ± 17.89 nm) and F9 (326.9 ± 16.74) have initial bursts equal to 47.89 ± 2.89, 41.21 ± 5.71 and 37.89 ± 4.45%, respectively. Additionally, more sustained effect with greater percent cumulative release is significantly demonstrated for At-Ca NPs with higher values of % EE. As it is shown in Figure 7A,B, F1 (% EE = 51.78 ± 1.89) and F4 (% EE = 74.52 ± 2.35) display total percent cumulative release equal to 69.82 ± 9.78 and 79.89 ± 8.79%, respectively. Whereas their initial bursts are equal to 31.75 ± 5.76 and 45.46 ± 1.74 % for F1 and F4, respectively. This could be attributed to increasing PVA concentrations which significantly enhances higher initial burst values (Figure 7A–C). F5 was chosen an optimum formula for further in vivo study because it offers the highest percent cumulative release (83.63 ± 3.71) with high % EE and a well-controlled particles size concurrently with proper values of PDI and zeta potential (Table 2).

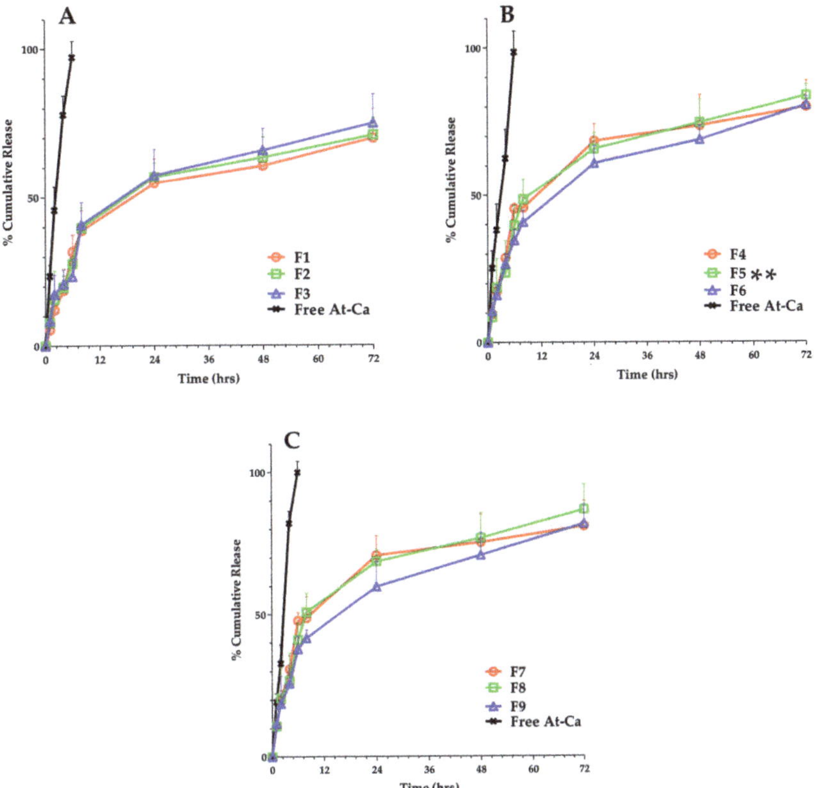

Figure 7. In vitro drug release profiles for At-Ca NPs (Formulae code shown in Table 1). For clarity, data are shown as mean + SD (n = 3). (**A–C**) patterns were for NPs formulations utilized PVA at concentrations of 0.5, 1 and 1.5 % w/v, respectively. At-Ca NPs (F5) has significantly higher initial burst and percent cumulative release (double asterisks, $p < 0.05$).

3.4. In Vivo Anti-Inflammatory Study

The optimized At-Ca NPs (F5) showed a significant ($p < 0.05$) decrease in paw oedema in all time intervals compared to untreated control, placebo NPs and free At-Ca as demonstrated in Figure 8. The anti-inflammatory effect of both placebo NPs and untreated control is statistically non-significant producing similar results of percent decrease in edema's weight. At 6 h time interval, it is obviously noted that free At-Ca has a significant reduction in edema's weight compared with untreated control and placebo NPs with a non-significant difference in the remaining time intervals till 72 h (Figure 8).

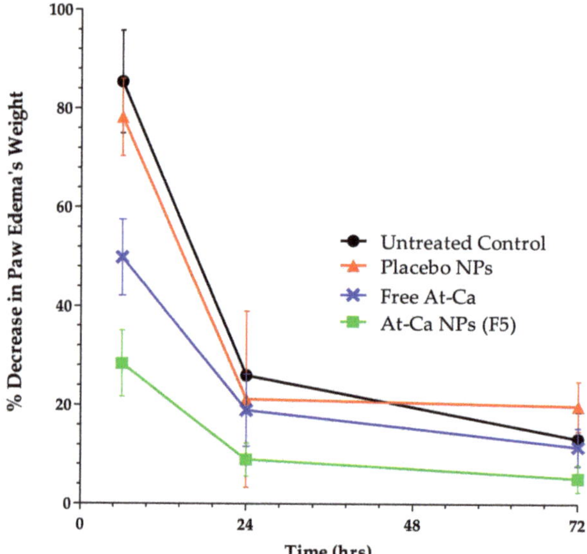

Figure 8. Percent decrease in paw edema's weight after carrageenan induced inflammation of untreated control (circles), placebo NPs (triangles), free At-Ca (cross) and At-Ca NPs, F5 (squares) in male Sprague–Dawley rats. Results are expressed as means ± SD. (n = 6 per group).

Microscopic examination of normal non inflamed paws tissue demonstrated typical histological features of skin layers including intact epidermal layer consisting of apparent well-organized keratinocytes, intact dermal layer containing abundant collagen fibers and minimal inflammatory cells infiltrates with normal density of subcutaneous muscles and vasculatures (Figure 9A,B).

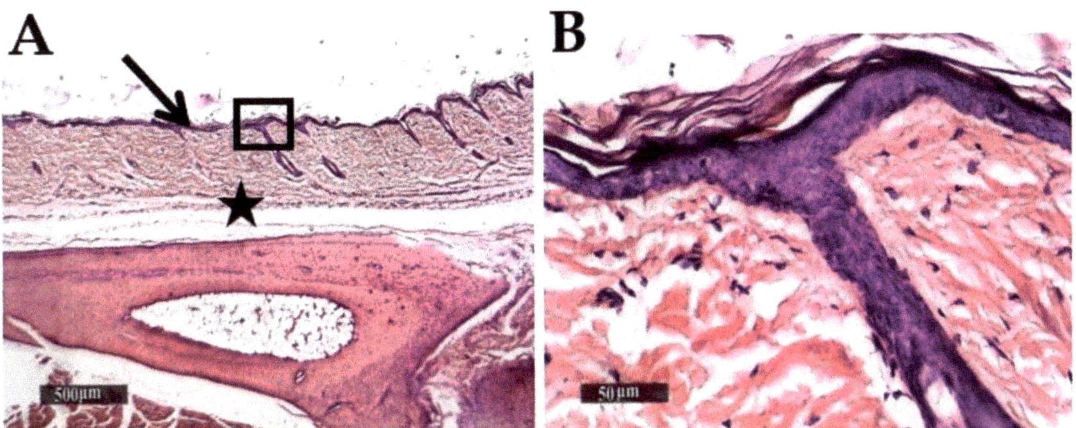

Figure 9. Normal morphological features of non-inflamed paws skin layers. H and E stain. (**A**) micrograph at ×40. Highlighted box in (**B**) micrograph at ×400. Black arrow and star show the epidermal layer and subcutaneous tissue, respectively.

Sever subcutaneous and intermuscular edema, wide dispersion of dermal collagen fibers, numerous congested subcutaneous blood vessels (BVs) and mild to moderate increase of lymphocytic and neutrophilic infiltrates (Figure 10A–D) were demonstrated in both untreated control and placebo NPs tissue sections (Table 4A) at 6 h interval after carrageenan injection. Moreover, lower cellular infiltrates with less accelerated influx records were showed in deep subcutaneous tissue of free AT-Ca and AT-Ca NPs (F5) samples with milder severity of subcutaneous edema (Figure 10E–H and Table 4A). More sever diffuse subcutaneous inflammatory cells infiltrates and edema extends to periosseous connective tissue with many necrotic subcutaneous muscles were observed in control untreated samples (Figure 11A,B) at 24 h after carrageenan injection. Extensive subcutaneous edema and necrotic muscle fibers with several congested deep BVs and moderate records of neutrophilic inflammatory cells infiltrates were shown in tissue samples treated with placebo NPs (Figure 11C,D and Table 4B). However, Figure 11E–H showed more preserved morphological features with mild focal records of degenerated muscle fibers for free AT-Ca and AT-Ca NPs (F5). Significant milder ($p < 0.05$) subcutaneous edema was observed in AT-Ca NPs (F5) compared to a moderate record for free AT-Ca samples accompanied with moderate mixed inflammatory cells infiltrates (Table 4B). More pronounced subcutaneous persistence of inflammatory reaction was shown at 72 h compared with 24 h post injection interval in both untreated control and placebo NPs samples (Figure 12A–D) with some dissolution of subcutaneous edema in placebo NPs samples and minimal records of congested subcutaneous BVs (Table 4C). Non-significant reduction of subcutaneous edema ($p > 0.05$) was observed in free AT-Ca samples (Table 4B,C) accompanied with persistent level of mixed inflammatory cells infiltrates (Figure 12E,F) resembling 24 h samples. Significant protective efficacy ($p < 0.05$) of morphological structures with more accelerated normalization of subcutaneous tissue components and cellular elements were recorded in AT-Ca NPs (F5) samples with sporadic few records of inflammatory cells infiltrates, minimal subcutaneous edema and normal vasculature with no congested blood vessels were shown in more than 80% of examined samples (Figure 12G,H, Table 4C).

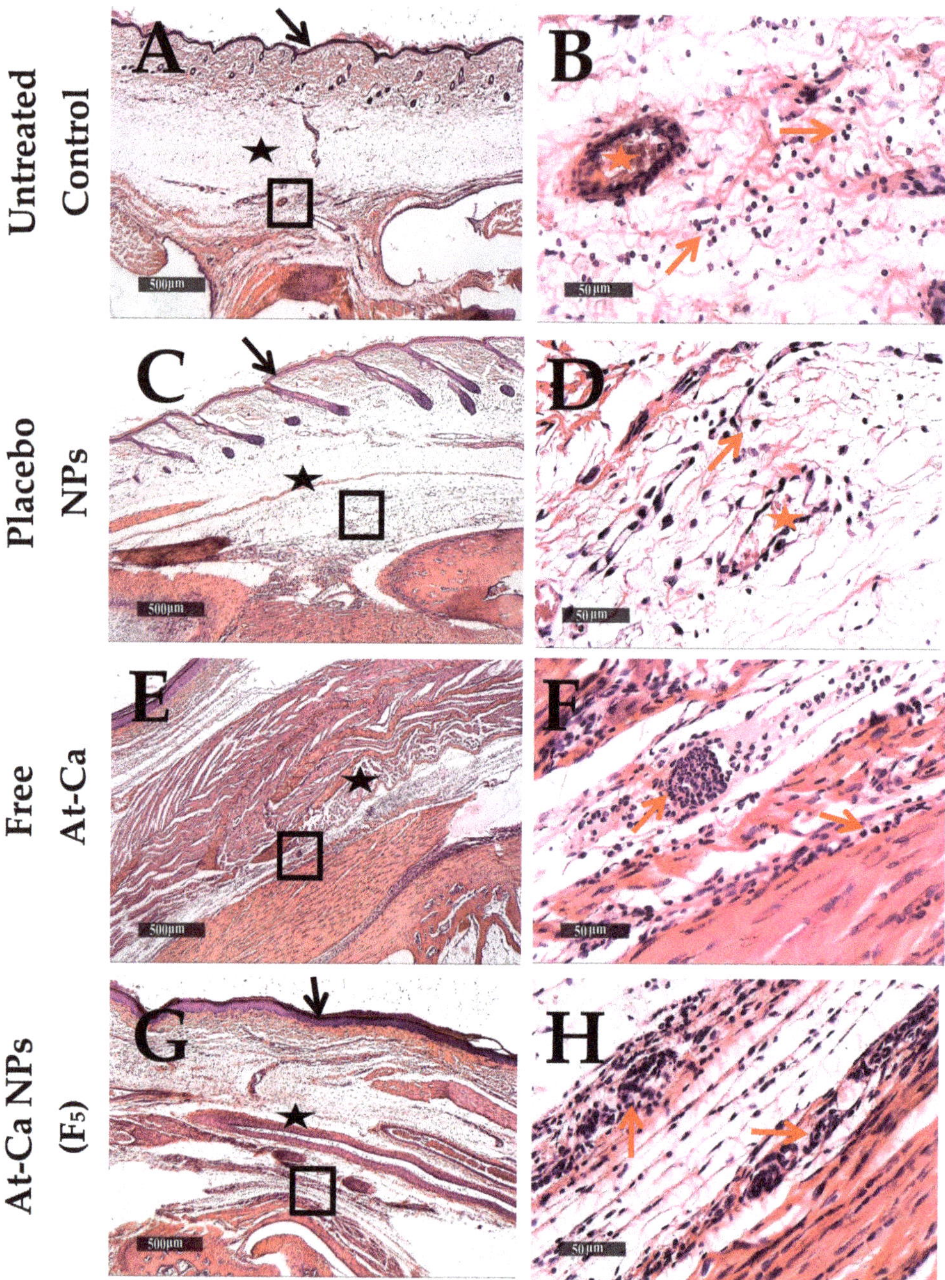

Figure 10. Histological changes of Paws skin layers in different groups (n = 6 per group), at 6 h time intervals after carrageenan injection. (**A,B**): untreated control, (**C,D**): Placebo NPs, (**E,F**): Free AT-Ca and (**G,H**): AT-Ca NPs (F5). H and E stain.   (**A,C,E,G**) micrographs at ×40. Highlighted boxes (**B,D,F,H**) micrographs at ×400. Black arrows and stars show the epidermal layer and subcutaneous tissue, respectively. Red arrows and stars show inflammatory cells infiltrates and congested BVs, respectively.

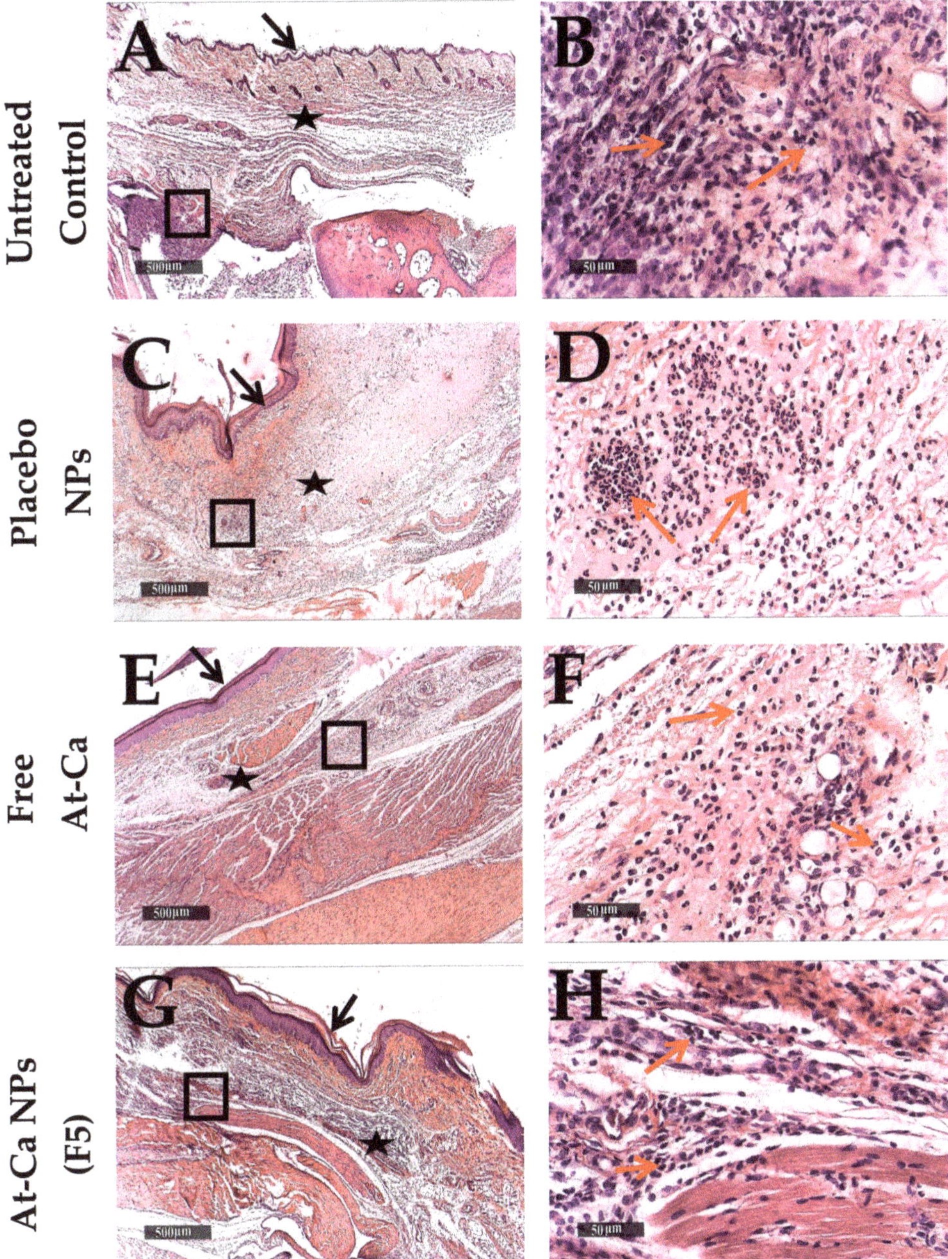

Figure 11. Histological changes of Paws skin layers in different groups (n = 6 per group), at 24 h time intervals after carrageenan injection. (**A,B**): untreated control, (**C,D**): Placebo NPs, (**E,F**): Free AT-Ca and (**G,H**): AT-Ca NPs (F5). H and E stain. (**A,C,E,G**) micrographs at ×40. Highlighted boxes (**B,D,F,H**) micrographs at ×400. Black arrows and stars show the epidermal layer and subcutaneous tissue, respectively. Red arrows show inflammatory cells infiltrates.

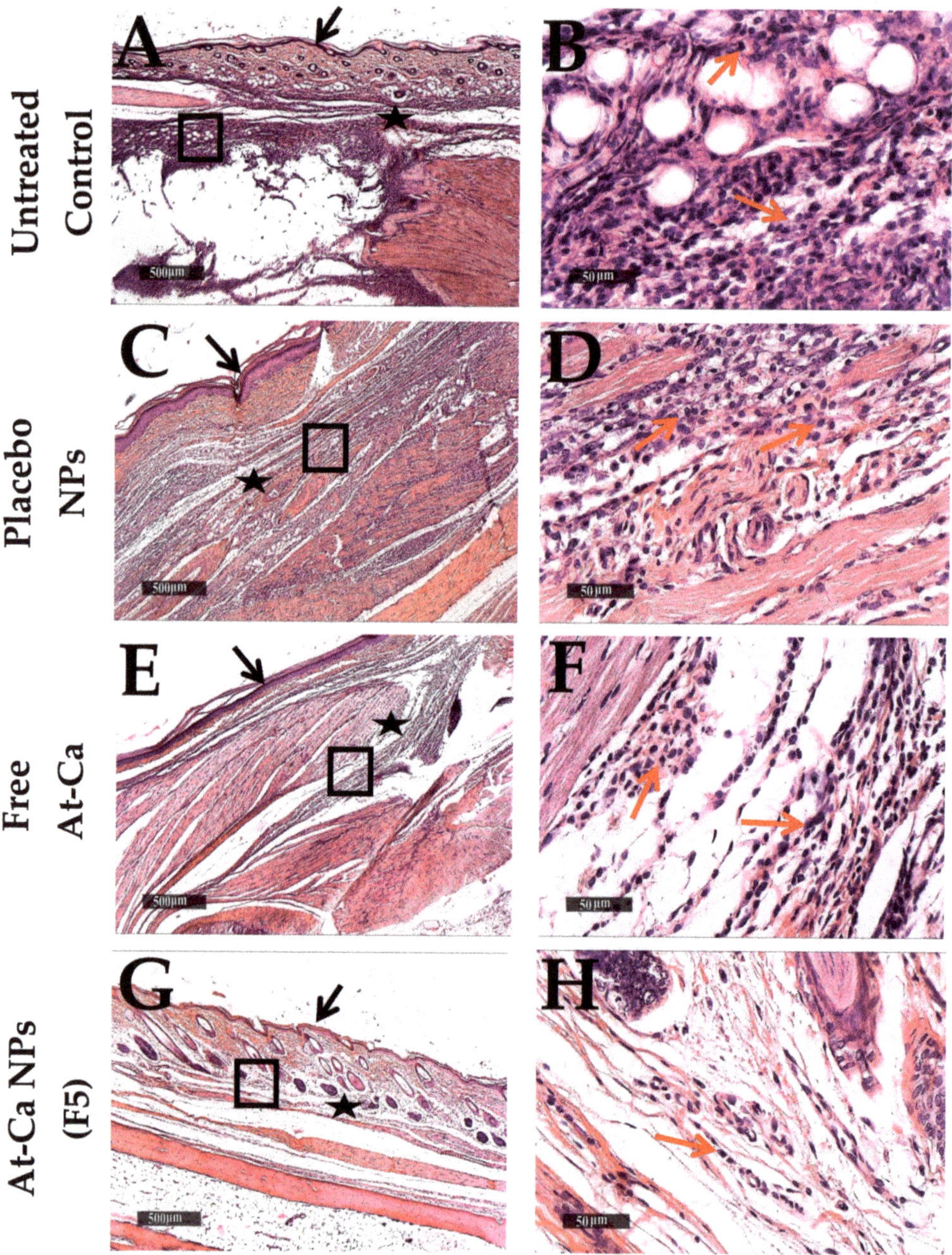

Figure 12. Histological changes of Paws skin layers in different groups (n = 6 per group), at 72 h time intervals after carrageenan injection. (**A,B**): untreated control, (**C,D**): Placebo NPs, (**E,F**): Free AT-Ca and (**G,H**): AT-Ca NPs (F5). H and E stain.   (**A,C,E,G**) micrographs at ×40. Highlighted boxes (**B,D,F,H**) micrographs at ×400. Black arrows and stars show the epidermal layer and subcutaneous tissue, respectively. Red arrows show inflammatory cells infiltrates.

Table 4. Score records of histological examined lesions after carrageenan injection at different time intervals. A, B and C for 6, 24 and 72 h, respectively.

A	Untreated Control	Placebo NPs	Free AT-Ca	AT-Ca NPs (F5)
Inflammatory cells infiltrates	+++	++	++	++
Subcutaneous oedema	+++	+++	++	++
Congested BVs	++	++	++	++
B	**Untreated Control**	**Placebo NPs**	**Free AT-Ca**	**AT-Ca NPs (F5)**
Inflammatory cells infiltrates	+++	++	++	++
Subcutaneous oedema	+++	+++	++	+
Congested BVs	++	++	+	+
C	**Untreated Control**	**Placebo NPs**	**Free AT-Ca**	**AT-Ca NPs (F5)**
Inflammatory cells infiltrates	+++	+++	++	+
Subcutaneous oedema	++	++	++	+
Congested BVs	++	+	+	-

- Nil (no lesions were demonstrated). + Mild lesion recorded in less than 15% of examined tissue sections. ++ Moderate lesion recorded in 16–35% of examined tissue sections. +++ Sever lesion recorded in more than 35% of examined tissue sections.

4. Discussion

PLGA utilized in At-Ca NPs fabrication is acid terminated functionalized with carboxylic acid groups. Generally, PLGA could be synthesized to be acid or ester terminated [38]. In our study we employed acid terminated PLGA because ester terminated PLGA could produce larger particle size and less stable NPs [39]. Acid terminated PLGA is more hydrophilic with negatively charged COOH and hence, more tendency to attract water molecules producing smaller particle size [39]. In addition, the interparticle electrostatic repulsion between similar negatively charged carboxylate groups might inhibit any possible aggregation between individually dispersed NPs leading to homogenous NPs distribution, better dispersibility parallel with increased stability [39]. PLGA concentration is a key controlling factor that significantly affects on particle size of At-Ca NPs. Higher concentration of PLGA increases the viscosity of the organic phase with greater oily droplet consistency which requires more shear force for proper emulsification. Coarser emulsion will be formulated because the homogenization power and stirring time are fixed at constant values throughout the study leading to NPs with bigger particle size [12]. Inadequate homogenization and poor drug dispersion occurred concurrently with increasing PLGA concentration which negatively affect on emulsification efficiency lead to non-uniformly particle size with high deviation in size distribution and PDI values [40]. The acidic contents of PLGA imparts a negative charge shielding NPs surface which is highly dependent on PLGA concentration, so increasing PLGA results in higher negative charge covering NPs surface offering greater electrostatic repulsion with better dispersibility [12,41]. % EE is an important in vitro parameter should be highly considered during evaluation of NPs because it mainly affects on release pattern of NPs and also, their in vivo implementation. Highly viscous organic phase obtained with increasing PLGA concentration leads to preventing drug molecules from diffusion out to the external surface with better incorporation into PLGA core which results in significant increase in % EE [12,42].

Increasing PVA concentration results in bigger NPs size. This could be attributed to formation of hydrophobic bonds between the hydrocarbon branches of PVA and the surface of PLGA NPs synchronously with spontaneous heavy inter/intra molecular hydrogen bonds constructed due to excessive hydration of PVA hydroxyl groups [12]. Higher concentration of PVA surfactant in the aqueous phase should enhance NPs hydrodynamic stability preventing their aggregation, therefore smaller deviation in size distribution and lower PDI values will be obtained [43]. Three washing cycles were carried out during NPs preparation to remove extra amount of PVA with micro levels might be attached on NPs surface which have negligible effect on zeta potential of NPs [12]. At lower PVA concentration equal to 0.5 and 1% w/v, PVA has a significant influence to increase % EE because the hydrophobic drug will be tightly encapsulated into PLGA internal core and its diffusion to the outer PVA surface will be reduced to the minimum. At higher PVA concentration of 1.5% w/v, the solubilizing power of PVA surfactant starts to improve drug solubility enhancing the drug leakage to the outer surface which leads to counteract drug desirability to be remained centralized in nanovesicle's core [12,44].

PEG is nonionic polymer, universally employed during fabrication of NPs to enhance their residence time in blood circulation leading to achieve better absorption and hence enhanced pharmacokinetic properties of nano formulations [45]. There are different methods used to incorporate PEG into polymeric PLGA NPs. Chemically synthesized coblock polymers is the main approach [45] and also, physically comixing PLGA with PEG is considered a promising strategy which efficiently affects on NPs physicochemical features [12,19]. Here in this study, we added comixed PEG for surface modification because PEG could form a periphery layer surrounding NPs. This layer acts as biological barrier protecting the nano system from macrophage engulfment which leads to higher stability and delayed degradation [19]. PEG plays a crucial role in improving in vitro characterization of NPs. Association of polymeric networks modified after PEG addition might produce NPs with smaller diameter [12,19,46]. As our model drug At-Ca is highly hydrophobic, its entrapment will efficiently increase into PLGA hydrophobic core with no tendency for diffusion out to the external hydrophilic PEG surface [47]. Additionally, PEG controls the prolonged release of At-Ca NPs for 72 h due to reducing drug release rates and slowing down its diffusion outside the nanovesicle [47]. PEG aids to exhibit sustained anti-inflammatory action of At-Ca PLGA NPs.

FTIR is as important analytical study should be fully investigated during NPs preparation to exclude any physical and chemical interaction might occur between the polymeric material and drug, also, to ensure the compatibility between all the components of NPs. FTIR aids to examine the effectiveness of drug encapsulation into polymeric core of NPs [7,45]. As illustrated in Section 3.3.1, FTIR spectrum of AT-Ca NPs shows no loss of the functional groups of free drug and native PLGA negating any chemical interaction occurred between the drug and polymer. This results proves that our fabrication method used for NPs preparation is a successful technique protecting the drug stability and integrity [7]. Most of detection peaks regarding to PEG spectrum were not clearly observed on At-Ca NPs spectrum. The spectral bands of PLGA (major component of polymeric matrix) might mask the spectral peaks of PEG existed in trace concentration comixed with PLGA [48]. DSC thermogram of At-Ca NPs elucidates that there is no drug in crystalline form after encapsulation into PLGA core. This ensures that At-Ca might be in amorphous form or dispersed homogeneously in polymeric matrix [7,34]. Additionally, this results demonstrate that At-Ca was efficiently dissolved in the organic phase during NPs preparation with no tendency to precipitate after solvent evaporation due to the polymeric matrix sufficiently surrounding drug molecules and preventing their aggregation and/or crystallization [7].

Biphasic release pattern of PLGA NPs could be explained by the initial burst firstly emerged due to the presence of loosely bound drug molecules on the outer surface that is highly enhanced by efficient wettability offered by PVA [49]. The second sustained release phase manifested by PLGA polymer showed a delayed release extended for three days [12,49]. In vitro release At-Ca from PLGA NPs is highly dependent on particle size. Smaller particles with larger surface area exposed to release media facilitate drug diffusion which results in higher initial bursts values [7,12]. Even though At-Ca PLGA NPs were orally administrated, pH of the release medium was adjusted to 7.4. Intact NPs would be absorbed directly into blood circulation (pH 7.4) via payer patches' pathway. All in vitro release experiments were performed in pH 7.4 simulating the in vivo condition [50].

The results of in vivo study revealed the sustained anti-inflammatory effect of At-Ca PLGA NPs [19,51] extended for approximately three days compared with free At-Ca which has rapid onset of action lasted for only six hours. Histological examination showed that the paw skin tissue could restore its normal histological structure of epidermal/dermal layers after oral administration of AT-Ca NPs within 72 h. Placebo NPs have similar results of untreated control which negates any anti-inflammatory effect for PLGA and PEG forming the main matrix of the polymeric nanoarchitecture.

5. Conclusions

The anti-inflammatory effect of At-Ca is considered a novel pharmacological action that could be widely employed in the near future. Polymeric PLGA NPs loaded with At-Ca provide excellent in vitro features offering a sustained release platform for 72 h. Normal density of subcutaneous muscles, minimal inflammatory cells, healthy blood vessels and no subcutaneous oedema were clearly observed in Paw' tissues of male Sprague–Dawley rats after orally administration of At-Ca PLGA NPs comparing with placebo NPs (contain no medication) and free drug.

Author Contributions: Conceptualization, D.H.A.; methodology, D.H.A. and A.K.A.; software, D.H.A.; investigation, D.H.A. and A.K.A.; resources, D.H.A., B.N.A. and A.S.A.; data curation, D.H.A. and A.K.A.; writing—original draft preparation, D.H.A.; writing—review and editing, D.H.A. and A.S.A.; visualization, M.A.K.; project administration, D.H.A., B.N.A. and A.S.A.; funding acquisition, B.N.A. and A.S.A. All authors have read and agreed to the published version of the manuscript.

Funding: This research received no external funding.

Institutional Review Board Statement: All animal experiments including handling, study design and euthanization were carried out in accordance with the rules approved by Faculty of Pharmacy and the University of Tanta's Animal Ethics Committee guidelines (PT00020, approved in March 2021).

Informed Consent Statement: Not applicable.

Data Availability Statement: All the data of this research are available upon request.

Conflicts of Interest: The authors declare no conflict of interest.

References

1. Koushki, K.; Shahbaz, S.K.; Mashayekhi, K.; Sadeghi, M.; Zayeri, Z.D.; Taba, M.Y.; Banach, M.; Al-Rasadi, K.; Johnston, T.P.; Sahebkar, A. Anti-inflammatory Action of Statins in Cardiovascular Disease: The Role of Inflammasome and Toll-Like Receptor Pathways. *Clin. Rev. Allergy Immunol.* **2021**, *60*, 175–199. [CrossRef]
2. Gilbert, R.; Al-Janabi, A.; Tomkins-Netzer, O.; Lightman, S. Statins as anti-inflammatory agents: A potential therapeutic role in sight-threatening non-infectious uveitis. *Porto Biomed. J.* **2017**, *2*, 33–39. [CrossRef]
3. Macin, S.M.; Perna, E.R.; Farías, E.F.; Franciosi, V.; Cialzeta, J.R.; Brizuela, M.; Medina, F.; Tajer, C.; Doval, H.; Badaracco, R. Atorvastatin has an important acute anti-inflammatory effect in patients with acute coronary syndrome: Results of a randomized, double-blind, placebo-controlled study. *Am. Heart J.* **2005**, *149*, 451–457. [CrossRef]
4. Bu, D.X.; Griffin, G.; Lichtman, A.H. Mechanisms for the anti-inflammatory effects of statins. *Curr. Opin. Lipidol.* **2011**, *22*, 165–170. [CrossRef] [PubMed]
5. Frostegård, J.; Zhang, Y.; Sun, J.; Yan, K.; Liu, A. Oxidized Low-Density Lipoprotein (OxLDL)-Treated Dendritic Cells Promote Activation of T Cells in Human Atherosclerotic Plaque and Blood, Which Is Repressed by Statins: MicroRNA let-7c Is Integral to the Effect. *J. Am. Heart Assoc.* **2016**, *5*, 1–15. [CrossRef] [PubMed]
6. Schönbeck, U.; Libby, P. Inflammation, immunity, and HMG-CoA reductase inhibitors: Statins as antiinflammatory agents? *Circulation* **2004**, *109*. [CrossRef]
7. Shaker, M.A.; Elbadawy, H.M.; Al Thagfan, S.S.; Shaker, M.A. Enhancement of atorvastatin oral bioavailability via encapsulation in polymeric nanoparticles. *Int. J. Pharm.* **2021**, *592*. [CrossRef]
8. Mahmoud, M.O.; Aboud, H.M.; Hassan, A.H.; Ali, A.A.; Johnston, T.P. Transdermal delivery of atorvastatin calcium from novel nanovesicular systems using polyethylene glycol fatty acid esters: Ameliorated effect without liver toxicity in poloxamer 407-induced hyperlipidemic rats. *J. Control. Release* **2017**, *254*, 10–22. [CrossRef] [PubMed]
9. Iqbal, A.; Hossain, M.S.; Shamim, M.A.; Islam, M.; Siddique, M.A.T. Formulation, in vitro evaluation and characterization of atorvastatin solid dispersion. *Trop. J. Pharm. Res.* **2020**, *19*, 1131–1138. [CrossRef]
10. Al-kazemi, R.; Al-basarah, Y.; Nada, A. Atorvastatin Cocrystals: Tablet Formulation and Stability. *Asian J. Pharm.* **2020**, *14*, 578–595.
11. Gardouh, A.R.; Nasef, A.M.; Mostafa, Y.; Gad, S. Design and evaluation of combined atorvastatin and ezetimibe optimized self-nano emulsifying drug delivery system. *J. Drug Deliv. Sci. Technol.* **2020**, *60*, 102093. [CrossRef]
12. Abdelkader, D.H.; El-Gizawy, S.A.; Faheem, A.M.; McCarron, P.A.; Osman, M.A. Effect of process variables on formulation, in-vitro characterisation and subcutaneous delivery of insulin PLGA nanoparticles: An optimisation study. *J. Drug Deliv. Sci. Technol.* **2018**, *43*, 160–171. [CrossRef]
13. Faheem, A.M.; Abdelkader, D.H. Novel drug delivery systems. In *Engineering Drug Delivery Systems*; Woodhead Publishing: Sawston, UK, 2020; ISBN 9780081025482.

14. Zhang, Y.; Zhang, R.; Illangakoon, U.E.; Harker, A.H.; Thrasivoulou, C.; Parhizkar, M.; Edirisinghe, M.; Luo, C.J. Copolymer composition and nanoparticle configuration enhance in vitro drug release behavior of poorly water-soluble progesterone for oral formulations. *Int. J. Nanomed.* **2020**, *15*, 5389–5403. [CrossRef] [PubMed]

15. Li, Z.; Tao, W.; Zhang, D.; Wu, C. The studies of PLGA nanoparticles loading atorvastatin calcium for oral administration in vitro and in vivo. *Asian J. Pharm. Sci.* **2017**, *12*, 285–291. [CrossRef]

16. Kurakula, M.A.T. Co-Delivery of Atorvastatin Nanocrystals in PLGA based in situ Gel for Anti-Hyperlipidemic Efficacy. *Curr. Drug Deliv.* **2016**, *13*, 211–220. [CrossRef]

17. Şimşek, S.; Eroğlu, H.; Kurum, B.; Ulubayram, K. Brain targeting of Atorvastatin loaded amphiphilic PLGA-b-PEG nanoparticles. *J. Microencapsul.* **2012**, 1–11. [CrossRef] [PubMed]

18. Kizilbey, K. Optimization of Rutin-Loaded PLGA Nanoparticles Synthesized by Single-Emulsion Solvent Evaporation Method. *ACS Omega* **2019**, *4*, 555–562. [CrossRef]

19. Abdelkader, D.H.; Osman, M.A.; El-Gizawy, S.A.; Hawthorne, S.J.; Faheem, A.M.; McCarron, P.A. Effect of poly(ethylene glycol) on insulin stability and cutaneous cell proliferation in vitro following cytoplasmic delivery of insulin-loaded nanoparticulate carriers—A potential topical wound management approach. *Eur. J. Pharm. Sci.* **2018**, *114*, 372–384. [CrossRef] [PubMed]

20. Weng, J.; Tong, H.H.Y.; Chow, S.F. In vitro release study of the polymeric drug nanoparticles: Development and validation of a novel method. *Pharmaceutics* **2020**, *12*, 732. [CrossRef]

21. Rivat, C.; Laulin, J.-P.; Corcuff, J.-B.; Célèrier, E.; Pain, L.; Simonnet, G. Fentanyl Enhancement of Carrageenan-induced Long-lasting Hyperalgesia in Rats. *Anesthesiology* **2002**, *96*, 381–391. [CrossRef]

22. Whiteley, P.E.; Dalrymple, S.A. Models of Inflammation: Carrageenan-Induced Paw Edema in the Rat. *Curr. Protoc. Pharmacol.* **2001**, 4–6. [CrossRef]

23. Culling, C.F.A. *Handbook of Histopathological and Histochemical Techniques*, 3rd ed.; Elsevier: London, UK, 2013; ISBN 9781483164793.

24. Abdelkader, D.H.; Osman, M.A.; El-Gizawy, S.A.; Faheem, A.M.; McCarron, P.A. Characterisation and in vitro stability of low-dose, lidocaine-loaded poly(vinyl alcohol)-tetrahydroxyborate hydrogels. *Int. J. Pharm.* **2016**, *500*, 326–335. [CrossRef] [PubMed]

25. Abdelkader, D.H.; Osman, M.A.; El-Gizawy, S.A.; Faheem, A.M.; McCarron, P.A. Adhesiveness of poly(vinyl alcohol) tetrahydroxyborate hydrogels to the stratum corneum and their applicability as topical drug delivery systems. *Int. J. Cosmet. Sci.* **2015**, *37*, 141–163.

26. Singh, G.; Kaur, T.; Kaur, R.; Kaur, A. Recent biomedical applications and patents on biodegradable polymer-PLGA. *Int. J. Pharmacol.* **2014**, *1*, 30–42.

27. Singh, R.; Kesharwani, P.; Mehra, N.K.; Singh, S.; Banerjee, S.; Jain, N.K. Development and characterization of folate anchored Saquinavir entrapped PLGA nanoparticles for anti-tumor activity. *Drug Dev. Ind. Pharm.* **2015**, *41*, 1888–1901. [CrossRef]

28. Xiao, H.; Wang, L. Effects of X-shaped reduction-sensitive amphiphilic block copolymer on drug delivery. *Int. J. Nanomed.* **2015**, *10*, 5309–5325. [CrossRef]

29. Asadi, H.; Rostamizadeh, K.; Salari, D.; Hamidi, M. Preparation of biodegradable nanoparticles of tri-block PLA-PEG-PLA copolymer and determination of factors controlling the particle size using artificial neural network. *J. Microencapsul.* **2011**, *28*, 406–416. [CrossRef]

30. Xiang, H.; Wang, S.; Wang, R.; Zhou, Z.; Peng, C.; Zhu, M. Synthesis and characterization of an environmentally friendly PHBV/PEG copolymer network as a phase change material. *Sci. China Chem.* **2013**, *56*, 716–723. [CrossRef]

31. Zhang, X.; Li, N.; Liu, Y.; Ji, B.; Wang, Q.; Wang, M.; Dai, K.; Gao, D. On-demand drug release of ICG-liposomal wedelolactone combined photothermal therapy for tumor. *Nanomed. Nanotechnol. Biol. Med.* **2016**, *12*, 2019–2029. [CrossRef]

32. Shaker, M.A.; Elbadawy, H.M.; Shaker, M.A. Improved solubility, dissolution, and oral bioavailability for atorvastatin-Pluronic®solid dispersions. *Int. J. Pharm.* **2020**, *574*, 118891. [CrossRef]

33. Shaker, M.A. Dissolution and bioavailability enhancement of Atorvastatin: Gelucire semi-solid binary system. *J. Drug Deliv. Sci. Technol.* **2018**, *43*, 178–184. [CrossRef]

34. Sun, S.B.; Liu, P.; Shao, F.M.; Miao, Q.L. Formulation and evaluation of PLGA nanoparticles loaded capecitabine for prostate cancer. *Int. J. Clin. Exp. Med.* **2015**, *8*, 19670–19681. [PubMed]

35. Valor, D.; Montes, A.; Monteiro, M.; García-Casas, I.; Pereyra, C.; de la Ossa, E.M. Determining the optimal conditions for the production by supercritical co2 of biodegradable plga foams for the controlled release of rutin as a medical treatment. *Polymers* **2021**, *13*, 1645. [CrossRef]

36. Dabbagh, A.; Mahmoodian, R.; Abdullah, B.J.J.; Abdullah, H.; Hamdi, M.; Abu Kasim, N.H. Low-melting-point polymeric nanoshells for thermal-triggered drug release under hyperthermia condition. *Int. J. Hyperth.* **2015**, *31*, 920–929. [CrossRef] [PubMed]

37. Kuru, A.; Aksoy, S.A. Cellulose–PEG grafts from cotton waste in thermo-regulating textiles. *Text. Res. J.* **2014**, *84*, 337–346. [CrossRef]

38. Öztürk, A.A.; Kırımlıoğlu, G.Y. Preparation and in vitro characterization of lamivudine loaded nanoparticles prepared by acid and/or ester terminated PLGA for effective oral anti-retroviral therapy. *J. Res. Pharm.* **2019**, *23*, 897–913. [CrossRef]

39. Gebreel, R.M.; Edris, N.A.; Elmofty, H.M.; Tadros, M.I.; El-Nabarawi, M.A.; Hassan, D.H. Development and Characterization of PLGA Nanoparticle-Laden Hydrogels for Sustained Ocular Delivery of Norfloxacin in the Treatment of Pseudomonas Keratitis: An Experimental Study. *Drug Des. Devel. Ther.* **2021**, *15*, 399–418. [CrossRef] [PubMed]

40. Patel, R.R.; Chaurasia, S.; Khan, G.; Chaubey, P.; Kumar, N.; Mishra, B. Cromolyn sodium encapsulated PLGA nanoparticles: An attempt to improve intestinal permeation. *Int. J. Biol. Macromol.* **2016**, *83*, 249–258. [CrossRef]
41. Sharma, D.; Maheshwari, D.; Philip, G.; Rana, R.; Bhatia, S.; Singh, M.; Gabrani, R.; Sharma, S.K.; Ali, J.; Sharma, R.K.; et al. Formulation and Optimization of Polymeric Nanoparticles for Intranasal Delivery of Lorazepam Using Box-Behnken Design:In Vitro and In Vivo Evaluation. *Biomed. Res. Int.* **2014**, *2014*, 1–14.
42. Sharma, N.; Madan, P.; Lin, S. Effect of process and formulation variables on the preparation of parenteral paclitaxel-loaded biodegradable polymeric nanoparticles: A co-surfactant study. *Asian J. Pharm. Sci.* **2016**, *11*, 404–416. [CrossRef]
43. Mohan, L.J.; McDonald, L.; Daly, J.S.; Ramtoola, Z. Optimising PLGA-PEG nanoparticle size and distribution for enhanced drug targeting to the inflamed intestinal barrier. *Pharmaceutics* **2020**, *12*, 1114. [CrossRef]
44. Seju, U.; Kumar, A.; Sawant, K.K. Development and evaluation of olanzapine-loaded PLGA nanoparticles for nose-to-brain delivery: In vitro and in vivo studies. *Acta Biomater.* **2011**, *7*, 4169–4176. [CrossRef]
45. Mares, A.G.; Pacassoni, G.; Marti, J.S.; Pujals, S.; Albertazzi, L. Formulation of tunable size PLGA-PEG nanoparticles for drug delivery using microfluidic technology. *PLoS ONE* **2021**, *16*, e0251821. [CrossRef] [PubMed]
46. Beletsi, A.; Panagi, Z.; Avgoustakis, K. Biodistribution properties of nanoparticles based on mixtures of PLGA with PLGA–PEG diblock copolymers. *Int. J. Pharm.* **2005**, *298*, 233–241. [CrossRef]
47. Samkange, T.; D'Souza, S.; Obikeze, K.; Dube, A. Influence of PEGylation on PLGA nanoparticle properties,hydrophobic drug release and interactions with humanserum albumin.pdf. *J. Pharm. Pharmacol.* **2019**, *71*, 1497–1507. [CrossRef] [PubMed]
48. Chan, K.L.A.; Kazarian, S.G. Detection of trace materials with Fourier transform infrared spectroscopy using a multi-channel detector. *Analyst* **2006**, *131*, 126–131. [CrossRef]
49. Panda, B.P.; Wei, M.X.; Shivashekaregowda, N.K.H.; Patnaik, S. Design, Fabrication and Characterization of PVA/PLGA Electrospun Nanofibers Carriers for Improvement of Drug Delivery of Gliclazide in Type-2 Diabetes. *Proceedings* **2020**, *78*, 14. [CrossRef]
50. Bachhav, S.S.; Dighe, V.D.; Devarajan, P. V Exploring Peyer's Patch Uptake as a Strategy for Targeted Lung Delivery of Polymeric Rifampicin Nanoparticles Exploring Peyer ' s Patch Uptake as a Strategy for Targeted Lung Delivery of Polymeric Rifampicin Nanoparticles National Center for Preclinical. *Mol. Pharm.* **2018**, 1–38.
51. Abdelkader, D.H.; Murtaza, T.; Mitchell, C.A.; Osman, M.A.; El-Gizawy, S.A.; Faheem, A.M.; Mohamed, E.-T.; McCarron, P.A. Enhanced cutaneous wound healing in rats following topical delivery of insulin-loaded nanoparticles embedded in poly(vinyl alcohol)-borate hydrogels. *J. Drug Deliv. Transl. Res.* **2018**, *8*, 1053–1065. [CrossRef]

pharmaceutics

Article

Optimisation of a Microfluidic Method for the Delivery of a Small Peptide

Felicity Y. Han [1,2,*], Weizhi Xu [1,2], Vinod Kumar [1], Cedric S. Cui [1], Xaria Li [1], Xingyu Jiang [3,4], Trent M. Woodruff [1], Andrew K. Whittaker [2,5] and Maree T. Smith [1]

[1]   Faculty of Medicine, School of Biomedical Sciences, The University of Queensland, Brisbane, QLD 4072, Australia; weizhi.xu@uq.net.au (W.X.); v.kumar1@uq.edu.au (V.K.); s.cui@uq.edu.au (C.S.C.); xaria.li@uq.edu.au (X.L.); t.woodruff@uq.edu.au (T.M.W.); maree.smith@uq.edu.au (M.T.S.)
[2]   Australian Institute for Bioengineering and Nanotechnology, The University of Queensland, Brisbane, QLD 4072, Australia; a.whittaker@uq.edu.au
[3]   National Center for Nanoscience and Technology, Beijing 100190, China; xingyujiang@nanoctr.cn
[4]   Department of Biomedical Engineering, Southern University of Science and Technology, Shenzhen 518055, China
[5]   ARC Centre of Excellence in Convergent Bio Nano Science and Technology, The University of Queensland, Brisbane, QLD 4072, Australia
*   Correspondence: f.han@uq.edu.au; Tel.: +61-7-3346-3807

Abstract: Peptides hold promise as therapeutics, as they have high bioactivity and specificity, good aqueous solubility, and low toxicity. However, they typically suffer from short circulation half-lives in the body. To address this issue, here, we have developed a method for encapsulation of an innate-immune targeted hexapeptide into nanoparticles using safe non-toxic FDA-approved materials. Peptide-loaded nanoparticles were formulated using a two-stage microfluidic chip. Microfluidic-related factors (i.e., flow rate, organic solvent, theoretical drug loading, PLGA type, and concentration) that may potentially influence the nanoparticle properties were systematically investigated using dynamic light scattering and transmission electron microscopy. The pharmacokinetic (PK) profile and biodistribution of the optimised nanoparticles were assessed in mice. Peptide-loaded lipid shell-PLGA core nanoparticles with designated size (~400 nm) and a sustained in vitro release profile were further characterized in vivo. In the form of nanoparticles, the elimination half-life of the encapsulated peptide was extended significantly compared with the peptide alone and resulted in a much higher distribution into the lung. These novel nanoparticles with lipid shells have considerable potential for increasing the circulation half-life and improving the biodistribution of therapeutic peptides to improve their clinical utility, including peptides aimed at treating lung-related diseases.

Keywords: drug delivery system; nanoparticles; poly (lactic-*co*-glycolic acid) (PLGA); microfluidic; pharmacokinetics (PK) and biodistribution

Citation: Han, F.Y.; Xu, W.; Kumar, V.; Cui, C.S.; Li, X.; Jiang, X.; Woodruff, T.M.; Whittaker, A.K.; Smith, M.T. Optimisation of a Microfluidic Method for the Delivery of a Small Peptide. *Pharmaceutics* **2021**, *13*, 1505. https://doi.org/10.3390/pharmaceutics13091505

Academic Editors: Oya Tagit and Alyssa Panitch

Received: 29 June 2021
Accepted: 14 September 2021
Published: 18 September 2021

1. Introduction

Polymeric nanoparticles have been widely studied for decades as a strategy to produce sustained release of encapsulated drugs as well as enhanced therapeutic effects [1,2]. Poly (lactic-*co*-glycolic acid) (PLGA) is superior to many other polymers used for fabricating nanoparticles due to its biodegradability and biocompatibility [3]. Importantly, PLGA is approved for clinical applications involving systemic dosing routes by many international regulatory agencies including the United States Food and Drug Administration (FDA) [4].

In the last two decades, considerable effort has been directed at improving formulation methods for producing polymeric nanoparticles. Versatile methods used to fabricate PLGA nanoparticles include nanoprecipitation, emulsion and hydrogel template methods, the use of supercritical CO_2, spray drying, coacervation, microfluidics, and the PRINT technique (particle in non-wetting templates) [5–9]. Compared with other methods, the microfluidic method has significant advantages, including the precise control over particle parameters (e.g., size, morphology, and charge) [10,11], monodispersed particles, automated and single-step formulation, as well as relatively high encapsulation efficiency [12–14]. This relatively novel formulation method provides homogeneous mixing conditions by precisely controlling micro flows, which cannot be achieved in conventional bulk methods [15,16]. Using microfluidics, the physicochemical properties of nanoparticles, including size, shape, structure, rigidity, and surface modification, can be controlled [16,17]. Additionally, lipid shell coated PLGA nanoparticles fabricated by a microfluidic method further extended the circulation half-life [18] and enhanced tumour accumulation and therapeutic effect [19].

Particle size and distribution are key parameters of nanoparticle-based drug delivery systems, because they affect multiple parameters including the blood circulation half-life [20,21], cellular uptake [21] and tumour penetration [22,23], toxicity, in vivo targeting, and metabolism [1,21]. The particle size needs to be optimized for different biomedical applications [22,24,25]. A recent study using a microfluidic method reported that PLGA particle size influences the drug release characteristics, cellular uptake, and in vivo clearance of the particles thus produced [10]. Our two-stage polydimethylsiloxane (PDMS) microfluidic chip was developed for the production of lipid-polymeric nanoparticles with varying water content to regulate the rigidity of nanoparticles [26]. This microfluidic chip had a maximum flow rate of up to 410 mL/h, enabling high-throughput synthesis of PLGA nanoparticles [27]. Additionally, this microfluidic chip can generate PLGA nanoparticles of various sizes by changing the flow rate and the flow rate ratio in the channel. We have also shown that increasing the total flow rate from 41 mL/h to 246 mL/h in the microfluidic method tends to form smaller particles [15]. However, particle size will also be influenced by a range of other factors, including drug loading, polymer chemistry, polymer concentration, additives, and other formulation conditions [6,28]. The innovation or optimization of microfluidic systems will be facilitated with the development of advanced manufacturing techniques [29–31]. Hence, it is important to systematically investigate the chemical and physical parameters that influence particle size using the microfluidic method.

The hexapeptide, phenylpropanoic acid-[ornithine-proline-d-cyclohexylalanine-tryptophan-arginine] (HC-[OPdChaWR]) is an antagonist of the C5a receptor 1 (C5aR1) of the innate immune complement cascade [32,33]. It has been shown to have efficacy in several animal disease models, such as inflammatory bowel disease, amyotrophic lateral sclerosis, allergic asthma, and spinal cord injury [34–37]. Typical of most peptides, however, it has a short plasma half-life in vivo (i.e., 0.17 h in mice following intravenous (IV) injection) and low oral bioavailability [38–40]. Utilising a nanoparticle drug delivery system has the potential to extend the circulation half-life of peptides, such as the C5aR1 antagonist investigated herein.

In this study, C5aR1 antagonist peptide-loaded lipid shell-PLGA core nanoparticles were fabricated using our two-stage microfluidic chip, and various formulation parameters were optimized to give the desired particle size distribution as well as a sustained in vitro release profile. Parameters in the microfluidic method that have the potential to influence particle size distribution were investigated. These included microfluidic flow rate, type of organic solvent utilized, theoretical drug loading, as well as PLGA types and concentration. The best performing nanoparticle formulations were identified for in vitro release investigation. The optimised peptide-loaded lipid shell-PLGA core nanoparticles were further characterized for drug incorporation efficiency and morphology. The in vivo pharmacokinetics (PK) and biodistribution of the optimised nanoparticle formulation were also assessed in mice.

2. Materials and Methods

2.1. Materials

The C5aR1 antagonist peptide (HC-[OPdChaWR]) was synthesised as previously described [41]. Poly (lactic-*co*-glycolic acid) (PLGA) (50:50, acid terminated, MW 24,000~38,000 Da) was purchased from Sigma–Aldrich Pty Ltd. (Sydney, NSW, Australia). Methoxy poly (ethylene glycol)-*b*-poly (lactide-*co*-glycolide) MW ~5000:55,000 Da (mPEG5K-PLGA55K) was purchased from PolySciTech (Akina, West Lafayette, IN, USA). Additionally, 1,2-Dihexadecanoyl-*sn*-glycero-3-phosphocholine (DPPC), 1,2-distearoylsn-glycero-3-phospho-ethanolamine-N-[methoxy (polyethylene glycol (DSPE-PEG) and cholesterol were obtained from Avanti Pty Ltd. (Tonawanda, NY, USA). Phosphate-buffered saline (PBS) tablets were from Astral Scientific (Sydney, NSW, Australia). Dichloromethane (DCM) and ethanol (EtOH) were from Biolab Pty Ltd. (Melbourne, VIC, Australia). Acetonitrile (ACN) was from Ajax Finechem Pty Ltd. (Brisbane, QLD, Australia). Methanol (MeOH) was bought from RCI Labscan Ltd. (Sydney, NSW, Australia). Ethyl acetate (EA), trifluoroethanol (TFE), and dimethylformamide (DMF) were from Sigma Aldrich (Sydney, NSW, Australia). All solvents used in the study were of analytical and/or high-performance liquid chromatography (HPLC) grade. Zolazepam and xylazine were obtained from Provet, Brisbane, QLD, Australia.

2.2. Preparation of Peptide-Loaded Nanoparticles

Peptide-loaded nanoparticles were fabricated using an optimised microfluidic method [42,43]. The two-stage microfluidic chip (Figure 1) was designed and fabricated [15] in the laboratory of Professor Xingyu Jiang. Briefly, the microfluidic chip was fabricated on a SU8-2100 master mould based on standard soft lithography. The two-stage microfluidic chip was composed of a first stage (three inlets and a straight mixing channel, 100 µm in width and 60 µm in height) and a second stage (a centre inlet and a mixing spiral channel, 300 µm in width and 60 µm in height). To obtain an appropriate ratio of thickness to height of the channels (60 µm), SU8-2100 photoresist (MicroChem Corp, Westborough, MA, USA) was spin-coated on a 4 inch silicon wafer at 500 rpm for 10 s and then at 3400 rpm for 60 s. The spin-coated SU8-2100 photoresist was baked at 65 °C for 5 min and at 95 °C for 10 min. The spin-coated SU8-2100 photoresist was exposed to ultraviolet light (150 mJ·cm^{-2}) through a photomask containing the pattern of channels. The exposed photoresist was baked at 60 °C for 6 min and 110 °C for 8 min. SU-8 developer (MicroChem Corp) was used to dissolve the unexposed photoresist. The method for fabricating the microfluidic chip with polydimethylsiloxane (PDMS) (Sylgard 184, Dow Corning Inc., Midland, MI, USA) and the master mould was the same as the previous work [27]. A peptide–polymer solution was prepared by dissolving peptide and PLGA in an organic solvent. The lipid solution was DPPC (4.55 mg/mL), DSPE-PEG (0.85 mg/mL), and cholesterol (0.48 mg/mL) in ethanol. The peptide–polymer solution, lipid solution, and water were injected into the microfluidic chip. The peptide–polymer solution and lipid solution were injected through the middle inlet and central inlet with the same flow rate, respectively. Water was injected from two side inlets with the same flow rate. The flow rate for each flow inlet was precisely controlled by syringe pumps (PHD ULTRA CP, SDR Scientific Pty Ltd., Sydney, NSW, Australia). The mixture containing peptide-loaded nanoparticles was collected from the outlet side of the microfluidic chip.

To remove the organic solvent, fabricated peptide-loaded nanoparticles were stirred on ice (200 rpm) for 1 h and then placed under vacuum for 30 min using a rotary evaporator (Rotavapor R-210, BUCHI Ltd., Newmarket, UK) at room temperature (~25 °C). Peptide-loaded nanoparticles were washed twice with Milli-Q water, and centrifuged at 20,000× *g* (Avanti JE, Beckman Coulter, Indianapolis, CA, USA) for 15 min. Particles were then resuspended in a small volume of PBS or 5% glucose for subsequent experiments.

Particle size and peptide release profiles were optimised by investigating a range of organic solvents (DCM, EA, TEF + DMF (3:7, *v/v*) [15], ACN or acetone), multiple sides (mL/h)/centre (mL/h) flow rate ratios (sides/centre, 60/4.5 [43], 120/3 [15] and 120/1), several theoretical drug loadings (5, 10, and 20%), as well as various PLGAs (PLGA or PLGA + PEG-PLGA) and polymer concentrations (1 or 2%). The flow rates of the two side (water) inlets were controlled at 60 mL/h, while the centre (polymer + drug) inlet flow rate was 4.5 mL/h. When the flow rates of the two side (water) inlets were increased to 120 mL/h, the centre inlet flow rate was reduced to either 3 or 1 mL/h. Detailed information on the parameters for all formulations prepared is summarised in

Supplementary Table S1. The flow rates were specified for various total flow rates and flow rate ratios (water in the side inlets to solvent in the centre inlet) calculated using the equations listed in Supplementary Table S2.

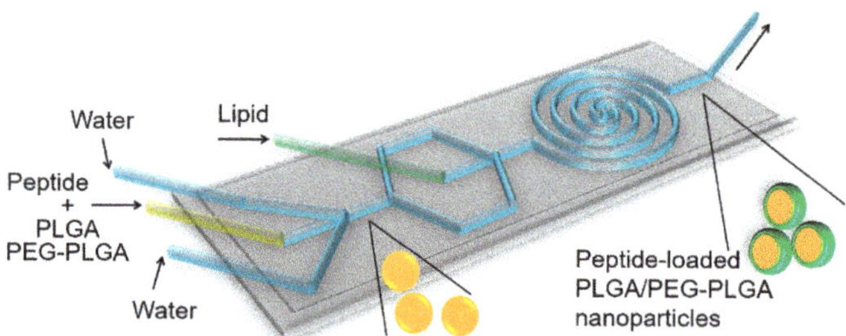

Figure 1. Schematic illustration of the microfluidic chip and the nanoparticle formulation process. There are four inlets and one outlet in the microfluidic chip. Peptide-PLGA solution is injected through the middle inlet while water is injected through two side inlets. The peptide-loaded PLGA nanoparticles were formed at this stage. Lipid injected through the central inlet forms the lipid shell structure of the PLGA nanoparticles. Fabricated peptide-loaded lipid shell-polymer core nanoparticles are collected from the outlet.

2.3. Particle Size, Polydispersity Index (PDI) and Zeta Potential

The size, size distribution, PDI, and zeta-potential of the nanoparticles were determined using dynamic light scattering (DLS) at a scattering angle of 173° (Zetasizer Nano ZS, Malvern Instruments, Worcestershire, UK).

2.4. Incorporation Efficiency

The peptide-loaded nanoparticle suspension (200 μL) was freeze-dried ($n = 3$) and then completely dissolved in 1 mL of acetonitrile, and the peptide concentration was quantified using LC-MS/MS [38]. The actual drug loading (%, w/w) and the drug incorporation efficiency, expressed as encapsulation efficiency (EE, %, w/w) were calculated using the equations below. The individual values for three replicate determinations and their mean values are reported.

$$\text{Actual drug loading } (\%) = w(\text{drug})/[w(\text{drug}) + w(\text{polymer})] \tag{1}$$

$$\text{EE } (\%) = w(\text{drug in particles})/w(\text{drug added}) \tag{2}$$

2.5. In Vitro Release Profile

The peptide-loaded nanoparticle suspension (200 μL) was transferred into dialysis tubes (MWCO 10 kDa). Each dialysis tube was sealed, placed into a capped container containing 5 mL PBS, and then placed into an incubator maintained at 37 °C and shaken horizontally at 100 shakes min^{-1} (IKA® KS130 shaker; Sigma–Aldrich Pty Ltd., Sydney, NSW, Australia). On each testing occasion (i.e., 1, 3, 6, 24, 30, 48, 54, and 72 h), the dialysis tubes were rinsed with 1 mL of PBS and transferred to 5 mL of fresh PBS. Concentrations of HC-[OPdChaWR] were determined in triplicate by LC-MS/MS [38] at each time of assessment, and the data are presented.

2.6. Measurement of Peptide by Liquid Chromatography–Tandem Mass Spectrometry (LC–MS/MS)

The C5aR1 peptide antagonist concentrations were quantified by a previously developed and validated LC–MS/MS method [38]. Briefly, the LC–MS/MS system consisted of an API 3200 (AB SCIEX) triple quadrupole LC–MS/MS mass spectrometer coupled with an Agilent 1200 series HPLC system (Agilent Technologies, Santa Clara CA, USA). For sample processing, 10 μL of the solution containing the C5aR1 peptide antagonist was mixed with

10 μL of internal standard (AcF-[OPdChaWR]) dissolved in solvent (ACN) followed by sample processing, extraction, supernatant isolation, drying, and reconstitution in a 75% methanol-in-water solution. A small volume of processed sample (approximately 50 μL) was used for the LC–MS measurement. The autosampler was set at 4 °C and the column oven temperature was maintained in the range of 25 ± 1 °C. The system control and data acquisition were executed by Analyst software (AB SCIEX, Applied Biosystems Inc., Framingham, MA, USA, version 1.5.1). Chromatographic separation was implemented using a Kinetex EVO C18 analytical column (100 mm × 2.1 mm, 100 Å, 5 μm, Phenomenex Inc., Torrance, CA, USA) under binary gradient conditions using mobile phase A (milliQ water containing 0.1% formic acid) and mobile phase B (acetonitrile containing 0.1% formic acid) with a variable flow rate, as described in Kumar et al. [38].

2.7. Morphology

The morphology of nanoparticles was observed under a transmission electron microscope (TEM, Hitachi HT7700, Hitachi High-Technologies Corporation, Tokyo, Japan) at 100 kV. Briefly, 10 μL of diluted nanoparticle suspension (1 mg/mL) was dropped onto a carbon-coated copper grid and incubated for 10 min to allow nanoparticles to attach to the membrane. Next, samples were negatively stained with 1% uranyl acetate and viewed.

2.8. Pharmacokinetics Study and Biodistribution

All experimental procedures involving animals were performed following approval from the animal ethics committee of the University of Queensland (The approval number is 241/18, and the date of approval is 21 August 2018). All experiments were conducted as per the National Health and Medical Research Council of Australia policies and guidelines for the care and use of animals for scientific purposes (8th Edition, 2013). Wild-type C57BL/6J mice (male, 10–12 weeks old) were purchased from the Animal Resources Centre (Perth, WA, Australia). All animals were housed within the University of Queensland Biological Resources animal facility in a pathogen-free environment with a 12 h dark/12 h light cycle and free access to food and water.

The plasma pharmacokinetic (PK) profile and biodistribution of peptide-loaded nanoparticles (1 mg/kg equivalent dose of free peptide) vs. free peptide were assessed in mice after intravenous (IV; via tail vein) dosing. Serial blood samples were collected via a tail vein using a microsampling technique [40]. Blood samples were collected at 0.04, 0.25, 0.5, 0.75, 1, 1.5, 1, 4, 6, and 24 h for both groups and then 48, 72, 96, and 120 h for the nanoparticle group. The blood samples were centrifuged at 2000× *g* for 10 min at 4 °C, and the separated plasma samples (10 μL) were collected, diluted with 100 μL of milliQ water and 10 μL of internal standard, filtered through Microcon®-10 centrifugal filters (Merck, Melbourne, VIC, Australia) to separate nanoparticles from released peptide, and stored at −80 °C until time of analysis. For tissue distribution, on the terminal day, i.e., day 1 for native peptide and day 5 for the nanoparticle group, animals were transcardially perfused using PBS solution to remove circulatory peptide/formulation, and tissue samples were collected, processed as described in our previously published method [38], and filtered as mentioned above.

On the day of analysis, samples were processed for C5aR1 peptide antagonist concentrations using the above-mentioned LC–MS/MS method as described previously [38]. The IV route of drug administration was used for this study to achieve 100% bioavailability of peptide or peptide-loaded nanoparticles in the systemic circulation of the mice.

2.9. Data Analysis

The data are presented as mean (±SEM) unless otherwise specified, and plasma concentration versus time curves were prepared using GraphPad Prism (GraphPad Software Inc., La Jolla, CA, USA, version 7). The statistical significance criterion was $p \leq 0.05$. LC–MS/MS data processing and analysis were performed using Analyst software (AB

SCIEX, Applied Biosystems Inc., Framingham, MA, USA, version 1.5.1) and Multiquant software (AB SCIEX, USA, version 2.0).

PK parameters were derived from the plasma concentration versus time profiles using a two-compartmental model as implemented in the WinNonlin software package (Certara, L.P, St. Louis, MO, USA, v8.1). The linear trapezoidal rule was applied for the calculation of area under the curve (AUC $_{0-t}$) of the plasma concentration vs. time profiles.

3. Results and Discussion

3.1. Optimisation of Particle Size

3.1.1. Effect of Flow Rate (Side/Centre)

Changing the side (mL/h)/centre (mL/h) flow rate from 60/4.5 to 120/3 then to 120/1 showed similar effects on the mean size of nanoparticles fabricated by the various organic solvents assessed (Figure 2). The mean size and PDI of particles for all formulations are summarised in Table S1. By increasing the total flow rate and flow rate ratio (water in side inlets to solvent in centre inlet) (Table S2), the size of nanoparticles fabricated using DCM decreased from ~400 to ~150 nm, the size of nanoparticles using ACN decreased from ~200 to <100 nm, and the size of nanoparticles using TFE + DMF (3:7, v/v) decreased from 250 to ~30 nm. It is worth noting that increasing the sides/centre flow rate ratio had a more significant effect on the PDI of nanoparticles formulated with DCM, resulting in a sharply decreased PDI, from 0.8 to approximately 0.3. In comparison, using a higher sides/centre flow rate ratio only caused a small effect on the PDI of nanoparticles from the TFE + DMF (3:7, v/v) formulation and no significant effect on nanoparticles from the ACN formulation. Generally, a higher total flow rate and higher flow rate ratio (water to solvent) are believed to provide a higher shear stress, which could produce smaller organic droplets and subsequently form smaller nanoparticles through precipitation [44]. Subsequently, the higher side/centre flow rate ratio results in smaller nanoparticles with greater size homogeneity (lower PDI) [15,42]. Additionally, the higher side/centre flow rate ratio results in more effective and rapid interfacial deposition, thus generating nanoparticles with more uniform size [15,42].

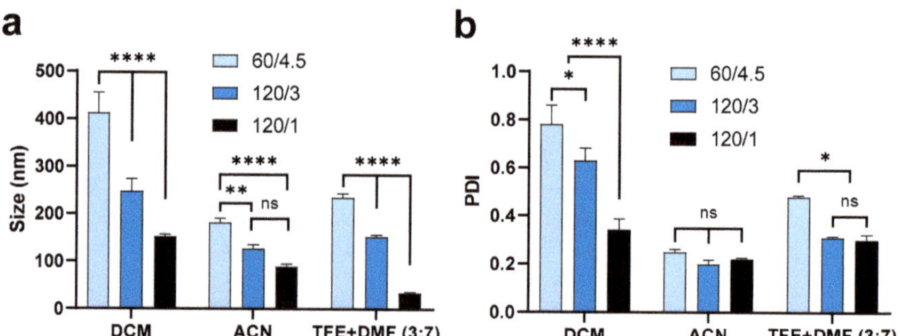

Figure 2. Mean (±SEM) (*n* = 3) size (**a**) and PDI (**b**) of 'blank' PLGA nanoparticles formulated using 1% PLGA in DCM, ACN, and TFE + DMF (3:7, v/v) under a range of side/centre flow rate conditions (water inside inlets to solvent in centre inlet, 60/4.5 = 60 to 4.5 mL/h, 120/3 = 120 to 3 mL/h, 120/1 = 120 to 1 mL/h). Using the same solvent, increasing flow rate and flow rate ratios significantly decreased the size of nanoparticles. Increasing flow rate or flow rate ratio caused a dramatic decrease in the PDI of nanoparticles formulated using DCM but did not significantly affect nanoparticles formulated using ACN or TFE + DMF (3:7). Two-way ANOVA analysis followed by Tukey test for multiple comparisons (**** $p < 0.0001$; ** $p < 0.01$; * $p < 0.05$; and ns, no significant difference).

3.1.2. Effect of Organic Solvent

As can be seen in Figure 2, the organic solvent utilized had a significant influence on the size of nanoparticles, even if all other factors were the same. Therefore, the mean size and PDI of peptide-loaded nanoparticles (5% theoretical loading) were compared between different organic solvents. As shown in Figure 3, using ACN resulted in the smallest size (200 nm) and PDI (~0.2) under all three different side/centre flow rate ratios. The nanoparticles fabricated under a low side/centre flow rate ratio (60/4.5) using acetone and the combined solvent, TFE + DMF (3:7, v/v) had a slightly larger size (200–250 nm) and relatively high PDI (0.5 and 0.3, respectively). On the other hand, nanoparticles fabricated using DCM were the largest size (~400 nm) and had the broadest size distribution (PDI 0.8) followed by nanoparticles fabricated using EA (size 300 nm, PDI 0.5). The same trend was observed under high flow rate conditions (side/centre at 120/3), but the difference between solvents was attenuated (Figure 3c,d). Peptide-loaded nanoparticles performed similarly to the 'blank' nanoparticles (Figure 3e,f), such that particle sizes were 400, 350, and 150 nm, both with and without peptide for nanoparticles formulated using DCM, EA, and ACN, respectively. However, in comparison to the blank (without peptide) nanoparticles (PDI 0.3–0.8), the PDI of nanoparticles containing encapsulated peptide was decreased (PDI 0.2–0.3).

In brief, ACN and TFE + DMF (3:7, v/v) always led to the fabrication of smaller sized particles, characterized by a narrower size distribution (lower PDI), whereas DCM and EA resulted in larger particle sizes and a broader size distribution (higher PDI). This fact could be explained by the differential miscibility of the various organic solvents in water [45]. DCM and EA are water-immiscible solvents, whereas TEF, DMF, ACN, and acetone are miscible with water. In agreement with others, larger particles were fabricated using DCM compared with partially/fully water-miscible solvents [45]. Similarly, increasing the ratio of DCM in a mixed solvent resulted in increased particle size and PDI [46]. Concerning the geometry of the microfluidic chip, the PLGA dissolved in a water-miscible solvent generated a more efficient shearing force that was beneficial for self-assembly via nanoprecipitation subsequently resulting in smaller and mono-dispersed nanoparticles [47]. Additionally, liquid viscosity, density, and surface tension also have a dynamic effect on the droplet formation and particle size distribution [47,48]. However, it was noticed that the microfluidic chip was blocked easily when using ACN as the organic solvent, and this caused substantial drug loss, so this method is not feasible for producing particles in a large quantity. Additionally, the PDI of nanoparticles fabricated using DCM was smaller when the peptide was encapsulated in comparison with 'blank' nanoparticles (Figure 3d,f). For further investigation, a range of drug loadings were assessed for their impact on particle size distribution. DCM was used for further formulation development, as its lower boiling point means that it can be fully removed under ambient temperature and pressure. Another consideration was that our peptide is soluble in water (1 mg/mL), and so water miscible solvents are unsuitable for this purpose.

3.1.3. Effect of Theoretical Drug Loading

The size of nanoparticles increased with increased theoretical drug loading (from 400 nm to 600 nm) (Figure 4). However, the nanoparticles loaded with peptide had a narrower size distribution (PDI 0.35–0.45) compared with the blank nanoparticles (PDI ~0.8). The presence of peptide aided the formation of homogenous nanoparticles, although the mechanism behind this phenomenon is unclear and needs further investigation. It is known that, during formulation, nanoparticles undergo nucleation, growth, and aggregation [49]. The C5aR1 peptide antagonist contains a hydrocinnamate residue and is sufficiently lipophilic to have a positive effect on nucleation or particle growth [32,50]. It is plausible that a hydrophobic core may form more readily by adding peptide compared with particles comprised of pure PLGA when DCM is used as the solvent in the microfluidic method. Therefore, a lower PDI was observed for particles with encapsulated peptide (Figure 4b).

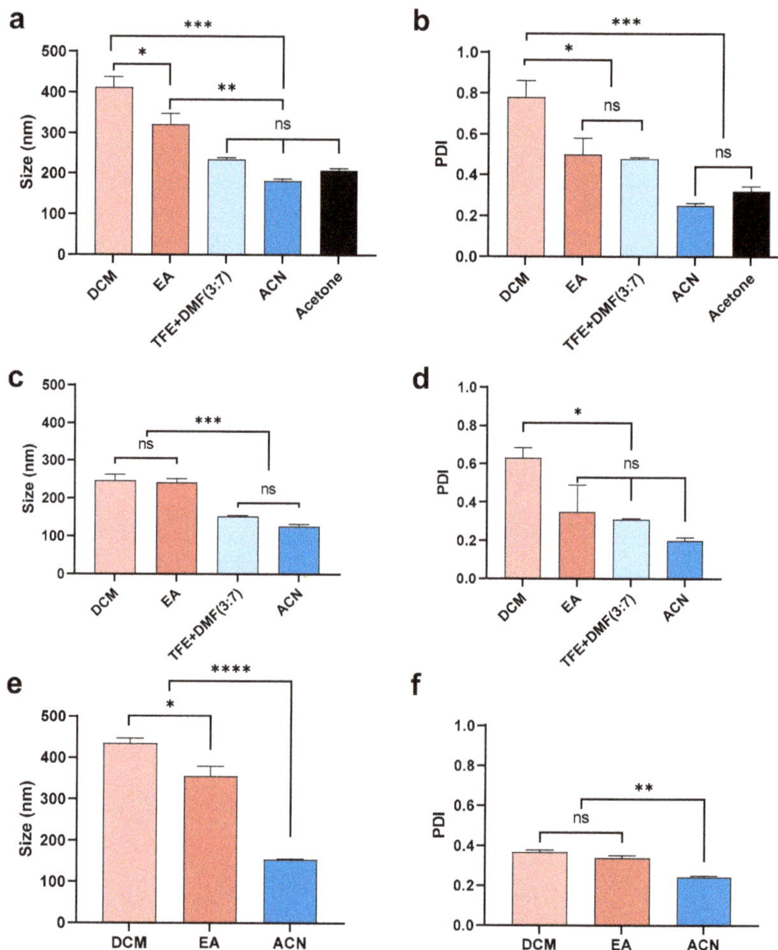

Figure 3. Mean (±SEM) (*n* = 3) size (**a**) and PDI (**b**) of blank nanoparticles formulated using a low sides/centre flow rate ratio (60/4.5) using 1% PLGA in a range of organic solvents (DCM, EA, TFE + DMF (3:7, *v*/*v*), ACN, and acetone). Nanoparticles fabricated using DCM had the largest size (400 nm) and the highest PDI (0.8). By comparison, use of ACN resulted in particles of the smallest size (<200 nm) and lowest PDI (~0.2). Mean (±SEM) (*n* = 3) size (**c**) and PDI (**d**) of blank nanoparticles formulated using a high sides/centre flow rate ratio (120/3) using 1% PLGA in different organic solvents (DCM, EA, TFE + DMF (3:7, *v*/*v*), and ACN). Both DCM and EA produced particles of a larger size (~250 nm) compared with ACN (~100 nm). DCM provided the highest PDI (0.6) while ACN led to the lowest PDI (<0.2). Mean (±SEM) (*n* = 3) size (**e**) and PDI (**f**) of 5% theoretical drug-loaded nanoparticles formulated with a low side/centre flow rate ratio (60/4.5) using 1% PLGA in different organic solvents (DCM, EA, and ACN). DCM provided a larger particle size (>400 nm) compared with EA (~350 nm) and ACN (150 nm). Solvents did not show a significant difference in PDI (0.2~0.3). One-way ANOVA analysis followed by Tukey test for multiple comparisons (**** $p < 0.0001$; *** $p < 0.001$; ** $p < 0.01$; * $p < 0.05$; and ns, no significant difference).

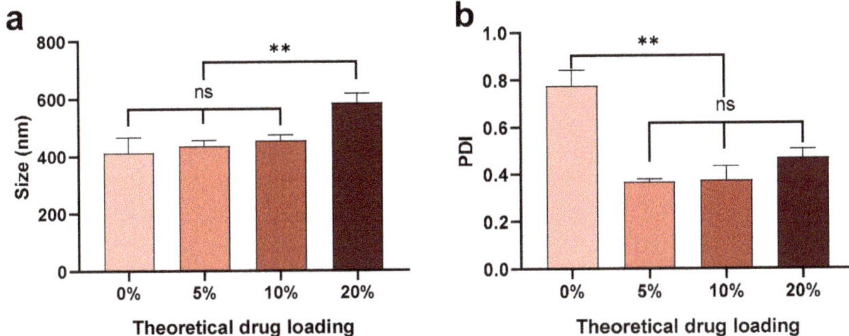

Figure 4. Mean (±SEM) (*n* = 3) size (**a**) and PDI (**b**) of nanoparticles with a range of theoretical drug loadings (0, 5, 10, and 20%) fabricated using 1% PLGA in DCM using a low side/centre flow rate ratio (60/4.5). The mean size of nanoparticles increased with increased drug loading, with this effect most marked at 20% peptide loading (0 < 5 < 10 < 20%). The PDI was significantly decreased for particles containing 5% peptide (<0.4) compared with blank nanoparticles (0.8). PDI increased with increased drug loading (5 < 10 < 20%). One-way ANOVA analysis followed by Tukey test for multiple comparisons (** $p < 0.01$; and ns, no significant difference).

3.1.4. Effect of Polymer

PLGA (1%) was compared with the mixture of 1%PLGA + 1%PEG-PLGA (Figure 5). There was no significant difference in size and PDI between nanoparticles fabricated using 1% PLGA or 1% PLGA + 1% PEG-PLGA when using DCM as the solvent and 5% theoretical drug loading (Figure 5a,b). This finding suggested that the addition of an extra 1% PEG-PLGA did not greatly affect particle size distribution. However, adding PEG-PLGA significantly decreased the Z-potential of nanoparticles (−10 mV) compared with PLGA alone (−40 mV) (Figure 5c). Our finding is consistent with others who have reported that PEGylation always results in a dramatic decrease in the charge of PLGA nanoparticles [51–53]. As reported by Garinot et al., the decreased zeta potential close to neutrality may result from the shielding effect of the PEG chains [53].

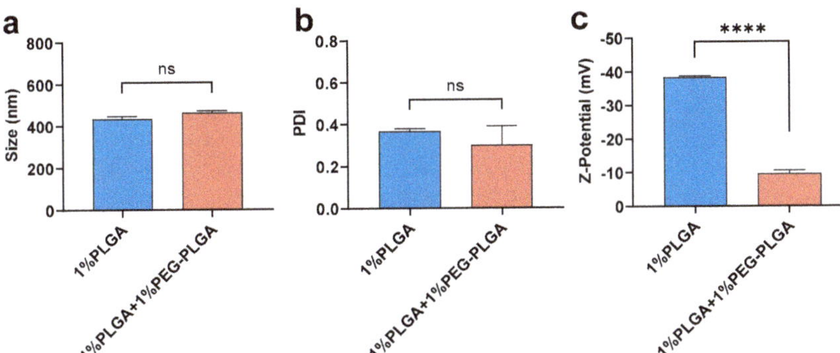

Figure 5. Mean (±SEM) (*n* = 3) size (**a**), PDI (**b**), and Z-potential (**c**) of peptide-loaded nanoparticles (5% theoretical drug loading) fabricated using 1% PLGA in DCM or 1% PLGA + 1% PEG-PLGA in DCM under low side/centre flow rate ratio conditions (60/4.5). The extra 1% PEG-PLGA showed a slight increment in size distribution but significantly lower Z-potential. One-way ANOVA analysis followed by Tukey test for multiple comparisons (**** $p < 0.0001$; and ns, no significant difference).

To produce drug-loaded polymeric nanoparticles, nanoprecipitation and emulsion-based methods are commonly used. Nano-precipitation is based upon the interfacial deposition of polymeric nanoparticles when water-miscible solvents containing dissolved polymer are mixed with water [54]. To generate uniform and size-controlled nanoparticles by nano-precipitation, rapid and sufficient mixing is generally required [55] and often leads to a wide range size distribution. In agreement with work by others [55,56], our data show that, using a rapid mixing microfluidic method to co-precipitate the drug and polymer, more uniform and size-controlled nanoparticles are produced.

3.2. Optimisation of In Vitro Release Profile and Further Characterization

As shown in Figure 6a, peptide-loaded PLGA/PEG-PLGA nanoparticles achieved a maximum cumulative release of ~25% within 72 h, whereas the release from peptide-loaded PLGA nanoparticles was only ~13%. Additionally, PEG-PLGA was reported to form stealth nanoparticles, which have a longer in vivo circulation half-life [52]. However, more polymer concentrations and other types of PLGA with different L:G ratios, different molecular weights, and different terminal groups (e.g., ester group) remain for future investigation to achieve the optimal release profile [1].

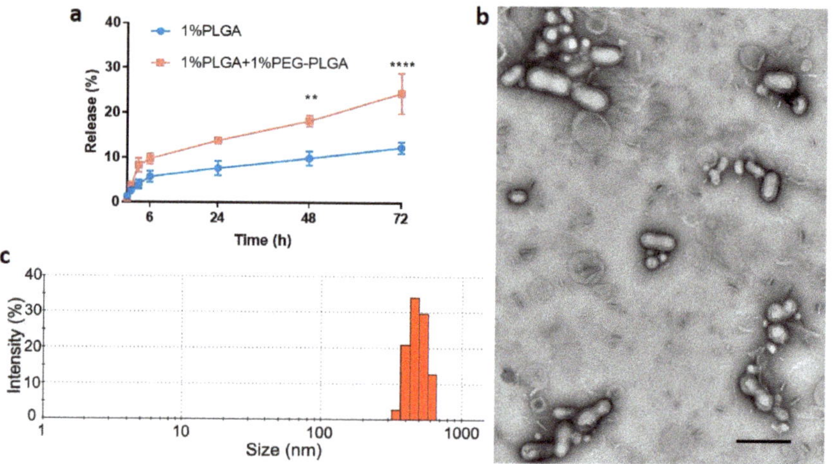

Figure 6. (**a**) Mean (±SEM) in vitro C5aR1 peptide antagonist release profiles ($n = 3$) for nanoparticles formulated using 1% PLGA in DCM or 1% PLGA + 1% PEG-PLGA in DCM using a low side/centre flow rate ratio (60/4.5). The percentage drug release was calculated based on actual drug loading. Two-way ANOVA analysis followed by Sidak's test for multiple comparisons (**** $p < 0.0001$; and ** $p < 0.01$). Nanoparticles prepared using 1% PLGA + 1%PEG-PLGA in DCM were characterized for morphology (**b**) using transmission electron microscopy (TEM) and size distribution (intensity%) (**c**) using DLS. The TEM image indicates that peptide nanoparticles were successfully coated by a lipid shell as the foggy boundary could be observed. The shadowy particles in the background are components that did not coat polymer particles. The scale bar is 400 nm.

The modified formulation of 1% PLGA + 1% PEG-PLGA in DCM and 5% theoretical drug loading using a low side/centre flow rate ratio (60/4.5) was further investigated before undertaking the in vivo study. The actual drug loading for this formulation was $2.0 \pm 0.5\%$ (mean ± SD, $n > 3$) with EE of $38.2 \pm 10.6\%$ (mean ± SD, $n > 3$). Note that the formulation with 1% PLGA alone has an actual drug loading of approximately 1.8% with the encapsulation efficiency at approximately 36%. The core–shell structure of these nanoparticles was confirmed using TEM (Figure 6b). The size distribution is shown in Figure 6c, and it is consistent with those previously determined (Figure 5). Although a few lipid components could be observed in the freshly prepared formulation, they can be readily removed through washing. The mean size of the particles was relatively large at 400 nm, consistent with previous work that showed relatively large particles had less initial burst release compared with smaller particles [57]. Additionally, particles of this size (<500 nm) are unlikely to be cleared by phagocytosis [58] before releasing their drug payload. Thus, the precise fabrication of particles at ~400 nm was achieved by using the microfluidic method and it was selected for the in vivo study.

Previously, we have reported on the release of linear and cyclic peptides from polyurethane films with different hard and soft segments. Solvent casting of the polyurethane at room temperature mixed with the various peptides resulted in reproducible efflux profiles with no evidence of drug degradation. The organic solvent used was DCM. The C5aR1 antagonist peptide retained its biological activity during casting of the polyurethane films and following release from the films [59].

3.3. In Vivo Pharmacokinetics Study and Bio-Distribution

The innate immune targeted C5aR1 antagonist peptide encapsulated within the biodegradable nanoparticles had a significantly reduced elimination half-life in mice compared to the peptide itself (Figure 7a). Specifically, the plasma peptide levels after IV dosing with free peptide were unmeasurable after 6 h post-dosing whereas peptide from the nanoparticles was measurable in plasma samples for up to five days after a single IV dose administration. The terminal elimination half-life ($t_{1/2}$) after dosing of mice with free peptide was 0.84 h, which is in agreement with the short $t_{1/2}$ values of 0.17 h and 0.33 h reported previously [38,40]. The half-life of peptide in the form of nanoparticles was extended significantly (24.9 h). Similarly, peptide-loaded nanoparticles have a much longer mean residence time (MRT) (32.61 h) compared with that of the free peptide (0.76 h). The systemic exposure (AUC_{0-t}) of peptide released from peptide-loaded nanoparticles was nearly five times larger than that for the free peptide (AUC for free peptide = 3825.7 ng.h/mL; AUC for peptide-loaded nanoparticle = 18,665.6 g.h/mL). Thus, our findings from the mouse pharmacokinetic study have confirmed that the peptide-loaded nanoparticles significantly increased the systemic exposure of this peptide when administered via the optimised nanoparticle formulation.

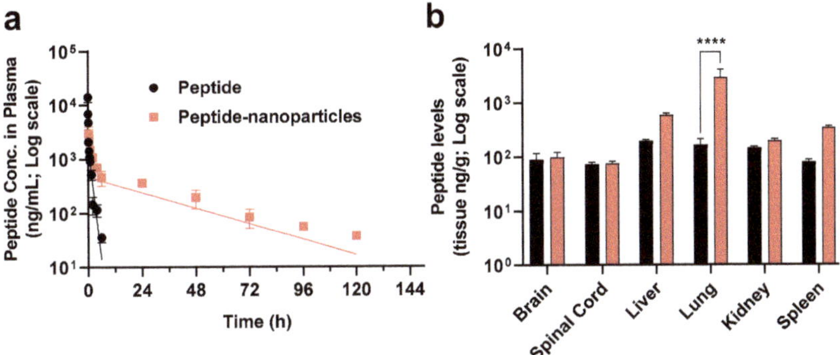

Figure 7. (**a**) Plasma concentration versus time profile (two-compartmental model) (mean ± SEM, *n* = 4) of the C5aR1 peptide antagonist administered by bolus intravenous (IV) injection at a dose of 1 mg/kg either as free compound (black) or in the form of lipid-shell PLGA nanoparticles (pink). (**b**) Biodistribution (mean ± SEM, *n* = 4) of peptide released from nanoparticles in different organs (brain, spinal cord, liver, lung, kidney, spleen) (ng/g). Tissues were extracted on day one and day five after a single IV injection of 1 mg/kg either as free peptide (black on day one due to the short half-life of this peptide based on our previous PK studies for the free peptide) or in the form of nanoparticles (pink on day five), respectively. Statistical analysis was conducted by two-way ANOVA, followed by Sidak's test for multiple comparisons (**** $p < 0.0001$).

Additionally, the biodistribution of peptide in mice administered a single IV injection of the optimised nanoparticle formulation, showing that 18-fold higher concentrations of peptide were present in the lung tissue of the same animals for at least five days compared with animals dosed with free peptide (Figure 7b). In support of this notion, previous work by others showed that C5aR1 inhibition alleviated influenza virus-induced lung injury in mice [60]. Recently, other work showed that anti-C5a monoclonal antibody treatment improved the deteriorating health of patients with COVID-19 [61], and targeting C5aR1 with anti-C5aR1 monoclonal antibodies also has demonstrated mechanistic merit [62]. Compared with peptides, antibodies are expensive, potentially lack tissue penetrance, and some patients may also develop auto-immune reactions. Thus, alternative drug modalities that can block the activation of C5aR1 and overcome the limitations of therapeutic antibodies are urgently needed. The C5aR1 antagonist peptide-loaded lipid-shell nanoparticles described in this study may therefore have potential for more effectively treating lung inflammation, and this remains an area for future investigation.

4. Conclusions

In summary, our findings provide valuable information on how to precisely control the fabrication of peptide-encapsulated nanoparticles to achieve a designated size using a microfluidic device, including flow rate, organic solvent, theoretical drug loading, as well as PLGA type and concentration. Small peptide-encapsulated nanoparticles (200–400 nm) were successfully produced. This study demonstrated a marked increase in the systemic exposure and a significant increase in the terminal half-life of the same IV dose of a C5aR1 peptide antagonist administered to mice in the form of peptide-loaded nanoparticles relative to the corresponding parameters for the free peptide. We also demonstrated that the lipid-shell-PLGA core nanoparticles significantly increased the biodistribution of peptide in the lung. Further development of our method has the potential to be developed into a safe, effective delivery platform for small molecule peptide drugs in general and pave the way for their progression to clinical development.

Supplementary Materials: The following are available online at https://www.mdpi.com/article/10.3390/pharmaceutics13091505/s1, Table S1: Formulations of peptide-loaded lipid shell-PLGA core nanoparticles, Table S2: Parameters of flow rate applied in the study.

Author Contributions: Conceptualization, F.Y.H.; Data curation, F.Y.H., W.X., V.K., C.S.C. and X.L.; Formal analysis, F.Y.H., W.X., V.K. and X.L.; Funding acquisition, F.Y.H., A.K.W. and M.T.S.; Investigation, F.Y.H. and W.X.; Methodology, F.Y.H., W.X., V.K., X.L. and X.J.; Project administration, F.Y.H. and W.X.; Resources, F.Y.H., X.J., T.M.W. and A.K.W.; Software, F.Y.H. and W.X.; Supervision, F.Y.H., T.M.W., A.K.W. and M.T.S.; Validation, F.Y.H., W.X. and V.K.; Visualization, F.Y.H., W.X., V.K., C.S.C. and M.T.S.; Writing—original draft, F.Y.H. and W.X.; Writing—review & editing, F.Y.H., W.X., V.K., C.S.C., X.J., T.M.W., A.K.W. and M.T.S. All authors have read and agreed to the published version of the manuscript.

Funding: W.X. and F.Y.H. were supported financially by a National Health and Medical Research Council (NHMRC) Grant [APP1107723]. We thank Richard J. Clark for providing the C5aR1 antagonist peptide used in this study. T.M.W. acknowledges support from the National Health and Medical Research Council (APP1118881 and APP1105420). Financial support from the Australian Research Council (LE0775684, LE110100028, LE110100033, LE140100087, LE160100168) is also acknowledged. The authors acknowledge the Queensland Government Smart State Research Facilities Programme for supporting Centre for Integrated Preclinical Drug Development (CIPDD; School of Biomedical Sciences) research infrastructure. CIPDD is also supported by Therapeutic Innovation Australia (TIA). TIA is supported by the Australian Government through the National Collaborative Research Infrastructure Strategy (NCRIS) program.

Institutional Review Board Statement: All experimental procedures involving animals were performed following approval from the animal ethics committee of the University of Queensland (The approval number is 241/18, and date of approval is 21 August 2018).

Conflicts of Interest: T.M.W. has previously consulted to Alsonex Pharmaceuticals Pty Ltd., who are commercially developing HC-[OPdChaWR]. He holds no stocks, shares, or other commercial interest in this company. All other authors have no conflicts of interest to declare in association with this work.

References

1. Allahyari, M.; Mohit, E. Peptide/protein vaccine delivery system based on PLGA particles. *Hum. Vaccin Immunother.* **2016**, *12*, 806–828. [CrossRef] [PubMed]
2. Operti, M.C.; Bernhardt, A.; Grimm, S.; Engel, A.; Figdor, C.G.; Tagit, O. PLGA-based nanomedicines manufacturing: Technologies overview and challenges in industrial scale-up. *Int. J. Pharm.* **2021**, *605*, 120807. [CrossRef] [PubMed]
3. Gentile, P.; Chiono, V.; Carmagnola, I.; Hatton, P.V. An overview of poly(lactic-co-glycolic) acid (PLGA)-based biomaterials for bone tissue engineering. *Int. J. Mol. Sci.* **2014**, *15*, 3640–3659. [CrossRef] [PubMed]
4. Han, F.Y.; Thurecht, K.J.; Whittaker, A.K.; Smith, M.T. Bioerodable PLGA-Based Microparticles for Producing Sustained-Release Drug Formulations and Strategies for Improving Drug Loading. *Front. Pharmacol.* **2016**, *7*, 185. [CrossRef] [PubMed]
5. Zambaux, M.F.; Bonneaux, F.; Gref, R.; Maincent, P.; Dellacherie, E.; Alonso, M.J.; Labrude, P.; Vigneron, C. Influence of experimental parameters on the characteristics of poly(lactic acid) nanoparticles prepared by a double emulsion method. *J. Control Release* **1998**, *50*, 31–40. [CrossRef]
6. Bisht, R.; Rupenthal, I.D. PLGA nanoparticles for intravitreal peptide delivery: Statistical optimization, characterization and toxicity evaluation. *Pharm. Dev. Technol.* **2018**, *23*, 324–333. [CrossRef]
7. Han, F.Y.; Whittaker, A.; Howdle, S.M.; Naylor, A.; Shabir-Ahmed, A.; Smith, M.T. Sustained-Release Hydromorphone Microparticles Produced by Supercritical Fluid Polymer Encapsulation. *J. Pharm. Sci.* **2019**, *108*, 811–814. [CrossRef] [PubMed]
8. Wan, F.; Yang, M. Design of PLGA-based depot delivery systems for biopharmaceuticals prepared by spray drying. *Int. J. Pharm.* **2016**, *498*, 82–95. [CrossRef]
9. Morikawa, Y.; Tagami, T.; Hoshikawa, A.; Ozeki, T. The Use of an Efficient Microfluidic Mixing System for Generating Stabilized Polymeric Nanoparticles for Controlled Drug Release. *Biol. Pharm. Bull.* **2018**, *41*, 899–907. [CrossRef]
10. Operti, M.C.; Dölen, Y.; Keulen, J.; van Dinther, E.A.W.; Figdor, C.G.; Tagit, O. Microfluidics-Assisted Size Tuning and Biological Evaluation of PLGA Particles. *Pharmaceutics* **2019**, *11*, 590. [CrossRef]
11. Operti, M.C.; Fecher, D.; van Dinther, E.A.W.; Grimm, S.; Jaber, R.; Figdor, C.G.; Tagit, O. A comparative assessment of continuous production techniques to generate sub-micron size PLGA particles. *Int. J. Pharm.* **2018**, *550*, 140–148. [CrossRef] [PubMed]
12. Xu, Q.; Hashimoto, M.; Dang, T.T.; Hoare, T.; Kohane, D.S.; Whitesides, G.M.; Langer, R.; Anderson, D.G. Preparation of monodisperse biodegradable polymer microparticles using a microfluidic flow-focusing device for controlled drug delivery. *Small* **2009**, *5*, 1575–1581. [CrossRef]
13. Hung, L.H.; Teh, S.Y.; Jester, J.; Lee, A.P. PLGA micro/nanosphere synthesis by droplet microfluidic solvent evaporation and extraction approaches. *Lab Chip* **2010**, *10*, 1820–1825. [CrossRef]
14. Valencia, P.M.; Basto, P.A.; Zhang, L.; Rhee, M.; Langer, R.; Farokhzad, O.C.; Karnik, R. Single-Step Assembly of Homogenous Lipid−Polymeric and Lipid−Quantum Dot Nanoparticles Enabled by Microfluidic Rapid Mixing. *ACS Nano* **2010**, *4*, 1671–1679. [CrossRef] [PubMed]
15. Feng, Q.; Zhang, L.; Liu, C.; Li, X.; Hu, G.; Sun, J.; Jiang, X. Microfluidic based high throughput synthesis of lipid-polymer hybrid nanoparticles with tunable diameters. *Biomicrofluidics* **2015**, *9*, 052604. [CrossRef]
16. Valencia, P.M.; Pridgen, E.M.; Rhee, M.; Langer, R.; Farokhzad, O.C.; Karnik, R. Microfluidic platform for combinatorial synthesis and optimization of targeted nanoparticles for cancer therapy. *ACS Nano* **2013**, *7*, 10671–10680. [CrossRef] [PubMed]
17. Zhang, L.; Chen, Q.; Ma, Y.; Sun, J. Microfluidic Methods for Fabrication and Engineering of Nanoparticle Drug Delivery Systems. *ACS Appl. Bio Mater.* **2020**, *3*, 107–120. [CrossRef]
18. Bose, R.J.; Lee, S.-H.; Park, H. Lipid-based surface engineering of PLGA nanoparticles for drug and gene delivery applications. *Biomater. Res.* **2016**, *20*, 34. [CrossRef]
19. Zhang, L.; Feng, Q.; Wang, J.; Zhang, S.; Ding, B.; Wei, Y.; Dong, M.; Ryu, J.Y.; Yoon, T.Y.; Shi, X.; et al. Microfluidic Synthesis of Hybrid Nanoparticles with Controlled Lipid Layers: Understanding Flexibility-Regulated Cell-Nanoparticle Interaction. *ACS Nano* **2015**, *9*, 9912–9921. [CrossRef]
20. Faraji, A.H.; Wipf, P. Nanoparticles in cellular drug delivery. *Bioorganic Med. Chem.* **2009**, *17*, 2950–2962. [CrossRef] [PubMed]
21. Hoshyar, N.; Gray, S.; Han, H.; Bao, G. The effect of nanoparticle size on in vivo pharmacokinetics and cellular interaction. *Nanomedicine* **2016**, *11*, 673–692. [CrossRef]
22. Albanese, A.; Tang, P.S.; Chan, W.C. The effect of nanoparticle size, shape, and surface chemistry on biological systems. *Annu. Rev. Biomed Eng.* **2012**, *14*, 1–16. [CrossRef]
23. Chauhan, V.P.; Stylianopoulos, T.; Martin, J.D.; Popović, Z.; Chen, O.; Kamoun, W.S.; Bawendi, M.G.; Fukumura, D.; Jain, R.K. Normalization of tumour blood vessels improves the delivery of nanomedicines in a size-dependent manner. *Nat. Nanotechnol.* **2012**, *7*, 383–388. [CrossRef]
24. Li, M.; Al-Jamal, K.T.; Kostarelos, K.; Reineke, J. Physiologically based pharmacokinetic modeling of nanoparticles. *ACS Nano* **2010**, *4*, 6303–6317. [CrossRef]
25. Ferguson, R.M.; Minard, K.R.; Krishnan, K.M. Optimization of nanoparticle core size for magnetic particle imaging. *J. Magn. Magn. Mater.* **2009**, *321*, 1548–1551. [CrossRef] [PubMed]
26. Sun, J.; Zhang, L.; Wang, J.; Feng, Q.; Liu, D.; Yin, Q.; Xu, D.; Wei, Y.; Ding, B.; Shi, X.; et al. Tunable rigidity of (polymeric core)-(lipid shell) nanoparticles for regulated cellular uptake. *Adv. Mater.* **2015**, *27*, 1402–1407. [CrossRef] [PubMed]
27. Wang, J.; Chen, W.; Sun, J.; Liu, C.; Yin, Q.; Zhang, L.; Xianyu, Y.; Shi, X.; Hu, G.; Jiang, X. A microfluidic tubing method and its application for controlled synthesis of polymeric nanoparticles. *Lab Chip* **2014**, *14*, 1673–1677. [CrossRef] [PubMed]

28. Silva, A.L.; Rosalia, R.A.; Sazak, A.; Carstens, M.G.; Ossendorp, F.; Oostendorp, J.; Jiskoot, W. Optimization of encapsulation of a synthetic long peptide in PLGA nanoparticles: Low-burst release is crucial for efficient CD8$^+$ T cell activation. *Eur. J. Pharm. Biopharm.* **2013**, *83*, 338–345. [CrossRef]

29. Kanai, T.; Tsuchiya, M. Microfluidic devices fabricated using stereolithography for preparation of monodisperse double emulsions. *Chem. Eng. J.* **2016**, *290*, 400–404. [CrossRef]

30. Zhou, Z.; Kong, T.; Mkaouar, H.; Salama, K.N.; Zhang, J.M. A hybrid modular microfluidic device for emulsion generation. *Sens. Actuators A Phys.* **2018**, *280*, 422–428. [CrossRef]

31. Liu, Y.; Li, Y.; Hensel, A.; Brandner, J.J.; Zhang, K.; Du, X.; Yang, Y. A review on emulsification via microfluidic processes. *Front. Chem. Sci. Eng.* **2020**, *14*, 350–364. [CrossRef]

32. March, D.R.; Proctor, L.M.; Stoermer, M.J.; Sbaglia, R.; Abbenante, G.; Reid, R.C.; Woodruff, T.M.; Wadi, K.; Paczkowski, N.; Tyndall, J.D.; et al. Potent cyclic antagonists of the complement C5a receptor on human polymorphonuclear leukocytes. Relationships between structures and activity. *Mol. Pharm.* **2004**, *65*, 868–879. [CrossRef]

33. Woodruff, T.M.; Pollitt, S.; Proctor, L.M.; Stocks, S.Z.; Manthey, H.D.; Williams, H.M.; Mahadevan, I.B.; Shiels, I.A.; Taylor, S.M. Increased potency of a novel complement factor 5a receptor antagonist in a rat model of inflammatory bowel disease. *J. Pharm. Exp.* **2005**, *314*, 811–817. [CrossRef] [PubMed]

34. Jain, U.; Woodruff, T.M.; Stadnyk, A.W. The C5a receptor antagonist PMX205 ameliorates experimentally induced colitis associated with increased IL-4 and IL-10. *Br. J. Pharmacol.* **2013**, *168*, 488–501. [CrossRef]

35. Brennan, F.H.; Gordon, R.; Lao, H.W.; Biggins, P.J.; Taylor, S.M.; Franklin, R.J.; Woodruff, T.M.; Ruitenberg, M.J. The Complement Receptor C5aR Controls Acute Inflammation and Astrogliosis following Spinal Cord Injury. *J. Neurosci.* **2015**, *35*, 6517–6531. [CrossRef] [PubMed]

36. Staab, E.B.; Sanderson, S.D.; Wells, S.M.; Poole, J.A. Treatment with the C5a receptor/CD88 antagonist PMX205 reduces inflammation in a murine model of allergic asthma. *Int. Immunopharmacol.* **2014**, *21*, 293–300. [CrossRef] [PubMed]

37. Lee, J.D.; Kumar, V.; Fung, J.N.; Ruitenberg, M.J.; Noakes, P.G.; Woodruff, T.M. Pharmacological inhibition of complement C5a-C5a1 receptor signalling ameliorates disease pathology in the hSOD1(G93A) mouse model of amyotrophic lateral sclerosis. *Br. J. Pharmacol.* **2017**, *174*, 689–699. [CrossRef]

38. Kumar, V.; Lee, J.D.; Clark, R.J.; Woodruff, T.M. Development and validation of a LC-MS/MS assay for pharmacokinetic studies of complement C5a receptor antagonists PMX53 and PMX205 in mice. *Sci. Rep.* **2018**, *8*, 8101. [CrossRef]

39. Strachan, A.J.; Shiels, I.A.; Reid, R.C.; Fairlie, D.P.; Taylor, S.M. Inhibition of immune-complex mediated dermal inflammation in rats following either oral or topical administration of a small molecule C5a receptor antagonist (vol 134, pg 579, 2001). *Brit. J. Pharm.* **2002**, *135*, 579–580. [CrossRef]

40. Kumar, V.; Lee, J.D.; Clark, R.J.; Noakes, P.G.; Taylor, S.M.; Woodruff, T.M. Preclinical Pharmacokinetics of Complement C5a Receptor Antagonists PMX53 and PMX205 in Mice. *ACS Omega* **2020**, *5*, 2345–2354. [CrossRef]

41. Li, X.X.; Lee, J.D.; Massey, N.L.; Guan, C.; Robertson, A.A.B.; Clark, R.J.; Woodruff, T.M. Pharmacological characterisation of small molecule C5aR1 inhibitors in human cells reveals biased activities for signalling and function. *Biochem. Pharm.* **2020**, *180*, 114156. [CrossRef]

42. Feng, Q.; Sun, J.S.; Jiang, X.Y. Microfluidics-mediated assembly of functional nanoparticles for cancer-related pharmaceutical applications. *Nanoscale* **2016**, *8*, 12430–12443. [CrossRef]

43. Zhu, M.; Whittaker, A.K.; Jiang, X.; Tang, R.; Li, X.; Xu, W.; Fu, C.; Smith, M.T.; Han, F.Y. Use of Microfluidics to Fabricate Bioerodable Lipid Hybrid Nanoparticles Containing Hydromorphone or Ketamine for the Relief of Intractable Pain. *Pharm. Res.* **2020**, *37*, 211. [CrossRef] [PubMed]

44. Zhang, L.; Chan, J.M.; Gu, F.X.; Rhee, J.-W.; Wang, A.Z.; Radovic-Moreno, A.F.; Alexis, F.; Langer, R.; Farokhzad, O.C. Self-Assembled Lipid−Polymer Hybrid Nanoparticles: A Robust Drug Delivery Platform. *ACS Nano* **2008**, *2*, 1696–1702. [CrossRef] [PubMed]

45. Song, K.C.; Lee, H.S.; Choung, I.Y.; Cho, K.I.; Ahn, Y.; Choi, E.J. The effect of type of organic phase solvents on the particle size of poly (d, l-lactide-co-glycolide) nanoparticles. *Colloids Surf. A Physicochem. Eng. Asp.* **2006**, *276*, 162–167. [CrossRef]

46. Sahana, D.; Mittal, G.; Bhardwaj, V.; Kumar, M.R. PLGA nanoparticles for oral delivery of hydrophobic drugs: Influence of organic solvent on nanoparticle formation and release behavior in vitro and in vivo using estradiol as a model drug. *J. Pharm. Sci.* **2008**, *97*, 1530–1542. [CrossRef]

47. Martins, J.P.; Torrieri, G.; Santos, H.A. The importance of microfluidics for the preparation of nanoparticles as advanced drug delivery systems. *Expert Opin. Drug Deliv.* **2018**, *15*, 469–479. [CrossRef] [PubMed]

48. Utada, A.S.; Fernandez-Nieves, A.; Stone, H.A.; Weitz, D.A. Dripping to jetting transitions in coflowing liquid streams. *Phys. Rev. Lett.* **2007**, *99*, 094502. [CrossRef]

49. Lince, F.; Marchisio, D.L.; Barresi, A.A. Strategies to control the particle size distribution of poly-epsilon-caprolactone nanoparticles for pharmaceutical applications. *J. Colloid Interface Sci.* **2008**, *322*, 505–515. [CrossRef]

50. Proctor, L.M.; Woodruff, T.M.; Sharma, P.; Shiels, I.A.; Taylor, S.M. Transdermal pharmacology of small molecule cyclic C5a antagonists. *Adv. Exp. Med. Biol.* **2006**, *586*, 329–345. [CrossRef]

51. Suk, J.S.; Xu, Q.G.; Kim, N.; Hanes, J.; Ensign, L.M. PEGylation as a strategy for improving nanoparticle-based drug and gene delivery. *Adv. Drug Deliv. Rev.* **2016**, *99*, 28–51. [CrossRef]

52. Li, Y.P.; Pei, Y.Y.; Zhang, X.Y.; Gu, Z.H.; Zhou, Z.H.; Yuan, W.F.; Zhou, J.J.; Zhu, J.H.; Gao, X.J. PEGylated PLGA nanoparticles as protein carriers: Synthesis, preparation and biodistribution in rats. *J. Control. Release* **2001**, *71*, 203–211. [CrossRef]
53. Garinot, M.; Fievez, V.; Pourcelle, V.; Stoffelbach, F.; des Rieux, A.; Plapied, L.; Theate, I.; Freichels, H.; Jerome, C.; Marchand-Brynaert, J.; et al. PEGylated PLGA-based nanoparticles targeting M cells for oral vaccination. *J. Control. Release* **2007**, *120*, 195–204. [CrossRef] [PubMed]
54. Sun, J.; Xianyu, Y.; Li, M.; Liu, W.; Zhang, L.; Liu, D.; Liu, C.; Hu, G.; Jiang, X. A microfluidic origami chip for synthesis of functionalized polymeric nanoparticles. *Nanoscale* **2013**, *5*, 5262–5265. [CrossRef] [PubMed]
55. Liu, Y.; Cheng, C.; Prud'Homme, R.K.; Fox, R. Mixing in a multi-inlet vortex mixer (MIVM) for flash nano-precipitation. *Chem. Eng. Sci.* **2008**, *63*, 2829–2842. [CrossRef]
56. Johnson, B.K.; Prud'homme, R.K. Chemical processing and micromixing in confined impinging jets. *Aiche J.* **2003**, *49*, 2264–2282. [CrossRef]
57. Kumari, A.; Yadav, S.K.; Yadav, S.C. Biodegradable polymeric nanoparticles based drug delivery systems. *Colloids Surf. B Biointerfaces* **2010**, *75*, 1–18. [CrossRef]
58. Aderem, A.; Underhill, D.M. Mechanisms of phagocytosis in macrophages. *Annu. Rev. Immunol.* **1999**, *17*, 593–623. [CrossRef]
59. Zhang, J.; Woodruff, T.M.; Clark, R.J.; Martin, D.J.; Minchin, R.F. Release of bioactive peptides from polyurethane films in vitro and in vivo: Effect of polymer composition. *Acta Biomater.* **2016**, *41*, 264–272. [CrossRef] [PubMed]
60. Song, N.; Li, P.; Jiang, Y.; Sun, H.; Cui, J.; Zhao, G.; Li, D.; Guo, Y.; Chen, Y.; Gao, J.; et al. C5a receptor1 inhibition alleviates influenza virus-induced acute lung injury. *Int. Immunopharmacol.* **2018**, *59*, 12–20. [CrossRef] [PubMed]
61. Gao, T.; Hu, M.; Zhang, X.; Li, H.; Zhu, L.; Liu, H.; Dong, Q.; Zhang, Z.; Wang, Z.; Hu, Y.; et al. Highly pathogenic coronavirus N protein aggravates lung injury by MASP-2-mediated complement over-activation. *medRxiv* **2020**. [CrossRef]
62. Carvelli, J.; Demaria, O.; Vely, F.; Batista, L.; Chouaki Benmansour, N.; Fares, J.; Carpentier, S.; Thibult, M.L.; Morel, A.; Remark, R.; et al. Association of COVID-19 inflammation with activation of the C5a-C5aR1 axis. *Nature* **2020**, *588*, 146–150. [CrossRef] [PubMed]

pharmaceutics

Article

Pharmacokinetics and Toxicity Evaluation of a PLGA and Chitosan-Based Micro-Implant for Sustained Release of Methotrexate in Rabbit Vitreous

Soumyarwit Manna [1], Anna M. Donnell [2], Rafaela Q. Caixeta Faraj [3], Blanca I. Riemann [3], Christopher D. Riemann [3], James J. Augsburger [3], Zelia M. Correa [3,4] and Rupak K. Banerjee [1,*]

[1] Department of Mechanical and Materials Engineering, University of Cincinnati, Cincinnati, OH 45221, USA; soumyarwit_m@hotmail.com
[2] Department of Chemistry, University of Cincinnati, Cincinnati, OH 45221, USA; donnelaa@ucmail.uc.edu
[3] Department of Ophthalmology, University of Cincinnati, Cincinnati, OH 45221, USA; rafaqcaixeta@hotmail.com (R.Q.C.F.); briemann@fuse.net (B.I.R.); criemann@cincinnatieye.com (C.D.R.); james.augsburger@uc.edu (J.J.A.); correazm@gmail.com (Z.M.C.)
[4] Department of Ophthalmology, Bascom Palmer Eye Institute, University of Miami, Miami, FL 33146, USA
* Correspondence: rupak.banerjee@uc.edu; Tel.: +1-5134772124

Citation: Manna, S.; Donnell, A.M.; Faraj, R.Q.C.; Riemann, B.I.; Riemann, C.D.; Augsburger, J.J.; Correa, Z.M.; Banerjee, R.K. Pharmacokinetics and Toxicity Evaluation of a PLGA and Chitosan-Based Micro-Implant for Sustained Release of Methotrexate in Rabbit Vitreous. *Pharmaceutics* **2021**, *13*, 1227. https://doi.org/10.3390/pharmaceutics13081227

Academic Editors: Jaehwi Lee and Bruno Sarmento

Received: 29 June 2021
Accepted: 5 August 2021
Published: 9 August 2021

Publisher's Note: MDPI stays neutral with regard to jurisdictional claims in published maps and institutional affiliations.

Abstract: The present research investigates the pharmacokinetics and toxicity of a chitosan (CS) and poly(lactic-co-glycolic) acid (PLGA)-based methotrexate (MTX) intravitreal micro-implant in normal rabbit eyes. PLGA and CS-based micro-implants containing 400 µg of MTX were surgically inserted in the vitreous of twenty-four New Zealand rabbits using minimally invasive procedures. The PLGA-coated CS-MTX micro-implant and the placebo micro-implant were inserted in the right eye and in the left eye, respectively, of each rabbit. The intravitreal MTX concentration was evaluated on Days 1, 3, 7, 14, 28 and 56. A therapeutic concentration of MTX (0.1–1.0 µM) in the rabbit vitreous was observed for 56 days. The release of MTX in the therapeutic release phase followed first-order kinetics. Histopathologic evaluation on Days 14, 28 and 56 of the enucleated eyes demonstrated no signs of toxicity or any anatomical irregularity in the vitreoretinal domain. Additionally, the micro-implants were stationary at the position of their implantation throughout the duration of the study. The PLGA-coated CS-MTX micro-implant can serve as a potential alternative to the current treatment modality of intravitreal MTX injections based on its performance, thereby avoiding associated complications and the treatment burden of multiple injections.

Keywords: methotrexate; chitosan; PLA/PLGA; sustained release; micro-implant; animal model; minimally invasive

1. Introduction

Methotrexate (MTX), an antimetabolite chemotherapeutic drug, has gained a substantial reputation in treating various types of cancer and autoimmune diseases [1]. In the domain of ophthalmology, intravitreal injections of MTX have been administered to treat various vitreoretinal (VR) diseases such as: (1) primary central nervous system lymphoma (PCNSL) with intraocular (vitreoretinal) involvement [2–4], also known as primary intraocular lymphoma (PIOL); (2) severe and recalcitrant intraocular inflammation (uveitis) [5–8]; and (3) certain cases of rhegmatogenous retinal detachment associated with elevated risk for proliferative vitreoretinopathy (PVR) [9,10]. Furthermore, in clinical practice, there has been a gradual incorporation of the off-label use of this drug, as described hereafter.

Various administration regimens have been employed when treating with intravitreal MTX injection. A representative dosing regimen would be either two intravitreal injections every week until apparent clinical improvement is demonstrated or one intravitreal injection every week thereafter until complete clinical regression in the eye(s) is observed. Alternatively, one intravitreal injection every month for at least three additional months is administered. Typically, this regimen accumulates to 20 to 25 injections or more [11]. A single intravitreal injection of MTX at the 200 to 400 µg/0.1 mL dose has shown good tolerance. Nevertheless, all intravitreal injections bear a potential risk of exogenous endophthalmitis, induced rhegmatogenous retinal detachment and elevated intraocular pressure [12]. The cumulative risk of these potential side effects and complications combined with the inconvenience and discomfort of multiple invasive procedures cannot be ignored when multiple intravitreal MTX injections are administered. Several cases of corneal epithelial toxicity [13], induced retinal cotton wool spots and macular edema [14] have been reported even at a dose of 400 µg/0.1 mL.

The evaluation of in vitro cytotoxic activity of MTX on 63 different cell lines showed the therapeutic levels of MTX range from 0.1 µM to 1 µM, with a mean IC_{50} of 0.32 µM [15]. Accordingly, a concentration range of 0.1–1 µM of MTX is considered to be therapeutic. When administered by intravitreal injection, the hydrophilic MTX drug (log p = -1.85) undergoes rapid clearance in the intravitreal domain (half-life of 14.3 h) [15]. Previously intravitreal MTX concentrations of 0.1–1 µM from one injection have been observed to have therapeutic effectiveness for a limited duration of 48–72 h [15,16]. It is expected that a prolonged therapeutic efficacy can be attained by delivering MTX through a sustained-release intravitreal drug delivery device (micro-implant) in the intravitreal domain, without causing any collateral complications. Currently, there are several FDA-approved intravitreal micro-implants, which deliver antiviral medications and corticosteroids [17], which are relatively hydrophobic in nature. These FDA-approved intravitreal implants along with other intravitreal drug delivery systems have been based on lipophilic (or hydrophobic) biodegradable polymers such as lactic and glycolic acid-based matrices such as poly-lactic acid (PLA), poly-glycolic acid (PGA), their copolymers and derivatives poly(lactic-co-glycolic) acid (PLGA) [18–20]. The issue with the existing polymer matrices (PLA, PGA and PLGA) is that they are lipophilic in nature and do not blend well with hydrophilic drugs such as MTX. Another disadvantage of these lipophilic polymer matrices is that they degrade very slowly even after the drug has been released, causing local toxicity [21]. At present, there is no sustained drug delivery platform for hydrophilic drugs, such as MTX, to treat VR diseases in commercial practice, which poses major challenges to develop a satisfactory micro-implant device.

Chitosan (CS) was chosen as the polymer matrix to fabricate the micro-implant, as it has been previously used as a delivery vehicle for MTX (hydrophilic drug) in other drug delivery formulations due to its similar hydrophilic nature [22–26]. CS is a copolymer of N-acetylglucosamine and glucosamine, which is a fully or partially N-deacetylated (DA) derivative of the natural polymer chitin. CS has been recognized as a Generally Recognized as Safe (GRAS) material by the FDA [27]. It has been reported by de la Fuente et al. [27] that CS-based ocular drug delivery formulations have shown good tolerance and acceptability in terms of toxicity. However, to restrict the rapid release of the hydrophilic MTX from the hydrophilic CS matrix, a lipophilic coating was imperative. Because PLA and PLGA have been extensively used in FDA-approved intraocular implants [17], they were chosen as the coating polymer to the hydrophilic CS-MTX matrix.

Our group developed a biodegradable intravitreal micro-implant platform that delivers a sustained therapeutic dose of MTX to the VR domain for several months [11,28–32]. Initially, a CS and PLA-based MTX intravitreal micro-implant device was developed that was able to maintain a sustained release of MTX between 0.2 and 2 µg/day in vitro for more than 50 days [29] and a therapeutic concentration of 0.1–1 µM MTX for ~30 days in vivo [11]. Non-invasive methodologies such as electroretinogram (ERG) [28] and B-scan ultrasound (US), as well as histopathology [30], have suggested favorable safety and toxicity of this micro-implant in normal rabbit eyes.

In a subsequent study, a PLGA and CS-based MTX intravitreal micro-implant was developed, which demonstrated an improved sustained MTX release rate of 0.2–2 µg/day for ~3–5 months in vitro [32]. In this in vitro study, the influence of (a) molecular weight of the PLA/PLGA and (b) the composition of the PLGA polymer on the biodegradation characteristics of the micro-implant, including swelling and disintegration of the micro-implant, was analyzed. The molecular weight of CS used was 50,000–190,000 with a degree of deacetylation (DA%) $\geq$ 75%. (Sigma Aldrich, St. Louis, MO, USA). The evaluated molecular weight of ester-terminated PLGA 5050 (PLA/PGA copolymer ratio in PLGA is 50:50) (Lactel Biodegradable Polymers, AL) with an inherent viscosity of 0.82 in hexafluoroisopropanol at 30 °C was Mw = 287,000 with a polydispersity index of 2.78 [32]. The safety and toxicity of this improved PLGA and CS-based MTX micro-implant were validated in normal rabbit eyes using *non-invasive* techniques such as ERG, US, slit-lamp biomicroscopy (SLB), fundoscopy and intraocular pressure (IOP) measurement for a period of approximately 2 months [31]. Non-invasive evaluation proved the micro-implant to be safe and well tolerated in rabbit eyes for the whole duration of the study. In the same pre-clinical trial, we evaluated the pharmacokinetics and gross morphology of the micro-implant concurrently using minimally invasive techniques for improved knowledge of the safety of the micro-implant in the vitreoretinal domain.

In our pilot in vivo studies [11,28,30], the PLA-coated CS-MTX micro-implant had a different composition of the coating polymer, which resulted in a shorter duration of release (~1 month in vivo). Additionally, the scope of the pilot studies was limited due to the limited number of rabbits (n = 8). Subsequently, a recent study involving the improved PLGA-coated CS-MTX micro-implant was focused on non-invasive parameters only (ERG, US, SLB, fundoscopy and IOP) [31]. This study is novel because it presents findings on the invasive parameters, such as pharmacokinetics and toxicity evaluation of the PLGA-coated CS-MTX micro-implant, in thirty rabbit eyes. Additionally, this improved micro-implant design resulted in a significant increase in the release duration (~2 months in vivo), as presented hereafter.

2. Materials and Methods

Fabrication of the Micro-implant: In this in vivo study, the PLGA 5050 (PLA/PGA ratio = 50:50)-coated CS-MTX micro-implant was used. It was fabricated based on the method described in our earlier study [29,32]. Briefly, CS-MTX fibers were obtained by freeze-drying the mixture of CS and MTX in 0.1 N HCl in Tygon® tubing (Saint-Gobain, Malvern, PA, USA) of a 1/16-inch internal diameter. Subsequently, these fibers were cut to desired lengths under a microscope, and then a ~200 μm coating of PLGA 5050 was applied on the surface using dip-coating techniques. As described earlier [29,32], the microstructure and the morphology of the micro-implant were evaluated using optical microscopy (Keyence Digital Microscope, VHX-600, Osaka, Japan) and scanning electron microscopy (SEM) (FEI XL 30-FEG, FEI, Hillsboro, OR, USA) using an accelerating voltage of 15 KV. The morphology and microstructure of the PLGA-coated CS-MTX micro-implant are shown in Figure 1.

The final dimensions of the PLGA-coated CS-MTX micro-implant containing 40% *w/w* MTX were ~4.3 mm in length and ~1.2 mm in diameter. The weight of the micro-implant containing 40% *w/w* MTX was ~1 mg, which translated to a drug loading of ~400 μg of MTX. The placebo micro-implant was fabricated following the same procedure without the drug. The rationale behind using the 40% *w/w* MTX loading was to compare the same dosage of MTX (400 μg of MTX), as administered in an intravitreal injection used in clinical practice. The micro-implants were sterilized under UV radiation for 20 min prior to surgical procedures.

Design of the Non-invasive In Vivo Experiment: Thirty immune-competent New Zealand albino rabbits, each weighing 2–3 kg, were used in this study. A sterilized MTX-loaded micro-implant and a sterilized placebo micro-implant (without MTX) were surgically implanted in the right eye and in the left eye, respectively, as per the study plan shown in Table 1. All procedures for the rabbit surgery were in accordance with the Institutional Animal Care and Use Committee protocol (IACUC No. 12-09-13-01, University of Cincinnati, dated: 21 November, 2012) and followed the ARVO Statement for the Use of Animals in Ophthalmic and Vision Research.

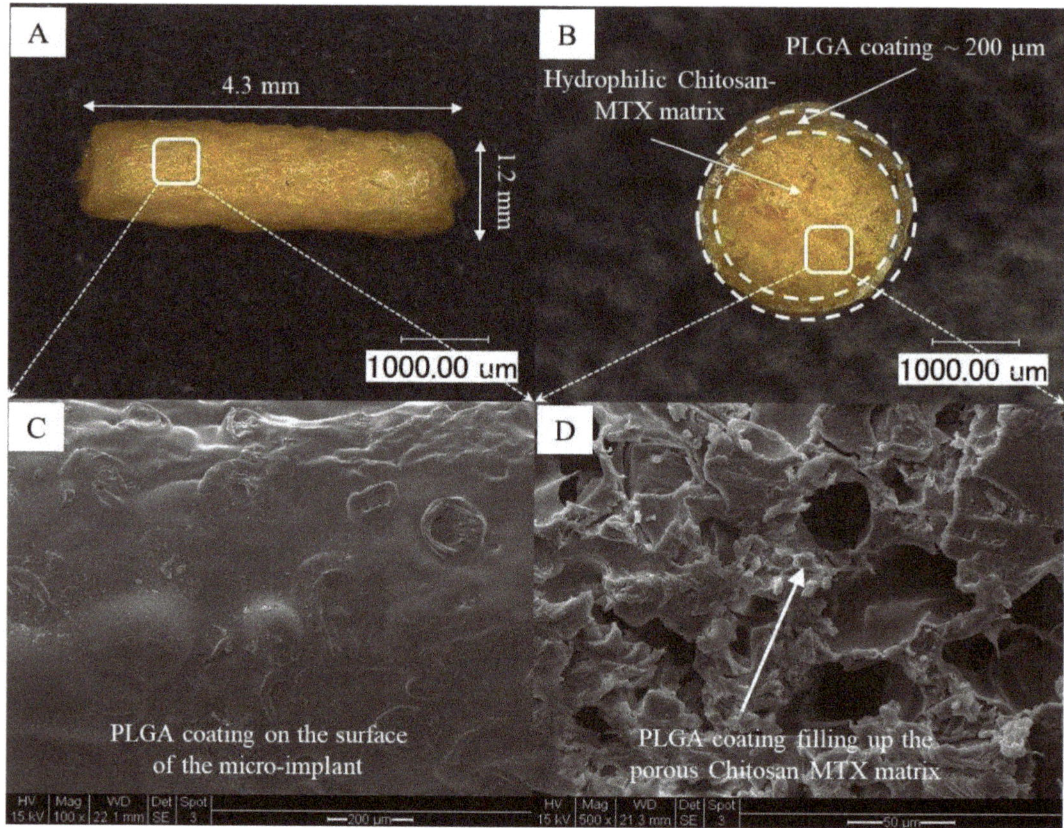

Figure 1. Optical microscopy images of the PLGA-coated CS-MTX micro-implant showing the top view (**A**) and the cross-sectional view (**B**). The microstructure of the micro-implant can be observed in the SEM images showing the top view (**C**) and the cross-sectional view (**D**).

Table 1. Design of the minimally invasive study.

Time-point	Number of Rabbits	Pharmacokinetic Evaluation	Gross Inspection of the Cross-Section of the Eye Globe	Histopathology ^
Day 1	3 + 1 (control *)	3 rabbits	None	None
Day 3	3 + 1 (control)	3 rabbits	None	None
Day 7	3 + 1 (control)	3 rabbits	None	None
Day 14	5 + 1 (control)	3 rabbits	2 rabbits	2 rabbits
Day 28	5 + 1 (control)	3 rabbits	2 rabbits	2 rabbits
Day 56	5 + 1 (control)	3 rabbits	2 rabbits	2 rabbits

Nomenclature: * For control rabbits, no pharmacokinetic evaluation, gross inspection of the eye globe and histopathology were conducted; ^ No pharmacokinetic evaluation was conducted in the eyes used for histopathology evaluation.

Surgery: The details of the surgery for this study are reported in our recent study [31] involving the non-invasive evaluation of safety and toxicity of the same micro-implant. Briefly, each animal was evaluated for the vital signs (heart rate, respiration rate, temperature) prior to surgery (PS) and verified to be within normal limits. Each rabbit was anesthetized with a mixture of xylazine hydrochloride (5 mg/kg) and ketamine hydrochloride (45 mg/kg) by intramuscular injection. Buprenorphine hydrochloride (0.02 mg/kg (Sub-Q)) was administered as an analgesic, and glycopyrrolate (0.006 mg/kg (IM)) was administered to stabilize the heart rate. For further supplementation of the anesthesia, tetracaine hydrochloride (0.05%) eye drops were administered topically to the eyes. The body temperature of the rabbit was maintained at 37 °C. An eyelid speculum was inserted to expose the right eye. A full-thickness eye wall incision was carried out parallel to the limbus at a distance measured 6 mm from the limbus in the superotemporal quadrant using a 1.1 mm side port paracentesis surgical knife. Subsequently, a PLGA-coated CS-MTX micro-implant was inserted through the eye wall incision using McPherson forceps (Figure 2). The conjunctival–scleral wound was immediately closed using a full-thickness mattress suture of 7-0 polyglactin 910 (suture material: Ethicon, Cincinnati, OH, USA). Then, the same procedure was repeated on the left eye except for the insertion of a placebo micro-implant (without MTX loading) in that eye. Immediately after suturing the conjunctival–scleral wound, Meloxicam 0.2 mg/kg (Sub-Q) was administered in each eye along with 2 drops of admixed neomycin sulfate (3.5 mg/mL)-polymyxin B (10,000 units/mL)-dexamethasone (1 mg/mL). In addition to standard animal care, each animal was given one drop of admixed neomycin sulfate (3.5 mg/mL)-polymyxin B (10,000 units/mL)-dexamethasone (1 mg/mL) eye drops to each eye twice daily for the first five post-surgical days.

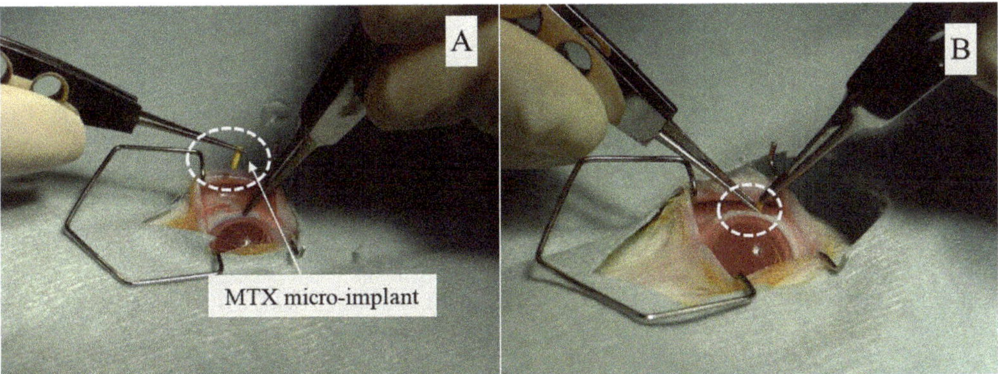

Figure 2. (**A**) MTX micro-implant before surgical insertion; (**B**) surgical insertion of the MTX micro-implant.

Euthanasia: On Day 1, Day 3, and Day 7, four rabbits were euthanized, and on Day 14, Day 28, and Day 56, six rabbits were euthanized by administration of sodium pentobarbital (150–200 mg/kg) IV.

Procurement of Ocular Specimens: Both eyes of each animal were removed immediately after euthanasia at each specified time-point. The eye globes were inspected to evaluate the insertion site of the micro-implant and observe any evident abnormalities of that site. For pharmacokinetic evaluation of the MTX concentration in the vitreous, the vitreous (~1 mL) was extracted manually from the eyes of the rabbits designated for pharmacokinetic evaluation and collected in a sterile transport vial. The vials containing the vitreous specimens were stored in dry ice and then kept frozen until the date of pharmacokinetic testing (see Pharmacokinetics Study below). The eye globes of the animals designated for histopathological evaluation were immersed immediately in 10% neutral buffered formalin for subsequent pathological processing and histopathological analysis (see Histopathological Study below).

Pharmacokinetics Study: The concentration of MTX in the vitreous samples obtained from the eyes receiving the MTX micro-implant and the placebo micro-implant for each time-point was analyzed using high-performance liquid chromatography (HPLC). The HPLC method was carried out as described in the United States Pharmacopeia assay for MTX [33]. This method has been previously described in our previous publication on the in vivo study of a similar PLA-coated CS-MTX micro-implant [11]. Briefly, the Agilent® 1100 HPLC system (Agilent Technologies, Santa Clara, C, USAA) with a diode array detector was used for the HPLC analysis. A C-18 column measuring 150 mm × 4.6 mm with a pore size of 80 Å was used. The column temperature was set at 23 °C. Acetonitrile and phosphate/citrate buffer (pH 6.0) mixed in the ratio of 10:90 was used as the mobile phase. A flow rate of 1 mL/min of the mobile phase was used with an injection volume of 10 μL. The characteristic UV wavelength of 302 nm was used for the detection of MTX. For each time-point, (a) n = 3 vitreous samples obtained from the eye receiving the MTX micro-implant and (b) n = 3 vitreous samples obtained from the eye receiving the placebo micro-implant were analyzed.

Histopathological Study: Similar to our previous study, the globes were grossed and sectioned to display the micro-implant and surgical wound in pupil–optic nerve (P-ON) sections [11]. The globes were then processed and stained as previously described [11]. The stained histopathology slides were then evaluated by one of the authors (Z.M.C.) to evaluate any potential toxicity or complications.

3. Results

There were no complications during the surgical insertion of the micro-implants. The gross inspection of the position of the micro-implant relative to the crystalline lens and the eye wall incision showed (a) there was no migration of the micro-implants from their respective initial implantation site; (b) the gross appearance of the micro-implant and the vitreous surrounding the micro-implant was not appreciably different in the eyes that received an MTX-containing micro-implant and those that received a placebo micro-implant; and (c) there were signs of swelling of the micro-implant in the eyes on the Day 28 and Day 56 time-points (Figure 3).

Pharmacokinetics Study: The concentration profile of MTX in the vitreous of the eye post-implantation on Days: 1, 3, 7, 14 and 56 is shown in Figure 4. The drug concentration is observed to be within the therapeutic window (0.1–1 μM) from Day 3 to Day 56. No MTX concentration was detected by the HPLC in the vitreous samples obtained from the eyes containing the placebo micro-implants. The MTX concentration data of our study have been compared with prior clinical study results [34] and the pre-clinical study results [15] obtained from a 400 μg MTX intravitreal injection and also with our prior in vitro study involving the same micro-implant as used in this study [32] (Figure 4).

A peak MTX concentration (Cmax-MTX) of 360 μM and 400 μM was observed from the clinical study [34] and the pre-clinical study [15], respectively, and the total duration of MTX release lasted for 2–6 days. In comparison, the MTX release profile observed in this current in vivo study showed a Cmax-MTX of 28.88 μM on the first day and subsequent delivery of MTX in the therapeutic window (0.1–1 μM) for about 2 months (56 days). The MTX concentration, obtained from our prior in vitro study [32] involving the same micro-implant, is comparatively higher than that of the MTX concentration observed in this in vivo study. The higher concentration of MTX in the in vitro study can be potentially attributed to the lack of realistic representation of the physiological convective–diffusive transfer of the vitreous in the in vitro environment. Additionally, phosphate-buffered saline (PBS) of pH 7.4 at ~38 °C used as the dissolution media in the in vitro study may not accurately replicate the solubility, diffusivity and convective clearance of MTX in the in vivo vitreous.

A regression value of $R^2 \sim 0.86$ is obtained by fitting the in vivo vitreous concentration of MTX in the therapeutic release phase to the characteristic first-order equation model. The regression value worsens ($R^2 \sim 0.31$) by fitting the in vivo vitreous concentration of MTX of the whole duration of release to the characteristic first-order equation model. This indicates the release mechanism of MTX in the therapeutic release phase follows first-order kinetics. Similarly, the half-life ($t_{1/2}$) of MTX is 40.76 days, as obtained from the first-order fit (first-order constant $k = 0.017$) in the therapeutic release phase. The $t_{1/2}$ of MTX worsens to ~14.43 days, as obtained from the first-order fit (first-order constant k = 0.048) of the whole duration of release.

Histopathology Study: The details of the histopathological findings have been reported in the non-invasive study of toxicity and performance of the same micro-implant on the same rabbits, as used in this study [31]. Briefly, the issues identified in the histopathology analysis were associated with surgical procedures. Otherwise, the micro-implant showed no signs of toxicity and appeared to be safe for application in the VR domain. Histopathology findings include focal vitreoretinal traction without any apparent predominance between MTX and placebo micro-implants (Figure 5). The retina appeared to be anatomically healthy in all the eyes.

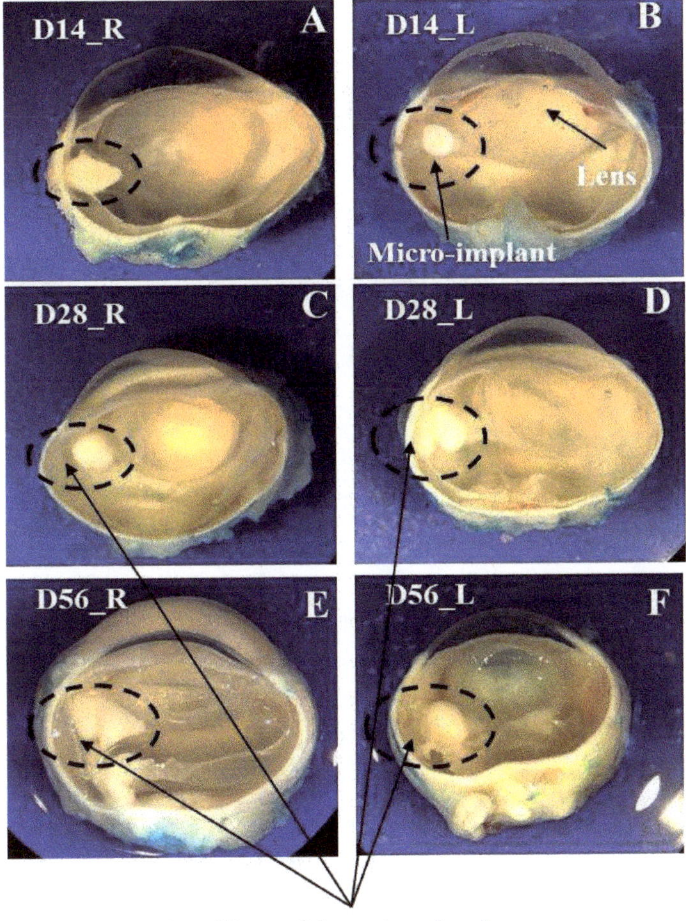

Figure 3. The position of the micro-implant relative to the lens can be seen in all the eye globes. The vitreous is clear in all the eye globes, indicating no toxicity caused by the MTX micro-implant and placebo micro-implant on the 14th day ((**A**,**B**), respectively); on the 28th day ((**C**,**D**), respectively); and on the 56th day ((**E**,**F**), respectively). Note the swelling of the micro-implants on the Day 28 and Day 56 time-points. Nomenclature of the eye globes—Dxx_y, where xx = time-points (14, 28, 56), and y = eye (R—right eye receiving the MTX micro-implant; L—left eye receiving the placebo micro-implant).

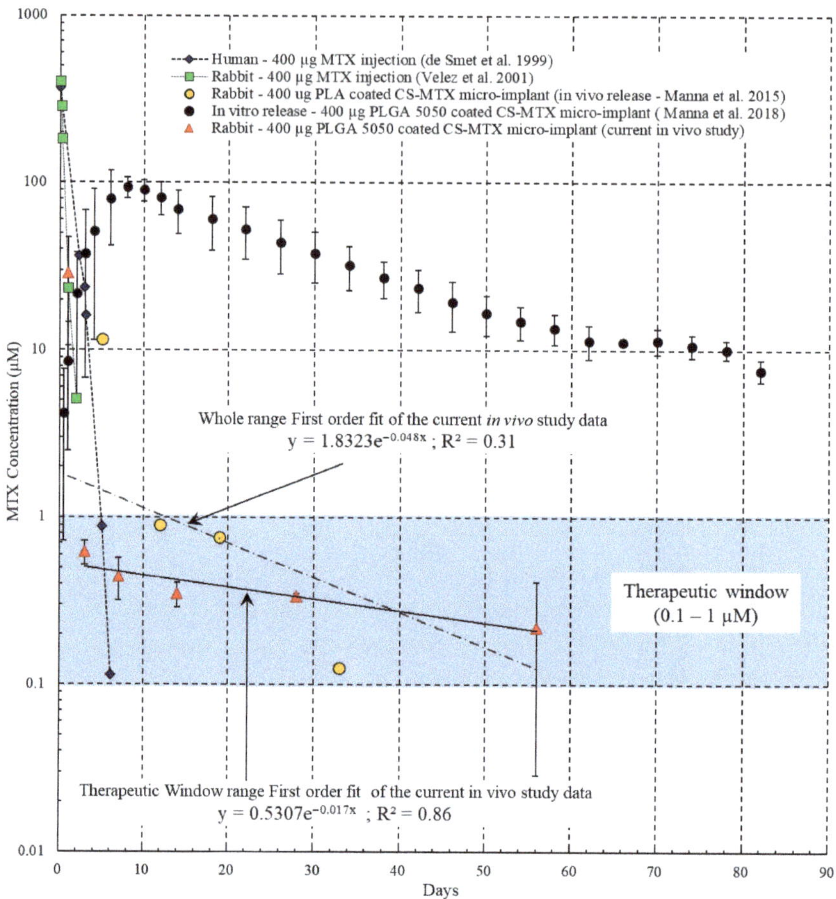

Figure 4. Comparison of concentration of MTX in the vitreous (therapeutic window—shaded region).

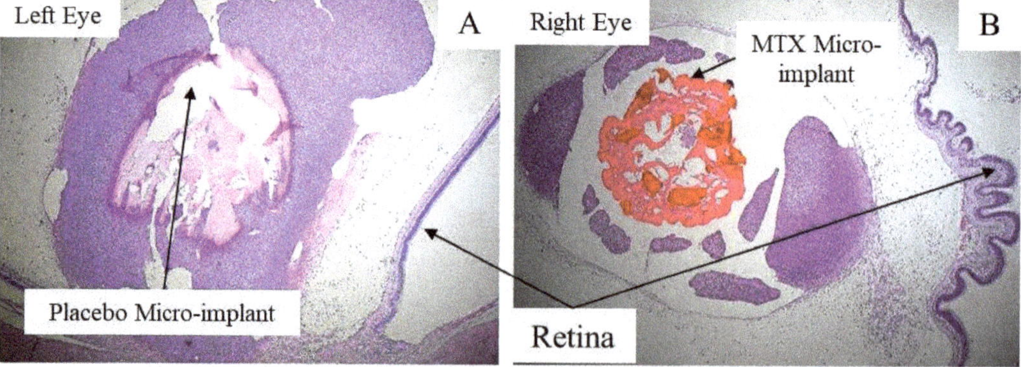

Figure 5. Histopathology findings using H&E staining. (**A**) microslide (4× shows the placebo micro-implant (Day 28) surrounded by a more pronounced inflammatory capsule; (**B**) microslide (4×) shows the MTX micro-implant (Day 28) with mild non-granulomatous inflammatory infiltrate surrounding the micro-implant and localized vitreous traction causing wrinkling of the retina. The retina appears anatomically healthy.

4. Discussion

Oral administration of MTX is used to treat uveitis (typical dose of 7–25 mg once weekly), often with supplemental oral folic acid to reduce systemic toxicity [7]. Additionally, administration of a single intravitreal MTX injection (typical dose of 200 to 400 µg/0.1 mL) or a series of injections over a several-month period has been used as an off-label treatment for severe or recalcitrant uveitis [8]. In selected cases of PIOL, intravitreal MTX (usually at the 400 µg/0.1 mL dose) [35] has been administered off-label to one or both eyes in addition to systemic MTX-based chemotherapy. Additionally, in patients without evidence of active brain–cerebrospinal fluid involvement, the intravitreal MTX injections are administered to the affected eye(s) alone [3].

This development of a sustained-release MTX intravitreal micro-implant stems from our motivation to create an alternative clinical approach to the current treatment modality of intravitreal MTX injections to treat VR diseases such as PIOL, uveitis and to prevent PVR. The FDA-approved sustained-release intraocular implants only deliver hydrophobic drugs (e.g., corticosteroids, ganciclovir, dexamethasone, fluocinolone acetonide) [17]. In comparison, this micro-implant platform offers a major advantage of delivering a sustained therapeutic concentration of hydrophilic drugs with a short half-life, such as MTX, over a prolonged duration. This study demonstrated the PLGA-coated CS-MTX micro-implants were able to deliver a therapeutic concentration of MTX (0.1–1 µM) over a period of 56 days in normal rabbit eyes without any signs of potential toxicity.

The PLGA-coated CS-MTX micro-implant showed a peak MTX concentration (Cmax-MTX) of 28.88 µM in the vitreous on the first-day time-point. This Cmax-MTX obtained from the PLGA-coated CS-MTX micro-implant is ~92% lower than the Cmax-MTX observed in intravitreal MTX injections (360–400 µM) [34]. Although the Cmax-MTX of 28.88 µM is beyond the therapeutic window (0.1–1 µM), it is anticipated to have an insignificant toxic effect, as it is still ~92% less than the Cmax-MTX of the intravitreal injections. Furthermore, the PLGA-coated CS-MTX micro-implant releases the same quantity of MTX (400 µg) over a duration of 56 days as compared to clearing after 2–3 days after intravitreal injection. Therefore, it can be expected that the PLGA-coated CS-MTX micro-implant will be well tolerated in the VR domain.

The $t_{1/2}$ of our MTX micro-implant, as obtained from the first-order fit of the whole duration of release, is ~14.43 days. This is very similar to the $t_{1/2}$ of MTX (13.1 days), as observed in the same PLGA-coated CS-MTX micro-implant in the in vitro study [32]. Additionally, the $t_{1/2}$ of our MTX micro-implant is prolonged to 40.76 days when the first-order fit is applied to the concentration data points in the therapeutic release phase (excluding the Cmax-MTX on Day 1). Therefore, it is likely that the PLGA coating further extends the duration of release in addition to reducing the peak MTX vitreous concentration (Cmax).

In our previous in vivo study on the PLA-coated CS-MTX micro-implant (400 µg of MTX), a therapeutic concentration of MTX (0.1–1 µM) was observed for 33 days without any signs of toxicity in rabbit eyes (Figure 4). Based on (a) our experience of limited time-points from that study (Days: 5, 12, 19 and 33) [11] and (b) the in vitro release profile of the MTX and swelling profile of the PLGA 5050-coated CS-MTX micro-implant [32], we introduced more time-points in this study (Days: 1, 3, 7 14, 28 and 56) for improved knowledge of the drug release profile in the intravitreal domain. Additionally, due to prior experience from the in vivo study involving the PLA-coated CS-MTX micro-implant, histopathology was conducted only at later time-points (Days 14, 28 and 33), as insignificant changes due to either toxicity or inflammation in the intraocular domain were expected at early time-points (Days: 1, 3 and 7).

The swelling of the micro-implants, as observed in cross-sections of the eye globes (Figure 3), on Day 28 and Day 56 is consistent with the findings of the in vitro study [32] and the findings of the SLB, fundoscopy and US in the non-invasive in vivo study on the same micro-implant [31]. Additionally, in the in vitro study [32], the PLGA-coated micro-implant showed peak swelling (6.2 times its original weight) around 30 days, which is similar to swelling observed on Day 28 of this study, and finally disintegrated around

66 days in vitro, which is approximately the duration of this study (~56 days). Furthermore, the ultrasound study also showed that (a) none of the micro-implants migrated in the vitreous; (b) the overall shape of each of the micro-implants was stable during the study barring the swelling towards the end of the experiment; and (c) fully attached peripheral retina and no evident abnormalities of the peripheral vitreous adjacent to the micro-implant in any of the evaluated eyes.

This study did not manifest any evidence of histological retinal deterioration at any postsurgical time-point related either to the PLGA-coated CS-MTX micro-implants or to the placebo micro-implants. This corroborates the findings of the electroretinogram (ERG) evaluation on the same rabbits across the duration of this study [31]. Statistical evaluation of the ERG analysis showed there was no change in the retinal functional integrity due to the micro-implants over the entire duration of the study.

Lastly, the histopathological evaluation corroborated the absence of toxicity such as thinning of the inner retina, photoreceptor atrophy and vascular events at later time-points [31], indicating observed changes are associated specifically with the implantation procedure and not caused by any toxicity associated with the micro-implants. Additionally, the materials used to fabricate the micro-implant are considered to be biodegradable. PLGA is known to be metabolized under physiological conditions to produce carbon dioxide and water, and CS is expected to be degraded by lysozyme, an enzyme present in the vitreous to form amino sugars [36].

While this study provides an improved evaluation of the pharmacokinetics and toxicity performance of this micro-implant, further studies are required in the future such as (a) cell invasion evaluation at later time-points; (b) assessment of MTX concentration in the aqueous humor; (c) improved sample size; (d) inclusion of more time-points for an improved resolution of the release concentration; and (e) immunohistochemistry.

5. Conclusions

Our investigation showed that the PLGA-coated CS-MTX micro-implant delivered a therapeutic concentration of MTX in rabbit eyes for ~56 days. Histopathology analysis demonstrated that the micro-implant appeared safe and well tolerated in the rabbit eyes. Our PLGA-coated CS-MTX micro-implant may serve as a potential platform for sustained release of a therapeutic concentration of MTX or other similar hydrophilic drugs in the VR domain. Additionally, it can be used as a promising alternative to the current treatment modality of recurrent intravitreal MTX injections, thereby avoiding the complications, discomfort and treatment burden associated with the intravitreal MTX injections.

Author Contributions: S.M.—Conceptualization, Methodology, Investigation, Data Curation, Writing—Original Draft Preparation, Writing—Review and Editing, A.M.D.—Methodology, Investigation, Data Curation, Writing—Review and Editing, R.Q.C.F.—Surgery, Writing—Review and Editing, B.I.R.—Surgery, Writing—Review and Editing, C.D.R.—Conceptualization, Investigation, Writing—Review and Editing, J.J.A.—Conceptualization, Investigation, Writing—Review and Editing, Z.M.C.—Conceptualization, Surgery, Investigation, Writing—Review and Editing, R.K.B.—Conceptualization, Methodology, Investigation, Data Curation, Writing—Original Draft Preparation, Writing—Review and Editing. All authors have read and agreed to the published version of the manuscript.

Funding: This study was supported by Accelerator Funding from the State of Ohio and University of Cincinnati-Ohio Department of Development, Cincy Tech 008431.

Institutional Review Board Statement: All procedures for the rabbit surgery were in accordance with the Institutional Animal Care and Use Committee protocol (IACUC No. 12-09-13-01, University of Cincinnati, dated: 21 November, 2012) and followed the ARVO Statement for the Use of Animals in Ophthalmic and Vision Research.

Informed Consent Statement: Not Applicable.

Data Availability Statement: The data presented in this study are available on request from the corresponding author.

Acknowledgments: The authors would like to acknowledge the support of Laboratory Animal Medical Services, University of Cincinnati, in assisting with the rabbit experiments.

Conflicts of Interest: The authors declare no conflict of interest.

References

1. Yu, W.J.; Huang, D.X.; Liu, S.; Sha, Y.L.; Gao, F.H.; Liu, H. Polymeric nanoscale drug carriers mediate the delivery of methotrexate for developing therapeutic interventions against cancer and rheumatoid arthritis. *Front. Oncol.* **2020**, *10*, 1734. [CrossRef]
2. Chan, C.C.; Wallace, D.J. Intraocular lymphoma: Update on diagnosis and management. *Cancer Control* **2004**, *11*, 285–295. [CrossRef] [PubMed]
3. Frenkel, S.; Hendler, K.; Siegal, T.; Shalom, E.; Pe'er, J. Intravitreal methotrexate for treating vitreoretinal lymphoma: 10 years of experience. *Br. J. Ophthalmol.* **2008**, *92*, 383–388. [CrossRef]
4. Hearne, E.; Netzer, O.T.; Lightman, S. Learning points in intraocular lymphoma. *Eye* **2021**, *35*, 1815–1817. [CrossRef]
5. Munoz-Fernandez, S.; Garcia-Aparicio, A.M.; Hidalgo, M.V.; Platero, M.; Schlincker, A.; Bascones, M.L.; Pombo, M.; Morente, P.; Sanpedro, J.; Martin-Mola, E. Methotrexate: An option for preventing the recurrence of acute anterior uveitis. *Eye* **2009**, *23*, 1130–1133. [CrossRef] [PubMed]
6. Taylor, S.R.J.; Habot-Wilner, Z.; Pacheco, P.; Lightman, S.L. Intraocular methotrexate in the treatment of uveitis and uveitic cystoid macular edema. *Ophthalmology* **2009**, *116*, 797–801. [CrossRef] [PubMed]
7. Khalatbari, D.; Mccallum, R.M.; Jaffe, G.J. Methotrexate. A Patient Education Monograph prepared for the American Uveitis Society. 2003. Available online: http://www.uveitissociety.org/pages/diseases/methotrexate.html (accessed on 25 March 2015).
8. Hardwig, P.W.; Pulido, J.S.; Erie, J.C.; Baratz, K.H.; Buettner, H. Intraocular methotrexate in ocular diseases other than primary central nervous system lymphoma. *Am. J. Ophthalmol.* **2006**, *142*, 883–885. [CrossRef]
9. Sadaka, A.; Sisk, R.A.; Osher, J.M.; Toygar, O.; Duncan, M.K.; Riemann, C.D. Intravitreal methotrexate infusion for proliferative vitreoretinopathy. *Clin. Ophthalmol.* **2016**, *10*, 1811–1817. [CrossRef]
10. Benner, J.D.; Dao, D.; Butler, J.W.; Hamill, K.I. Intravitreal methotrexate for the treatment of proliferative vitreoretinopathy. *BMJ Open Ophthalmol.* **2019**, *4*, e000293. [CrossRef]
11. Manna, S.; Banerjee, R.K.; Augsburger, J.J.; Al-Rjoub, M.F.; Donnell, A.; Correa, Z.M. Biodegradable chitosan and polylactic acid based intraocular micro-implant for sustained release of methotrexate into vitreous: Analysis of pharmacokinetics and toxicity in rabbit eyes. *Graefe's Arch. Clin. Exp. Ophthalmol.* **2015**, *253*, 1297–1305. [CrossRef]
12. Sampat, K.M.; Garg, S.J. Complications of intravitreal injections. *Curr. Opin. Ophthalmol.* **2010**, *21*, 178–183. [CrossRef]
13. Gorovoy, I.; Prechanond, T.; Abia, M.; Afshar, A.R.; Stewart, J.M. Toxic corneal epitheliopathy after intravitreal methotrexate and its treatment with oral folic acid. *Cornea* **2013**, *32*, 1171–1173. [CrossRef]
14. Kuroiwa, N.; Abematsu, N.; Matsuo, Y.; Nakao, K.; Sakamoto, T. A case of intraocular lymphoma having retinal adverse events associated with intravitreal methotrexate. *Nippon Ganka Gakkai Zasshi* **2011**, *115*, 611–616. [PubMed]
15. Velez, G.; Yuan, P.; Sung, C.; Tansey, G.; Reed, G.F.; Chan, C.C.; Nussenblatt, R.B.; Robinson, M.R. Pharmacokinetics and toxicity of intravitreal chemotherapy for primary intraocular lymphoma. *Arch. Ophthalmol.* **2001**, *119*, 1518–1524. [CrossRef]
16. Palakurthi, N.K.; Krishnamoorthy, M.; Augsburger, J.J.; Correa, Z.M.; Banerjee, R.K. Investigation of kinetics of methotrexate for therapeutic treatment of intraocular lymphoma. *Curr. Eye Res.* **2010**, *35*, 1105–1115. [CrossRef] [PubMed]
17. Lee, S.; Hughes, P.; Ross, A.; Robinson, M. Biodegradable implants for sustained drug release in the eye. *Pharm. Res.* **2010**, *27*, 2043–2053. [CrossRef] [PubMed]
18. Kunou, N.; Ogura, Y.; Hashizoe, M.; Honda, Y.; Hyon, S.-H.; Ikada, Y. Controlled intraocular delivery of ganciclovir with use of biodegradable scleral implant in rabbits. *J. Control. Release* **1995**, *37*, 143–150. [CrossRef]
19. Kunou, N.; Ogura, Y.; Yasukawa, T.; Kimura, H.; Miyamoto, H.; Honda, Y.; Ikada, Y. Long-term sustained release of ganciclovir from biodegradable scleral implant for the treatment of cytomegalovirus retinitis. *J. Control. Release* **2000**, *68*, 263–271. [CrossRef]
20. Fialho, S.L.; Silva Cunha, A.d. Manufacturing techniques of biodegradable implants intended for intraocular application. *Drug Deliv.* **2005**, *12*, 109–116. [CrossRef]
21. Kimura, H.; Ogura, Y. Biodegradable polymers for ocular drug delivery. *Ophthalmologica* **2001**, *215*, 143–155. [CrossRef] [PubMed]
22. Yang, X.; Zhang, Q.; Wang, Y.; Chen, H.; Zhang, H.; Gao, F.; Liu, L. Self-aggregated nanoparticles from methoxy poly(ethylene glycol)-modified chitosan: Synthesis; characterization; aggregation and methotrexate release in vitro. *Colloids Surf. B Biointerfaces* **2008**, *61*, 125–131. [CrossRef] [PubMed]
23. Kumar, S.C.; Satish, C.S.; Shivakumar, H.G. Formulation and evaluation of chitosan-gellan based methotrexate implants. *J. Macromol. Sci. Part A Pure Appl. Chem.* **2008**, *45*, 643–649. [CrossRef]
24. Sun, Y.; Cui, F.; Shi, K.; Wang, J.; Niu, M.; Ma, R. The effect of chitosan molecular weight on the characteristics of spray-dried methotrexate-loaded chitosan microspheres for nasal administration. *Drug Dev. Ind. Pharm.* **2009**, *35*, 379–386. [CrossRef] [PubMed]
25. Singh, U.V.; Udupa, N. In vitro characterization of methotrexate-loaded poly(lactic acid) microspheres of different molecular weights. *Drug Deliv.* **1998**, *5*, 57–61. [CrossRef]
26. Seo, D.H.; Jeong, Y.-I.; Kim, D.G.; Jang, M.J.; Jang, M.K.; Nah, J.W. Methotrexate-incorporated polymeric nanoparticles of methoxy poly (ethylene glycol)-grafted chitosan. *Colloids Surf. B Biointerfaces* **2009**, *69*, 157–163. [CrossRef] [PubMed]

27. de la Fuente, M.; Ravina, M.; Paolicelli, P.; Sanchez, A.; Seijo, B.; Alonso, M.J. Chitosan-based nanostructures: A delivery platform for ocular therapeutics. *Adv. Drug Deliv. Rev.* **2010**, *62*, 100–117. [CrossRef] [PubMed]
28. Manna, S.; Augsburger, J.J.; Correa, Z.M.; Al-Rjoub, M.F.; Rao, M.B.; Banerjee, R.K. Non-invasive electroretinography assessment of intravitreal sustained-release methotrexate micro-implants in rabbit eyes. *J. Ocul. Pharm.* **2016**, *32*, 583–594. [CrossRef]
29. Manna, S.; Augsburger, J.J.; Correa, Z.M.; Landero, J.A.; Banerjee, R.K. Development of chitosan and polylactic acid based methotrexate intravitreal micro-implants to treat primary intraocular lymphoma: An in vitro study. *J. Biomech. Eng.* **2014**, *136*, 021018. [CrossRef]
30. Manna, S.; Banerjee, R.K.; Augsburger, J.J.; Al-Rjoub, M.F.; Correa, Z.M. Ultrasonographical assessment of implanted biodegradable device for long-term slow release of methotrexate into the vitreous. *Exp. Eye Res.* **2016**, *148*, 30–32. [CrossRef]
31. Manna, S.; Caixeta Faraj, R.Q.; Riemann, B.; Rao, M.B.; Nair, V.; Riemann, C.D.; Augsburger, J.J.; Correa, Z.M.; Banerjee, R.K. Non-invasive evaluation of toxicity in vitreoretinal domain following insertion of sustained release methotrexate micro-implant. *Exp. Eye Res.* **2021**, *205*, 108505. [CrossRef]
32. Ricker, R.; Manna, S.; Donnell, A.M.; Kaval, N.; Al-Rjoub, M.F.; Augsburger, J.J.; Banerjee, R.K. Improved design and characterization of PLGA/PLA-coated Chitosan based micro-implants for controlled release of hydrophilic drugs. *Int. J. Pharm.* **2018**, *547*, 122–132. [CrossRef] [PubMed]
33. Ricker, R. *USP Method for HPLC Analysis of Methotrexate*; The United States Pharmacopeial Convention: Rockville, MD, USA, 2010.
34. de Smet, M.D.; Vancs, V.S.; Kohler, D.; Solomon, D.; Chan, C.C. Intravitreal chemotherapy for the treatment of recurrent intraocular lymphoma. *Br. J. Ophthalmol.* **1999**, *83*, 448–451. [CrossRef] [PubMed]
35. Nakauchi, Y.; Takase, H.; Sugita, S.; Mochizuki, M.; Shibata, S.; Ishiwata, Y.; Shibuya, Y.; Yasuhara, M.; Miura, O.; Arai, A. Concurrent administration of intravenous systemic and intravitreal methotrexate for intraocular lymphoma with central nervous system involvement. *Int. J. Hematol.* **2010**, *92*, 179–185. [CrossRef] [PubMed]
36. Stainer, G.A.; Peyman, G.A.; Berkowitz, R.; Tessler, H.H. Intraocular lysozyme in experimental uveitis in rabbits: Aqueous and vitreous assay. *Investig. Ophthalmol. Vis. Sci.* **1976**, *15*, 312–315.

pharmaceutics

Article

Preparation of Poly-Lactic-Co-Glycolic Acid Nanoparticles in a Dry Powder Formulation for Pulmonary Antigen Delivery

Regina Scherließ [1,*] and Julia Janke [1,2]

1   Department of Pharmaceutics and Biopharmaceutics, Kiel University, 24118 Kiel, Germany;
    julia_janke@gmx.de
2   Merck Healthcare KGaA, 64289 Darmstadt, Germany
*   Correspondence: rscherliess@pharmazie.uni-kiel.de; Tel.: +49-431-880-1330

Abstract: One of the key requirements for successful vaccination via the mucosa is particulate antigen uptake. Poly-lactic-co-glycolic acid (PLGA) particles were chosen as well-known model carriers and ovalbumin (OVA) as the model antigen. Aiming at application to the respiratory tract, which allows direct interaction of the formulation with the mucosal immune system, this work focuses on the feasibility of delivering the antigen in a nanoparticulate carrier within a powder capable of pulmonary delivery. Further requirements were adequate antigen encapsulation in order to use the characteristics of the particulate carrier for (tunable) antigen release, and capability of the production process for industrialisation (realisation in industry). For an effective particulate antigen uptake, nanoparticles with a size of around 300 nm were prepared. For this, two production methods for nanoparticles, solvent change precipitation and the double emulsion method, were evaluated with respect to antigen incorporation, transfer to a dry powder formulation, redispersion and antigen release characteristics. A spray drying step was included in the production procedure in order to obtain a respirable powder with an aerodynamic particle size of between 0.5 and 5 μm. The dried products were characterised for particle size, dispersibility and aerodynamic behaviour, as well as for immune response and cytotoxicity in cell culture models. It could be shown that the double emulsion method is suitable to prepare nanoparticles (270 nm) and to incorporate the antigen. By modifying the production method to prepare porous particles, it was possible to obtain an acceptable antigen release while maintaining an antigen load of about 10%. By the choice of polyvinyl alcohol as a stabiliser, nanoparticles could be dried and redispersed without further excipients and the production steps were capable of realisation in industry. Aerodynamic characteristics were good with a mass median aerodynamic diameter of 3.3 μm upon dispersion from a capsule-based inhaler.

Keywords: double-emulsion method; dry powder inhalation; antigen release; porous PLGA particles; microfluidics

Citation: Scherließ, R.; Janke, J. Preparation of Poly-Lactic-Co-Glycolic Acid Nanoparticles in a Dry Powder Formulation for Pulmonary Antigen Delivery. *Pharmaceutics* **2021**, *13*, 1196. https://doi.org/10.3390/pharmaceutics13081196

Academic Editor: Oya Tagit

Received: 7 June 2021
Accepted: 20 July 2021
Published: 3 August 2021

1. Introduction

As a target for vaccination, the respiratory tract offers the advantage that most pathogens enter the body via this pathway. Hence, this route allows direct interaction with the mucosal immune system [1]. This is attractive not only for the efficient prevention of infections such as SARS-CoV-2, which threatens the world with its pandemic spread which started 2019 [2], but also in the course of therapeutic vaccination [3]. Mucosal vaccination is based on the idea that a particulate vaccine formulation is taken up and processed locally. This requires efficient particle uptake by endocytosis or phagocytosis. These processes are clearly size-dependent; the smaller the particles, the better they can be taken up. Hence, it is obvious to use nanoparticles rather than microparticles as antigen carriers for mucosal vaccination, despite the higher amount of antigen per carrier which can be formulated in microparticles. Within the nanometre range, the particles should be large enough not to be drained via the lymphatic route (i.e., larger than 100 nm) and small enough not to be cleared by macrophages to a larger extent. Macrophages generally take up particles of 500 nm and above [4]. For respiratory administration, it is reported that alveolar macrophages predominantly take up particles between 3 µm and 6 µm [5]. Hence, nanoparticulate vaccine carriers should best have a size between 100 nm and 500 nm (possibly up to 3 µm). For the present work, a target size of 300 nm was determined [6]. Nanoparticles can be prepared by a range of different techniques using "top down" technologies, such as milling, or "bottom up" technologies, such as precipitation, ionic gelation or double emulsion/solvent evaporation techniques [7]. All these processes take place in liquid phase and result in a suspension of nanoparticles in the respective dispersion media, mostly an aqueous system. Nonetheless, physical and thermal stability may be greatly increased when the product can be dried. A drying process shall not harm the antigen, and it also has to retain the nanoparticle being formulated in the preceding step. This comprises retention of primary nanoparticle size and redispersion capability. Spray drying (or freeze drying) is a feasible technique, but further efforts need to be made to stabilise the nanoparticles and allow redispersion. If a nanoparticle suspension is spray dried without excipients, the nanoparticles will probably aggregate to a great extent due to their large surface. Depending on the carrier material, they may also coalesce to larger, undefined material and completely lose their initial small size during wet storage or drying without excipients. This is detrimental for nanoparticulate uptake of antigen and also with respect to pulmonary delivery, where an aerodynamic particle size between 0.5 µm and 5 µm is aimed at [8]. To facilitate redispersion, the nanoparticles are hence embedded into a microparticulate matrix during spray drying, resulting in Nano-in-Microparticles (NiMs). With this, nanoparticle agglomeration and aggregation are prevented and the product is easier to handle. If the matrix is water soluble, such as a carbohydrate, the matrix can dissolve upon water contact (e.g., deposition on the wet mucosa) and release the embedded nanoparticulate vaccine carrier, which can then be taken up.

Ideally, a polymer for vaccination should be degradable in the body to allow the release of antigen and clearance of the drug. It should be biocompatible, of low toxicity and without intrinsic immune response. It needs to be water insoluble to allow particulate uptake, but sufficiently soluble (under non-physiological conditions) in (non-chlorinated) aqueous/organic solvents to allow formulation. The polymer needs to be compatible with proteins and has to be solid at room temperature and up to about 40 °C to allow formulation and storage as a dry powder. Poly-lactic-co-glycolic acid (PLGA), which was used as the model polymer in this study, is a synthetic polymer, which is only soluble in organic solvents. The polymer can be degraded hydrolytically over time [9], resulting in the monomers lactic acid and glycolic acid. Degradation is a function of polymer length (molecular weight) and composition and may take up to several months, making it suitable for sustained-release dosage forms [10]. The shortest degradation time is described for a 50:50 (lactic acid to glycolic acid) polymer, which is used in this work. The polymer can further be functionalised [11,12]. It has been proved that nanoparticles made of, e.g., PLGA can be used as carriers for the antigen to increase the elicited adaptive response by means of increased antigen uptake, processing and presentation [13,14]. PLGA particles have been shown to be effective vaccine carriers upon respiratory administration [15,16]. However, due to their slow degradation, PLGA particles appear to be more suitable for sustained release or depot formulations [10], whereas for vaccination, an immediate and sustained presence of the antigen is required. It had been shown that a more rapid release from PLGA derivatives was associated with better immune response [17], but the use of novel polymers goes along with safety issues. Thus, the well-known PLGA was used as model and nanoparticle preparation was modified to allow less-sustained release of the antigen.

The aim of this work was to design a formulation which can be administered by dry powder inhalation, and which can be prepared using industry-realisable and scalable processes.

2. Materials and Methods

Nanoprecipitation

For solvent-change nanoprecipitation, 1.6% (m/V) polylactic-co-glycolic acid (PLGA, Resomer 503 H, Evonik, Darmstadt, Germany) was dissolved in a mixture of 9 + 1 acetone (J.T.Baker, Deventer, The Netherlands) and methanol (J.T.Baker, Deventer, The Netherlands). This organic phase was added dropwise in an aqueous phase containing 0.5% (w/w) HPMC (Metolose 60 SH 50, Shin-Etsu, Tokyo, Japan), or other stabilisers such as 0.1% Polysorbate 80 or 0.1% Poloxamer 188 in ultrapure water. The mixture was stirred with an Ultra-Turrax (IKA, Staufen, Germany) at 9500 rpm. When the addition of organic phase was completed, the speed was raised to 13,500 rpm for three minutes. Afterwards, the dispersion was stirred with a magnetic stirrer (IKA, Staufen, Germany) at 600 rpm in order to evaporate the organic solvents. After the evaporation process, the dispersion was sonicated in an ultrasonic bath (Sonorex Super RK 106, Bandelin electronic GmbH, Berlin, Germany) for 10 min in order to disperse agglomerates. To determine the suitability of micromixers (interdigital slit mixer and impinging jet mixer, IMM, Mainz, Germany), the two phases were pumped by HPLC pumps (Knaur, Berlin, Germany) into the respective mixing device. The resulting nanosuspension was collected in a glass and was stirred to remove the organic solvent as described above. To load the antigen to the particles, ovalbumin (Sigma, St. Louis, MO, USA) as the model antigen was adsorbed to the particle surface. An ovalbumin stock solution (2 mg/mL) was produced by dissolving ovalbumin in phosphate buffered solution at pH 7.4. For adsorption, the nanoparticle dispersion was mixed with the ovalbumin stock solution at a ratio of 2 + 1 (dispersion + ovalbumin stock solution) and shaken for 4 h at 37 °C prior to spray drying.

Double emulsion method

For NP preparation via the double emulsion method, 2.5% (m/V) PLGA was dissolved in ethyl acetate (Merck, Darmstadt, Germany) as organic phase. This organic phase was homogenised for two minutes at 20,500 rpm by Ultra-Turrax (IKA, Staufen, Germany) with a first hydrophilic phase containing 4% (m/V) ovalbumin (Sigma-Aldrich, St. Louis, MO, USA) in phosphate buffered solution of pH 7.4, resulting in a W/O emulsion. Afterwards, a second hydrophilic phase containing 5% (m/V) PVA (Mowiol 4-88, Hoechst, Frankfurt, Germany) dissolved in ultrapure water was added and homogenised for two minutes. The produced W/O/W-emulsion was transferred into a stabilising solution with 1% PVA in water and stirred with a magnetic stirrer (IKA, Staufen, Germany) at 400 rpm in order to harden the produced particles and to evaporate the organic solvent. Afterwards, the particles were washed to remove free ovalbumin. For this, the dispersion was centrifuged at 14,000 rpm (Centrifuge 5430 R, Eppendorf, Hamburg, Germany), leading to a clear supernatant. The supernatant with free ovalbumin was removed and replaced by a fresh 1% (m/V) PVA solution (and in some experiments 0.01% (m/V) l-leucine) for redispersion of the nanoparticles.

Porous nanoparticles

Porous particles were produced by the double emulsion technique. Here, the first hydrophilic phase was prepared by dissolving 4% (m/V) ovalbumin as well as trehalose (5% to 50% (m/V), British sugar plc, Peterborough, UK) in phosphate buffered solution with pH 7.4. Preparation of W/O and W/O/W emulsion and washing was performed as described above. The supernatant was replaced by a solution containing 1% (m/V) PVA in water for redispersion.

Dry powder formulation by spray drying

The nanoparticle dispersion was spray dried using the Mini-Büchi B-290 (Büchi, Flawil, Switzerland) equipped with a two-fluid nozzle and a high-performance cyclone. No further matrix excipients were added prior to spray drying. Drying was performed at an inlet temperature of 100 °C and an outlet temperature of about 42 °C in order to not melt the polymer.

Particle size distribution and morphology

Nanoparticle size was characterised by dynamic light scattering (Zetasizer, Malvern Instruments, Malvern, UK) before spray drying and after redispersion of the dry powder in water to determine the particle diameter and polydispersity index (PDI). For this, every sample was assayed in triplicate. Particle size of the dry powder after spray drying was characterised by laser diffraction using dry dispersion at 3 bar (Helos with Rodos module, Sympatec GmbH, Clausthal-Zellerfeld, Germany). To visualise particle size and morphology, SEM pictures of the respective formulations were taken (Smart SEM Supra 55VP or Zeiss DSM 940, Zeiss, Oberkochen, Germany).

Antigen content

For antigen quantification, the micro-BCA assay (Thermo Scientific, Rockford, IL, USA) with an OVA calibration was used. When antigen content of intact nanoparticles was determined, PLGA particles were degraded in 0.1N NaOH and samples were neutralised with 0.1 N HCl prior to protein quantification.

Dispersion behaviour and aerodynamic characterisation

For device dispersion experiments and aerodynamic characterisation of the formulations, two capsule-based inhalers, the Unihaler (Figure 1) and the Cyclohaler (Figure 2), were used.

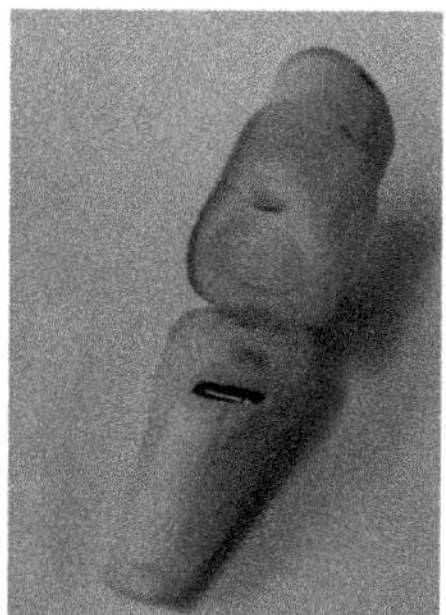

Figure 1. Unihaler.

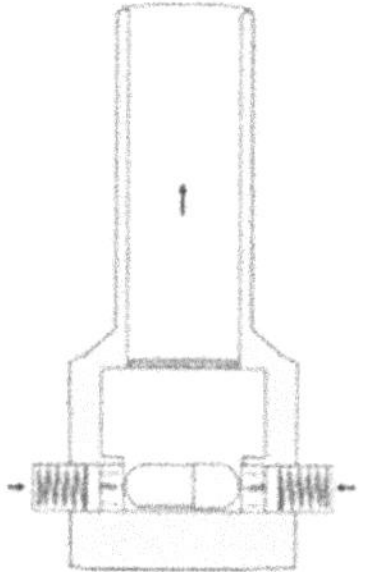

Figure 2. Schematic of the Cyclohaler.

Capsules (HPMC capsules, size 3, Qualicaps Europe, Alcobendas, Spain) were filled manually with 20 mg of the respective powder. Particle size distribution upon device dispersion was measured with the INHALER module (Helos, Sympatec GmbH, Clausthal-Zellerfeld, Germany) at the respective flow rate, creating a pressure drop of 4 kPa over the device (59 L/min for the Unihaler and 100 L/min for the Cyclohaler). Aerodynamic characterisation was performed with the Next Generation Pharmaceutical Impactor (NGI, Copley Scientific, Nottingham, UK). One capsule was used per run. The measurements were performed in a conditioned environment at 21 °C and 45% rH. All stages were coated with a mixture of propylene glycol and isopropanol 50:50 to minimise particle bounce. Deposited powder was dissolved in 0.1 M NaOH. After neutralisation with 0.1 M HCl, the samples were analysed for ovalbumin content with a Micro BCA Protein Assay Kit (Thermo Scientific, Rockford, IL, USA). As reference, 20 mg of powder was dissolved and analysed with the BCA assay. All samples were analysed four times. The fine particle fraction and mass median aerodynamic diameter of the delivered dose were calculated using the Copley Inhaler testing data analysis software (Citdas, Version 3.10, Copley, Nottingham, UK).

Antigen release

For release studies, 10 mg of the spray dried powder was weighed in 2 mL centrifuge tubes in duplicate for every time point. Afterwards, 1 mL of release medium (phosphate buffered solution at pH 7.4 or pH 5.5, prepared according to Ph. Eur. 7.0) was added. Samples were agitated in a water bath with 40 rpm at 37 °C. At predefined time points over 24 h, the respective samples were removed and centrifuged at 14,000 rpm for 10 min to separate any particles from the supernatant. The supernatant was analysed for ovalbumin content. All samples were analysed four times.

Formulation toxicity by MTT assay and endotoxin content

Formulation toxicity in vitro on Calu-3 cells and endotoxin content by an LAL test were assessed. For cytotoxicity testing, 3×10^4 cells were used per well and the test was performed after 3 days of cell growth. In every well, 200 µL of sample was added and incubated for 4 h at 37 °C and 5% CO_2. The samples contained spray dried formulation suspended in Hanks balanced salt solution (HBSS) with 30 mM HEPES, a negative control (HBSS) or a positive control (5 mM SDS in HBSS). Every sample was determined 4-fold. After 4 h of incubation, the samples were replaced by 25 µL MTT solution and incubated for 2 h. Afterwards, 100 µL lysis solution (5% SDS in 50:50 DMF:water with pH 4.7) was added into each well. Then, the absorbance of every well was determined by using a plate reader (Spectra Thermo Reader with Software easyWINfitting V6.0a, Männedorf, Switzerland) at 570 nm (reference wavelength of 690 nm). By using the absorbance of the negative and positive control, the cell viability in each well was calculated.

For the determination of endotoxin content, 20 mg of spray dried formulation was redispersed with 5 mL LAL reagent water (Acila, Weiterstadt, Germany) and lightly shaken for 60 min. Afterwards, the particles were removed by centrifugation, 45 min at 7800 rpm, and the endotoxin content in the supernatant was determined by using LAL reagent Limusate® (sensitivity 0.03 E.U./mL, Waku Chemicals, Richmond, VA, USA). Therefore, different dilutions of the supernatant were prepared, and the LAL reagent was added. After 60 min of motionless incubation at 37 °C, the samples were tested for gelling. From sample dilution and sensitivity of reagent, the endotoxin content was determined semi-quantitatively. These tests were performed to ensure the prepared nanoparticles and resulting dry powders were not harmful to the cells and did not induce an unspecific immune response during further in vitro testing.

Storage stability

Storage stability of the dry powder was assessed over time (0, 1, 3 months, 6 months, ongoing) at ambient conditions (21 ± 1.3 °C and $32 \pm 3\%$ rH). At the respective time points, a sample was analysed with respect to redispersion and nanoparticle size, protein content, dry powder particle size, dispersion and aerodynamic characteristics.

3. Results and Discussion

The absence of chlorinated solvents was one prerequisite for the formulation of PLGA nanoparticles due to toxicological considerations. Therefore acetone, methanol and ethyl acetate were used as solvents for PLGA, unlike in many other studies which use methylene chloride. An advantage of acetone and methanol is their fast evaporation, which eases removal from the formulation. Ethyl acetate is more challenging to remove and is further the most toxic solvent of the aforementioned. Hence, special attention needs to be given to possible residues in the final formulation. Both evaluated formulation processes, nanoprecipitation and the double emulsion technique, resulted in solid nanoparticles of the targeted size range with an acceptable low-size distribution as seen from the PDI (Table 1). It had been assessed earlier how preparation parameters influence the particle characteristics of the resulting particles [18], and parameters had been optimised to achieve the targeted size. For nanoprecipitation, different stabilisers had been screened (polysorbate, Poloxamer 188, HPMC and PVA), with PVA resulting in the smallest nanoparticles below the target size. HPMC, being a macromolecule of moderate surface activity, was able to increase the size concentration-dependently. As HPMC is also favourable if the product is to be dried afterwards, this has been used for nanoprecipitation. For the double emulsion method, HPMC could not be used as the concentration needed for stabilisation, resulting in an unmanageable viscosity. Here, the nanoparticle size was governed by the emulsion droplet size and as such by the homogenisation, whereas an efficient stabiliser (namely PVA) was required to quickly cover interfaces and prevent droplets from growing.

Table 1. Characteristics of solid nanoparticles produced with the different techniques (manual process).

	Solvent Change Precipitation	**Double Emulsion**
Mean particle size	250 ± 7 nm	270 ± 20 nm
PDI	0.11 ± 0.02	0.13 ± 0.03
Redispersibility	good	good
OVA loading	by adsorption	incorporation in first hydrophilic phase
Loading capacity (OVA)	4% (w/w)	10% (w/w)
SEM image of nanoparticles		

Nanoprecipitation is a process which can easily be automated and may run in a semi-continuous mode by the use of a central mixing device which is constantly supplied with the polymer solution and non-solvent by two pumps providing constant flow rates [19,20]. This has been evaluated in this study for an interdigital slit mixer and an impinging jet mixer. The advantage of an impinging jet mixer is that precipitation takes place externally. With this, the device is less prone to blocking. For the interdigital slit mixer, resulting particles were comparable to the manual process, but showed a broader distribution (PDI of 0.3) and tended to block the mixing chamber. Initial experiments with the impinging jet mixer resulted in smaller particles of about 100 nm (PDI 0.1), which would be too small for the intended use as an antigen carrier, but the size was shown to be tuneable by parameter optimisation [18]. Industrialisation of a micromixing setup can be performed by "numbering up" (simultaneous use of many micromixing devices) without changing the dimensions of the individual mixing element. The micromixers were used at a flow rate of 24 mL/min, resulting in a solid output of 424 mg/min. For the double emulsion method, focused ultrasound was evaluated by a partner and was found to be a suitable preparation technique, which can be scaled to industrial size [21]. Utilising the same equipment, namely a focused ultrasound system (S220x, LGC-KBioscience, Teddington, UK), batch sizes between 1 mg and 2.5 g could be prepared uniformly.

One important difference in the two preparation methods (precipitation vs. emulsion) is the incorporation of antigen. Nanoprecipitation results in solid PLGA particles, where the antigen needs to be adsorbed to in a second step. If the antigen is present in the non-solvent during precipitation, it might be incorporated into the polymer matrix in parts but will mostly be present at the surface due to its surface activity. Adsorption on the surface will further be hindered by stabiliser molecules which also adsorb to the interface. Hence, comparably low amounts (4% *w/w* in this study) were associated with the particles and release of the antigen is predominantly guided by desorption and diffusion. This leads to soluble antigen being present around the particles as soon as they are deposited on wet mucosal surfaces. This effect is unwanted as particulate uptake of the antigen is a prerequisite for local processing. For this reason, nanoprecipitation is not the best technique for the formulation of antigen-carrying nanoparticles with sustained release. If solvent-change precipitation can be used to prepare nanocapsules by coating antigen crystals, the product could have superior characteristics. This approach was not followed further in this work, but it focussed on the double emulsion method, which incorporates the antigen being dissolved in the inner phase of the primary emulsion into the polymer matrix. Here, loading capacities of 10% and a size of 270 nm with narrow distribution were reproducibly achieved (Table 1).

The formulations were transferred to a dry powder by spray drying to increase storage stability and allow direct application of the dry powder by inhalation. Here, only HPMC and PVA were suitable as stabilisers of the nanosuspension as the stabiliser cannot be removed completely prior to spray drying. If a liquid surfactant, such as Polysorbate 80, was used as stabiliser, the resulting product after spray drying was highly aggregated and sticky, and hence unsuitable for dry powder dispersion. HPMC and PVA as stabilisers resulted in a dry powder with good bulk characteristics and dispersion behaviour without the need of further matrix components. During spray drying, the nanoparticles were incorporated into the polymeric stabiliser, being present in excess and forming microparticles (Figure 3). The particle size of the spray dried powder is governed by atomisation and solid content of the feed liquid [22].

Figure 3. SEM pictures of spray dried powder. **Left**: PLGA nanoparticles loaded with OVA prepared by solvent change evaporation and 0.5% HPMC as stabiliser/matrix. **Right**: PLGA nanoparticles with OVA prepared by the double emulsion technique and 1% PVA as stabiliser/matrix.

Both dry powder formulations were easily redispersible in water and nanoparticle size was unchanged (data not shown). The powders were well dispersible in dry air and resulting particle size distribution was monomodal with a mean size below 3 µm (Figure 4). As nanoparticles prepared by the double emulsion technique reveal preferable characteristics for vaccination in terms of antigen incorporation, this formulation approach was used for more detailed characterisations. When the dry powder was dispersed from a capsule-based device, particle size distribution is almost identical to 3 bar pressurised air dispersion as seen in Figure 5. This shows that the powder is excellently dispersible to individual primary microparticles by both tested devices. A slight difference between the two tested devices can be seen, which is probably due to differences in device setup, but this effect does not translate to the aerodynamic assessment.

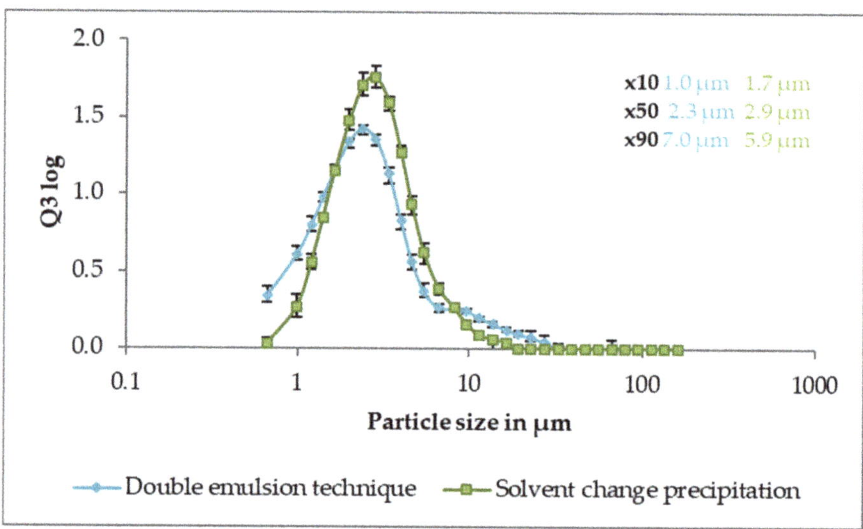

Figure 4. Particle size distribution of spray dried powders with nanoparticles prepared via solvent change precipitation (green) or double emulsion technique (blue) as measured by laser diffraction (3 bar pressurised air dispersion). Data is average of n = 7, error bars show min–max.

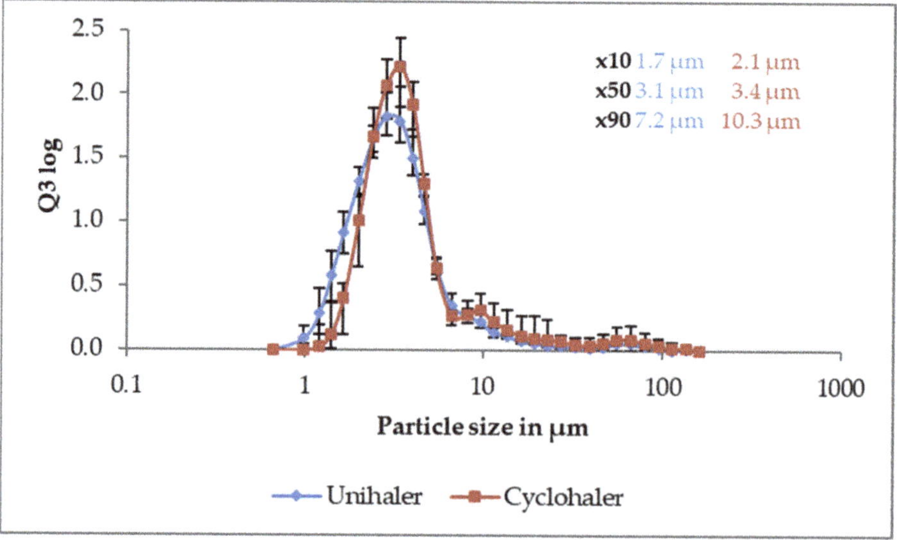

Figure 5. Particle size distribution of spray dried powder with nanoparticles prepared via double emulsion technique dispersed from the Cyclohaler (red) or the Unihaler device (blue) as measured by laser diffraction. Data is average of n = 10, error bars show min–max.

Aerodynamic assessment with the NGI revealed a good distribution profile in the NGI (Figure 6), resulting in a fine particle fraction of about 50% of the loaded dose. Nonetheless, it could be observed that a proportion of the formulation remains in the capsule and the device, which is unwanted. Moreover, it would be favourable to maximise FPF further. An excipient which could decrease capsule and device retention and could maximise FPF is leucine, which has already been assessed as a dispersion modifier and surface coating in dry powder inhalation formulations [23,24].

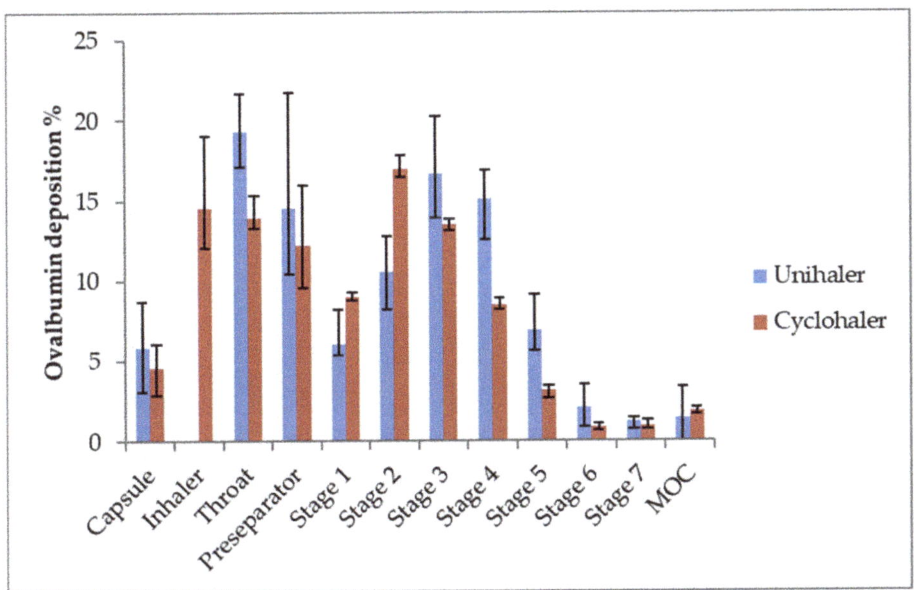

Figure 6. Deposition profile of the NGI of spray dried powder with nanoparticles prepared via double emulsion technique dispersed from the Cyclohaler (red) or the Unihaler (blue) device. n = 3, error bars show min–max.

An addition of 0.01% leucine in the hydrophilic stabiliser phase prior to spray drying resulted in a slightly decreased particle size of the spray dried powder and lower capsule/device retention, but did not improve FPF further (Table 2).

Table 2. Mean particle size from laser diffraction measurements (3 bar), mass median aerodynamic diameter (MMAD) and fine particle fraction (FPF) < 5 µm from aerodynamic characterisation in the NGI utilising two capsule-based devices for different dry powder formulations (average of n = 3). n.d. = not determined.

Formulation	Mean Particle Size (x_{50}, 3 Bar)	Cyclohaler		Unihaler	
		FPF	MMAD	FPF	MMAD
solvent change precipitation, HPMC	2.3 µm	n.d.	n.d.	n.d.	n.d.
double emulsion, PVA	2.9 µm	51%	3.3 µm	49%	3.1 µm
double emulsion, PVA + 0.01% leucine	2.6 µm	45%	3.5 µm	42%	3.2 µm

One important measure for PLGA as a sustained-release polymer is the release of the antigen from the formulation over time. Upon uptake of PLGA nanoparticles, they are routed into intracellular compartments which are acidified [25]. Here, they need to release the antigen, which should then leave the endosome to be processed and presented.

Hence, release studies at a reduced pH of 5.5 representing the endosome [26] are relevant to estimate the possibility of the antigen to be released at these conditions.

The immune effect of solid antigen-loaded PLGA particles is limited, as shown in in-vitro models [17], and this finding goes along with a low release of the antigen over the first 24 h (Figure 7, black line). To increase antigen release, porous nanoparticles were developed (Figure 8).

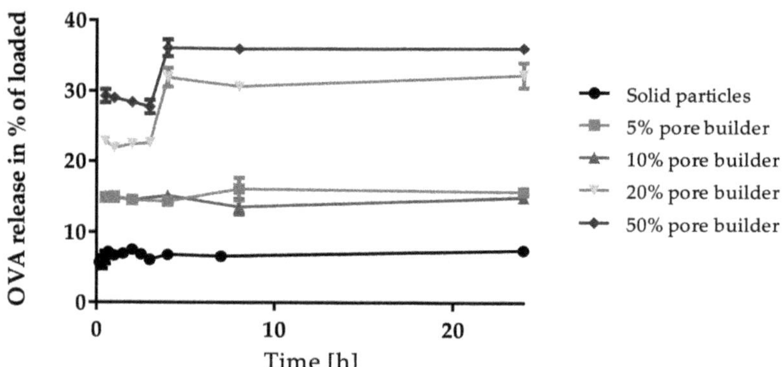

Figure 7. Release of OVA from OVA-loaded PLGA nanoparticles produced by double emulsion technique (without (solid) and with different % amounts of pore builder) in % of loaded OVA over 24 h. n = 3, error bars are standard deviation.

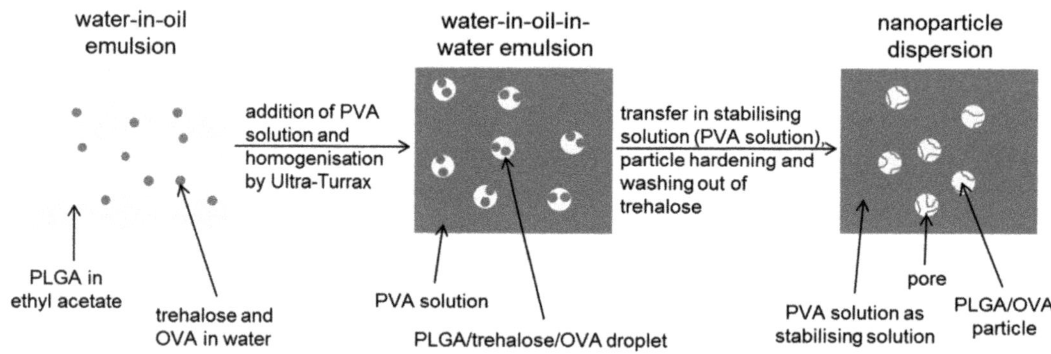

Figure 8. Preparation steps for the formulation of porous PLGA nanoparticles.

Depending on the percentage of pore builder, an increased amount of OVA was released over 24 h (Figure 7). Nonetheless, the release profile remained similar: antigen release was dominated by an initial burst release followed by a plateau. This is due to the antigen being incorporated closely to the surface of the particles as PLGA will not degrade that fast. If release is followed over a much longer time period, PLGA degradation leads to further release, but for a vaccination setup, antigen released over the first 24 h after uptake is most interesting [27]. With increasing amounts of pore builder, loading capacity of the nanoparticles decreased and a maximum of 1498 µg OVA released per 100 mg of polymer was observed when pores were formed from 10% trehalose in the first hydrophilic phase (Table 3). When compared to the data from [17], even a low pore builder concentration of 5% should be sufficient to increase the release to be sufficient to provoke a sound immune response while maintaining a good loading capacity.

Table 3. Effect of trehalose addition on OVA loading capacity and OVA release.

Formulation	Loading Capacity, %	Released OVA after 24 h Normalised to 100 mg Polymer, μg
50% trehalose	3.6	1024
20% trehalose	4.6	1120
10% trehalose	9.8	1498
5% trehalose	8.6	1283
solid particles (no pore builder)	11.5	789

Endotoxin contamination can play an important role in immunological setups, as endotoxins can activate the immune system themselves, leading to false positive results in in vitro experiments and to unforeseeable reactions in vivo. Endotoxin content of the spray dried powders was between 0.37 and 0.75 I.U./mg depending on the formulation components, which is in an acceptable range [28]. Cell culture experiments revealed no acute toxicity of the dry formulation in concentrations up to 20 mg powder/mL (data not shown). If biocompatibility should be shown, the assay would need to be performed for longer. However, the general biocompatibility of PLGA is well known. The purpose of this study was to assess possible acute toxic effects, which would exclude the prepared particles from further in vitro and in vivo examinations.

Stability of the dry powder formulation (solid PLGA NP prepared by double emulsion) was assessed for storage at room temperature. Particles were redispersible after storage and mean nanoparticle size increased marginally (264 nm $\pm$ 5 nm vs. 278 nm $\pm$ 9 nm) due to a slightly increased PDI (0.109 vs. 0.180). Dry powder dispersion and aerodynamic characteristics, which started at a lower level than the batches tested before, were improved over storage, resulting in a higher FPF of 34% after 3 months compared to 27% directly after preparation.

4. Conclusions

As the double emulsion technique allowed the removal of non-incorporated OVA prior to spray drying with a washing step, and with this, minimisation of free OVA in the formulation was possible, this preparation technique is preferred for the preparation of nanoparticulate PLGA systems for mucosal vaccination. The primary nanoparticles of a size between 250 nm and 300 nm incorporated about 10% OVA as model antigen. Particles were formulated to a dry powder which is easily dispersible by capsule-based dry powder inhalers, resulting in a good fraction of fine particles < 5 μm (aerodynamic particle size) which can enter the lung upon oral inhalation. Immunological evaluation of PLGA particles showed that they are capable of antigen delivery and can provoke an antigen-specific immune response.

Author Contributions: Conceptualization, R.S. and J.J.; methodology, J.J., R.S.; formal analysis, J.J.; investigation, J.J.; resources, R.S.; data curation, J.J.; writing—original draft preparation, J.J., R.S.; writing—review and editing, R.S.; visualization, J.J., R.S.; supervision, R.S.; project administration, R.S.; funding acquisition, R.S. All authors have read and agreed to the published version of the manuscript.

Funding: This work was performed as part of the BMBF-funded project PeTrA (grant number 13N11455). The authors would like to thank BMBF and Merck KGaA (Darmstadt, Germany) for the financial support of this project.

Data Availability Statement: Not applicable.

Conflicts of Interest: The authors declare no conflict of interest. The company had no role in the design of the study; in the collection, analyses, or interpretation of data; in the writing of the manuscript, and in the decision to publish the results.

References

1. Shahiwala, A.; Vyas, T.K.; Amiji, M.M. Nanocarriers for Systemic and Mucosal Vaccine Delivery. *Recent Pat. Drug Deliv. Formul.* **2007**, *1*, 1–9. [CrossRef]
2. Moreno-Fierros, L.; García-Silva, I.; Roseales-Mendoza, S. Development of SARS-CoV-2 vaccines: Should we focus on mucosal immunity? *Expert Opin. Biol. Ther.* **2020**, *20*, 831–836. [CrossRef] [PubMed]
3. Saxena, M.; van der Burg, S.H.; Melief, C.J.M.; Bhardwaj, N. Therapeutic cancer vaccines. *Nat. Rev. Cancer* **2021**, *21*, 360–378. [CrossRef]
4. Fifis, T.; Gamvrellis, A.; Crimeen-Irwin, B.; Pietersz, G.A.; Li, J.; Mottram, P.L.; McKenzie, I.F.C.; Plebanski, M. Size-Dependent Immunogenicity: Therapeutic and Protective Properties of Nano-Vaccines against Tumors. *J. Immunol.* **2004**, *173*, 3148–3154. [CrossRef] [PubMed]
5. Blank, F.; Stumbles, P.; von Garnier, C. Opportunities and challenges of the pulmonary route for vaccination. *Expert Opin. Drug Deliv.* **2011**, *8*, 547–563. [CrossRef]
6. Joshi, V.B.; Geary, S.M.; Salem, A.K. Biodegradable Particles as Vaccine Delivery Systems: Size Matters. *AAPS J.* **2013**, *15*, 85–94. [CrossRef] [PubMed]
7. Nicolas, J.; Mura, S.; Brambilla, D.; Mackiewicz, N.; Couvreur, P. Design, functionalization strategies and biomedical applications of targeted biodegradable/biocompatible polymer-based nanocarriers for drug delivery. *Chem Soc. Rev.* **2012**. [CrossRef]
8. Depreter, F.; Pilcer, G.; Amighi, K. Inhaled proteins: Challenges and perspectives. *Int. J. Pharm.* **2013**, *447*, 251–280. [CrossRef]
9. Soppimath, K.S.; Aminabhavi, T.M.; Kulkarni, A.R.; Rudzinski, W.E. Biodegradable polymeric nanoparticles as drug delivery devices. *J. Control. Release* **2001**, *70*, 1–20. [CrossRef]
10. Mukherjee, B.; Satra, K.; Pattnaik, G.; Ghosh, S. Preparation, characterization and in-vitro evaluation of sustained release protein-loaded nanoparticles based on biodegradable polymers. *Int J. Nanomed.* **2008**, *3*, 487–496. [CrossRef]
11. Babiuch, K.; Gottschaldt, M.; Werz, O.; Schubert, U.S. Particulate transepithelial drug carriers: Barriers and functional polymers. *RSC Adv.* **2012**. [CrossRef]
12. Nie, Y.; Zhang, Z.-R.; He, B.; Gu, Z. Investigation of PEG-PLGA-PEG Nanoparticles-based Multipolyplexes for Il-18 Gene Delivery. *J. Biomat. Appl.* **2012**, *26*, 893–916. [CrossRef] [PubMed]
13. Morales-Cruz, M.; Flores-Fernández, G.M.; Morales-Cruz, M.; Orellano, E.A.; Rodriguez-Martinez, J.A.; Ruiz, M.; Griebenow, K. Two-step nanoprecipitation for the production of protein-loaded PLGA nanospheres. *Results Pharma Sci.* **2012**, *2*, 79–85. [CrossRef] [PubMed]
14. Silva, A.L.; Soema, P.C.; Slütter, B.; Ossendorp, F.; Jiskoot, W. PLGA particulate delivery systems for subunit vaccines: Linking particle properties to immunogenicity. *Hum. Vaccines Immunother.* **2016**, *12*, 1056–1069. [CrossRef] [PubMed]
15. Slütter, B.; Bal, S.; Keijzer, C.; Mallants, R.; Hagenaars, N.; Que, I.; Kaijzel, E.; van Eden, W.; Augustijns, P.; Löwik, C.; et al. Nasal vaccination with N-trimethyl chitosan and PLGA based nanoparticles: Nanoparticle characteristics determine quality and strength of the antibody response in mice against the encapsulated antigen. *Vaccine* **2010**, *28*, 6282–6291. [CrossRef]
16. Singh, J.; Pandit, S.; Bramwell, V.W. and Alpar, H.O.; Diphtheria toxoid loaded poly-(E-caprolactone) nanoparticles as mucosal vaccine delivery systems. *Methods* **2006**, *38*, 96–105. [CrossRef]
17. Rietscher, R.; Schröder, M.; Janke, J.; Czaplewska, J.A.; Gottschaldt, M.; Scherließ, R.; Hanefeld, A.; Schubert, U.S.; Schneider, M.; Knolle, P.; et al. Antigen delivery via hydrophilic PEG-b-PAGE-b-PLGA nanoparticles boosts vaccination induced T cell immunity. *Eur. J. Pharm. Biopharm.* **2016**, *102*, 20–31. [CrossRef] [PubMed]
18. Janke, J. *PLGA Nanoparticles for Inhalative Therapeutic Vaccination [Monograph in German]*; Kiel University: Kiel, Germany, 2014.
19. Bally, F.; Garg, D.K.; Serra, C.A.; Hoarau, Y.; Anton, N.; Brochon, C.; Parida, D.; Vandamme, T.; Hadziioannou, G. Improved size-tunable preparation of polymeric nanoparticles by microfluidic nanoprecipitation. *Polymer* **2012**, *53*, 5045–5051. [CrossRef]
20. Gradl, J.; Schwarzer, H.-C.; Schwertfirm, F.; Manhart, M.; Peukert, W. Precipitation of nanoparticles in a T-mixer: Coupling the particle population dynamics with hydrodynamics through direct numerical simulation. *Chem. Eng. Phys.* **2006**, *45*, 908–916. [CrossRef]
21. Schiller, S.; Hanefeld, A.; Schneider, M.; Weigandt, M.; Lehr, C.-M. Focused Ultrasound as a Contact-Free and Scalable Method to Manufacture Protein-Loaded PLGA Nanoparticles. In Proceedings of the AAPS Annual Meeting and Exposition, San Antonio, TX, USA, 10–14 November 2013.
22. Vehring, R.; Foss, W.R.; Lechuga-Ballesteros, D. Particle formation in spray drying. *J. Aerosol Sci.* **2007**, *38*, 728–746. [CrossRef]
23. Minne, A.; Boireau, H.; Horta, M.J.; Vanbever, R. Optimization of the aerosolization properties of an inhalation dry powder based on selection of excipients. *Eur. J. Pharm. Biopharm.* **2008**, *70*, 839–844. [CrossRef] [PubMed]
24. Raula, J.; Thielmann, F.; Naderi, M.; Lehto, V.-P.; Kauppinen, E.I. Investigations on particle surface characteristics vs. dispersion behaviour of l-leucine coated carrier-free inhalable powders. *Int. J. Pharm.* **2010**, *385*, 79–85. [CrossRef] [PubMed]
25. Baleeiro, R.B.; Rietscher, R.; Diedrich, A.; Czaplewska, J.A.; Lehr, C.M.; Scherließ, R.; Hanefeld, A.; Gottschaldt, M.; Walden, P. Spatial separation of the processing and MHC class I loading compartments for cross-presentation of the tumor-associated antigen HER2/neu by human dendritic cells. *OncoImmunology* **2015**, *4*, e1047585. [CrossRef]

26. Geisow, M.J.; Evans, W.H. pH in the Endosome. *Exp. Cell Res.* **1984**, *150*, 36–46. [CrossRef]
27. Allahyari, M.; Mohit, E. Peptide/protein vaccine delivery system based on PLGA particles. *Hum. Vaccines Immunother.* **2016**, *12*, 806–828. [CrossRef] [PubMed]
28. Brito, L.A.; Singh, M. Acceptable Levels of Endotoxin in Vaccine Formulations During Preclinical Research. *AAPS J. Pharm Sci.* **2011**, *100*, 34–37. [CrossRef] [PubMed]

pharmaceutics

Article

Fluorescently Labeled PLGA Nanoparticles for Visualization In Vitro and In Vivo: The Importance of Dye Properties

Vasilisa Zhukova [1], Nadezhda Osipova [1], Aleksey Semyonkin [1], Julia Malinovskaya [1], Pavel Melnikov [2], Marat Valikhov [2], Yuri Porozov [3,4], Yaroslav Solovev [5], Pavel Kuliaev [6], Enqi Zhang [7], Bernhard A. Sabel [7], Vladimir Chekhonin [2], Maxim Abakumov [1,8], Alexander Majouga [1], Jörg Kreuter [9], Petra Henrich-Noack [7,10], Svetlana Gelperina [1] and Olga Maksimenko [1,*]

1 Drug Delivery Systems Laboratory, D. Mendeleev University of Chemical Technology of Russia, Miusskaya pl. 9, 125047 Moscow, Russia; vassaz@list.ru (V.Z.); kompacc@yandex.ru (N.O.); semyonkin.aleksey@gmail.com (A.S.); j.malinowskaya@gmail.com (J.M.); abakumov1988@gmail.com (M.A.); rector@muctr.ru (A.M.); svetlana.gelperina@gmail.com (S.G.)
2 Department of Neurobiology, V. Serbsky Federal Medical Research Centre of Psychiatry and Narcology of the Ministry of Health of the Russian Federation, Kropotkinskiy per. 23, 119034 Moscow, Russia; proximopm@gmail.com (P.M.); marat.valikhov@gmail.com (M.V.); chekhoninnew@yandex.ru (V.C.)
3 World-Class Research Center "Digital Biodesign and Personalized Healthcare", I.M. Sechenov First Moscow State Medical University, Trubetskaya ul. 8, 119048 Moscow, Russia; yuri.porozov@gmail.com
4 Department of Computational Biology, Sirius University of Science and Technology, Olympic Ave 1, 354340 Sochi, Russia
5 Laboratory of Bioinformatics Approaches in Combinatorial Chemistry and Biology, Department of Functioning of Living Systems, Institute of Bioorganic Chemistry, Russian Academy of Sciences, Miklukho-Maklaya ul. 16/10, 119991 Moscow, Russia; solovev@ibch.ru
6 TheoMAT Group, ITMO University, Kronverkskiy pr. 49, 197101 Saint Petersburg, Russia; kulyaevp@gmail.com
7 Institute of Medical Psychology, Otto-von-Guericke University, Leipziger Str. 44, 39120 Magdeburg, Germany; enqi.zhang@med.ovgu.de (E.Z.); bernhard.sabel@med.ovgu.de (B.A.S.); phnoack@gmail.com (P.H.-N.)
8 Department of Medical Nanobiotecnology, Pirogov Russian National Research Medical University, Ostrovityanova ul 1, 117997 Moscow, Russia
9 Institute of Pharmaceutical Technology, Biocenter, Goethe University, Max-von-Laue-Str. 9, 60438 Frankfurt am Main, Germany; Kreuter@em.uni-frankfurt.de
10 Clinic of Neurology with Institute of Translational Neurology, University Clinic Muenster, Mendel Str. 7, 48149 Muenster, Germany
* Correspondence: omnews@mail.ru

Citation: Zhukova, V.; Osipova, N.; Semyonkin, A.; Malinovskaya, J.; Melnikov, P.; Valikhov, M.; Porozov, Y.; Solovev, Y.; Kuliaev, P.; Zhang, E.; et al. Fluorescently Labeled PLGA Nanoparticles for Visualization In Vitro and In Vivo: The Importance of Dye Properties. *Pharmaceutics* **2021**, *13*, 1145. https://doi.org/10.3390/pharmaceutics13081145

Academic Editor: Oya Tagit

Received: 19 June 2021
Accepted: 21 July 2021
Published: 27 July 2021

Publisher's Note: MDPI stays neutral with regard to jurisdictional claims in published maps and institutional affiliations.

Abstract: Fluorescently labeled nanoparticles are widely used for evaluating their distribution in the biological environment. However, dye leakage can lead to misinterpretations of the nanoparticles' biodistribution. To better understand the interactions of dyes and nanoparticles and their biological environment, we explored PLGA nanoparticles labeled with four widely used dyes encapsulated (coumarin 6, rhodamine 123, DiI) or bound covalently to the polymer (Cy5.5.). The DiI label was stable in both aqueous and lipophilic environments, whereas the quick release of coumarin 6 was observed in model media containing albumin (42%) or liposomes (62%), which could be explained by the different affinity of these dyes to the polymer and lipophilic structures and which we also confirmed by computational modeling (log PDPPC/PLGA: DiI—2.3, Cou6—0.7). The importance of these factors was demonstrated by in vivo neuroimaging (ICON) of the rat retina using double-labeled Cy5.5/Cou6-nanoparticles: encapsulated Cou6 quickly leaked into the tissue, whereas the stably bound Cy.5.5 label remained associated with the vessels. This observation is a good example of the possible misinterpretation of imaging results because the coumarin 6 distribution creates the impression that nanoparticles effectively crossed the blood–retina barrier, whereas in fact no signal from the core material was found beyond the blood vessels.

Keywords: PLGA nanoparticles; fluorescent labeling; DiI; coumarin 6; rhodamine 123; Cy5.5; quantum yield; brightness; stability of fluorescent label; confocal microscopy; intracellular internalization; in vivo neuroimaging

1. Introduction

Polymeric nanoparticles hold promise as carriers for the targeted delivery of diagnostics or drugs. By altering the biodistribution of drugs, nanoparticles can improve their therapeutic efficacy and reduce adverse side effects. Therefore, comprehensive knowledge of a nanocarrier biodistribution profile is of paramount importance for the successful development of delivery systems [1]. Radioactive labeling is probably an unsurpassable technique for the quantitative analysis of bulk nanoparticle uptake in the organs, using, for example, single-photon emission computed tomography (SPECT) or positron emission tomography (PET). However, besides the extra efforts necessary for synthesizing and handling the radioactive marker substances, this approach offers little information regarding the precise localization of nanocarriers in target tissues and cells due to the intrinsically limited spatial resolution [2,3]. This information can, however, be obtained by fluorescence imaging. Because of the availability of highly sensitive imaging techniques, such as laser confocal fluorescent microscopy, flow cytometry and systems for the in vivo imaging of fluorescence, fluorescently labeled nanoparticles have become increasingly important in biomedical research (reviewed in several papers [4–9]).

Polymeric nanocarriers used for drug delivery are usually not fluorescent per se, and therefore their visualization in a biological environment requires labeling with fluorescent dyes. According to Fili and Toseland, an ideal "fluorescent label should be small, bright and stable, without causing any perturbation to the biological system [10]." The original phrase refers to the labeling of biological molecules; however, the same principle can be applied to the labeling of nanoparticles. These requirements can be fulfilled by certain small-molecule organic dyes. Obviously, fluorescently labeled nanoparticles serving as a prototype for preclinical evaluation of the nanoparticulate formulation or delivery system should be as similar as possible to the final nanoparticle system. Along with the necessary similarity of their basic parameters such as size and charge, the essential parameter defining the suitability of fluorescent nanoparticles as a prototype model is its stable dye retention. However, fluorescent nanosystems can be used for different purposes: (i) using the fluorescent dye as a model for a drug, which should be delivered by the nanocarrier to a target organ, (ii) developing a stable dye-nanoparticle unit for diagnostic purposes or (iii) following the distribution of the nanocarrier itself. In this respect it is of utmost importance to differentiate between the distribution patterns of the label and the carrier and to know when they exist as one entity and when they are separated.

Widely used approaches to prepare fluorescently labeled nanoparticles from prefabricated non-fluorescent polymers are the physical entrapment of a dye in nanoparticles, achieved by encapsulation during the nanoparticle formation, and the covalent binding of a dye to a core polymer [11]. Both approaches have their advantages and limitations. The encapsulation of a dye in nanoparticles is a relatively simple method based on an emulsification–solvent evaporation technique or nanoprecipitation. After solvent removal, the dye remains physically entrapped/adsorbed in the polymer matrix. One potential limitation of this technique is the dependence of adsorption or encapsulation efficiency on the physicochemical properties of both dye and polymer. Moreover, the encapsulation of a dye might change the properties of nanoparticles and the fluorescence properties of a dye, for example, by inducing fluorescence quenching or, on the contrary, enhancing emission [12]. As mentioned above, another potential drawback for nanoparticle tracking is that the dye might be prone to leakage from the nanoparticles under in vivo conditions [13]. This would eventually decrease their brightness and may create a background signal preventing proper determination of nanoparticle locations.

In contrast, the "covalent" approach enables the preparation of labeled nanoparticles with more predictable parameters that are sufficiently stable under physiological conditions [14,15]. This approach is only applicable, however, when both the polymer and the fluorophore have suitable groups for conjugation.

Stability of labeling has been evaluated previously in a number of studies (for example [13,16–21]); however, these studies employed different nanoparticles and experimental techniques for individual dyes. Based on a non-systematic pool of data, it is not easy to select the optimal experimental strategy for the preparation of the labeled particles with the desired fluorescence properties.

The objective of the present study was to study the influence of the essential parameters, such as the nanoparticle preparation procedure and the dyes' physicochemical properties, on the label performance and properties of the resulting fluorescent nanoparticles. PLGA nanoparticles labeled with four common fluorescent dyes—DiI [22,23], coumarin 6 [24–27], rhodamine 123 [28–30] and Cy5.5 [31,32] were employed.

The labeled PLGA nanoparticles were compared with regard to how a dye influences their distribution, brightness (which, in turn, depends on the dye loading efficiency and quantum yield) and stability of the dye retention. Computer simulation and empirical measurements were applied for the modeling of dye leakage from the nanoparticles and the interaction of the dye molecules with lipid membranes. The stability of the encapsulated label in vitro and in vivo was then monitored using the double-labeled nanoparticles containing both the conjugated Cy5.5 and one of the encapsulated dyes. Furthermore, the nanoparticle–cell interactions were studied using confocal microscopy and in vivo neuroimaging to verify their potential value for medical applications.

2. Experimental

2.1. Materials

Poly(lactic-co-glycolic acid) (PLGA, Resomer RG 502 H, LA/GA ratio 50:50, acid terminated, Mw 7–17 kDa, η = 0.21 dL/g) was purchased from Evonik Nutrition & Care GmbH, Germany. The cyanine5.5 amine (Cy5.5 analogue) was obtained from Lumiprobe Life Science solutions (Cockeysville, MD, USA). Rhodamine 123 hydrochloride (Rh123) was purchased from Acros Organics (New Jersey, NJ, USA). Block-copolymer PLGA-PEG (PEG MW 2000 Da, PLGA MW 11,500 Da), DiI (1,1′-dioctadecyl-3,3,3′,3′-tetramethylindocarbocyanine perchlorate), coumarin 6 (Cou6), poly(vinyl alcohol) (PVA, 9 10 kDa, 80% hydrolyzed), poloxamer 188 (Kolliphor P188, BASF), N-hydroxysuccinimide (NHS), N,N-diisopropylethylamine (DIPEA), N-(3-dimethylaminopropyl)-N-ethylcarbodiimide hydrochloride (EDC), bis(2-ethylhexyl)sulfosuccinate sodium salt (AOT) and other reagents were purchased from Sigma-Aldrich (St. Louis, MO, USA). Gel filtration media Sepharose CL-2B and Sephadex G-25 were from GE Healthcare BioSciences (Uppsala, Sweden). L-α-phosphatidylcholine (egg, chicken) were purchased from Avanti Polar Lipids Inc. (Alabaster, AL, USA). All organic solvents were of analytical grade.

2.2. Preparation of Fluorescently Labeled Nanoparticles

PLGA nanoparticles labeled with DiI or coumarin 6 (Cou6). The PLGA nanoparticles labeled with the hydrophobic dyes DiI and Cou6 were produced by a high-pressure emulsification–solvent evaporation technique. Briefly, the polymer and the dye were dissolved in dichloromethane (DCM). The dye-to-polymer mass ratios ranged from 1:2353 to 1:39 in the case of DiI and from 1:300 to 1:25.5 in the case of Cou6. Then the organic phase was mixed with a 1% aqueous PVA solution (aqueous-to-organic phase ratio = 5:1, *v/v*), and the mixture was emulsified using a high shear homogenizer (Ultra-Turrax T18 digital, IKA, Staufen im Breisgau, Germany) followed by high-pressure homogenization at 1000 bar (Microfluidizer M-110P, Microfluidics, Newton, MA, USA) to obtain an *o/w* emulsion. After the removal of the organic solvent under vacuum, the resulting suspension was filtered through a glass-sintered filter (pore size 90–150 μm). Free dye was washed from the nanosuspension by two-step centrifugation (48,384× *g*, 5 °C, 30 min at the first step and 40 min at the second step, Avanti JXN-30, Beckman Coulter, Brea, CA, USA); each time the precipitates were resuspended in distilled water. Additionally, AOT, a bulky hydrophobic counterion, was added in the equimolar ratio to DiI during the nanoparticle preparation process (PLGA-DiI-AOT NPs) in order to decrease the self-quenching of the dye inside the nanoparticles [33]. Finally, D-mannitol was added to the suspension in an amount sufficient to produce 1% (*w/v*) concentration to act as a cryoprotectant, and the nanoparticle suspension was frozen at $-70\ °C$ and freeze-dried.

The PEG-PLGA nanoparticles labeled with coumarin 6 (PEG-PLGA-Cou6) were prepared using blends of PLGA and PLGA-PEG with varying mass ratio (PLGA:PLGA-PEG = 10:1 and PLGA:PLGA-PEG = 5:1, *w/w*). The amount of Cou6 varied from 0.3 to 3.7% (*w/w*). Homogenization was carried out as described above (15,000–20,000 psi for 1 min). A coarse nanoparticle fraction with a size >1000 nm was separated by centrifugation (15,000× *g* at 18 °C for 5 min, Avanti JXN-30, Beckman Coulter, Brea, CA, USA), and then the supernatant was additionally filtered through a 0.45 μm membrane. These conditions were sufficient to sediment the aggregated dye but not sufficient to cause significant loss of the PLGA nanoparticles in suspension. Afterwards, the nanoparticles were freeze-dried with addition of 10% (*w/v*) of mannitol. Nanoparticles were washed from the free dye as described above. Two types of the reference preparations were produced for comparison by a similar high-pressure emulsification solvent evaporation procedure: non-labeled nanoparticles (placebo PLGA nanoparticles) produced by the same technique, without adding the dye and without subsequent washing and/or coarse fraction separation, and a dye nanosuspension stabilized with PVA (polymer-free control), produced by the described techniques without adding the polymer to an organic phase and also without washing and/or fraction separation.

PLGA nanoparticles were labeled with rhodamine 123. The PLGA-Rh123 nanoparticles were produced employing a high-pressure double emulsion solvent evaporation technique as follows. An aqueous solution of Rh123 was added to the PLGA solution in DCM at a ratio of 1:48.4 (*w/w*) and emulsified using an Ultra-Turrax T18 (23,600 rpm, 1 min). Then, the obtained emulsion (*w/o*) was added to a 1% aqueous PVA solution (9–10 kDa) and passed through a multi-step homogenization process as described above. Non-encapsulated (free) dye was removed from the nanosuspension by gel filtration chromatography using a Sephadex G-25 column. The nanoparticles were freeze-dried with 1% mannitol.

Preparation of nanoparticles using PLGA conjugated with cyanine5.5 amine (PLGA-Cy5.5). The cyanine5.5 amine was covalently bound to a carboxylic terminal group of PLGA via a carbodiimide coupling reaction as described earlier [15]. For a detailed description of the procedure, see the electronic support information. The PLGA-Cy5.5 nanoparticles were produced by an *o/w* emulsion solvent evaporation technique. PLGA and PLGA-Cy5.5 (total amount 200 mg, content of pre-modified polymer 0, 50, 100, 150, 200 mg) were dissolved in 4 mL of DCM. The obtained solution was added to 20 mL of a 1% aqueous solution of PVA, mixed, and then passed through a multi-step homogenization process as described above. The organic solvent was removed using a rotary evaporator. The nanoparticles were freeze-dried with 1% mannitol.

Preparation of double-labeled PLGA nanoparticles. The double-labeled nanoparticles, i.e., the Cy5.5-PLGA nanoparticles labeled additionally with DiI, coumarin 6 or rhodamine 123 were prepared as described above using a 1:1 (*w/w*) mixture of PLGA-Cy5.5 and PLGA polymers. Thus, the PLGA-Cy5.5/DiI NPs were prepared similarly to the PLGA-DiI nanoparticles. PLGA (100 mg), PLGA-Cy5.5 (100 mg) and DiI (0.8 mg) were dissolved in 4 mL of DCM and added to a 1% aqueous solution of PVA (9–10 kDa). Cy5.5 and DiI were used in an equimolar ratio (8.6×10^{-7} mol). The PLGA-Cy5.5/Cou6 NPs were prepared by the same procedure as the PLGA-PEG-Cou6 nanoparticles using a blend of PLGA and PLGA-PEG polymer (5:1, *w/w*). The PLGA-Cy5.5/Rh123 nanoparticles were produced similarly to the preparation of PLGA-Rh123 nanoparticles.

2.3. Physicochemical Characterization

Particle size and zeta potential characterization. Average particle sizes (hydrodynamic diameter) and polydispersity indices (PDI) of the nanoparticles were measured by dynamic light scattering (DLS) using a Zetasizer Nano ZS particle size analyzer (Malvern Panalytical Ltd., Malvern, UK, equipped with He-Ne with 633 nm laser, maximum power 4 mW) at a scattering angle of 173° and a temperature of 25 °C. Samples were diluted to a concentration of 0.2 mg/mL polymer with milli-Q water. Zeta potentials were measured using the same instrument by electrophoretic light scattering (ELS) in a U-shaped disposable folded capillary cell either without dilution or diluted the same way as for the size measurements. All measurements were repeated 4 times. Sizes were evaluated immediately after preparation, before lyophilization, after lyophilization and several hours after resuspension to assess colloidal stability.

Dye content and encapsulation efficiency. The dye encapsulation efficiency ($EE_\%$) was calculated as a ratio of the dye amount entrapped in the nanoparticles to the total amount of a dye found in a sample:

$$EE_\% = \frac{m_{total} - m_{unbound}}{m_{total}} \cdot 100\% \tag{1}$$

The total amount of a dye was determined spectrophotometrically after dissolving samples in DMSO. Each measurement was performed in triplicate.

PLGA content. The amount of PLGA in the samples was analyzed by the amount of lactate formed after the nanoparticle destruction by hydrolysis in 1 M NaOH (37 °C, 24 h, constant shaking at 200 rpm) using capillary electrophoresis (CAPEL 105M, Lumex, St. Petersburg, Russia) as described in our previous work [15].

Scanning electron microscopy. Images of the PLGA-Cou6 nanoparticles were obtained using a scanning electronic microscope, JSM 6510LV (Jeol, Tokyo, Japan). Resolution HV mode was up to 5 nm; magnification/zoom was up to 300,000. Before the microscopical investigation, the nanoparticle samples were washed from stabilizer (PVA) and cryoprotectant (D-mannitol) by centrifugation. A drop of the resulting suspension was placed on a glass slide, dried in air and then sputter-coated with platinum for 6–30 s using an Auto Fine Coater JFC-1600 (JEOL, Tokyo, Japan). Studies were performed using the equipment of D. Mendeleev University's Center for Collective Use equipment.

Evaluation of physical labeling stability. Stability of the nanoparticle labeling was evaluated by the assessment of the amount of the dye released from nanoparticles at chosen time points. The freeze-dried nanoparticles labeled with Rh123, Cou6 and DiI were resuspended and diluted 25-fold with the release media (1% solution of poloxamer 188 in 0.15 M PBS, pH 7.4). Then the resulting suspensions were incubated at 37 °C under continuous shaking. At chosen intervals, 1.5 mL aliquots of the suspensions were sampled, and the nanoparticles were separated by centrifugation at $48,380 \times g$ for 30 min at 18 °C (Avanti JXN-30, Beckman, USA). The centrifugation conditions were preliminarily optimized for the residual polymer content in the supernatant, measured by the capillary electrophoresis, as described in 2.2 (data not shown). Thus, after centrifugation of a 1.5 mL sample at $48,380 \times g$ at 5 °C, the PLGA content in the supernatant did not exceed 5% of the total amount. In addition, the presence of the particles was not detected by the DLS method.

The amount of the released dye in supernatants was determined spectrophotometrically upon dilution with ethanol to 50%. The experiment was performed in triplicate. The release profiles of DiI and Cou6 were additionally assessed using 1% solution of human serum albumin (HSA) in 0.15 M PBS (pH 7.4) as the release medium.

The coumarin 6 release profile from the PLGA-PEG nanoparticles was also evaluated using the dialysis technique with frequent medium replacement (Float-A-Lyzer dialysis device, cellulose ester membrane, MWCO 50 kDa, Spectrum Laboratories Inc., Rancho Dominquez, CA, USA. The release medium was a 1% solution of poloxamer 188 in 0.15 M PBS at pH 7.4. A nanosuspension of coumarin 6 was used as a non-polymeric control. At the end of the experiment, the membrane samples from the dialysis bags were dissolved in DMSO, and the amount of coumarin 6 adsorbed on the membrane was measured spectrophotometrically. The experiment was performed in triplicate.

Fluorescence measurements. Fluorescence quantum yields (QY) were determined using an adaptation of the comparative method described by Rhys Williams et al. [34]

and the measurement conditions recommended by IUPAC [35]. The method is based on measurements of the fluorescence spectrum of the sample and comparison of its integrated intensity with the same quantity of a reference system, i.e., a certain standard with a known absolute QY and similar maxima of absorption and emission.

The general procedure was as follows. A series of standard solutions with varied concentrations and a series of suspensions of varied dilutions were prepared. Standards were chosen, when possible, based on the data of Brouwer [35]. The following standard solutions were used: rhodamine 6G solution in water for the PLGA-DiI nanoparticles, coumarin 153 solution in ethanol for the PLGA-Cou6 nanoparticles and Rh123 in ethanol for PLGA-Rh123, Cy5.5 in 96% ethanol for PLGA-Cy5.5 nanoparticles. Then, absorbance was measured for each sample at the relevant excitation wavelengths (488 nm for PLGA-DiI NPs, 450 nm for PLGA-Cou6 NPs, 480 nm for PLGA-Rh123 NPs, 684 nm for PLGA-Cy5.5 NPs). Absorbance values used were ≤ 0.25 in order to minimize the reabsorption effects. The fluorescence spectra were recorded for the same suspensions and solutions using the aforementioned wavelengths for excitation (RF-6000, Shimadzu, Japan). Optical path length was 10 mm, and slit widths for excitation and emission were held constant (5 nm). Integrated intensities were obtained for each sample. To calculate the relative quantum yields (QY_R), the intensities were plotted against absorbances of the corresponding samples. The slopes of the acquired graphs were used in the following equation:

$$QY_r = QY_{st} \cdot \frac{tg\alpha_x}{tg\alpha_{st}} \cdot \frac{n_x^2}{n_{st}^2} \tag{2}$$

where QY_{st} is the absolute quantum yield of the reference (0.90 for rhodamine 6G [36], 0.20 for cyanine5.5 amine (manufacturer's data) and 0.53 ± 0.02 for coumarin 153 [37]); $tg\alpha_x$ and $tg\alpha_{st}$ are the slopes of the graphs of nanoparticles and standards, respectively; n_x and n_{st} are the refractive indexes of solvents (1.3333 for water and 1.3617 for ethanol). The specific brightness of fluorescence was calculated accordingly as:

$$\text{Specific brightness} = \varepsilon \cdot QY \cdot n \tag{3}$$

where ε is the molar extinction coefficient of the dye loaded in nanoparticles calculated for a given sample calculated using the Beer–Lambert law, l/mol·cm; QY is the quantum yield, dimensionless (fraction of 1); n is the amount of substance of fluorophore molecules per mg of PLGA, mol/mg.

For nanoparticles labeled with both DiI and Cy5.5, FRET efficiency was calculated as described in the work of Majoul et al. [38] (See Supplementary Materials.)

Dye partitioning in liposomes. The liposomes were prepared by lipid membrane hydration. Briefly, L-α-phosphatidylcholine (20.4 mg) and cholesterol (4.8 mg) were dissolved in 5 mL of chloroform. The organic solvent was removed under vacuum at 40 °C until the lipid film was formed. Further, 20 mL of 0.01 M PBS (pH 7.4) was added, and the mixture was sonicated in an ice bath for 2 min at 100% power (Sonopuls HD2070, Bandelin, Berlin, Germany). The resulting suspension was filtered through a glass porous filter (pore size 100–160 μm).

For evaluation of the dye leakage, the liposomes (2–5 mL) were mixed with the fluorescent PLGA nanoparticles resuspended previously in 1 mL of 20% sucrose solution in 0.01 M PBS (pH 7.4). The control samples were prepared without the addition of liposomes. All samples were incubated for 1 h at 37 °C at constant shaking (200 rpm). Then the nanoparticles and the lipid fraction were separated by centrifugation (5 °C, 30 min, 15,000× g). The supernatants containing liposomes were removed, and the precipitates (nanoparticles) were resuspended, frozen and freeze-dried. The amount of the dye in precipitates was measured spectrophotometrically after their dissolution in DMSO, as described above. The total amount of the dye transferred to liposomes was calculated as the difference between the concentrations in control and in the sample ($c_{contr} - c_{lipo}$). The amount of a free dye was taken into consideration through encapsulation efficiency. To calculate the fraction of a dye transferred to liposomes (T), the following equation was applied:

$$T = \frac{\left(c_{contr} - c_{lipo}\right) \cdot EE}{c_{contr}} \cdot 100\%,\qquad(4)$$

where c_{contr} is the total amount of the dye, measured in precipitate of the control sample; c_{lipo} is the residual amount of the dye in the nanoparticles after incubation with liposomes; EE is the encapsulation efficiency.

In this experiment, two types of DiI samples were used: in one case, the free dye was removed by centrifugation (5 °C, 48,380× g, 30 min), in another by means of GPC (Sepharose CL-2B), followed by centrifugation as described above.

Differential scanning calorimetry. Glass transition temperature (T_g) was measured for the Cou6-PLGA and DiI-PLGA nanoparticles. Non-loaded PLGA nanoparticles and bulk polymer were used as control. The samples of Cou6-PLGA and DiI-PLGA nanoparticles used for the DSC analysis were prepared as described above at the dye/polymer weight ratio of 1:20 and were not subjected to washing procedure. The calorimetric measurements were performed using differential scanning calorimeters TA-4000 equipped with a DSC-30 heating cell and DSC823e (Mettler Toledo GmbH, Greifensee, Switzerland) at a heating rate of 10 °C/min under argon. The samples were measured in aluminum pans. The temperature range was −20 °C to 100 °C. A blank aluminum pan was used as reference. All glass transition temperatures (T_g) were reported as the onset of the transition.

Modeling of dye interaction with lipid membranes. The model lipid membranes consisted of either dipalmitoylphosphatidylcholine (DPPC) or dimyristoylphosphatidylcholine (DMPC) bilayers. The analysis of the dyes' conformational isomers (conformers) – their structure and stability—was carried out using a composite hierarchical approach. Molecular mechanics (MM) was employed to perform the initial screening of conformers with the OPLS3e force field [39] as implemented in the conformer advanced search script [40]. At this step, for each type of the dye molecules, 100 lowest-energy conformers were selected, the structures and energetics of which were further refined in semi-empirical electronic structure calculations at the PM6-D3H4 level [41] using the MOPAC2016 program package (MOPAC2016, supported by J.P. Stewart, Stewart Computational Chemistry, Colorado Springs, CO, USA). Next, the results were clustered with RMSD and energy to determine the most stable conformers, which were further used for the DFT calculations with the solvation model. Finally, the COSMO-RS approach was used to figure out the Boltzmann distribution of conformers for each dye, which allows us to find solubilities and partition coefficients more accurately. DFT calculations were carried out using the Turbomole program package (TURBOMOLE V7.2 2017, development of University of Karlsruhe and Forschungszentrum Karlsruhe GmbH) at the BP86/TZVPD level of theory [42–45] as a recommended procedure for COSMO-RS calculations [46]. The solubility of dyes and their partition coefficients in bi-phase systems were then computed using a hybrid COSMO-RS solvation model [46,47] with CosmoTherm 16 program [48]. Polymers (PLA and PLGA) were simulated with CosmoTherm 16 program according to the recommended approach [49].

2.4. Biological Studies

In vitro visualization. The 4T1 murine breast carcinoma cells were purchased from the American Type Culture Collection (ATCC, Manassas, VA, USA). The cells were cultured at a seeding density of around 50,000/cm² in Ham's F-12K medium (2.5% fetal bovine serum, 15% horse serum and 1% penicillin) and DMEM (10% newborn calf serum, 100 U/mL penicillin and 100 mg/mL streptomycin), respectively, at 37 °C in 5% CO_2 for 24 h to allow cell attachment. Before use, nanoparticles were additionally coated with a surfactant, which was accomplished by resuspending the samples in a 1% poloxamer 188 aqueous solution and incubating them for 30 min. Then, nanoparticles at a final concentration of 100 µg/mL were added to the fresh medium, and the cells were incubated from 0 to 20 min. After the incubation, the cells were gently washed three times with PBS and further incubated with 2.5 µg/mL of Hoechst, 50 nM/mL of Lysotracker Green DND26 for 20 min and then again washed three times with PBS. The cells were observed using a confocal laser scanning microscope (inverted multiphoton confocal microscope Nikon A1 MP, Nikon Instruments Inc., Tokyo, Japan) for 20 min. FRET imaging was applied to the double-stained nanoparticles after their incubation with the cells for 20 min.

In vivo confocal neuroimaging (ICON). All procedures were performed with ethical approval according to the requirement of the German National Act on the use of experimental animals (Ethic committee Referat Verbraucherschutz, Veterinärangelegenheiten; Landesverwaltungsamt Sachsen-Anhalt, Halle, approval code: 203.6.3-42502-2-1469 UniMD_G/09.07.2018). Adult male Lister hooded rats (Crl: LIS strain; Charles River) were housed on a 12 h light/12 h dark cycle under standard environment conditions at ambient temperature of 22 °C at 50–60% humidity. Animals had access to food and water ad libitum, except on the day before the induction of narcosis, when food was removed. Animals were kept at least one week for adaption in group cages and were handled before starting the experiment to reduce stress.

Cy5.5-PLGA-Cou6 NPs for ICON imaging were dispersed in 0.5 mL of a sterile 1% poloxamer 188 solution in deionized distilled water (dd water). Fluorescent signals of Cy5.5-PLGA-Cou6 NPs were observed as described in previous studies [50,51]. Briefly, the animals were anesthetized by an intraperitoneal injection of ketamine (75 mg/kg) and medetomidine (0.5 mg/kg) and kept on a heating pad. A cannula was inserted into the tail vein for application of the NP solution. The irises of the eyes of anaesthetized rats were dilated with neosynephrine-POS 5%. A contact lens was adjusted onto the cornea and immersed in Vidisic eye gel. Then, the rats were fixed under a confocal scanning microscope with the eye positioned in working distance underneath the objective of a Zeiss LSM 880 microscope, and in vivo confocal neuroimaging (ICON) of the retina was performed. The images were captured at time point 0 min (baseline before NP injection) and post-injection at 40 min (excitation/emission wavelengths used for coumarin 6 and Cy5.5 were 488/502 nm and 633/708 nm, respectively). The rats were kept on the heating pad throughout the in vivo imaging process. All the images were obtained with the same resolution (5× magnification and same image size).

Statistical analysis. Student's *t*-test and ANOVA on ranks were used to investigate the differences statistically. The differences were considered significant at values of $p < 0.05$. All values were expressed as mean ± SD.

3. Results & Discussion

The dyes investigated in this study are widely used to visualize nanoparticles in vitro and in vivo. However, data regarding their properties, especially the values of partition coefficients characterizing the distribution of the dyes in various media, differ from one source to another or are absent in the literature. This is mostly due to the variety of software used for calculation based on various algorithms or, in the case of experimental values, it is because these dyes are so poorly soluble that it is hard to obtain a reliable value. Nevertheless, one can with certainty claim that the dyes' hydrophobicity increases in the following order: Rh123—Cou6—DiI and that DiI is significantly more hydrophobic than Cou6 and Rh123 (Table S1). Those properties (namely, poor solubility in aqueous media and hydrophobicity) in many cases determine the choice of dye used in the visualization of nanoparticles. Researchers believe that these properties allow the dyes to be easily encapsulated into nanoparticles and to remain in the nanoparticles during in vitro/in vivo experiments. Moreover, for the same reasons, coumarin 6 is used not only as a bright and convenient label for the visualization of various nanocarriers [25,26] but is also considered a suitable model of a hydrophobic drug [52,53]. In order to further investigate the acceptability and reliability of the aforementioned dyes as labels for nanoparticles, we have conducted this study concerning various aspects that are important for fluorescent nanoparticles: colloidal stability, brightness, dye retention, interaction with biological media, etc.

3.1. Preparation and Basic Properties of the Labeled Nanoparticles

Due to their hydrophobic nature, DiI and coumarin 6 are insoluble in water but readily dissolve in dichloromethane (DCM), as does the polymer PLGA; therefore, these nanoparticles could be obtained by a simple (*o/w*) emulsification–solvent evaporation method. Similarly, for the polymer conjugated with Cy5.5, the same technique was employed. Rhodamine 123 hydrochloride is more soluble in water than in DCM; therefore, in this case the PLGA-Rh123 nanoparticles were obtained by a double (*w/o/w*) emulsification technique. Initial dye-to-polymer ratios were chosen based on preliminary experiments so that they could enable the visualization of the nanoparticles in the in vitro and in vivo experiments.

In general, the encapsulation of both DiI and rhodamine 123 was efficient and did not produce any considerable changes in the basic parameters of the nanoparticles: these nanoparticles had mean diameters of ~100 nm with relatively low PDI (<0.2) and negative zeta potentials (Table S2). In the case of the nanoparticles labeled with DiI, the final dye-to-polymer ratio varied in the range of 0.2–12.5 µg/mg (initial loadings were in the range of 0.4–25.6 µg/mg) due to losses during the preparation and washing procedures. In this range, no influence of the dye-to-polymer ratio on the encapsulation efficiency of DiI was observed; the average value was around 70%.

Resomer RG 502 H used in this study as a core polymer has a terminal carboxylic end group that enabled the labeling of the nanoparticles by its conjugation with the reactive amine derivative of Cy5.5. Labeling of PLGA with cyanine5.5 amine was performed using a carbodiimide method. The absence of fluorescence in the aqueous phase upon purification of the conjugate by washing signified the quantitative conjugation. The absence of unbound dye in the final preparation was evidenced by TLC. TLC also demonstrated that the label was stable after the nanoparticle preparation.

The PLGA-Cy5.5 nanoparticles were produced using different ratios of pre-modified PLGA-Cy5.5 polymer and the original polymer while keeping the total polymer amount constant. The presence of the modified polymer did not exert any influence on the nanoparticles' basic parameters: the average size of all PLGA-Cy5.5 nanoparticles was about 120 nm (PDI < 0.2), with a zeta potential of around −22 mV (Table S2).

In the double-labeled PLGA nanoparticles, containing one of the aforementioned dyes and covalently attached Cy5.5, the latter served as a marker, allowing for a more accurate analysis of the localization of both the particles and the physically entrapped dye, whether it is still associated with the nanoparticles or has leaked out. This approach is often a much-needed tool because otherwise it is hard to recognize the difference between a signal generated by a nanoparticle and a signal of the released dye, in part due to resolution limitations. The data obtained in this study indicate that double labeling represents a useful tool to verify the stability of the fluorescent dye incorporated into the nanoparticles (see below).

Among the nanoparticles prepared and investigated in this study, the PLGA-Cou6 nanoparticles labeled with coumarin 6 exhibited a distinctive behavior. These nanoparticles were characterized by a high encapsulation efficiency of 93–99% with a dye content of 12–17 µg/mg PLGA at the initial dye-to-polymer ratio of 1:170 (*w/w*). Interestingly, despite the higher solubility of coumarin 6 compared to DiI (Table S1) and its lower hydrophobicity (calculated logP 5.0 and 10.3, respectively see Section 3.5), the PLGA-Cou6 nanoparticles had a higher polydispersity index (PDI >0.2) and a less negative surface charge (−8.5 ± 0.6 mV) than the PLGA-DiI nanoparticles, which obviously was due to the dye adsorption on the nanoparticle surface leading to their increased hydrophobicity and the formation of aggregates. A tendency towards agglomeration with the formation of unusual clusters was also observed in the SEM images of these nanoparticles (Figure 1a,b).

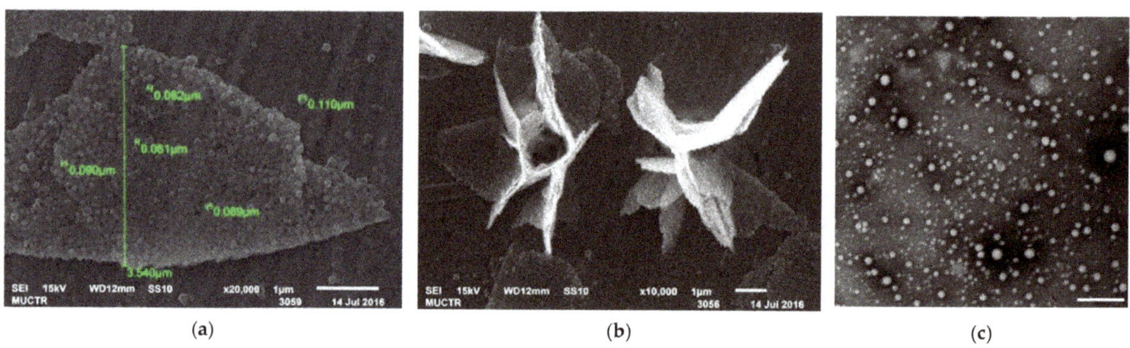

Figure 1. (**a**): SEM image of PLGA-Cou6 nanoparticles (non-PEGylated); (**b**): nanoparticles clusters (broader view); scale bar 1 μm; (**c**): PLGA-PEG-Cou6 nanoparticles (scale bar 500 nm).

This unwanted phenomenon could be avoided by using a mixture of PLGA and PEG-PLGA in nanoparticle preparation (Figure 1c), a technique frequently used to increase colloidal stability [54]. The hydrophilic brush on the nanoparticle surface created by the PEGylated polymer could considerably decrease agglomeration, and, accordingly, lead to the formation of smaller Cou6-labeled nanoparticles with lower polydispersity (see Supplementary Materials). However, the presence of the PEG moiety also changed the surface properties of these nanoparticles compared to the plain PLGA nanoparticles. Overall, it is hard to consider coumarin 6 a "neutral" fluorescent label. Therefore, it is advisable to use Cou6 in small concentrations as a fluorescent label; moreover, the nanoparticles should be thoroughly washed from the free dye, however, the washing procedure might not be possible if the target therapeutic nanoparticles preparation procedure does not imply it. At the same time, the encapsulation of DiI and Rh123 did not lead to significant changes in the nanoparticles' physicochemical properties.

The double-stained nanoparticles were obtained to get insight into the process of dye leakage in biological media. The physicochemical parameters of the double-labeled nanoparticles were similar to those of the respective single-labeled and non-labeled nanoparticles (Table S5). For the PLGA-Cy5.5/DiI nanoparticles, the encapsulation efficiency of DiI was on average higher compared to the PLGA-DiI nanoparticles.

3.2. Stability of the Nanoparticle Labeling

After having determined the optimal dye-to-polymer ratio, it is equally important to evaluate the stability of the fluorescent label in nanoparticles under conditions simulating the biological environment (lipophilic or hydrophilic or with different pH levels). However, it is challenging to mimic the physiological conditions in test tube conditions for release investigation, especially for hydrophobic substances [55]. As highlighted by Simonsson et al. [56], the classical methods based on employment of water or buffer as the release medium will not mimic the release in an in vivo environment, where multiple compartments are present. Various methods are used to evaluate the transfer of dye from nanocarriers to the cells: the incubation of the nanoparticles with cells at 4 °C to avoid the active transport [17]; the use of lipid acceptor compartments mimicking the cell membrane, such as liposomes, oils and other lipids [16,57]; and the application of the FRET phenomenon [20]. In this study, we applied three complementary approaches: common in vitro dye release studies in aqueous media, incubation with liposomes and computer simulations (prediction of dye interaction with lipid bilayer).

The stability of nanoparticle labeling was evaluated by an assessment of the dye fraction released after 2 h and 24 h of incubation (Table 1). The in vitro release of the encapsulated agents from the nanoparticles is generally analyzed in aqueous media simulating body fluids, such as, for example, phosphate buffered saline (PBS, pH 7.4). However, among the dyes used for encapsulation in this study, only rhodamine 123 was slightly soluble in PBS, others were practically insoluble. To prevent precipitation of the dyes during the continuous experiment, a 1% solution of poloxamer 188 in 0.15 M PBS (pH 7.4) was used as a release medium. However, even in the presence of the surfactant, the reliable results were obtained only within 2 h of incubation. At later time points, the dyes' precipitation interfered with the accuracy of measurements. Therefore, considering that studies on nanoparticle trafficking are mostly performed using plasma-containing media (i.e., in cell culture experiments or upon administration of the particles in the bloodstream), the stability of label retention was also evaluated in the presence of human serum albumin (HSA, 1% solution in PBS) as the most abundant plasma protein.

Table 1. Stability of labeling of the PLGA nanoparticles. In vitro release of DiI, Rh123, and Cou6 from PLGA nanoparticles (HSA, 1% solution in PBS, pH 7.4, 37 °C, $n = 3$).

Type of Nanoparticles	Fraction of Released Dye, %		
	1% Solution of Poloxamer 188	1% Solution of HSA	
	2 h	2 h	24 h
PLGA-DiI	9.8 ± 1.5	1.2 ± 0.1	42.3 ± 1.7
PLGA-Cy5.5/DiI	4.1 ± 0.3	<0.1	15.2 ± 0.7
PEG-PLGA-Cou6	17.0 ± 0.4	42.4 ± 2.1	65.0 ± 2.9
PLGA-Rh123	57.0 ± 3.5	n/d	n/d

All dyes demonstrated a different behavior in the release studies. Rhodamine 123 was quickly released from the nanoparticles. For PEG-PLGA-Cou6 nanoparticles, the release rate depended by the medium: Cou6 was released slowly in a 1% poloxamer 188 solution but much faster in a 1% HSA solution, as HSA enhanced its solubilization. In contrast, the initial release of DiI was much slower in an HSA solution compared to the poloxamer 188 solution. However, after 24 h incubation in the presence of HSA, the free dye concentration substantially increased (Table 1).

In the protein-free medium, DiI labeling was the most stable: approximately 10% of the dye was released from the PLGA-DiI nanoparticles after 2 h incubation. The PEG-PLGA-Cou6, nanoparticles released 17% of the dye content in these conditions. In contrast, the PLGA-Rho123 demonstrated a considerable burst-effect: 57% was released during the first 2 h (more likely desorbed from the surface); however, after this period the release was slow and, after 24 h, about 45% of the dye remained bound to the nanoparticles. There was no release of the conjugated Cy5.5 during 24 h incubation and up to 7 days.

Importantly, while the retention of DiI and coumarin 6 was seemingly sufficient in the protein-free medium, in the presence of HSA, the release rates of both dyes were enhanced considerably. This phenomenon is most probably explained by the known ability of HSA to form complexes with lipophilic molecules, which contributes to their more effective transfer from the nanoparticles into the aqueous medium. However, while the released fraction of coumarin 6 after 2 h of incubation was above 40%, only 1% of DiI was released from the nanoparticles at this time. Thus, in terms of label stability, the PLGA-DiI nanoparticles appear to be a reliable system for short-term imaging experiments in vitro and in vivo.

Both DiI and coumarin 6 are known for their ability to stain cell membranes, which is in line with their lipophilicity. Moreover, the rapid redistribution of coumarin 6 from nanocarriers into cell membranes was observed in the in vitro experiments [58,59]. The rate of coumarin 6 redistribution also depended on the type of the carrier and was higher in the case of the liposomes compared to the solid lipid nanoparticles. The role of the nanocarrier in the dye partitioning into the cell membranes probably explains the different results obtained for the lipid nanocapsules and PEG-polystyrene nanoparticles labeled with DiI: whereas the lipid-based carrier enabled a stable retention of the dye in the presence of cell membranes [56], in the case of polymeric nanoparticles, the release rate was strongly facilitated [60]

The use of dialysis for the separation of the nanoparticles from free coumarin 6 was unsuccessful due to the dye adsorption on the membrane (6.3%) and its slow diffusion through the membrane (after 96 h, the amount of the released dye in the medium was <1.6%). Similar results were obtained for the non-polymeric formulation of Cou6 (suspension). Hence, in the case of coumarin 6, the use of the dialysis technique may lead to the misinterpretation of the dye release rate from the nanoparticles.

No leakage of the covalently linked dye (Cy5.5) was observed for at least 24 h in various media (water, poloxamer 188 or HSA solution) (data not shown). Interestingly, labeling with DiI appeared to be more stable in the double-labeled PLGA-Cy5.5/DiI nanoparticles: after two weeks of incubation, 27.5% of DiI was released from the PLGA-DiI nanoparticles, whereas, in the case of the PLGA-Cy5.5/DiI nanoparticles, only 13.8% of the dye was released during this time (discussed below).

Covalently bound Cy5.5 label was not found in liposomes indicating that, under the experimental conditions, all of the dye remained associated with the nanoparticles. This result confirms that covalent conjugation of the dye with the core polymer is one of the most reliable ways to provide a stable staining of the nanoparticles. As shown in our previous study, the doxorubicin-loaded PLGA nanoparticles labeled with covalently bound Cy5.5; this approach yields accurate results on the nanoparticle uptake and localization within cells [15].

An interesting phenomenon of a higher encapsulation efficiency and slower release rate in the case of the double-stained PLGA-DiI/Cy5.5 nanoparticles was probably caused by the π-stacking between the molecules of these dyes, which contributed to the retention of DiI in the polymeric matrix due to electrostatic interaction [61]. A similar approach is described for polymeric micelles loaded with doxorubicin: a part of the total amount of doxorubicin was covalently bound to the carboxylic groups of poly(aspartate) micelles, whereas another part was physically loaded into the micelles. This approach enabled the better retention of doxorubicin in the polymeric micelles due to π-π interactions between free and covalently bound doxorubicin molecules [62]

3.3. Fluorescent Properties of the Labeled Nanoparticles

A very important characteristic of a fluorescent label is its brightness, which determines the sensitivity of its detection. In the case of suspension of polymeric nanoparticles labeled with a dye, the difficulty of brightness evaluation lies in the contribution of light scattering, which may interfere with the accuracy of measurements. In addition, the polymer environment and the dye–dye interaction, depending on the dye-to-polymer concentration, can influence the fluorescent properties (QY, ε) of dye molecules inside the nanoparticles. These factors complicate comparison of the properties of fluorescent nanoparticles in suspension with a solution of the dye itself. In this study, the specific brightness values were determined, because this parameter is very convenient for comparing different types of fluorescent particles [63]

3.3.1. PLGA Nanoparticles with a Single Label

The extent to which a dye-to-polymer mass ratio influenced the optical properties (quantum yield and brightness) of nanoparticles varied. In the case of the PLGA-DiI nanoparticles, the highest QY (14.91 ± 0.16%) was observed for the nanoparticles with a DiI content of 6.3 µg/mg PLGA (0.63 wt.%). (Figure 2). These particles also exhibited the maximal brightness (7.3×10^{-5} ℓ cm^{-1}/mg PLGA). As a result, the DiI content in the nanoparticles that provided the maximal brightness was found to be ~5 µg/mg PLGA.

A further increase of the dye concentration in the nanoparticles led to a decrease in the quantum yield and brightness, which is most likely associated with aggregation-caused quenching (ACQ). Interestingly, in contrast to our expectations, a bulky counter-ion, bis(2-ethylhexyl) sulfosuccinate (AOT) that was added in order to prevent fluorescence self-quenching of DiI (PLGA-DiI-AOT nanoparticles) led to a two-fold decrease in the fluorescence QY (from 13.72 ± 0.16% to 6.58 ± 0.03%); at the same time, their brightness increased by about 1.5 times (from 2.06×10^{-5} to $3.14 \times 10^{-5}\ \ell\ cm^{-1}/mg$ PLGA) due to the increased number of fluorophore molecules in the PLGA core. Possibly this decrease of the fluorescence QY is associated with the formation of a non-luminescent ion pair DiI-AOT. Moreover, the zeta potential of the PLGA-AOT-DiI nanoparticles was almost neutral, thus being substantially different from the negatively charged non-labeled and DiI-labeled nanoparticles prepared without AOT (-0.2 ± 0.2 mV versus -40.7 ± 2.1 mV, respectively).

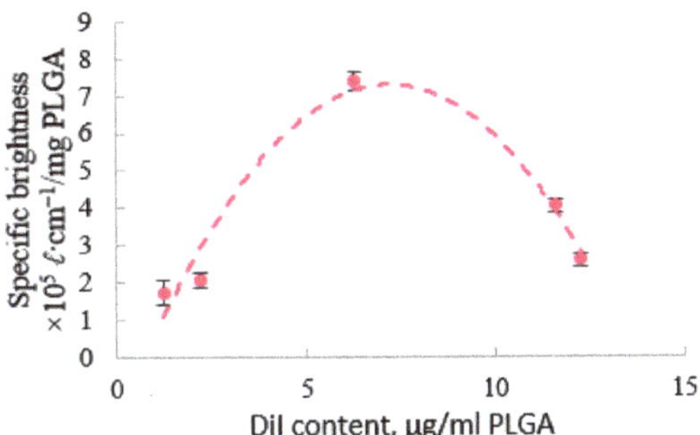
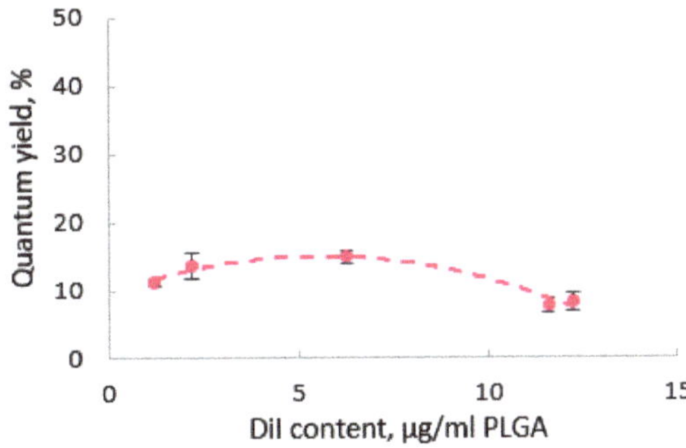

Figure 2. Influence of DiI content in PLGA nanoparticles on their specific brightness (**left**) and on quantum yield (λ_{EX} 488 nm) (**right**).

The optical properties of the PEG-PLGA-Cou6 nanoparticles also depended on the dye content (thoroughly washed nanoparticles were used for the measurements). The maximal QY ($91.9 \pm 0.2\%$) was observed for the nanoparticles containing 12.2 µg of dye per 1 mg of PLGA (Figure 3). These nanoparticles also demonstrated the highest specific brightness, ranging from 3.39×10^{-4} to $13.37 \times 10^{-4}\ \ell\ cm^{-1}/mg$ PLGA A, and the number of fluorophore molecules per mg PLGA was ten times higher than in the PLGA-DiI NPs. At higher dye concentrations, the QY of the fluorescence decreased due to concentration quenching.

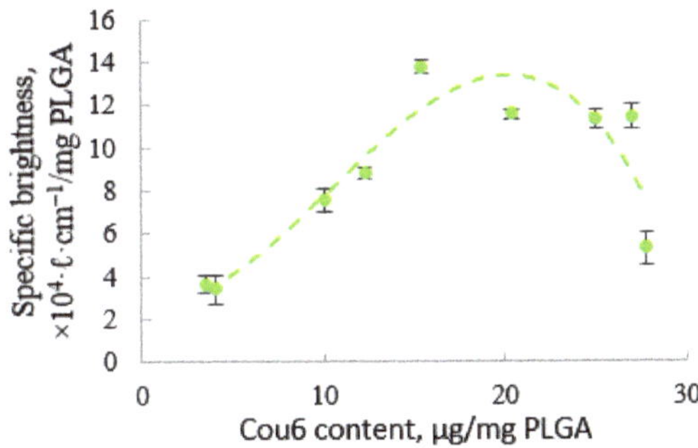
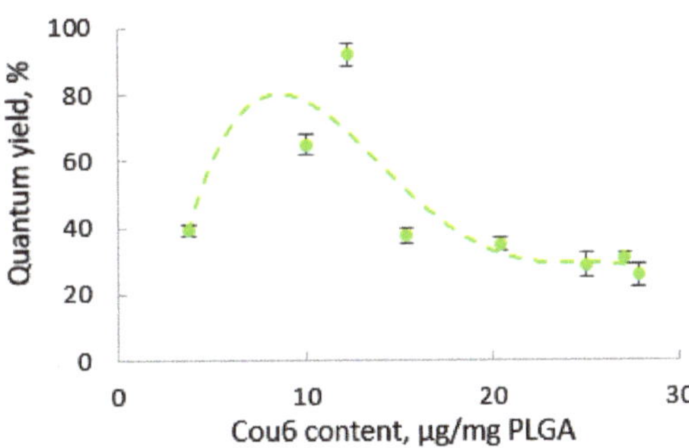

Figure 3. Influence of Cou6 content in PEG-PLGA nanoparticles on their specific brightness per mg of PLGA (**left**) and on quantum yield (**right**) (λ_{EX} 450 nm).

The Rh123-labeled nanoparticles had a QY of 46.8% and specific brightness of $2.02 \times 10^{-4}\ \ell\ cm^{-1}/mg$ PLGA that was 3-fold higher than that of the brightest PLGA-DiI NPs but lower compared to the PLGA-PEG-Cou6 nanoparticles. No further optimization of the Rh123-labeled nanoparticles was performed because of the fast release of this dye.

Quantum yield of Cy5.5 in the PLGA-Cy5.5 nanoparticles was significantly higher than that of a free dye in ethanolic solution (~50% versus 20%, respectively), which is probably due to the known increase in the fluorescence properties of indocyanine dyes in a viscous microenvironment [64,65]. In this case, no significant influence of the fluorophore amount on the quantum yield was observed in the range from 0.41 µg/mg to 1.66 µg/mg PLGA: in all cases, QY was about 50%. Brightness, however, increased along with the PLGA-Cy5.5 content in the nanoparticles (Figure 4).

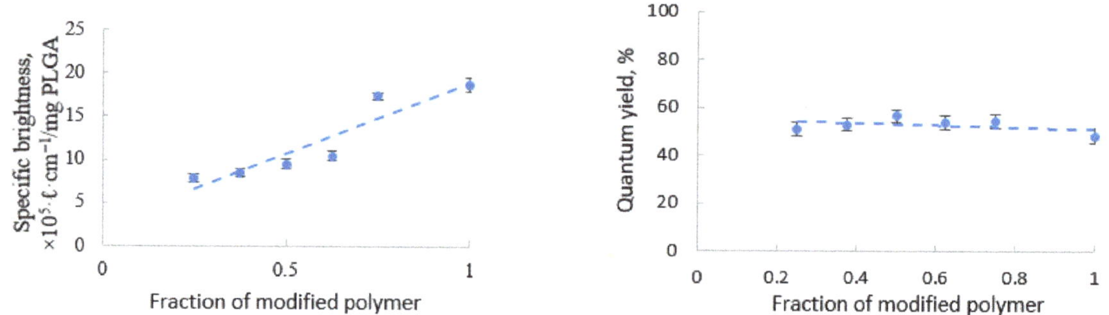

Figure 4. Influence of the PLGA-Cy5.5 content in nanoparticles on brightness (**left**) and quantum yield (**right**).

As shown in Figure 5, the PEG-PLGA-Cou6 nanoparticles were by far the brightest among the single-labeled nanoparticles.

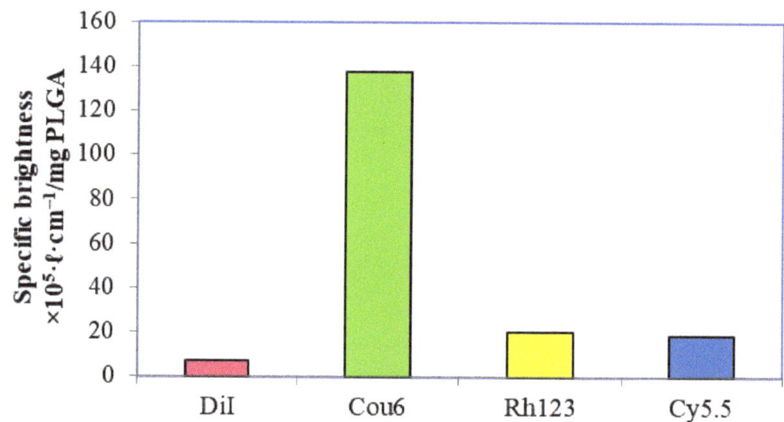

Figure 5. Maximal specific brightness values for the PLGA nanoparticles with a single label.

3.3.2. PLGA Nanoparticles with a Double Label

Emission spectra of the dyes, which were used in pairs, do not overlap significantly, and all dyes retained their characteristic emission peaks (Supplementary Materials, Figures S1–S4). The absence of overlap is advantageous for the signal discrimination in in vitro and in vivo experiments. On the other hand, the overlap between emission and excitation spectra of two dyes indicates that the pair will display Förster resonance energy transfer (FRET), where the magnitude of the effect depends on the area of overlap. Among the pairs used, such overlap was the greatest for Cy5.5 and DiI, followed by the Cy5.5–Rh123 and Cou6–Cy5.5 pairs.

Accordingly, the most considerable FRET phenomenon was observed for the DiI/Cy5.5 pair (not reported previously). The efficiency of FRET calculated according to Equation (4), using the quantum yields for DiI in the double-labeled (Cy5.5-PLGA-DiI) and DiI-labeled (PLGA-DiI) nanoparticles of 11.3% and 32.5%, respectively, was 65.2%, which correlated with the molar ratio of the dyes in the nanoparticles. FRET was also observed in the case of

DiI and Rh123. However, the efficiency of Rh123/Cy5.5 energy transfer was not sufficient for application in the in vitro and in vivo experiments.

A rather considerable overlap between the fluorescence spectrum of DiI and absorbance spectrum of Cy5.5 offers the possibility of using FRET as an additional method of monitoring nanoparticle stability. The fact that an intensive FRET signal was observed in the cells after 1 h of incubation confirms that no leakage occurs at that time, which correlates with the results of the release studies.

3.4. Evaluation of the Dyes' Partitioning in Liposomes

Both DiI and Cou6 exhibit affinity to plasma membranes and other lipid structures [66,67], which is in line with their lipophilicity. Thus, rapid redistribution of coumarin 6 from nanocarriers into cell membranes was observed in the in vitro experiments [58,68]. The rate of coumarin 6 redistribution also depended on the type of the carrier and was higher in the case of the liposomes as compared to the solid lipid nanoparticles. The role of the nanocarrier in the dye partitioning into the cell membranes probably explains also the different results obtained for the lipid nanocapsules and PEG-polystyrene nanoparticles labeled with DiI in the presence of cell membranes: whereas the lipid-based carrier enabled a stable retention of the dye [56], in the case of the polymeric nanoparticles, the release rate was strongly facilitated [60].

Therefore, for the proper evaluation of nanoparticle biodistribution, it is important to evaluate the affinity of the encapsulated dyes to the cell membranes. Liposomes can act as the model cell membranes mimicking the arrangement of the lipids in natural cell membranes and therefore are frequently used for this purpose. For this purpose, in order to evaluate the role that lipid components may play in the dye leakage from the nanoparticles, the dye retention was tested in the medium containing liposomes. The Z-average diameter and PDI of the liposomes employed in the present study were 307 ± 21 nm and 0.139 ± 0.010, respectively. The dye-loaded and washed nanoparticles were incubated with liposomes for 1 h; then, the liposomes and nanoparticles were separated by centrifugation in 20% sucrose, followed by analysis of both fractions. In the case of properly washed PLGA-DiI and PLGA-Cy5.5 nanoparticles, no dye transfer to the lipid layer was observed, whereas for the non-washed PLGA-PEG-Cou6 or PLGA-DiI nanoparticles, the fraction of a dye transitioned into liposomes was about 62% and 38%, respectively (Table 2).

Table 2. Dye partitioning in liposomes.

Type of Nanoparticles	Fraction of a Dye Transitioned to Liposomes, %	Amount of Dye Found in Liposomes, µg/mg
PLGA-Cou6	62.4 ± 7.1	0.93
PLGA-DiI	38.7 ± 0.6	1.45
PLGA-DiI *	0	0
PLGA-Rh123	8.1 ± 1.4	0.20
PLGA-Cy5.5	0	0

* Nanoparticles washed by centrifugation and subsequent GPC.

3.5. Modeling of Dyes' Interaction with Model Lipid Membranes

Thorough the removal of non-encapsulated DiI (preferably by GPC, followed by centrifugation) is important for a reliable interpretation of the imaging results in vitro and in vivo, since dye molecules that are loosely associated with nanoparticles easily distribute into the cell membranes, and, therefore, may create artifacts upon visualization. This pattern of the DiI behavior was further confirmed by computer simulation data (DFT). DiI is the most hydrophobic of the studied dyes and is practically insoluble in water (Table 3). It appears that the affinity of DiI to the model lipid membrane (DPPC/DMPC) is much higher than its affinity to the polymer. This relatively low affinity to the polymer explains the appearance of a considerable fraction of the dye adsorbed onto the surface of the nanoparticles, which is due to its tendency to form its own phase at the boundary of the emulsion droplet during the particle preparation process. In an aqueous medium, the molecule of that dye will "prefer" the lipid membrane to the polymeric matrix, given that there is a choice. In the absence of direct contact with the lipid layer, the probability that the dye molecule will transit from the nanoparticle surface into water and then to the membrane is negligible, since the solubility of DiI is very low. However, when the PLGA-DiI nanoparticles come in contact with the lipid membrane, the partition of the surface-adsorbed DiI into this membrane is a quick and efficient event, as was observed in the experiment wherein the PLGA-DiI nanoparticles were incubated with the liposomes. Indeed, in the case of the insufficiently washed PLGA-DiI nanoparticles, ~38% of DiI leaked into the liposomes. The results concerning the DiI affinity to the cell membrane in various publications are rather different [17,20,56]. Although lipophilic DiI is widely used as a cellular membrane tracer in biological studies [69], we observed that both the PLGA-DiI nanoparticles and non-polymeric DiI nanosuspension were internalized into the cells and distributed within the cytoplasm (Figure 6). No membrane staining was observed. Moreover, confocal images of the double-stained PLGA-Cy5.5/DiI nanoparticles confirmed the colocalization of DiI and Cy5.5 fluorescence signals. The stabilization of the dye in aqueous media could possibly prevent its diffusion into the cellular membrane. Indeed, according to several protocols recommended for staining of cellular membranes and for retrograde and anterograde neuronal labeling, the lipophilic dyes (DiI, DiO, DiD) should be first dissolved in DMSO and then applied to the samples in ethanolic solution [70].

Table 3. Calculated parameters of the dyes' distribution in different phases.

Parameter	DiI	Coumarin 6	Rhodamine 123
Topological polar surface area, Å^2 ("polarity")	6.25 *	42.43 *	91.48 *
$\log P_{\text{OCT/WAT}}$ ** (hydrophobicity)	10.3	5.0	0.04
$\log P_{\text{DPPC/WAT}}$ ** (affinity to a lipid bilayer)	8.6	5.2	−0.6
$\log P_{\text{DMPC/WAT}}$ ** (affinity to lipid bilayer)	8.7	5.2	−0.6
$\log P_{\text{PLGA/WAT}}$ ** (affinity to NPs' polymeric matrix)	5.3	1.2	−3.6
$\log P_{\text{DCM/PLGA}}$ **	0	0	0
$\log P_{\text{DPPC/PLGA}}$ **	2.3	0.7	2.6
$\log P_{\text{DMPC/PLGA}}$ **	1.6	0.7	2.1

* Data calculated using MarvinSketch software, ChemAxon Ltd.; ** quantum mechanical calculation (25 °C, pH 7).

Computer simulation is another useful approach for predicting the behavior of a fluorescent dye loaded in nanoparticles in vitro. To estimate the possibility for each of the dyes to exit the nanoparticles into the aqueous medium and to enter a model lipid membrane composed of either dipalmitoylphosphatidylcholine (DPPC) or dimyristoylphosphatidylcholine (DMPC), we have employed the DFT simulations in the frames of a COSMO-RS model. This model allows the calculation of the chemical potentials of substances depending on the temperature, pressure, concentrations and composition of the mixture. Thus, it is suitable for the computation of various physicochemical properties. This approach is a well-known method used for the evaluation of the solubilities and partition coefficients of small organic molecules in polymers [49,71]. Also, the chosen computational approach is quite sensitive for predicting the partition coefficients of the drug-like molecules, which was demonstrated in the blinded test challenges SAMPLE5 [72] and SAMPLE6 [73].

To investigate the activity of the studied dyes in complex multicomponent media, phase distributions for each compound in all examined systems were calculated. For each dye, only the most common conformers in vacuum, liquids and DPPC or DMPC bilayers were obtained. The final conformers of both DPPC and DMPC molecules were chosen as the components of a membrane lipid bilayer (see section of Supplementary Data

"Modeling of dye interaction with lipid membranes" and Supplementary Table S6). Since nanoparticle formation occurs at ambient temperature, the computations were conducted at 25 °C (298.15 K). At this temperature, the DMPC bilayer exists in the liquid crystalline state (gel-to-liquid crystalline transition temperature 24 °C), whereas the DPPC bilayer exists in the gel state (gel-to-liquid crystalline transition temperature 41 °C) [74].

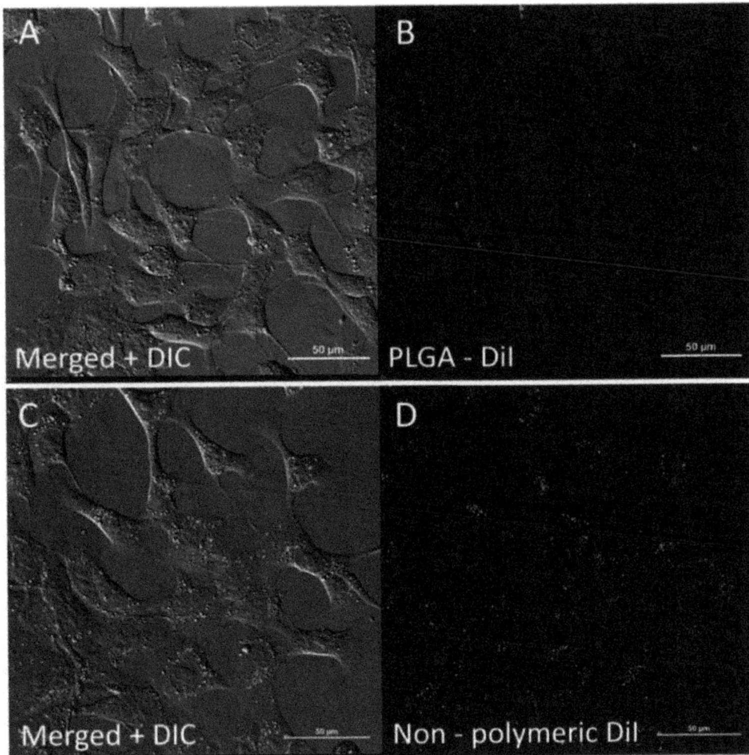

Figure 6. Laser scanning confocal microscopy of 4T1 cells after 40 min of incubation with DiI nanosuspension: PLGA-DiI NPs (**A,B**) and non-polymeric DiI formulation, stabilized with 1% PVA) (**C,D**). (**A,C**)—merged image and differential interference contrast (DIC); (**B**)—PLGA-DiI fluorescence; D—non-polymeric DiI fluorescence. Scale bar 50 μm.

According to the obtained data, rhodamine 123 hydrochloride is the only dye that dissolves well both in water and in hydrophobic media: $\log P_{DPPC/WAT}$ and $\log P_{DMPC/WAT}$ is equal to −0.58, and the calculated $\log P_{OCT/WAT}$ value is equal to 0.04. Both DiI and Cou6 are soluble in hydrophobic liquids only. For coumarin 6, the $\log P_{OCT/WAT}$ value is equal to 5 and both $\log P_{DPPC/WAT}$ and $\log P_{DMPC/WAT}$ values are equal to 5.2. Interestingly, the difference in the lipid bilayer composition and phase state did not affect the dyes' distribution between the membrane bilayers. Thus, we considered only the DPPC membrane. The results are presented in Table 3, featuring topological polar surface areas (polarity measure) calculated using MarvinSketch software, ChemAxon Ltd.

According to our in silico calculation, a rather small amount of coumarin 6 is able to saturate the membrane—1% (*w/w*) is the saturation level. Expectedly, the maximal amount of the dye transferred to liposomes, about 68%, was detected for PLGA-PEG-Cou6 NPs. Active dye leakage was also confirmed by confocal imaging for the double-stained PEG-PLGA-Cy5.5/Cou6 nanoparticles: the absence of co-localization for Cy5.5 and Cou6 in the

4T1 cells (see below, Section 3.7). Coumarin 6 is actively released from the nanoparticles and distributed in the cell membrane.

Rhodamine 123 also easily incorporates into the nanoparticles. However, it can just as easily leave it and then migrate through the membrane in both directions. Rho123 had little affinity to liposomes but was mostly transferred to the aqueous phase (Table 2). Rho 123 was the most unstable as a fluorescent label demonstrating a fast release and the subsequent staining of mitochondria (see Section 3.7.).

In the system of polylactide/dichloromethane, the affinity of any dye to PLGA is the same as its affinity to the solvent (log $P_{DCM/PLGA}$ = 0). That means that when the solvent is being removed the dye molecules are distributed evenly in the nanoparticle matrix, and the maximum loading of the substance is limited by its solubility in dichloromethane (Table 3).

By their hydrophobicity, the dyes are arranged in the following order: DiI > Cou6 > Rho123 and, by their polarity, they are arranged in the opposite one: DiI < Cou6 < Rho123. The dyes' affinity to polymers (polymer/water system, log $P_{PLGA/WAT}$) and their affinity to lipid bilayers (log $P_{DPPC/WAT}$) are arranged in the same order as hydrophobicity. The affinities of DiI to both polymers and lipids are much higher than those of coumarin 6, let alone rhodamine 123. In practice, this should mean that the probability of DiI leaking into water is rather low and is much lower compared to coumarin 6 and rhodamine 123.

3.6. Differential Scanning Calorimetry Assay of Interaction between Dyes and PLGA

The influence of DiI and coumarin 6 on PLGA thermal behavior was analyzed by differential scanning calorimetry (DSC). The dye-to-polymer ratio in the nanoparticle samples used for this assay was increased to 1:20 (*w/w*) to better reveal the influence of a dye on the polymer properties. As shown in Table 4, the glass transition temperatures (T_g) of both labeled nanoparticles were lower compared with that of non-loaded PLGA NPs and a bulk polymer, which indicates that coumarin 6 and DiI act as plasticizers for PLGA.

Table 4. Glass transition temperature (T_g) of PLGA (bulk) and PLGA nanoparticles, both non-loaded and loaded with DiI or coumarin 6.

Sample	T_g, °C
PLGA (bulk)	43.27
Non-loaded PLGA nanoparticles	44.90
Cou6-PLGA nanoparticles	43.91
DiI-PLGA nanoparticles	42.35

3.7. In Vitro Behavior of PLGA-Cou6 Nanoparticles

Among the nanoparticles produced, several samples possessing the optimal physicochemical qualities as outlined previously were used in further in vitro experiments. A brief characterization of them is given in Table 5. The non-polymeric DiI formulation was used in this experiment to mimic the behavior of free dye in aqueous media when the nanoparticles are not properly separated from unbound dye. Surprisingly, the intracellular distribution patterns of free DiI (non-polymeric DiI formulation) and DiI encapsulated in the nanoparticles (PLGA-DiI) were similar (Figure 6).

This phenomenon could be explained by the precipitation of the dye in the aqueous medium (PBS buffer), as was shown by molecular dynamic simulation. Interestingly, both, PLGA-DiI NP and non-polymeric DiI nanosuspension, appeared to be internalized into the cells without staining the cell membrane. Therefore, it was important to distinguish between the free dye nanoparticles and the PLGA nanoparticles to prevent the misinterpretation of the imaging data. Double-stained PLGA-Cy5.5/DiI nanoparticles were used for this purpose. However, as seen in Figure 7, both the DiI-labeled nanoparticles and the DiI nanosuspension displayed significant colocalization (Pearson's correlation coefficient is 0.68; Mander's overlap coefficient is 0.85) which suggests that, at this time, the DiI molecules are located mostly within the nanoparticles and, moreover, they are located close to the Cy5.5 molecules. The FRET signal in Figure 8 demonstrated that NP preserves their integrity during internalization process within 1 h of incubation. To determine the FRET signal, the samples were exited with a 561 nm laser (donor excitation wavelength), and the fluorescence was captured with both detectors simultaneously (both green and red fluorescent channels). Detectors from 570 to 620 nm and from 650 to 730 nm for DiI and Cy5.5 imaging, respectively, were used. Then the samples were exited with both 561 and 647 nm lasers, separately, for DiI and Cy5.5 excitation, respectively.

Table 5. Characteristics of fluorescent PLGA NPs used in in vitro experiments.

Type of NPs	Dye Content, μg/mg PLGA		EE, %	Mean Diameter Z-ave, nm	PDI	Zeta Potential, mV
PLGA-DiI	6.4		76.6	156 ± 5	0.156 ± 0.016	-27 ± 2
Non-polymeric DiI formulation	23.6 μg/mL *		-	333 ± 93	0.355 ± 0.077	-21 ± 6
PEG-PLGA/Cou6	27.2		99.7	139 ± 2	0.036 ± 0.028	-15 ± 3
Non-polymeric Cou6 formulation	32.6 μg/mL *		-	1086 ± 166	0.830 ± 0.111	-6.6 ± 0.1
PLGA-Cy5.5/DiI	DiI Cy5.5	6.2 0.8	77	137 ± 1	0.136 ± 0.022	-18 ± 1
PEG-PLGA-Cy5.5/Cou6 NPs	Cou6 Cy5.5	5.7 0.8	91	128 ± 1	0.137 ± 0.022	-22 ± 1
PLGA-Cy5.5/Rh123	Rh123 Cy5.5	6.8 0.8	96.6	98 ± 1	0.196 ± 0.031	-24 ± 4

* For non-polymeric forms the dye content is given per ml of a suspension.

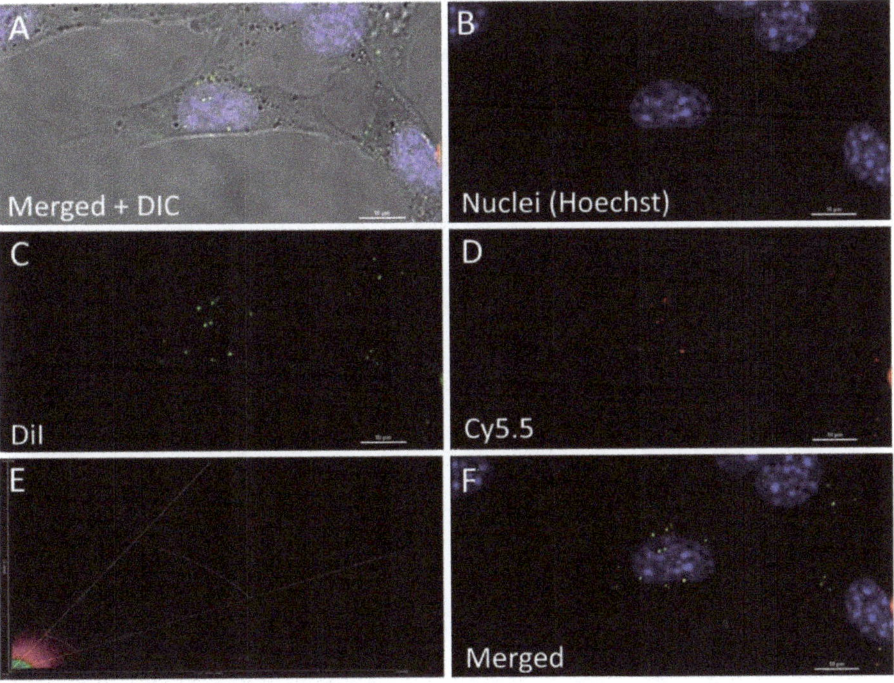

Figure 7. 4T1 murine mammary carcinoma cells after 40 min incubation with the PLGA-Cy5.5/DiI nanoparticles: maximum intensity projection of a z-stack (Laser Scanning Confocal Microscopy, Nikon A1 MP). (**A**)—merged image and differential interference contrast (DIC); (**B**)—cell nuclei (Hoechst). (**C**)—DiI fluorescence; (**D**)—PLGA-Cy5.5 fluorescence; €—Scatterplot representing colocalization between DiI and Cy5.5 fluorescent signals. (**F**)—Merged image. Scale bar 10 μm.

Coumarin 6 rapidly released from the double-stained nanoparticles and accumulated in the cells within the first minutes of incubation, while significant internalization of nanoparticles into the cells (traced by Cy5.5 signal) was observed only within 60 min of incubation (Figure 9).

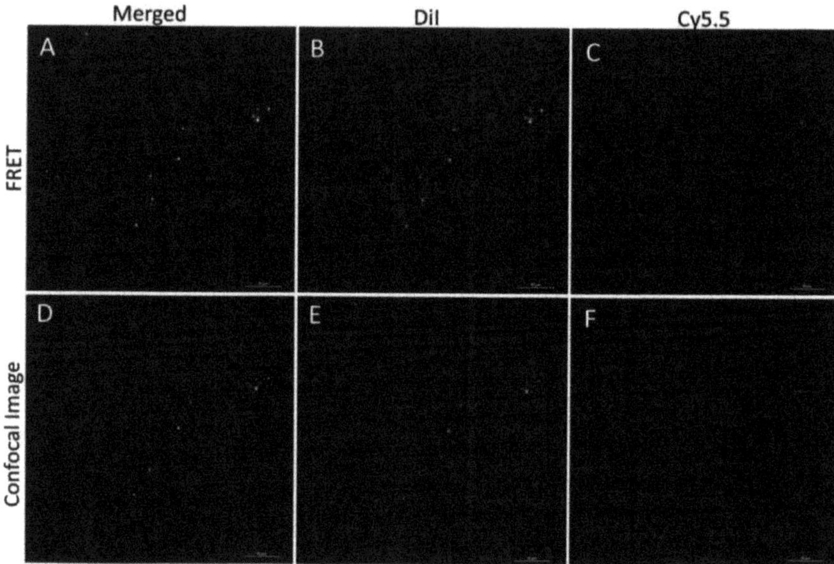

Figure 8. Imaging of 4T1 cells after 1 h of incubation with PLGA-Cy5.5/DiI NPs. Upper panels (**A–C**) represent the FRET signal at donor excitation wavelength (561 nm) -FRET phenomenon; lower panels represent DiI and Cy5.5 fluorescence separately (ex 561 and 647 nm); (**A,D**)—merged images; (**B,E**)—DiI fluorescence; (**C,F**)—Cy5.5 fluorescence. Laser scanning confocal microscopy (Nikon A1 MP). Scale bar 10 μm.

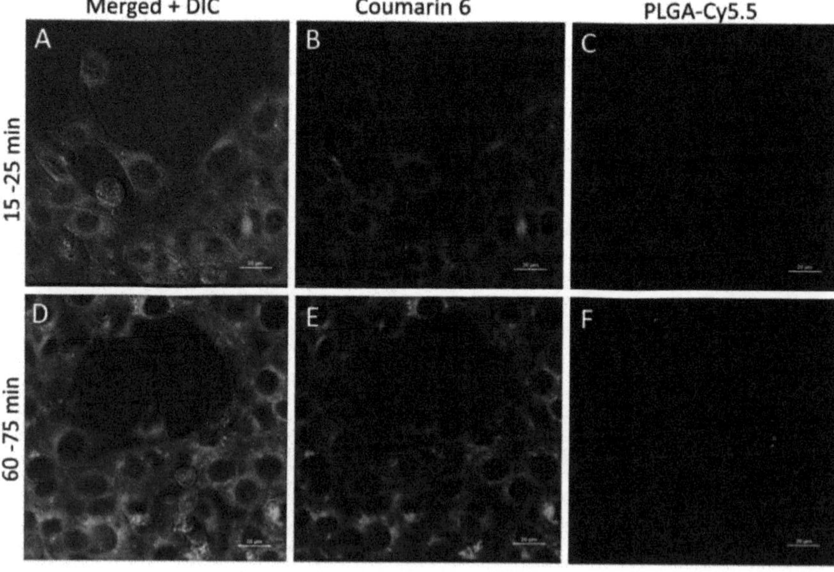

Figure 9. Internalization of PLGA-Cy5.5/Cou6 nanoparticles into 4T1 cells after 15–25 min (**A–C**) or 60–75 min (**D–F**) incubation (laser scanning confocal microscopy, Nikon A1 NP). (**A,D**)—merged image and differential interference contrast (DIC); (**B,E**)—coumarin 6 fluorescence; (**C,F**)—PLGA-Cy5.5 fluorescence. Scale bar 20 μm.

The same phenomenon was determined for rhodamine 123-loaded PLGA-Cy5.5 NP. However, rhodamine 123 penetrates into cells slower than coumarin 6, which correlates with computer simulation data (see Table 3). The dynamics of the nanoparticle and dye accumulation into the cells are shown in Figure 10. Moreover, rhodamine 123 was localized mainly in mitochondria (stained with MitoTracker Red, Life Technologies). The cells were imaged within 40 min of incubation with NPs. To track NP uptake by the cells and rhodamine 123 release and accumulation in real time, cells were not washed from the NP suspension. The graph in Figure 10M represents the increase in the fluorescence intensity of rhodamine 123 colocalized with mitochondria recognized in Figure 10K–O (applied blue channel). Rhodamine 123 accumulation was assessed in the cell mitochondria defined as regions of interest (ROI). ROI were obtained by thresholding based on the mitotracker fluorescence intensity profile.

3.8. In Vivo Behavior of PLGA-Cou6 Nanoparticles

To observe the in vivo distribution of Cou6 as the encapsulated label and Cy5.5 as a covalently linked label, the retinae of rats were imaged using in vivo confocal imaging after intravenous injection of the nanoparticles. As shown in Figure 11A, in the case of the Cy5.5-PLGA-Cou6 nanoparticles, the signal of Cy5.5 concentrated in the blood vessels (Figure 11A); however, the Cou6 signal was extremely bright in the tissue background of the retina, confirming the different distribution between the fluorescent labels encapsulated or covalently linked to the nanoparticles. The intensity analysis further confirmed that most of the Cy5.5 signals were visible inside the blood vessels, whereas the Cou6 signals were mainly distributed outside the blood vessel and in the tissue background.

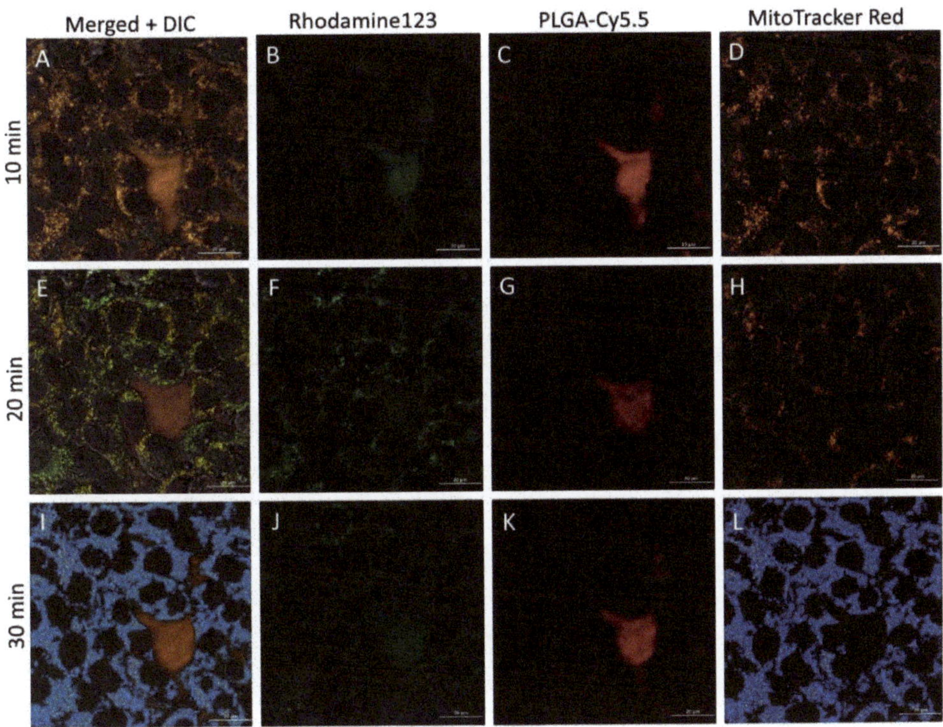

Figure 10. *Cont.*

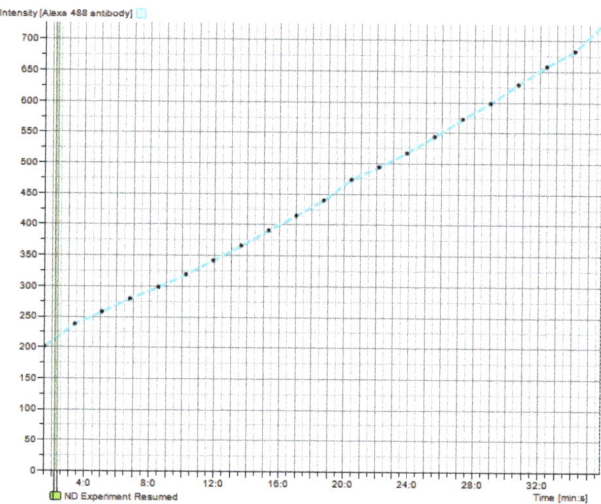

Figure 10. Imaging of 4T1 cells after 1 min (**A–D**) and 30 min (**E–H**) of incubation with PLGA-Cy5.5/Rh123 NPs. A,E—merged images and differential interference contrast (DIC); (**B,F,J**)—rhodamine 123 fluorescence; (**C,G,K**)—PLGA-Cy5.5 fluoresce; images (**I,L**) represent rhodamine 123 fluorescence intensity in the regions of mitochondrial fluorescence intensity (thresholding by mitochondrial fluorescence intensity of the images (**E–H**) was applied); (**M**)—Dynamic of rhodamine 123 accumulation within the mitochondria (ROIs are selected based on mitotracker fluorescence intensity—artificial blue channel (**I,L**)); 0–30 min of incubation. Laser scanning confocal microscopy (Nikon A1 NP). Scale bar 20 μm.

In the case of Cou6 the in vivo neuroimaging (ICON) data for the PLGA-Cy5.5/Cou6 nanoparticles corresponded to the data obtained above: the nanoparticle signal (Cy5.5) was found in the retinal vessels, whereas the Cou6 signal was distributed in the tissue background. Retinal vessels are characterized by the presence of the blood–retinal barrier, which in many ways is similar to the blood–brain barrier [75]. The obtained results can be easily misinterpreted, creating an impression that the nanoparticles effectively cross the blood–retina barrier, when in fact, no signal from the core material is found beyond the vessels. This was expected, because it was shown [76] that coumarin 6 dye stains liposomes and cells with no significant release into aqueous media, and therefore, it would most likely leak in vivo. The in vivo behavior of DiI and Rh123 was described in our previous study [77]: Preliminary tests for PLGA nanoparticles loaded with DiI showed that the results depended on the amount of loosely-bound dye: washed DiI nanoparticles did not stain liposomes and cells, while non-washed did. DiI and rhodamine 123 colocalized with the cyanine5.5 signal well at early stages of the experiment; however, the kinetics of the signal fading was different, especially for rhodamine 123, which leaked from the nanoparticles and was eliminated quickly. In contrast to the in vitro experiments with liposomes, the intravenously injected DiI-labeled nanoparticles provided long-term staining of the endothelial cell membranes, even when the nanoparticles were washed. This discrepancy may be explained by a more effective contact of nanoparticles with blood vessel endothelial cells achieved in vivo that could be due to blood pressure, compared to the contact that was achieved using liposomes.

Dyes, especially hydrophobic dyes, are sometimes used as model drugs for the evaluation of the targeting capability of various delivery systems [78,79], which is due to the fact that they can be easily visualized unlike most drug molecules, which are rarely fluorescent. In this respect, the most important difference between drugs and labels is the requirement of their retention by the carrier. Indeed, the drug is supposed to be released sooner or later, whereas the label should be retained by the carrier as long as possible. Thus, in a recent study conducted by our group [80], coumarin 6 served as a model drug that, in concert with the above observations, was very loosely associated with the PLGA nanoparticles.

However, being quickly released from the nanoparticles upon intravenous administration in mice, it was successfully delivered into the retina, whereas the nanoparticles remained associated with the blood capillary endothelial cells (Figure 11). Thus, while being not an optimal label, coumarin 6 helped to provide more evidence regarding the possibility of drug delivery by nanoparticles using a "kiss-and-run-mechanism", wherein the drug is delivered into the tissue upon nanoparticle contact with the cell membrane, and no particle entry is necessary [81].

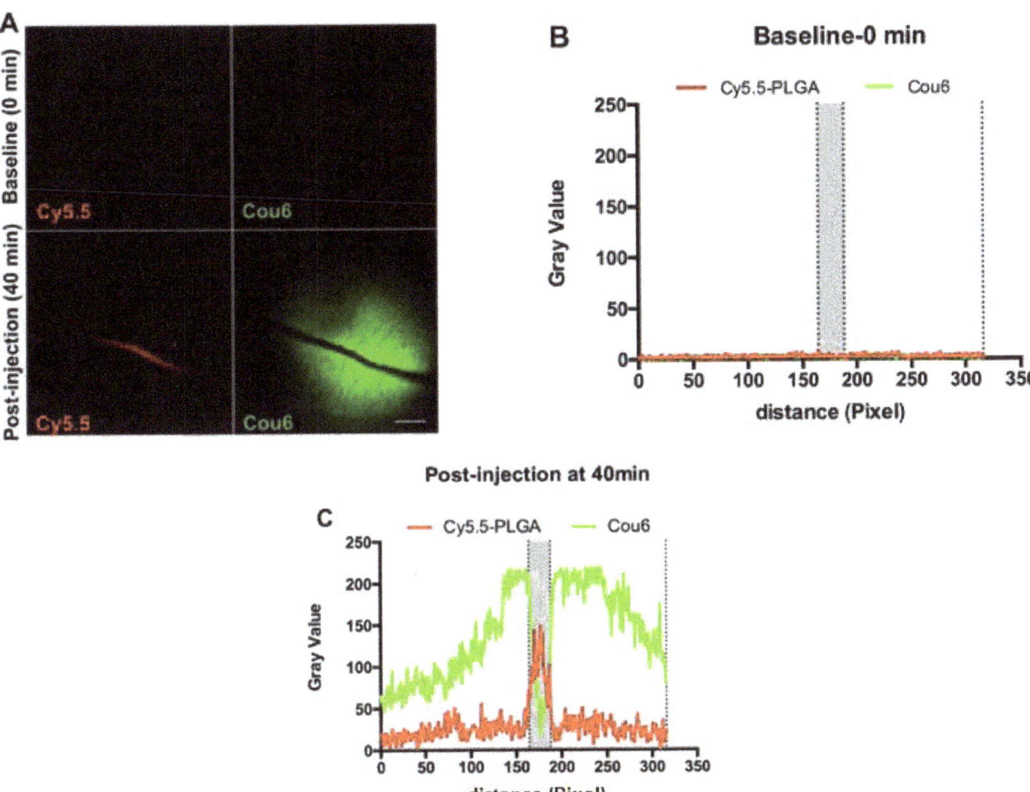

Figure 11. Biodistribution analysis of the intravenously injected Cy5.5-PLGA-Cou6 nanoparticles in rat retinal tissue. (**A**)—signal of Cou6 as encapsulated dye and Cy5.5 as covalently linked dye observed by the ICON technique before intravenous injection (baseline; upper row) or post injection at 40 min (lower row); scale bar: 400 µm. (**B**)—background intensity of signal of Cou6 and Cy5.5 before intravenous injection at baseline. (**C**)—intensity of signal of Cou6 and Cy5.5 40 min post injection. Area highlighted in gray in (**B**) and (**C**): blood vessel lumen.

4. Conclusions

The findings of this study highlight the importance of the careful design and evaluation of the fluorescently labeled nanoparticles intended for biodistribution studies. Unsurprisingly, the most reliable way to label the nanoparticles is by the conjugation of the fluorescent dye to the polymeric core. In cases when studies require particles with physically entrapped dye, in vitro or in vivo experiments with the simultaneous tracking of the colocalization of both the nanoparticle and the dye have to be used to ascertain the nanoparticle system's integrity and applicability.

We have shown that the usual threefold washing with water does not allow the removal of hydrophobic dyes from the surface of nanoparticles. In this case, the dye is practically not released into the usual model media but is being easily transferred from nanoparticles into the lipid bilayer of liposomes, which was also confirmed by the method of computer simulation.

Consequently, using the double-labeled nanoparticles as an example, it has been shown that, after the incubation of nanoparticles with coumarin 6, the latter is rapidly released from nanoparticles and stains the cell membranes of retinal vessels even after thorough preliminary washing of nanoparticles from free dye. Even very hydrophobic dyes included in the composition of polylactic nanoparticles and not covalently bound to their matrix can precipitate in a separate phase during the preparation of nanoparticles or be released from the surface of nanoparticles in biological media upon contact with lipid or protein components. The extent of this phenomenon depends on the affinity of the dye for the polymer matrix of the nanoparticles and the environmental conditions.

At this time we do not foresee that there is a single approach to traceable nanoparticles that can serve as a "golden standard" for all kinds of nanoparticle compositions and labels. Future studies should take into account the complex interferences of the fluorescent dye and the carrier prior to the use of fluorescently labeled nanoparticles.

Supplementary Materials: The following are available online at https://www.mdpi.com/article/10.3390/pharmaceutics13081145/s1, Table S1: Characteristics of fluorescent dyes; Table S2: Physico-chemical parameters of the PLGA nanoparticles labeled with DiI, rhodamine 123 and Cy5.5 (representative samples); Table S3: Influence of the PEG-PLGA/PLGA ratio on Cou6-labeled nanoparticle size and the size distribution at the dye-to-polymer mass ratio of 1:170; Table S4: Influence of the Cou6-to-polymer ratio (w/w) on the size distribution of PEG-PLGA-Cou6 nanoparticles (PEG-PLGA/PLGA = 1:5, w/w, number of measurements = 4; mean $\pm$ SD); Table S5: Representative physicochemical parameters of the double-labeled PLGA nanoparticles; Table S6: Conformer distribution in liquid phases at 298.15 K and pH 7; Figure S1: Normalized spectra of the possible FRET pairs with Cy5.5 as acceptor. Top to bottom: rhodamine 123 (orange line) —Cy5.5 (blue line), DiI (purple line Cy5.5 (blue line), coumarin 6 (green line)—Cy5.5 (blue line). Absorbance spectra are in solid lines, and fluorescence spectra are in dashed lines. The spectral overlap between a donor absorbance spectrum and an acceptor excitation spectrum is an indication of the FRET efficiency. The efficiency of FRET is shown with filling. The overlap between DiI/Cy5.5 is the greatest of the three pairs; Figure S2: Fluorescence (solid line) and absorbance (dashed line) spectra of the PLGA-DiI/Cy5.5 nanoparticles. The arrow shows the direction of FRET; Figure S3: Fluorescence (solid line) and absorbance (dashed line) spectra of PLGA-Rh123/Cy5.5 NPs; Figure S4: Fluorescence (solid lines) and absorbance (dashed line) spectra of the PLGA-Cou6/Cy5.5 nanoparticles.

Author Contributions: Conceptualization, O.M., S.G. and J.K., data curation: N.O., V.Z., A.S., J.M., O.M., E.Z. and P.H.-N., formal analysis: N.O., A.S., J.M., E.Z., Y.S. and P.K., funding acquisition: A.M., P.H.-N. and B.A.S., investigation: V.Z., N.O., A.S., J.M., P.M., M.V. and E.Z., methodology: O.M., V.C., A.M., M.A. and P.H.-N., software: Y.P., Y.S. and P.K., supervision: O.M., S.G., Y.P., B.A.S. and P.H.-N., validation: N.O., O.M., S.G. and M.A., visualization: J.M., M.V. and P.M., writing—original draft: A.S., N.O., V.Z., J.M. and O.M., writing—review and editing: O.M., S.G., B.A.S. and J.K. All authors have read and agreed to the published version of the manuscript.

Funding: This research was funded by Ministry of Science and Higher Education of the Russian Federation, project No 075-15-2020-792 (unique identifier RF-190220X0031). The APC was funded from the same source.

Institutional Review Board Statement: The study was conducted according to the requirement of the German National Act on the use of experimental animals (Ethic committee Referat Verbraucherschutz, Veterinärangelegenheiten; Landesverwaltungsamt Sachsen-Anhalt, Halle, approval code: 203.6.3-42502-2-1469 UniMD_G/09.07.2018).

Informed Consent Statement: Not applicable.

Data Availability Statement: Not applicable.

Acknowledgments: The authors would like to acknowledge contributions from Corbion Biomaterials to part of this work.

Conflicts of Interest: The authors report no declarations of interest.

References

1. Man, F.; Lammers, T.; de Rosales, R.T.M. Imaging Nanomedicine-Based Drug Delivery: A Review of Clinical Studies. *Mol. Imaging Biol.* **2018**, *20*, 683–695. [CrossRef]
2. Moses, W.W. Fundamental Limits of Spatial Resolution in PET. *Nucl. Instrum. Methods Phys. Res. Sect. A Accel. Spectrom. Detect. Assoc. Equip.* **2011**, *648*, S236–S240. [CrossRef]
3. Ostrowski, A.; Nordmeyer, D.; Boreham, A.; Holzhausen, C.; Mundhenk, L.; Graf, C.; Meinke, M.C.; Vogt, A.; Hadam, S.; Lademann, J.; et al. Overview about the Localization of Nanoparticles in Tissue and Cellular Context by Different Imaging Techniques. *Beilstein J. Nanotechnol.* **2015**, *6*, 263–280. [CrossRef]
4. Li, K.; Liu, B. Polymer-Encapsulated Organic Nanoparticles for Fluorescence and Photoacoustic Imaging. *Chem. Soc. Rev.* **2014**, *43*, 6570–6597. [CrossRef]
5. Thurner, G.C.; Debbage, P. Molecular Imaging with Nanoparticles: The Dwarf Actors Revisited 10 Years Later. *Histochem. Cell Biol.* **2018**, *150*, 733–794. [CrossRef]
6. Shang, L.; Nienhaus, G.U. Small Fluorescent Nanoparticles at the Nano-Bio Interface. *Mater. Today* **2013**, *16*, 58–66. [CrossRef]
7. Priem, B.; Tian, C.; Tang, J.; Zhao, Y.; Mulder, W.J.M. Fluorescent Nanoparticles for the Accurate Detection of Drug Delivery. *Expert Opin. Drug Deliv.* **2015**, *12*, 1881–1894. [CrossRef]
8. Boschi, F.; de Sanctis, F. Overview of the Optical Properties of Fluorescent Nanoparticles for Optical Imaging. *Eur. J. Histochem.* **2017**, *61*, 2830. [CrossRef]
9. Pratiwi, F.W.; Kuo, C.W.; Chen, B.C.; Chen, P. Recent Advances in the Use of Fluorescent Nanoparticles for Bioimaging. *Nanomedicine* **2019**, *14*, 1759–1769. [CrossRef]
10. Fili, N.; Toseland, C.P. Fluorescence and labelling: How to choose and what to do. In *Fluorescent Methods for Molecular Motors*; Springer: Berlin/Heidelberg, Germany, 2014; Volume 105, pp. 1–24. ISBN 9783034808569.
11. Reisch, A.; Klymchenko, A.S. Fluorescent Polymer Nanoparticles Based on Dyes: Seeking Brighter Tools for Bioimaging. *Small* **2016**, *12*, 1968–1992. [CrossRef]
12. Ma, X.; Sun, R.; Cheng, J.; Liu, J.; Gou, F.; Xiang, H.; Zhou, X. Fluorescence Aggregation-Caused Quenching versus Aggregation-Induced Emission: A Visual Teaching Technology for Undergraduate Chemistry Students. *J. Chem. Educ.* **2016**, *93*, 345–350. [CrossRef]
13. Abdel-Mottaleb, M.M.A.; Beduneau, A.; Pellequer, Y.; Lamprecht, A. Stability of Fluorescent Labels in PLGA Polymeric Nanoparticles: Quantum Dots versus Organic Dyes. *Int. J. Pharm.* **2015**, *494*, 471–478. [CrossRef]
14. Weiss, B.; Schaefer, U.F.; Zapp, J.; Lamprecht, A.; Stallmach, A.; Lehr, C.M. Nanoparticles Made of Fluorescence-Labelled Poly(L-Lactide-Co-Glycolide): Preparation, Stability, and Biocompatibility. *J. Nanosci. Nanotechnol.* **2006**, *6*, 3048–3056. [CrossRef]
15. Malinovskaya, Y.; Melnikov, P.; Baklaushev, V.; Gabashvili, A.; Osipova, N.; Mantrov, S.; Ermolenko, Y.; Maksimenko, O.; Gorshkova, M.; Balabanyan, V.; et al. Delivery of Doxorubicin-Loaded PLGA Nanoparticles into U87 Human Glioblastoma Cells. *Int. J. Pharm.* **2017**, *524*, 77–90. [CrossRef]
16. Bastiat, G.; Pritz, C.O.; Roider, C.; Fouchet, F.; Lignières, E.; Jesacher, A.; Glueckert, R.; Ritsch-Marte, M.; Schrott-Fischer, A.; Saulnier, P.; et al. A New Tool to Ensure the Fluorescent Dye Labeling Stability of Nanocarriers: A Real Challenge for Fluorescence Imaging. *J. Control. Release* **2013**, *170*, 334–342. [CrossRef] [PubMed]
17. Snipstad, S.; Hak, S.; Baghirov, H.; Sulheim, E.; Mørch, Ý.; Lélu, S.; von Haartman, E.; Bäck, M.; Nilsson, K.P.R.; Klymchenko, A.S.; et al. Labeling Nanoparticles: Dye Leakage and Altered Cellular Uptake. *Cytom. Part A* **2017**, *91*, 760–766. [CrossRef]
18. Tenuta, T.; Monopoli, M.P.; Kim, J.A.; Salvati, A.; Dawson, K.A.; Sandin, P.; Lynch, I. Elution of Labile Fluorescent Dye from Nanoparticles during Biological Use. *PLoS ONE* **2011**, *6*, e25556. [CrossRef]
19. Andreozzi, P.; Martinelli, C.; Carney, R.P.; Carney, T.M.; Stellacci, F. Erythrocyte Incubation as a Method for Free-Dye Presence Determination in Fluorescently Labeled Nanoparticles. *Mol. Pharm.* **2013**, *10*, 875–882. [CrossRef]
20. Chen, H.; Kim, S.; Li, L.; Wang, S.; Park, K.; Cheng, J.X. Release of Hydrophobic Molecules from Polymer Micelles into Cell Membranes Revealed by Förster Resonance Energy Transfer Imaging. *Proc. Natl. Acad. Sci. USA* **2008**, *105*, 6596–6601. [CrossRef] [PubMed]
21. Tosi, G.; Rivasi, F.; Gandolfi, F.; Costantino, L.; Vandelli, M.A.; Forni, F. Conjugated Poly(D,L-Lactide-Co-Glycolide) for the Preparation of in Vivo Detectable Nanoparticles. *Biomaterials* **2005**, *26*, 4189–4195. [CrossRef]
22. Chang, J.; Jallouli, Y.; Kroubi, M.; Yuan, X.; Feng, W.; Kang, C.; Pu, P.; Betbeder, D. Characterization of Endocytosis of Transferrin-Coated PLGA Nanoparticles by the Blood-Brain Barrier. *Int. J. Pharm.* **2009**, *379*, 285–292. [CrossRef] [PubMed]
23. Sahin, A.; Esendagli, G.; Yerlikaya, F.; Caban-Toktas, S.; Yoyen-Ermis, D.; Horzum, U.; Aktas, Y.; Khan, M.; Couvreur, P.; Capan, Y. A Small Variation in Average Particle Size of PLGA Nanoparticles Prepared by Nanoprecipitation Leads to Considerable Change in Nanoparticles' Characteristics and Efficacy of Intracellular Delivery. *Artif. Cells Nanomed. Biotechnol.* **2017**, *45*, 1657–1664. [CrossRef] [PubMed]
24. Panyam, J.; Zhou, W.; Prabha, S.; Sahoo, S.K.; Labhasetwar, V. Rapid Endo-lysosomal Escape of Poly(DL-lactide-Co Glycolide) Nanoparticles: Implications for Drug and Gene Delivery. *FASEB J.* **2002**, *16*, 1217–1226. [CrossRef]
25. Yan, F.; Zhang, C.; Zheng, Y.; Mei, L.; Tang, L.; Song, C.; Sun, H.; Huang, L. The Effect of Poloxamer 188 on Nanoparticle Morphology, Size, Cancer Cell Uptake, and Cytotoxicity. *Nanomed. Nanotechnol. Biol. Med.* **2010**, *6*, 170–178. [CrossRef]

26. Lü, J.M.; Wang, X.; Marin-Muller, C.; Wang, H.; Lin, P.H.; Yao, Q.; Chen, C. Current Advances in Research and Clinical Applications of PLGA-Based Nanotechnology. *Expert Rev. Mol. Diagn.* **2009**, *9*, 325–341. [CrossRef]
27. Zhang, J.; Shi, Y.; Zheng, Y.; Pan, C.; Yang, X.; Dou, T.; Wang, B.; Lu, W. Homing in on an Intracellular Target for Delivery of Loaded Nanoparticles Functionalized with a Histone Deacetylase Inhibitor. *Oncotarget* **2017**, *8*, 68242–68251. [CrossRef]
28. Kirthivasan, B.; Singh, D.; Bommana, M.M.; Raut, S.L.; Squillante, E.; Sadoqi, M. Active Brain Targeting of a Fluorescent P-Gp Substrate Using Polymeric Magnetic Nanocarrier System. *Nanotechnology* **2012**, *23*, 255102. [CrossRef]
29. Weinheimer, M.; Fricker, G.; Burhenne, J.; Mylius, P.; Schubert, R. The Application of P-Gp Inhibiting Phospholipids as Novel Oral Bioavailability Enhancers—An in Vitro and in Vivo Comparison. *Eur. J. Pharm. Sci.* **2017**, *108*, 13–22. [CrossRef]
30. Reimold, I.; Domke, D.; Bender, J.; Seyfried, C.A.; Radunz, H.E.; Fricker, G. Delivery of Nanoparticles to the Brain Detected by Fluorescence Microscopy. *Eur. J. Pharm. Biopharm.* **2008**, *70*, 627–632. [CrossRef] [PubMed]
31. Saito, E.; Kuo, R.; Pearson, R.M.; Gohel, N.; Cheung, B.; King, N.J.C.; Miller, S.D.; Shea, L.D. Designing Drug-Free Biodegradable Nanoparticles to Modulate Inflammatory Monocytes and Neutrophils for Ameliorating Inflammation. *J. Control. Release* **2019**, *300*, 185–196. [CrossRef] [PubMed]
32. Jiang, C.; Cano-Vega, M.A.; Yue, F.; Kuang, L.; Narayanan, N.; Uzunalli, G.; Merkel, M.P.; Kuang, S.; Deng, M. Dibenzazepine-Loaded Nanoparticles Induce Local Browning of White Adipose Tissue to Counteract Obesity. *Mol. Ther.* **2017**, *25*, 1718–1729. [CrossRef]
33. Yao, H.; Ashiba, K. Highly Fluorescent Organic Nanoparticles of Thiacyanine Dye: A Synergetic Effect of Intermolecular H-Aggregation and Restricted Intramolecular Rotation. *RSC Adv.* **2011**, *1*, 834–838. [CrossRef]
34. Rhys Williams, A.T.; Winfield, S.A.; Miller, J.N. Relative Fluorescence Quantum Yields Using a Computer-Controlled Luminescence Spectrometer. *Analyst* **1983**, *108*, 1067–1071. [CrossRef]
35. Brouwer, A.M. Standards for Photoluminescence Quantum Yield Measurements in Solution; IUPAC Technical Report. *Pure Appl. Chem.* **2011**, *83*, 2213–2228. [CrossRef]
36. Magde, D.; Wong, R.; Seybold, P.G. Fluorescence Quantum Yields and Their Relation to Lifetimes of Rhodamine 6G and Fluorescein in Nine Solvents: Improved Absolute Standards for Quantum Yields. *Photochem. Photobiol.* **2002**, *75*, 327–334. [CrossRef]
37. Grabolle, M.; Spieles, M.; Lesnyak, V.; Gaponik, N.; Eychmüller, A.; Resch-Genger, U. Determination of the Fluorescence Quantum Yield of Quantum Dots: Suitable Procedures and Achievable Uncertainties. *Anal. Chem.* **2009**, *81*, 6285–6294. [CrossRef]
38. Majoul, I.; Jia, Y.; Duden, R. Practical Fluorescence Resonance Energy Transfer or Molecular Nanobioscopy of Living Cells. In *Handbook of Biological Confocal Microscopy*, 3rd ed.; Springer: Boston, MA, USA, 2006; pp. 788–808. [CrossRef]
39. Harder, E.; Damm, W.; Maple, J.; Wu, C.; Reboul, M.; Xiang, J.Y.; Wang, L.; Lupyan, D.; Dahlgren, M.K.; Knight, J.L.; et al. OPLS3: A Force Field Providing Broad Coverage of Drug-like Small Molecules and Proteins. *J. Chem. Theory Comput.* **2016**, *12*, 281–296. [CrossRef] [PubMed]
40. Watts, K.S.; Dalal, P.; Murphy, R.B.; Sherman, W.; Friesner, R.A.; Shelley, J.C. ConfGen: A Conformational Search Method for Efficient Generation of Bioactive Conformers. *J. Chem. Inf. Model.* **2010**, *50*, 534–546. [CrossRef]
41. Řezáč, J.; Hobza, P. Advanced Corrections of Hydrogen Bonding and Dispersion for Semiempirical Quantum Mechanical Methods. *J. Chem. Theory Comput.* **2012**, *8*, 141–151. [CrossRef]
42. Becke, A.D. Density-Functional Exchange-Energy Approximation with Correct Asymptotic Behavior. *Phys. Rev. A* **1988**, *38*, 3098–3100. [CrossRef]
43. Vosko, S.H.; Wilk, L.; Nusair, M. Accurate Spin-Dependent Electron Liquid Correlation Energies for Local Spin Density Calculations: A Critical Analysis. *Can. J. Phys.* **1980**, *58*, 1200–1211. [CrossRef]
44. Perdew, J.P. Density-Functional Approximation for the Correlation Energy of the Inhomogeneous Electron Gas. *Phys. Rev. B* **1986**, *33*, 8822–8824. [CrossRef]
45. Schäfer, A.; Huber, C.; Ahlrichs, R. Fully Optimized Contracted Gaussian Basis Sets of Triple Zeta Valence Quality for Atoms Li to Kr. *J. Chem. Phys.* **1994**, *100*, 5829–5835. [CrossRef]
46. Klamt, A. The COSMO and COSMO-RS Solvation Models. *Wiley Interdiscip. Rev. Comput. Mol. Sci.* **2018**, *8*, 699–709. [CrossRef]
47. Eckert, F.; Klamt, A. Fast Solvent Screening via Quantum Chemistry: COSMO-RS Approach. *AIChE J.* **2002**, *48*, 369–385. [CrossRef]
48. Marenich, A.V.; Cramer, C.J.; Truhlar, D.G. Universal Solvation Model Based on Solute Electron Density and on a Continuum Model of the Solvent Defined by the Bulk Dielectric Constant and Atomic Surface Tensions. *J. Phys. Chem. B* **2009**, *113*, 6378–6396. [CrossRef]
49. Loschen, C.; Klamt, A. Prediction of Solubilities and Partition Coefficients in Polymers Using COSMO-RS. *Ind. Eng. Chem. Res.* **2014**, *53*, 11478–11487. [CrossRef]
50. Prilloff, S.; Fan, J.; Henrich-Noack, P.; Sabel, B.A. In Vivo Confocal Neuroimaging (ICON): Non-Invasive, Functional Imaging of the Mammalian CNS with Cellular Resolution. *Eur. J. Neurosci.* **2010**, *31*, 521–528. [CrossRef] [PubMed]
51. Sabel, B.A.; Engelmann, R.; Humphrey, M.F. In Vivo Confocal Neuroimaging (ICON) of CNS Neurons. *Nat. Med.* **1997**, *3*, 244–247. [CrossRef]
52. Finke, J.H.; Richter, C.; Gothsch, T.; Kwade, A.; Büttgenbach, S.; Müller-Goymann, C.C. Coumarin 6 as a Fluorescent Model Drug: How to Identify Properties of Lipid Colloidal Drug Delivery Systems via Fluorescence Spectroscopy? *Eur. J. Lipid Sci. Technol.* **2014**, *116*, 1234–1246. [CrossRef]

53. Rivolta, I.; Panariti, A.; Lettiero, B.; Sesana, S.; Gasco, P.; Gasco, M.R.; Masserini, M.; Miserocchi, G. Cellular Uptake of Coumarin-6 as a Model Drug Loaded in Solid Lipid Nanoparticles. *J. Physiol. Pharmacol.* **2011**, *62*, 45–53. [PubMed]

54. Guerrini, L.; Alvarez-Puebla, R.A.; Pazos-Perez, N. Surface Modifications of Nanoparticles for Stability in Biological Fluids. *Materials* **2018**, *11*, 1154. [CrossRef] [PubMed]

55. Abouelmagd, S.A.; Sun, B.; Chang, A.C.; Ku, Y.J.; Yeo, Y. Release Kinetics Study of Poorly Water-Soluble Drugs from Nanoparticles: Are We Doing It Right? *Mol. Pharm.* **2015**, *12*, 997–1003. [CrossRef] [PubMed]

56. Simonsson, C.; Bastiat, G.; Pitorre, M.; Klymchenko, A.S.; Béjaud, J.; Mély, Y.; Benoit, J.P. Inter-Nanocarrier and Nanocarrier-to-Cell Transfer Assays Demonstrate the Risk of an Immediate Unloading of Dye from Labeled Lipid Nanocapsules. *Eur. J. Pharm. Biopharm.* **2016**, *98*, 47–56. [CrossRef]

57. Shabbits, J.A.; Chiu, G.N.C.; Mayer, L.D. Development of an in Vitro Drug Release Assay That Accurately Predicts in Vivo Drug Retention for Liposome-Based Delivery Systems. *J. Control. Release* **2002**, *84*, 161–170. [CrossRef]

58. Pretor, S.; Bartels, J.; Lorenz, T.; Dahl, K.; Finke, J.H.; Peterat, G.; Krull, R.; Al-Halhouli, A.T.; Dietzel, A.; Büttgenbach, S.; et al. Cellular Uptake of Coumarin-6 under Microfluidic Conditions into HCE-T Cells from Nanoscale Formulations. *Mol. Pharm.* **2015**, *12*, 34–45. [CrossRef]

59. Tian, Y.; Halle, J.; Wojdyr, M.; Sahoo, D.; Scheblykin, I.G. Quantitative Measurement of Fluorescence Brightness of Single Molecules. *Methods Appl. Fluoresc.* **2014**, *2*, 035003. [CrossRef]

60. Zou, P.; Chen, H.; Paholak, H.J.; Sun, D. Noninvasive Fluorescence Resonance Energy Transfer Imaging of in Vivo Premature Drug Release from Polymeric Nanoparticles. *Mol. Pharm.* **2013**, *10*, 4185–4194. [CrossRef]

61. Wang, H.; Chen, J.; Xu, C.; Shi, L.; Tayier, M.; Zhou, J.; Zhang, J.; Wu, J.; Ye, Z.; Fang, T.; et al. Cancer Nanomedicines Stabilized by π-π Stacking between Heterodimeric Prodrugs Enable Exceptionally High Drug Loading Capacity and Safer Delivery of Drug Combinations. *Theranostics* **2017**, *7*, 3638. [CrossRef] [PubMed]

62. Nakanishi, T.; Fukushima, S.; Okamoto, K.; Suzuki, M.; Matsumura, Y.; Yokoyama, M.; Okano, T.; Sakurai, Y.; Kataoka, K. Development of the Polymer Micelle Carrier System for Doxorubicin. *J. Control. Release* **2001**, *74*, 295–302. [CrossRef]

63. Behnke, T.; Würth, C.; Laux, E.M.; Hoffmann, K.; Resch-Genger, U. Simple Strategies towards Bright Polymer Particles via One-Step Staining Procedures. *Dye. Pigment.* **2012**, *94*, 247–257. [CrossRef]

64. Sharafy, S.; Muszkat, K.A. Viscosity Dependence of Fluorescence Quantum Yields. *J. Am. Chem. Soc.* **1971**, *93*, 4119–4125. [CrossRef]

65. Hall, L.M.; Gerowska, M.; Brown, T. A Highly Fluorescent DNA Toolkit: Synthesis and Properties of Oligonucleotides Containing New Cy3, Cy5 and Cy3B Monomers. *Nucleic Acids Res.* **2012**, *40*, e108. [CrossRef] [PubMed]

66. Edetsberger, M.; Knapp, M.; Gaubitzer, E.; Miksch, C.; Gvichiya, K.E.; Köhler, G. Effective Staining of Tumor Cells by Coumarin-6 Depends on the Stoichiometry of Cyclodextrin Complex Formation. *J. Incl. Phenom. Macrocycl. Chem.* **2011**, *70*, 327–331. [CrossRef]

67. Li, Y.; Song, Y.; Zhao, L.; Gaidosh, G.; Laties, A.M.; Wen, R. Direct Labeling and Visualization of Blood Vessels with Lipophilic Carbocyanine Dye DiI. *Nat. Protoc.* **2008**, *3*, 1703–1708. [CrossRef] [PubMed]

68. Tian, B.-C.; Zhang, W.-J.; Xu, H.-M.; Hao, M.-X.; Liu, Y.-B.; Yang, X.-G.; Pan, W.-S.; Liu, X.-H. Further Investigation of Nanostructured Lipid Carriers as an Ocular Delivery System: In Vivo Transcorneal Mechanism and in Vitro Release Study. *Colloids Surf. B Biointerfaces* **2013**, *102*, 251–256. [CrossRef]

69. Jensen, K.H.R.; Berg, R.W. CLARITY-Compatible Lipophilic Dyes for Electrode Marking and Neuronal Tracing. *Sci. Rep.* **2016**, *6*, 32674. [CrossRef]

70. Poulain, F.E.; Gaynes, J.A.; Hörndli, C.S.; Law, M.Y.; Chien, C.B. Analyzing retinal axon guidance in zebrafish. *Methods Cell Biol.* **2010**, *100*, 2–26.

71. Goss, K.U. Predicting Equilibrium Sorption of Neutral Organic Chemicals into Various Polymeric Sorbents with COSMO-RS. *Anal. Chem.* **2011**, *83*, 5304–5308. [CrossRef] [PubMed]

72. Klamt, A.; Eckert, F.; Reinisch, J.; Wichmann, K. Prediction of Cyclohexane-Water Distribution Coefficients with COSMO-RS on the SAMPL5 Data Set. *J. Comput. Aided Mol. Des.* **2016**, *30*, 959–967. [CrossRef]

73. Loschen, C.; Reinisch, J.; Klamt, A. COSMO-RS Based Predictions for the SAMPL6 LogP Challenge. *J. Comput. Aided Mol. Des.* **2020**, *34*, 385–392. [CrossRef] [PubMed]

74. Biltonen, R.L.; Lichtenberg, D. The Use of Differential Scanning Calorimetry as a Tool to Characterize Liposome Preparations. *Chem. Phys. Lipids* **1993**, *64*, 129–142. [CrossRef]

75. Steuer, H.; Jaworski, A.; Elger, B.; Kaussmann, M.; Keldenich, J.; Schneider, H.; Stoll, D.; Schlosshauer, B. Functional Characterization and Comparison of the Outer Blood-Retina Barrier and the Blood-Brain Barrier. *Investig. Ophthalmol. Vis. Sci.* **2005**, *46*, 1047–1053. [CrossRef] [PubMed]

76. Pietzonka, P.; Rothen-Rutishauser, B.; Langguth, P.; Wunderli-Allenspach, H.; Walter, E.; Merkle, H.P. Transfer of Lipophilic Markers from PLGA and Polystyrene Nanoparticles to Caco-2 Monolayers Mimics Particle Uptake. *Pharm. Res.* **2002**, *19*, 595–601. [CrossRef] [PubMed]

77. Zhang, E.; Zhukova, V.; Semyonkin, A.; Osipova, N.; Malinovskaya, Y.; Maksimenko, O.; Chernikov, V.; Sokolov, M.; Grigartzik, L.; Sabel, B.A.; et al. Release Kinetics of Fluorescent Dyes from PLGA Nanoparticles in Retinal Blood Vessels: In Vivo Monitoring and Ex Vivo Localization. *Eur. J. Pharm. Biopharm.* **2020**, *150*, 131–142. [CrossRef]

78. Demuth, P.C.; Garcia-Beltran, W.F.; Ai-Ling, M.L.; Hammond, P.T.; Irvine, D.J. Composite Dissolving Microneedles for Coordinated Control of Antigen and Adjuvant Delivery Kinetics in Transcutaneous Vaccination. *Adv. Funct. Mater.* **2013**, *23*, 161–172. [CrossRef]

79. Tahara, K.; Karasawa, K.; Onodera, R.; Takeuchi, H. Feasibility of Drug Delivery to the Eye's Posterior Segment by Topical Instillation of PLGA Nanoparticles. *Asian J. Pharm. Sci.* **2017**, *12*, 394–399. [CrossRef]

80. Zhang, E.; Osipova, N.; Sokolov, M.; Maksimenko, O.; Semyonkin, A.; Wang, M.; Grigartzik, L.; Gelperina, S.; Sabel, B.A.; Henrich-Noack, P. Exploring the Systemic Delivery of a Poorly Water-Soluble Model Drug to the Retina Using PLGA Nanoparticles. *Eur. J. Pharm. Sci.* **2021**, 105905. [CrossRef]

81. Hofmann, D.; Messerschmidt, C.; Bannwarth, M.B.; Landfester, K.; Mailänder, V. Drug Delivery without Nanoparticle Uptake: Delivery by a Kiss-and-Run Mechanism on the Cell Membrane. *Chem. Commun.* **2014**, *50*, 1369–1371. [CrossRef]

 *pharmaceutics*

MDPI

Review

PLGA-Based Nanoparticles for Neuroprotective Drug Delivery in Neurodegenerative Diseases

Anthony Cunha [1,2], Alexandra Gaubert [1,*], Laurent Latxague [1,*] and Benjamin Dehay [2,*]

[1] Université de Bordeaux, INSERM U1212, CNRS UMR 5320, ARNA, ARN: Régulations Naturelle et Artificielle, ChemBioPharm, 146 rue Léo Saignat, F-33076 Bordeaux, France; anthony.cunha@u-bordeaux.fr
[2] Univ. Bordeaux, CNRS, IMN, UMR 5293, F-33000 Bordeaux, France
* Correspondence: alexandra.gaubert@u-bordeaux.fr (A.G.); laurent.latxague@u-bordeaux.fr (L.L.); benjamin.dehay@u-bordeaux.fr (B.D.)

Abstract: Treatment of neurodegenerative diseases has become one of the most challenging topics of the last decades due to their prevalence and increasing societal cost. The crucial point of the non-invasive therapeutic strategy for neurological disorder treatment relies on the drugs' passage through the blood-brain barrier (BBB). Indeed, this biological barrier is involved in cerebral vascular homeostasis by its tight junctions, for example. One way to overcome this limit and deliver neuroprotective substances in the brain relies on nanotechnology-based approaches. Poly(lactic-*co*-glycolic acid) nanoparticles (PLGA NPs) are biocompatible, non-toxic, and provide many benefits, including improved drug solubility, protection against enzymatic digestion, increased targeting efficiency, and enhanced cellular internalization. This review will present an overview of the latest findings and advances in the PLGA NP-based approach for neuroprotective drug delivery in the case of neurodegenerative disease treatment (i.e., Alzheimer's, Parkinson's, Huntington's diseases, Amyotrophic Lateral, and Multiple Sclerosis).

Keywords: PLGA nanoparticles; neurodegenerative diseases; drug delivery; central nervous system; neuroprotective drugs

Citation: Cunha, A.; Gaubert, A.; Latxague, L.; Dehay, B. PLGA-Based Nanoparticles for Neuroprotective Drug Delivery in Neurodegenerative Diseases. *Pharmaceutics* **2021**, *13*, 1042. https://doi.org/10.3390/pharmaceutics13071042

Academic Editor: Oya Tagit

Received: 3 June 2021
Accepted: 6 July 2021
Published: 8 July 2021

1. Introduction

Neurodegenerative diseases (NDD) represent a major societal issue. For example, in 2015, more than 46.8 million people worldwide were affected by dementia [1] and over 6 million by Parkinson's disease (PD) [2]. This number is increasing each year due to, in particular, the population aging. NDDs are a complex group of diseases with no curative treatments for the five most common and known of them: Alzheimer's disease (AD), Parkinson's disease, Huntington's disease (HD), Amyotrophic Lateral Sclerosis (ALS), and Multiple Sclerosis (MS) (Table 1). These incurable and incapacitating diseases lead to the progressive degeneration and death of nerve cells, causing troubles related to motor disabilities or mental dysfunction (called dementias), considerably worsening patient quality of life.

Therefore, the development of chronic treatments for NDDs both effective and easy to administer remains a major challenge due to the necessary blood-brain barrier (BBB) crossing to get drugs into the brain. This barrier filters and controls the passage of foreign blood substances, potentially dangerous molecules or pathogens, and prevents them from passing freely from the blood into the extracellular fluid of the brain's gray matter. In addition, the BBB protects the brain cells by regulating their environment (cerebral vascular homeostasis, pH) against hormone and neurotransmitter variations, for example. Thus, it offers strong resistance to the ion movement, allowing to maintain the brain chemical balance and adjusting this environment to guarantee a perfect transmission of signals between neurons. In addition, the existence of tight junctions (TJs) between each endothelial cell greatly limits the cerebral passage of numerous compounds (98% of the substances cannot cross it) [3,4].

Table 1. Non-exhaustive description of most known neurodegenerative diseases.

Disease	Prevalence (Per 100,000 Person)	Symptoms	Affected Areas
AD	511–690 [5]	Memory impairment, changes in thinking, judgment, language, behavioral changes, etc.	Cortex, hippocampus, brainstem
PD	100–200 [6]	Rest tremors, slowness of movement, decrease of spontaneous mobility, muscular stiffness, etc.	Basal ganglia, cortex
HD	5.96–13.7 [7]	Motor disorders, breathing difficulties, speech and swallowing disorders, etc.	Striatum and other basal ganglia regions
ALS	3.92–4.96 [8]	Progressive muscle paralysis, muscle atrophy, spasticity, breathing and swallowing disorders, etc.	Motor cortex, spinal cord
MS	35.87–35.95 [9]	Numbness in a limb, vision problems, electric shock sensations in a limb or the back, movement problems, etc.	Brain, spinal cord, optic nerve

One way to overcome these drawbacks relies on nanotechnologies. This approach presents various advantages such as drug protection, targeting, controlled release, or interest in NDD treatment, the BBB crossing. A wide range of pharmaceutical nanocarriers was developed, including liposomes, solid lipid nanoparticles, micelles, dendrimers, or nanoparticles (NPs) [10,11].

Thanks to their physicochemical characteristics, such as their size, biocompatibility, or low cytotoxicity, polymeric NPs represent very interesting tools for delivering neuroprotective agents for NDD treatments [12]. Indeed, their ability to cross biological barriers, their versatility (encapsulation of either hydrophilic or hydrophobic drugs), or their targeting ability (drug delivery to a specific body site decreasing the off-target toxicity) explains the increasing interest for this approach. Among the polymers used for NP formulation (natural or synthetic), the most widely used is the poly(lactic-*co*-glycolic) acid (PLGA). This polymer has a double interest in the NDD treatment for its acidifying [13], particularly for PD, and drug encapsulation properties [14].

In the following review, we propose a non-exhaustive summary of the last decade's advances in delivering neuroprotective substances via PLGA NPs for NDD treatment. A focus will be made on the existing neuroprotective drugs such as phytol, rhynchophylline, or curcumin to go to the potential new neuroprotective candidates, benefiting from the PLGA NP vectorization.

2. PLGA NPs

Nowadays, PLGA NPs are a booming topic, especially for the development of NDD treatments (more than 50 articles during the last 5 years on Scopus using "PLGA", "nanoparticles", "neurodegenerative disease" as keywords). PLGA, a commercially available synthetic copolymer obtained from lactic and glycolic acid, is approved by the US regulatory agency (Food and Drug Administration: FDA) and the European Medicine Agency (EMA). To date, not based on PLGA but poly-lactic acid (PLA), paclitaxel-loaded PEG-PLA micelles have reached the market in South Korea, India, and Indonesia (Genexol® PM). It is currently undergoing Phase III clinical trials for access to the EU and US markets. For the NPs, only one Phase II clinical trial based on PEG-PLGA/PLA-PEG NPs (BIND-014) for metastatic castration-resistant prostate cancer was reported [15]. However, in the case of NDDs treatment, no PLGA NPs are currently on the market or in clinical trials, but are currently only at the preclinical stage [16]. Indeed, several pre-clinical studies based on drug loaded nano-objects are in progress, including delivery of curcumin, levodopa, cholesterol or rapamycin for AD, PD, HD and MS treatment. The advantage of the PLGA NPs approach relies on their potential for drug encapsulation, excellent biocompatibility, and biodegradability [17]. PLGA degradation by hydrolysis of its ester bonds in aqueous media releases its two constitutive monomers, which are naturally produced under physiological conditions by several metabolic pathways [18,19]. Thanks to this property, PLGA was reported as an active substance, notably for the treatment of PD since lactic acid and glycolic acid decrease lysosomal pH [13,20]. For example, studies on pathological PD models demonstrated that PLGA NPs reacidified defective lysosomes to basal level (pH 4), restored lysosomal deleterious effects, due to PD-linked mutations, and overcame lysosomal acidification impairments. The acidification also supports mitochondrial membrane potential and neuronal survival in several models related to mitochondrial dysfunctions.

2.1. PLGA NP Formulation and Optimization

These last years, a wide variety of preparation techniques for PLGA NPs were reported in the literature, i.e., single and double emulsion solvent evaporation, nanoprecipitation, coacervation and spray-drying [14,21]. The single and double emulsion solvent evaporation methods are the most frequently used: PLGA is dissolved into an organic phase emulsified with an aqueous medium containing surfactants or stabilizers prior to solvent evaporation and NP formation, corresponding to a single emulsion solvent evaporation method [22]. The single method involves a two-phase process (oil in water O/W or water in oil W/O) while three-phase are used with the double emulsion solvent method (O/W/O or W/O/W). The main drawback of these two techniques is the use of shear stress during the homogenization step leading to low protein encapsulation efficiency. To overcome this drawback, the nanoprecipitation method may be beneficial. In that case, the organic solvent containing PLGA is added dropwise to the aqueous medium, leading to NPs formation by a rapid diffusion of the miscible solvent [23]. Nevertheless, the previous methods are not the most appropriate for industrial scale-up, unlike spray drying, which transforms liquid substances (sprayed in a thin stream of heated air) into powders [14].

Depending on the process used, amphiphilic, hydrophobic, or hydrophilic drugs can be either loaded inside the NP core, trapped among the polymer chains, or adsorbed on the NP surface (leading to burst release and potential off target effects). Even though a low encapsulation efficiency characterizes PLGA NPs, in particular for hydrophilic drugs, there is room for improvement by playing with the molecular weight (i.e., the higher the molecular weight, the larger the NPs) or the poly(D,L-lactic acid) (PLA)/poly (D,L-glycolyc acid) (PGA) ratio [24]. The latter parameter is directly related to the crystallinity degree and the melting point of PLGA copolymers, and drives the capacity of polymers to undergo hydrolysis. Indeed, PLA exhibiting methyl side groups is more hydrophobic than PGA, leading to a slower degradation of PLGA NPs possessing a high PLA ratio [17].

Depending on the biomedical application, other parameters must be considered for NP optimization such as particle size, polydispersity index (PdI), zeta potential (ζ potential), drug loading, or encapsulation efficiency. It was reported that a diameter lower than 200 nm enables an increase of the in vivo half-life and a better membrane passage, which are crucial for biomedical applications [25,26]. If the NP diameter is essential, the size distribution, characterized by the Polydispersity Index (PdI), is just as important: the smaller the PdI, the more homogeneous the population of NPs. Values of 0.2 and below are commonly deemed acceptable in practice for biomedical applications of polymer-based NPs [27]. To ensure a moderate colloidal stability of these nano-objects, a ζ potential below -30 mV or above $+30$ mV is required to avoid coagulation or flocculation phenomena [28]. Moreover, the NP charge also impacts their biological effects: positively charged NPs increase cellular uptake and cytotoxicity in nonphagocytic cells while negatively charged NPs enhance cytotoxicity in phagocytic cells [29].

Modulation of NP functionalization enables designing NP platforms to address many issues, especially in the diagnostic, the active targeting, or the shielding fields (Figure 1) [30,31]. Regarding the diagnostic approach, probes (containing fluorescent moieties, isotopes, etc.) are introduced inside the NP or at the surface, allowing the detection by different imaging techniques: near-infrared, fluorescence, positron-emission tomography, or single-photon emission computed tomography [30]. Another aspect is proper targeting to ensure the drug delivery directly on site. Ligands such as proteins, polysaccharides, peptides, aptamers, or small molecules are promising candidates in this approach [32]. For instance, tween 80, also known as polysorbate 80, can adsorb apolipoprotein E onto PLGA NPs enabling the binding of lipoprotein receptor-related proteins (LRPs), thus facilitating the BBB crossing [33]. This hydrophilic nonionic surfactant and emulsifier is usually used at 1% (w/v) to stabilize aqueous formulations for parenteral administration. It is speculated that weak interactions control the formation of an adsorbed monolayer of polysorbate 80 onto the nanoparticle surface [34]. As a coating, polysorbate 80 presents some clinical advantages such as decreasing the drug dosage (therefore reducing the potential side effects) or increasing its viability. Nevertheless, liver and renal toxicity, hypersensitivity or erythema need to be monitored to ensure treatment safety [35] About intranasal administration, the critical point relies on the residence time of the drug. This problem can be addressed by using mucoadhesive polymers, such as chitosan or specific ligands such as lectins [36]. A better binding to the nasoepithelial surface increases the residence time and thus the drug bioavailability. The last point to consider is the shielding of the nano-objects to reduce aggregation, opsonization, phagocytosis, and immune clearance, which may otherwise decrease their circulation time and their targeting efficiency. For this purpose, NPs are coated with several molecules such as carbohydrates, proteins, or lipids, knowing that PEGylation remains the main strategy to improve drug efficiency and gene delivery [37–39].

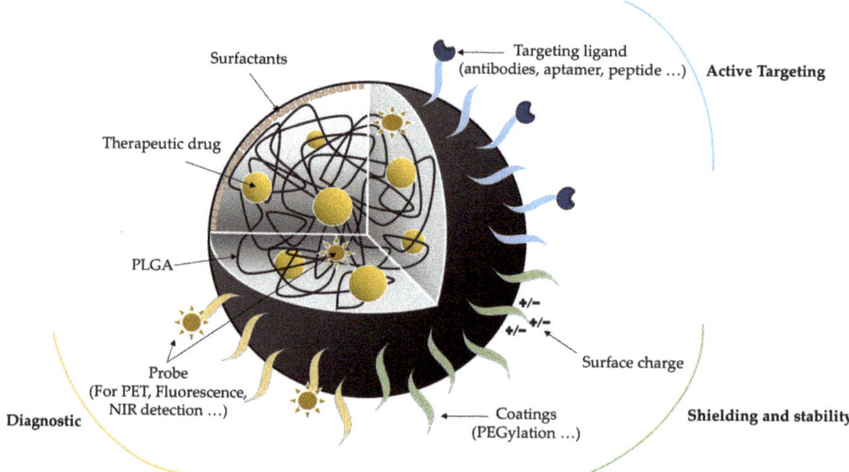

Figure 1. Schematic representation of PLGA NPs and their multiple potential functionalizations: for diagnostic, targeting, shielding, or stability purposes [30] (NIR: Near-infrared; PET: Positron-emission tomography).

2.2. Administration Routes

Due to their incredible modularity and small size, PLGA NPs present numerous advantages for NDD treatments via different available administration routes [40]. Stereotaxic surgery overpasses the BBB and thus directly delivers the drugs without peripheral drug inactivation [41]. The main advantage of this approach is to be as precise as possible on the target despite the technic invasiveness. Otherwise, other techniques such as enteral, parental, and intranasal routes represent good alternatives. The first two require crossing the BBB through various mechanisms (simple diffusion, receptor, adsorption mediated endocytosis, or carrier-mediated transport), depending on NP functionalization, with a first pass effect decreasing the drug concentration on site in the case of enteral route [42,43]. The last one (intraneuronal absorption by a direct nose to brain delivery of the NP platforms), allows a rapid and important active substance passage into the systemic circulation without first-pass metabolism avoiding drug degradation. It offers the advantage of a better patient compliance, due to the low invasiveness of the daily-based administrations, and therefore represents a promising option for NDD treatment [44].

Two types of strategies can be considered with PLGA NPs for NDD therapy: as active substances by themselves and as cargos for neuroprotective drugs.

3. PLGA NPs for Neuroprotective Drug Delivery in Neurological Disorder Therapy

Many neuroprotective medicines cannot reach the brain due to the lack of drug-specific transport systems through the BBB. The development of new strategies based on PLGA NPs to enhance brain drug delivery is of great interest in NDD therapy [45]. Indeed, even if different pathologies are included in NDD, they all share common mechanisms such as mitochondrial or oxidative injury, neuro-inflammation, apoptosis, and protein aggregation, which contribute to neuronal loss mainly by the intrinsic mitochondrial apoptotic pathway (Figure 2) [46,47]. The mitochondria complexes and mitochondrially located monoamine oxidase B (MAO-B) are involved in the nitrogen and reactive oxygen species (ROS) production. The dysfunction of these pathways is one characteristic of NDD as the neuro-inflammation for AD, PD, and MS. Neuro-inflammation is a defense mechanism that initially protects the brain by removing or inhibiting diverse pathogens. In the case of NDD, it involves abundant activated microglia inhibiting amyloid-beta (Aβ) phagocytosis contributing to plaque accumulation. These mechanisms are targets of choice for treating protein aggregation diseases (proteinopathies) such as AD, PD, HD, ALS, and MS.

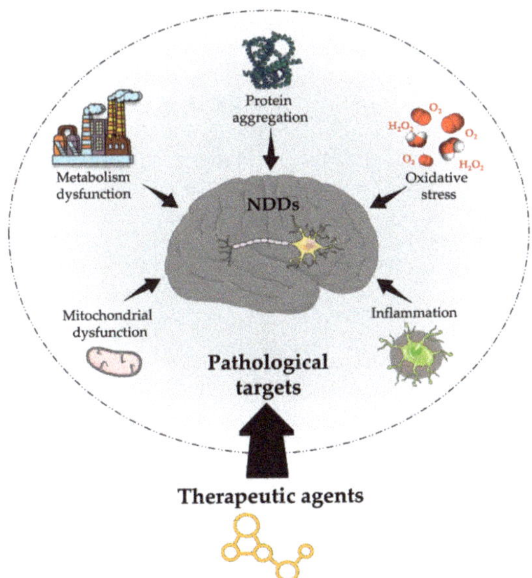

Figure 2. Pathological targets for active substances in the case of the most common NDDs: protein aggregation, oxidative stress, inflammation, mitochondrial, and metabolism dysfunction.

To act on the mechanisms involved in NDDs, the new trend relies on the encapsulation of specific active compounds (natural or synthetic) into PLGA NPs (Table 2).

3.1. Alzheimer's Disease

AD is the most common neurodegenerative disease responsible for a progressive decline in cognitive functions. The association of two neuropathological brain lesions characterizes this pathology: extracellular Aβ protein and intracellular hyperphosphorylated tau protein deposits [48,49]. Over time, neurodegeneration occurs in the brain regions related to memory and language, from the hippocampal region to the entire cerebral cortex, explaining the progression of the symptoms to aphasia, apraxia, visio-spatial, and executive function disorders [50]. The causes of protein aggregation, into amyloid deposits and neurofibrillary degeneration, are still unknown, but genetic and environmental factors may contribute to their appearance [51]. The protein aggregation may, in turn, induce oxidative stress (OS) leading to neuroinflammation [52]. OS results from an imbalance between the production of free radicals and their detoxification rates by the SuperOxide Dismutase enzyme (SOD) especially.

There is currently no cure for AD, only palliative strategies (tacrine, rivastigmine, etc.) to limit its progression and improve the patient's quality of life, opening the doors for the nanotechnology approach and more especially the PLGA NP one (Table 2).

3.1.1. Thymoquinone

As previously said, AD is linked to protein (i.e., Aβ) aggregation which may, in turn, induce oxidative stress leading to neuro-inflammation. Yusuf et al. designed PLGA NPs loaded with thymoquinone (TQ), a phytochemical compound, to act on this aspect in animal models due to its potent antioxidant and anti-inflammatory properties (Figure 3) [53,54]. Male albino mice administered with Streptozotocin (SZT) (mimicking AD oxidative stress by a decrease of SOD activity) were treated with TQ loaded PLGA NPs, coated with polysorbate 80 (P-80-TQN). These NP were prepared by the single-emulsion solvent-evaporation method, leading to spherical nano-objects with an average size of 226 $\pm$ 4.6 nm and -45.6 ± 2.6 mV as ζ-potential. In vitro, TQ release assessment was characterized by an initial burst for 2 h (drug diffusion from the NP surface and by the beginning of PLGA matrix erosion) followed by a longer sustained release (stabilized dipole-dipole interactions taking place between TQ and PLGA components). The P-80-TQ NPs successfully crossed the BBB by endocytosis through LDL receptors (mediated by the polysorbate coating). Once on-site, these systems significantly affected SOD activity (increase) from day 7 to day 28. Concurrently, an animal and cognition test (Despair test) was undertaken to highlight a meaningful improvement after NP injection, further confirming their beneficial role. Finally, histopathological examinations of brain hippocampal tissues, before and after P-80-TQ NP treatment (5 mg.kg^{-1} equivalent), showed a drastic size reduction of the protein aggregate generated by the STZ-induced AD in mice.

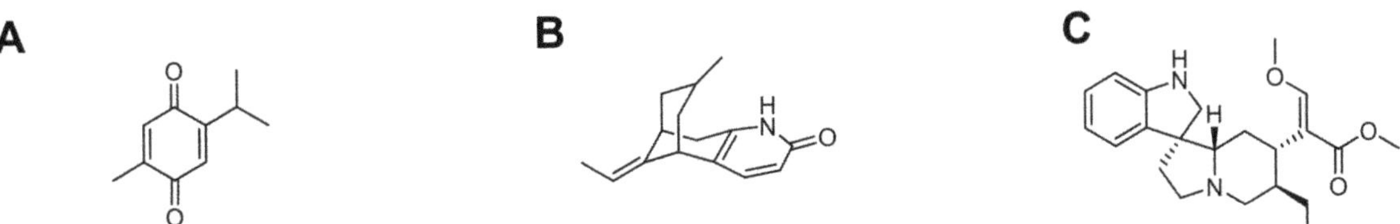

Figure 3. Chemical structures of the three described compounds encapsulated into PLGA NPs for AD treatment (Thymoquinone (**A**), Huperzine A (**B**), and Rhynchophylline (**C**)).

3.1.2. Huperzine A

Meng et al., developed an intranasal drug delivery system (DDS) as PLGA NPs loaded with Huperzine A (HupA) (Figure 3) co-modified with lactoferrin (Lf) to enhance nose-to-brain delivery and N-trimethylchitosan (TMC) for easier mucoadhesion [55]. HupA possesses neuroprotective effects especially in the treatment of AD. Still, it displays a poor brain selectivity resulting in several side effects (including nausea, diarrhea, vomiting, …), justifying the need for a more specific DDS [56]. NPs were formulated by the single-emulsion solvent-evaporation method with maleimide-TMC, allowing the further conjugation of thiolated Lf on the maleimide moiety leading to the Lf-TMC-NPs. No cytotoxicity was observed for free HupA and HupA-loaded NPs concentrations ranging from 12.5 to 25 μg.mL^{-1} against human bronchial epithelial cell lines (16HBE) used as a model of nasal mucosa cells. The mucoadhesive assessment was evaluated by mucin adsorption onto NPs with a binding efficiency up to 86% compared to unmodified PLGA NPs (32%). Lf-TMC NPs achieved a boost release of HupA in vitro within 4 h followed by a sustained release of 74.5% at 48 h. The cellular uptake of NPs was assessed using 16HBE and SH-SY5Y (human neuroblastoma cell line) cells as brain cell models, using nile red-loaded NPs. They were taken up in both cell lines with a significantly higher fluorescence intensity for Lf-TMC NPs than unmodified NPs or even TMC-NPs. The same trend was explicitly observed in vivo after intranasal administration of DiR-loaded NPs in mice. A strong signal was obtained with TMC-NPs in the brain, attributed to a better nose-to-brain drug delivery and mucoadhesion provided by TMC. Furthermore, an even larger Lf-TMC NPs was localized in the brain, suggesting a Lf-mediated transport mechanism. Finally, following the intranasal administration of HupA loaded NPs, pharmacokinetics exhibited a minimal elimination rate associated with a greater bioavailability of Lf-TMC NPs, particularly in the olfactory bulb and the memory-related hippocampus.

3.1.3. Rhynchophylline

Rhynchophylline (RIN) is a spirocyclic alkaloid isolated from Uncaria species that exhibits numerous pharmacological activities, including neuroprotective effect (Figure 3) [57]. In the case of AD, RIN inhibits soluble Aβ-induced hyperexcitability of hippocampal neurons. In 2020, the first study on the design and development of brain-targeting therapy for AD using RIN administration was published by Xu et al. [58]. Tween 80 coated methoxy poly (ethylene glycol) PLGA NPs were prepared by nanoprecipitation technique to enhance the pharmacological activity of RIN and targeting purposes. No hemolysis occurred using T80 coated RIN loaded PLGA NPs, suggesting that these NP suspensions were safe to use. Internalization into bEnd.3 cells was confirmed by confocal laser scanning microscopy, using DiD fluorophore-loaded PLGA NPs; T80 coated NPs exhibited higher permeation. To go further, an in-vitro BBB model, with bEnd.3 cells, was developed to study the crossing of such nano-systems and demonstrated their higher transport compared to free RIN or uncoated RIN loaded NPs. The benefits of using T80 for targeting brain was then highlighted in vivo in healthy C57BL/6 mice. Finally, survival rate of PC12 cells injured by Aβ_{25-35} was improved by incubation with T80 RIN NPs, and cell apoptosis was significantly reduced. These results confirmed that encapsulation did not affect RIN neuroprotective effects, demonstrating the great potential of loaded PLGA NPs for NDD therapy.

3.2. Parkinson's Disease

PD is the second most common NDD after AD. Three characteristic motor symptoms are associated with Parkinson's syndrome: tremor, akinesia (slowness of movement), and limb rigidity [59]. The disease hallmarks are defined by a selective neuronal dopaminergic loss in specific brain regions, and deposits of misfolded proteins through the presence of α-synuclein aggregates (Lewy bodies, LB). Other pathophysiological symptoms include altered dopamine metabolism, impaired mitochondrial function, oxidative stress, and inflammation. Nowadays, no cure is available for AD, but symptomatic treatments, such as L-dopa (dopamine precursor), can reduce the motor symptoms. This lack of therapeutic alternatives paves the way for the nanotechnology approach, especially using the PLGA system, since this compound was proved biologically active in PD treatment [13,20]. Combining PLGA with active drugs could thus constitute a good option to achieve a synergistic action on PD (Table 2).

3.2.1. Schisantherin A

Schisantherin A (SA) is a dibenzocyclooctadiene with neuroprotective activity against MPP(+) (or 1-methyl-4-phenyl pyridinium, a toxic molecule interfering with oxidative phosphorylation in mitochondria), thus a potential candidate for neuron loss treatment in PD (Figure 4) [60]. In 2017, Chen et al., prepared spherical SA-NPs (ø: 70.6 $\pm$ 2.2 nm), by flash nanoprecipitation method, stable for one week at room temperature [61]. For oral administration purposes, the in vitro drug release studies were conducted in a simulated gastric fluid with pepsin (SGF, pH 1.2), a simulated intestinal fluid with trypsin (SIF, pH 6.8), and PBS at physiological pH (7.4). All conditions gave an almost similar release pattern with a burst release (25%) in the initial 1 h followed by a sustained release reaching 90–94% at 48 h. DiO and DiI fluorophores were loaded into the NP to form FRET-NPs of similar sizes concerning SA-NPs and were stable for more than four weeks, which is required for cellular uptake studies. MDCK cells were incubated with the fluorescent NPs for 2 h and only a moderately decrease of the FRET signal was observed, indicating a slow intracellular release. The epithelial transport of NPs was also assessed with the same FRET-NPs and showed that a fraction of intact NPs was transported from the apical to the basolateral side. In vivo biodistribution studies in adult zebrafish after oral administration for 2 h showed a fast absorption of NPs into the blood circulation before reaching the brain. Similarly, after oral administration of SA-NPs to rats, they directly reached the systemic circulation with a decreased SA elimination rate in plasma (vs. control) and even more in the brain, suggesting specific brain targeting properties of this system. Finally, the neuroprotective effect of SA-NPs was evaluated using an MPTP-induced zebrafish model of PD; a decrease of the MPTP-induced neurotoxicity (vs. control), involving the Akt/Gsk3β signaling pathway activation, was observed.

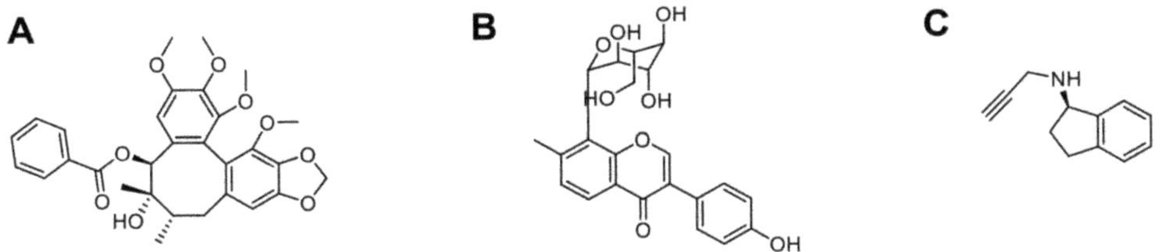

Figure 4. Chemical structures of the three described compounds encapsulated into PLGA NPs for PD treatment (Schisantherin A (**A**), Puerarin (**B**), and Rasagiline (**C**)).

3.2.2. Puerarin

Puerarin (PU) is a flavonoid glycoside, widely used in China, extracted from the root of the leguminous plant, Pueraria lobata, and Thomson Kudzuvine Root (Figure 4). Puerarin displays a series of beneficial activities notably neuroprotection, antioxidant, etc. [62]. In 2019, Chen et al., described the development of PU-loaded PLGA NPs to treat PD while enhancing half-life (pharmacokinetic study in rats) and bioavailability of the drug [63]. Homogeneous suspensions (ø: 88.4 ± 1.7 nm and PdI: 0.047 ± 0.007) of NPs were prepared by nanoprecipitation. Cytotoxicity and internalization of this formulation were investigated in MDCK cells after 24 h and 48 h. Overall, PU-loaded PLGA NPs exhibited no significant toxicity, and the uptake was confirmed. For oral administration purposes, the in vivo drug release studies were conducted in adult zebrafish; an effective brain uptake was observed using FRET-NPs (DiO and DiI fluorophores). In a dose-dependent manner (from 10 to 50 µM), PU NPs prevented cell death (SH-SY5Y cells) by decreasing MPP+ neurotoxicity and mitochondrial oxidative stress. Finally, PU-loaded NPs improved behavioral deficits, TH-positive neuron loss and associated mobility impairments (i.e., bradykinesia, fall latency, number of falls, average travel distance or speed), and restored dopamine level and MPTP-mediated neurotoxicity in mice.

3.2.3. Rasagiline

Rasagiline (RA) is a selective irreversible second-generation MAO-B inhibitor, which can be used both as monotherapy and as co-adjuvant therapy in combination with levodopa to treat PD (Figure 4) [64,65]. RA efficacy was evaluated using the Unified PD Rating Scale (UPDR). It showed that patients taking the lowest dose at the earliest possible time had slightly better benefic effects at 18 months than patients who received this dose only at 9 months [66]. To overcome poor solubility issues, low bioavailability and half-life, Bali and Salve designed transdermal films of gellan gum on which RA mesylate-loaded PLGA NPs were embedded [67]. This administration route was a novel approach for the sustained release of NPs. The nano-objects were prepared by double emulsion solvent evaporation technique and characterized by DLS (ø: 221.7 ± 5.7 nm) and field emission scanning electron microscopy (FE-SEM). A relatively high PdI (0.388 ± 0.86) was observed, suggesting a heterogeneous suspension. An in vivo study in Wistar rats highlighted an enhanced brain bioavailability using RA-PLGA NPs. Transdermal films slowed down the initial drug release rate followed by a constant release for more than 72 h compared to intravenous or oral administration. The sustained release of RA restored presynaptic depletion of dopamine, inhibited MAO-B enzyme, and prevented neuronal damage caused by oxidative stress in PD model. Finally, gamma scintigraphy study was performed to evaluate BBB permeation and targeting efficiency using ^{99m}TC-loaded PLGA NPs embedded transdermal film. It was observed that PLGA NPs are targeted from dorsal part toward the brain region of rats. Overall, this study appeared promising for the development of an easy-to-administer DDS able to cross the BBB and target the brain.

3.3. Huntington's Disease

Described in 1872 by George Huntington, HD is a genetic and hereditary disease, due to the mutation of the gene IT15, resulting in the repeat expansion of CAG trinucleotide, coding for the huntingtin protein (mHtt) [68]. Symptoms of HD often appear between the ages of 30 and 50 years old with the onset of progressive motor (involuntary jerking/writhing movements known as chorea, rigidity, muscle contracture, etc.), behavioral (dementia, difficulty of organization and learning) and psychiatric disorders (feelings of irritability, insomnia) [69]. These troubles are associated with the death of striatal neurons in the cortex, the striatum hippocampus and other brain regions such as the thalamus, subthalamic nucleus, and substantia nigra pars reticulate in advanced cases of HD [70,71]. The mutation of the gene IT15 leads to an accumulation of mHtt proteins and formation of toxic neuronal intranuclear inclusions (NIIs). The only drug approved by the regulatory agencies is the tetrabenazine (TBZ) for the symptomatic relief [72]. Research in the area is thus very interesting to expand the therapeutic arsenal, especially via the PLGA NP approach (Table 2).

3.3.1. Oligonucleotides QBP1, NT17, and PGQ$_9$P^2

Published in 2019, Joshi et al., designed peptide-loaded PLGA NPs coated with polysorbate 80 to act on mHtt aggregation. QBP1, NT17, and PGQ$_9$P^2 are oligonucleotides exhibiting great inhibitory potential against mutant huntingtin protein aggregation [73,74]. In this study, nanoprecipitation peptide-loaded PLGA NPs were prepared to obtain NPs from 158 to 180 nm in diameter with a low PdI (<0.100) and a ζ potential between −23.3 and −27.5 mV. Polysorbate 80 coating enabled an active targeting to the brain, while improving peptide bioavailability and half-life. Overall, in vitro studies performed using Madin-Darby canine kidney (MDCK) cells suggested that transcytosis was facilitated by absorption of apolipoprotein E (ApoE) and low-density lipoprotein (LDL) receptors. Neutral red PLGA NPs were used to conduct in vivo study in healthy mice and confirmed the higher concentration in the brain of polysorbate 80 coated NPs compared to uncoated NPs. The peptide-loaded PLGA NPs displayed dose dependent inhibition of mHtt aggregation in both Neuro 2A and PC12 cell models. Finally, using Drosophila model of HD, their ability to improve motor performance was evaluated; after feeding larvae and adult flies with high NT$_{17}$ and QBP1 concentrations; higher crawling and climbing activities were observed, compared to the control group.

3.3.2. Cholesterol

Cholesterol is a lipid of the sterol family produced by cells and found in food. It is an essential element for synthesizing many hormones, neurotransmitter release, membrane structure, vesicle assembly and fusion. Recently, cholesterol metabolism in the central nervous system of HD patients was described as significantly altered, leading to brain malformations and impaired cognitive functions (Figure 5) [75,76]. A study by Valenza et al., reported the delivery of cholesterol using PLGA NPs to treat HD [77]. Homogenous and monomodal suspensions (ø: 200–300 nm, PdI: 0.09–0.3, dose-dependent) of cholesterol-loaded NPs were prepared with various cholesterol concentrations (1, 5, to 10 µM). In vitro studies showed an initial burst release during the first days followed by a constant release (5–10 days), and an internalization into neuron and astrocyte models of HD. In vivo assays in R6/2 mice highlighted that cholesterol PLGA NPs crossed the BBB and restored electrophysical phenotypes such as synaptic transmission in striatal medium-sized spiny neuron impairment. Finally, cholesterol supplementation did not significantly improve behavioral issues (locomotion, resting time, rearing) but slowed down the disease progression. Using a novel object recognition test for cognitive task evaluation, the cholesterol treated mice group appeared to benefit from the drug with resolved memory deficits.

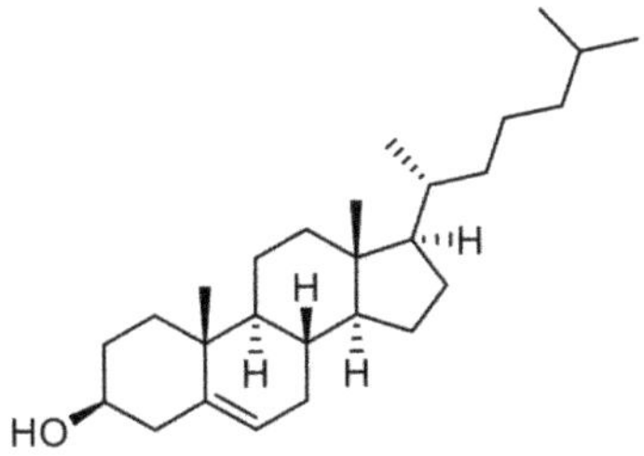

Figure 5. Chemical structure of cholesterol encapsulated into PLGA NPs for HD treatment.

Even if the therapeutic arsenal for HD treatment is limited to TBZ, the current studies open the way to new promising tools.

3.4. Amyotrophic Lateral Sclerosis

ALS, also known as Lou Gehrig's disease, is a NDD involving the motor neurons (motoneurons) located in the anterior horn of the spinal cord and the motor nuclei of the last cranial nerves. ALS results in a progressive paralysis of the muscles responsible for voluntary motor skills. The symptoms go from muscle twitching, cramping, stiffness, weakness, involuntary jerking movements, tremors, to phonation (the production of sounds) and swallowing [78]. No curative treatment is available, and only riluzole is used to delay the disease evolution [79]. New developments of therapeutics for ALS treatment would be very beneficial (Table 2).

PHA-767491

PHA-767491 (PHA) is the first cell division cycle seven kinase inhibitor described as a molecule preventing neurodegeneration, especially in the treatment of ALS (Figure 6) [80]. PHA prevents phosphorylation by CDC7 of TDP-43, a nuclear protein encoded by TARDBP gene-regulating several RNA processes as transcription, mRNA transport, and microRNA biosynthesis. To address low permeability, rapid metabolism, and unspecific distribution issues, Rojas-Prats et al., used PHA-loaded PLGA NPs [81]. They were prepared by nanoprecipitation method and observed by SEM microscopy using gold-coated NPs. The obtained nano-objects exhibited a 141–155 nm diameter with a low PdI ($\leq$0.15) characteristic of a monodisperse suspension. Encapsulation of PHA was controlled and confirmed by imaging PHA fluorescence using confocal laser scanning microscopy. An encapsulation efficiency of 12–18% and a drug loading of 2–4% were determined by HPLC. Using porcine lipid to emulate the human BBB, PHA crossing was enhanced, being loaded in the PLGA NPs compared to free PHA. Finally, in vitro assays, in the SH-SY5Y cell model of TDP-43 phosphorylation, highlighted a significant protection of neuronal cells from death and decreased TDP-43 phosphorylation.

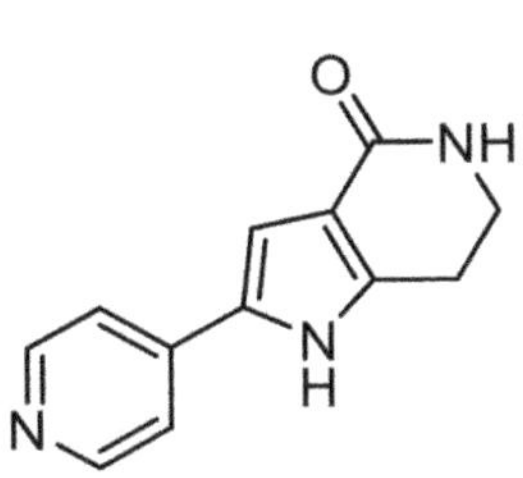

Figure 6. Chemical structure of PHA-767491 encapsulated into PLGA NPs for ALS treatment.

3.5. Multiple Sclerosis

Multiple sclerosis (MS) is a chronic autoimmune inflammatory disease and NDD that attacks the CNS, particularly the brain, nerves, and spinal cord [82]. It involves various pathogenic mechanisms leading to inflammatory infiltrates (T cells, B cells, macrophages) and alteration of the myelin, which forms a protective sheath around the nerve endings, and thus of nerve impulse transmission. Symptoms vary depending on the localization of the attacks: limb numbness, vision troubles, electric shock sensations in a limb or the back, movement dysfunctions, etc. Nowadays, over a dozen therapeutic agents reducing the number of attacks and delaying MS progression are approved by the regulatory agencies. A few of them may be vectorized by PLGA NPs to treat some MS symptoms (Table 2) [83].

3.5.1. Interferon-Beta-1a

Interferon-beta-1a (IFN β-1a) is a naturally occurring protein produced by various cell types including fibroblasts and macrophages. Binding specific receptors exert its biological effect on the human cell surface to slow down MS progression [84]. To decrease drug dosage and improve its controlled release, Fodor-Kardos et al., designed IFN β-1a-loaded PEG-PLGA NPs by double emulsion solvent evaporation method with great encapsulation efficiency (95.9%) [85]. According to the used technic: 165 nm (PdI: 0.093) by DLS or 50–120 nm by SEM, differences in the NP size were observed by SEM, linked to the hydration of the sample. BSA loaded NPs were also prepared as a model for release kinetics comparison: a significant slower release was observed in the case of PEG-PLGA NPs compared to simple PLGA NPs. Cyanine 5 amine fluorescent dye was conjugated to the NPs to confirm the cellular uptake and the absence of cytotoxicity in vitro in hepatocytes. Finally, in vivo toxicity studies were performed on healthy Wistar male rats. In the first step, the loaded NPs exerted some toxicity due to the fabrication process and potential remains of solvent, while the unloaded ones were safe to use. In the second step, the formulation process was modified to nanoprecipitation, however mild toxic signals were observed for blank PLGA and PEG-PLGA NPs. To go further, a new formulation process will be required to obtain IFN β-1a-loaded PEG-PLGA NPs without or with low in vivo toxicity allowing to investigate the neuroprotective effect of the system.

3.5.2. Leukemia Inhibitory Factor

Leukemia Inhibitory Factor (LIF) is a neural stem cell growth factor belonging to the interleukin-6 family of cytokines. LIF plays an important role in different biological processes, including differentiation of leukemia cells, inflammatory response, neuronal development, etc. [86]. Recent evidence has demonstrated a crucial role of LIF in neuroprotection, axonal regeneration as well as the prevention of demyelination [87]. Preliminary studies to design and prepared PLGA-NG2 NPs, by double emulsion solvent preparation method, as carrier for LIF were performed by Rittchen et al. [88]. Anti-NG2 antibodies, targeting oligodendrocyte precursor cells (OPCs), were attached on the avidin coated NPs. LIF-loaded PLGA NPs were characterized by scanning electron microscopy on a size (126 ± 50 nm) and morphological levels before performing biological assays. In vitro studies attested of the targeting efficiency toward OPCs, and of LIF activity. To evaluate the CNS remyelination properties of NG2-targeted LIF-NPs, in vivo assays were performed on a mouse brain model having received a stereotaxic injection of myelin toxin, lysophosphatidylcholine, into the Corpus callosum. Remyelination was quantified using electron microscopy by measuring the percentage of myelinated axons and myelin thickness. Overall, NG2-targeted LIF-PLGA NPs induced maturation of OPCs to myelin-competent oligodendrocytes within three days and promoted high-quality myelin repair.

3.5.3. Proteolipid Protein (PLP139–151)

In 2020, Ferreira Lima et al., designed antigen (PLP139–151) loaded PLGA NPs using the double emulsion solvent evaporation method to induce antigen-specific tolerance, regulate immune responses, as well as withstand multiple immunogenic challenges in a MS model. Indeed, this protein was identified as a potent tolerogenic molecule in MS disease. Immunization with this peptide induced experimental autoimmune encephalomyelitis (EAE) in an animal model of MS. Various PLP loaded NPs were prepared with an average diameter of 195.1 $\pm$ 10.1 nm for a drug loading of 20%. However, a low encapsulation efficiency (over 60% of PLP not encapsulated) was observed, as well as a complete drug release within the first hours requiring formulation improvements. NPs were incorporated into polymeric microneedles to design a minimal invasive administration system. However, no biological evaluation of the neuroprotective properties of PLP encapsulated PLGA NPs were performed, requiring further studies to confirm their interest.

Table 2. Non-exhaustive summary of neuroprotective drugs vectorized by the PLGA NP approach for NDD treatment (D.L.: Drug Loading, E.E.: Encapsulation Efficiency).

Active Substance	Encapsulation Method	NP Properties	Model of Use	Neuroprotective Activities	Ref.
			Alzheimer's Disease		
Rosmarinic acid	Emulsion/solvent diffusion	ø: 60 to 80 nm PdI: 0.047 ζ potential: 2 to 8 mV E.E.: 25 to 40%	In vitro: HBMECs, Has and Aβ-insulted SK-N-MC cells	Eliminate peroxynitrite anions Reduce inflammatory responses Disrupt the integrity of HBMEC cytoskeletal microfibrils Decrease the structural density of TJ Increase the permeability and reduce the TEER by targeting HBMECs with the use of 83-14 Mab	[89]
Curcumin *Curcuma longa*	Single emulsion/solvent evaporation	ø: 150 to 200 nm ζ Potential: −30 to −20 mV	In vitro: GI-1 glioma cells	Inhibit Aβ fibril formation Reduce brain amyloid level and plaque burden Free-radical scavenging activity	[90]
Rhynchophylline *Uncaria rhynchophylla*	Nanoprecipitation	ø: 145.2 nm PdI: 0.133 D.L.: 10.3% E.E.: 60%	In vitro: bEnd.3 and PC12 cells In vivo: healthy C57BL/6 mice	Inhibit soluble Aβ-induced hyperexcitability of hippocampal neurons Reduce cell apoptosis significantly	[58]
Quecertin	Double emulsion/solvent evaporation	ø < 145 nm	In vitro: human neuronal cells (SH-SY5Y) In vivo: BALB/c mice	Inhibit AChE and secretase enzymes Disrupt amyloid aggregates Prevent and inhibit Aβ_{1-42} fibrils formation Dissolve Aβ_{1-42} aggregates	[91]
iA β5	Nanoprecipitation	ø: 153 to 166 nm PdI: 0.090 to 0.100 ζ potential: −10.1 to −13 mV E.E.: 63 ± 9%	Porcine brain capillary endothelial cells	N/A	[92]
Phytol	Emulsion/solvent evaporation	ø: 177.4 ± 5.9 nm ζ potential: −32.8 ± 2.2 mV D.L.: 56% E.E.: 92%	In vitro: Neuro2a cells In vivo: healthy male Wistar rats, rats injected with scopolamine	Anti-cholinesterase and anti-oxidative Disrupt amyloid aggregates Attenuate the impairment of learning and memory induced by scopolamine in rat brain Prevent oxidation of proteins and lipids against scopolamine-induced oxidative damage	[93,94]

256

Table 2. *Cont.*

Active Substance	Encapsulation Method	NP Properties	Model of Use	Neuroprotective Activities	Ref.
Alzheimer's Disease					
Thymoquinone	Single emulsion/solvent evaporation	ø: 226.3 ± 4.6 nm PdI: 0.143 ζ potential: −45.7 ± 2.6 mV E.E.: 69.5 ± 3.0%	In vivo: STZ-induced Alzheimer model mice	Act on neurosis, ROS formation, free radical scavenging capacity, etc. Prevent Aβ aggregates and hyperphosphorylated τ-protein tangles	[53]
Nigella sativa oil and Plasmid DNA	Modified solvent diffusion	ø: 600 to 700 nm PdI: 0.229 ζ potential: + 35.6 mV D.L.: 13.0% E.E.: 73.8%	In vitro: murine neuroblastoma (N2a) cells	Inhibit radicals O_2 Promote neurite outgrowth and neuroregeneration	[95]
Anthocyanins Natural pigments	Emulsion/solvent evaporation	ø: 165 nm PdI: 0.400 ζ potential: −12 mV E.E.: 60%	In vitro: human neuronal cells (SH-SY5Y)	Inhibit $A\beta_{1-42}$-induced ROS generation Anti-oxidant	[96]
Huperzine A	Emulsion/solvent evaporation	ø: 78.1 to 153.2 nm PdI: 0.182 to 0.229 ζ potential: −21.2 to +35.6 mV D.L.: 5 to 25% E.E.: 83.2 to 73.8%	In vitro: 16HBE and SH-SY5Y cells In vivo: KM mice	Inhibit acetylcholinesterase Disrupt amyloid aggregates	[55]
Selegiline Methamphetamine derivative	Emulsion/solvent evaporation	ø: 217 to 246 nm PdI: 0.295 to 0.331 ζ potential: 34.7 to 38.0 mV E.E.: 3.3 to 11.7%	In vitro: fibrils ($fA\beta_{1-40}$ and $fA\beta_{1-42}$)	Inhibit MAO-B and Aβ fibrils formation Increase the destabilizing effects of the drug by increasing selegiline concentration and incubation time	[97]
Parkinson's Disease					
Schisantherin A	Flash nanoprecipitation	ø: 70.6 ± 2.2 nm PdI: 0.104 ± 0.012 ζ potential: −24.7 ± 3.5 mV D.L.: 28.0 ± 0.8% E.E.: 91.1 ± 2.6%	In vitro: SH-SY5Y and MDCK cells In vivo: larval zebrafish, MPTP-induced zebrafish and male Sprague-Dawley rats	Reduce tyrosine hydroxylase TH+ DA neuronal death Prevent MPTP-induced decrease in TH+ region Reverse locomotor deficiency Stronger neuroprotective effects in MPP+ -induced SH-SY5Y cell injury model	[98]

Table 2. *Cont.*

Active Substance	Encapsulation Method	NP Properties	Model of Use	Neuroprotective Activities	Ref.
Alzheimer's Disease					
Puearin	Antisolvent precipitation	ø: 88.4 ± 1.7 nm PdI: 0.047 ± 0.007 ζ potential: −18.9 ± 2.8 mV D.L.: 43.0 ± 1.6% E.E.: 89.5 ± 1.7%	In vitro: MDCK and SH-SY5Y cells In vivo: zebrafish and their embryos, male Sprague−Dawley (SD) rats, male C57BL/6 mice (with MPTP group)	Reduce tyrosine hydroxylase TH+ DA neuronal death Reduce behavioral deficits and associated mobility impairments in MPTP-mediated neurotoxicity in mice Reduce TH+ neuron loss and associated neurotoxicity	[63]
Levodopa	Double emulsion/solvent evaporation	ø: 329.0 to 383.7 nm PdI: 0.384 to 0.426 ζ potential: −4.5 to −20.8 mV E.E.: 50.5 to 73.0%	In vitro: PC12 cells In vivo: healthy CD57/BL6 mice (with MPTP group)	Improve locomotor activity over time Improve brain delivery by intranasal route Enhance dopamine concentration in the brain compared to the blood	[99]
Rasagiline	Double emulsion/solvent evaporation	ø: 221.7 ± 5.7 nm PdI: 0.388 ± 0.860 ζ potential: −36.1 ± 4.4 mV E.E.: 29.2 ± 1.8%	In vivo: healthy and Parkinson Wistar rats	Inhibit MAO-B enzyme Prevent neuronal damage caused by oxidative stress	[67]
Huntington's Disease					
Peptides QBP1, NT17, and PGQ$_9$P^2	Nanoprecipitation	ø: 158 to 180 nm PdI: 0.031 to 0.066 ζ potential: −23.3 to −27.5 mV	In vitro: MDCK, Neuro 2A and PC12 cell model of HD In vivo: healthy mice, Drosophila, larvae, and adult fly model of HD	Inhibit mHtt aggregation Restore motor activity	[74]
Cholesterol	Nanoprecipitation	ø: 200 to 300 nm PdI: 0.090 to 0.300 ζ potential: −8 to −12 mV D.L.: 0.7 to 2.5% E.E.: 48 to 68%	In vitro: neural stem (NS) cell model of HD In vivo: genotyping of R6/2 mouse	Restore synaptic alterations and delaying cognitive	[77]
Amyotrophic Lateral Sclerosis					

Table 2. *Cont.*

Active Substance	Encapsulation Method	NP Properties	Model of Use	Neuroprotective Activities	Ref.
Alzheimer's Disease					
PHA-767491	Nanoprecipitation	ø: 141 to 155 nm PdI ≤ 0.150 D.L.: 24% E.E.: 12 to 18%	In vitro: SH-SY5Y cell model of ALS	Reduce TDP-43 phosphorylation	[81]
Multiple Sclerosis					
Interferon-β-1a	Double emulsion/solvent evaporation	ø: 150 to 165 nm PdI: 0.081 to 0.093 ζ potential: 17.7 to 18.8 mV D.L.: 0.7 to 2.5% E.E.: 95.9%	In vitro: human blood plasma, primary hepatocytes from male Wistar rats In vivo: Wistar male rats	N/A	[85]
LIF	Double emulsion/solvent evaporation	ø: 126 ± 50 nm	In vitro: oligodendrocyte precursor cells from neonatal Sprague–Dawley rats	Induce maturation of OPCs to myelin-competent oligodendrocytes within 3 days Promote high-quality myelin repair	[88]
Proteolipid Protein	Double emulsion/solvent evaporation	ø: 195.1 ± 10.1 nm PdI: 0.220 ζ potential: −20 mV D.L.: 20% E.E.: 40%	N/A	N/A	[100]

4. New Candidates for PLGA NPs Vectorization

Many active substances with neuroprotective activities remain to be evaluated for the PLGA NP approach to improve their brain delivery [101–115]. Some of them, such as trehalose, exhort numerous properties ranging from anti-inflammatory to anti-oxidant, which can potentially be applied for out of current applications and NDDs treatment. Trehalose is a non-reducing disaccharide that has been widely used in the food industry (Figure 7). Still, it has also recently shown many unique properties, indicating its potential use in preventing neurodegeneration [105]. It may act as a potent stabilizer of proteins, preserving protein structural integrity and, as an mTOR-independent autophagy inducer, improving the clearance of the mutant proteins that act as autophagy substrates when aberrant protein deposition occurs. Trehalose treatment reduced the level of toxic protein aggregates, which in turn improve behavioral symptoms and survival in animal models of NDDs including AD [106], PD [107,108], and HD [109]. However, trehalose can be cleaved by trehalase, an enzyme observed in different mammalian organs [110]. Therefore, PLGA NPs could be interesting candidates to protect trehalose from degradation and improve its bioavailability.

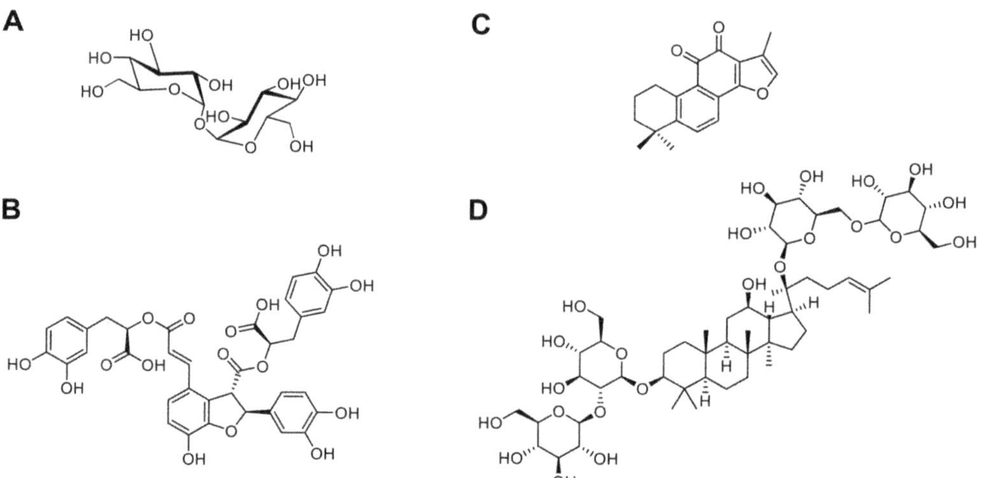

Figure 7. Chemical structures of the four described compounds for AD and PD treatment (Trehalose (**A**), Salvianolic acid B (**B**), Tanshinone IIA (**C**), and Panax notoginsenoside Rb1 (**D**)).

Another good example is the association of salvianolic acid B (Sal B), tanshinone IIA (TS IIA), and panax notoginsenoside (PNS) (Figure 7), which demonstrated neuroprotective effects on cerebral ischemia reperfusion (I/R) injury. In 2013, Zhang and coworkers encapsulated them into PLGA NPs for brain delivery via the intratympanic route (inner ear administration) [111]. Homogeneous suspensions were obtained by double emulsion solvent evaporation method and composed of particles with a mean diameter of 154 nm. In vitro, loaded PLGA NPs exhibited an excellent sustained release of 76% and 93% of the drug after 72 h for Sal B and PNS, respectively. A slower release was observed for TS IIA (23.6% after 72 h), explained by the drug low solubility in the medium. Pharmacokinetic studies revealed that PLGA NPs improved drug distribution within the inner ear, the CSF, and the brain after intratympanic administration demonstrating the usefulness of this route to deliver drugs without BBB crossing. The anti-oxidant activity of such nanosystems was then evaluated since recent studies confirmed that oxidative stress (SOD, MDA, NOS, ROS and NO) plays a role in the pathogenesis of I/R injury. The lipid peroxidation evaluation showed that drug-loaded PLGA NPs exhibited neuroprotection by regulation of SOD,

MDA, and NOS levels, especially a decrease of MDA and NOS levels validating the antioxidant properties of Sal B, TS IIA, and PNS. Combining these compounds with PLGA NPs may be a very interesting new tool for brain disease treatment, such as AD and PD thanks to their neuroprotective activity. Indeed, Young Woo Lee and coworkers demonstrated that Sal B improved memory impairment caused by $A\beta_{25-35}$ peptide-induced neuronal damage in AD [103] while Zeng et al., published its ability to protect cells against MPP-induced apoptosis for PD treatment [112]. TS IIA inhibited transcription and translation of iNOS, MMP-2, and NF-κBp65, and suppressed expression of NADPH oxidase and iNOS, respectively, involved in AD [113] and PD [114] development. Finally, in AD, PNS Rb1 suppressed the phosphorylated tau protein expression and upregulated the expression levels of brain-derived neurotrophic factor (BDNF) [115].

5. Conclusions

BBB crossing or the potential degradation of neuroprotective substances remain strategic points needing to be overcome for NDD treatments. In this context, PLGA NPs are a promising approach for new therapeutics. Being both biocompatible and biodegradable, they allow for drug protection, enhance their bioavailability, and possess active targeting capabilities facilitating the delivery to a precise site. These beneficial effects were used for active substance administration (i.e., thymoquinone, curcumin, etc.) and delivery directly into the brain without losing their neuroprotective properties. Moreover, they also demonstrated therapeutic activities such as restoration of the lysosomal pH in the case of PD. Some challenges remain regarding the biological (i.e., biological barrier breaching), technological (i.e., scale-up synthesis), and study design (i.e., data treatment) levels, which impact the potential clinical trials of PLGA NPs. They need to be addressed for the successful development of such therapeutics [116]. However, new guidelines are established to facilitate the transposition from bench to clinic and ensure the product quality and safety [21,117].

Even if these researches are often in the early stages of development, nanovectorization of neuroprotective substances announce future great advances in the pharmaceutical field for NDD treatment.

Author Contributions: A.C. prepared and wrote the original draft. A.G., L.L. and B.D. collectively corrected the draft of the Review, including text and figures. All authors have read and agreed to the published version of the manuscript.

Funding: This work was supported by Fondation de France, grant number 00066525 and by IDEX Emergence Grant number OPE-2018-410 (B.D.). A.C. is a recipient of a MSER fellowship (France). We thank Le collège des écoles doctorales (appel doctorat interdisciplinaire 2018) for Financial support.

Institutional Review Board Statement: Not applicable.

Informed Consent Statement: Not applicable.

Data Availability Statement: Not applicable.

Acknowledgments: We apologize to the authors of several high-quality scientific articles that contributed significantly to the development of the field, which could not be cited due to space limits.

Conflicts of Interest: The authors declare no conflict of interest.

References

1. Scott, A.M. Global Burden of Neuropsychiatric Disorders. *Alzheimer's Dis. Int. World's Alzheimer's Rep.* **2017**, *2015*, 10.
2. Feigin, V.L.; Krishnamurthi, R.V.; Theadom, A.M.; Abajobir, A.A.; Mishra, S.R.; Ahmed, M.B.; Abate, K.H.; Mengistie, M.A.; Wakayo, T.; Abd-Allah, F.; et al. Global, regional, and national burden of neurological disorders during 1990–2015: A systematic analysis for the Global Burden of Disease Study 2015. *Lancet Neurol.* **2017**, *16*, 877–897. [CrossRef]
3. Furuse, M. Molecular Basis of the Core Structure of Tight Junctions. *Cold Spring Harb. Perspect. Biol.* **2010**, *2*, a002907. [CrossRef]
4. Pardridge, W.M. Drug transport across the blood-brain barrier. *J. Cereb. Blood Flow Metab.* **2012**, *32*, 1959–1972. [CrossRef] [PubMed]

5. Nichols, E.; Szoeke, C.E.I.; Vollset, S.E.; Abbasi, N.; Abd-Allah, F.; Abdela, J.; Aichour, M.T.E.; Akinyemi, R.O.; Alahdab, F.; Asgedom, S.W.; et al. Global, regional, and national burden of Alzheimer's disease and other dementias, 1990–2016: A systematic analysis for the Global Burden of Disease Study 2016. *Lancet Neurol.* **2019**, *18*, 88–106. [CrossRef]

6. Tysnes, O.-B.; Storstein, A. Epidemiology of Parkinson's disease. *J. Neural Transm.* **2017**, *124*, 901–905. [CrossRef] [PubMed]

7. Baig, S.S.; Strong, M.; Quarrell, O.W. The global prevalence of Huntington's disease: A systematic review and discussion. *Neurodegener. Dis. Manag.* **2016**, *6*, 331–343. [CrossRef]

8. Xu, L.; Liu, T.; Liu, L.; Yao, X.; Chen, L.; Fan, D.; Zhan, S.; Wang, S. Global variation in prevalence and incidence of amyotrophic lateral sclerosis: A systematic review and meta-analysis. *J. Neurol.* **2020**, *267*, 944–953. [CrossRef] [PubMed]

9. Walton, C.; King, R.; Rechtman, L.; Kaye, W.; Leray, E.; Marrie, R.A.; Robertson, N.; La Rocca, N.; Uitdehaag, B.; van der Mei, I.; et al. Rising prevalence of multiple sclerosis worldwide: Insights from the Atlas of MS, third edition. *Mult. Scler. J.* **2020**, *26*, 1816–1821. [CrossRef] [PubMed]

10. Naqvi, S.; Panghal, A.; Flora, S.J.S. Nanotechnology: A Promising Approach for Delivery of Neuroprotective Drugs. *Front. Neurosci.* **2020**, *14*, 1–26. [CrossRef]

11. Brambilla, D.; Le Droumaguet, B.; Nicolas, J.; Hashemi, S.H.; Wu, L.P.; Moghimi, S.M.; Couvreur, P.; Andrieux, K. Nanotechnologies for Alzheimer's disease: Diagnosis, therapy, and safety issues. *Nanomed. Nanotechnol. Biol. Med.* **2011**, *7*, 521–540. [CrossRef]

12. Leyva-Gómez, G.; Cortés, H.; Magaña, J.J.; Leyva-García, N.; Quintanar-Guerrero, D.; Florán, B. Nanoparticle technology for treatment of Parkinson's disease: The role of surface phenomena in reaching the brain. *Drug Discov. Today* **2015**, *20*, 824–837. [CrossRef] [PubMed]

13. Bourdenx, M.; Daniel, J.; Genin, E.; Soria, F.N.; Blanchard-Desce, M.; Bezard, E.; Dehay, B. Nanoparticles restore lysosomal acidification defects: Implications for Parkinson and other lysosomal-related diseases. *Autophagy* **2016**, *12*, 472–483. [CrossRef] [PubMed]

14. Essa, D.; Kondiah, P.P.D.; Choonara, Y.E.; Pillay, V. The Design of Poly(lactide-co-glycolide) Nanocarriers for Medical Applications. *Front. Bioeng. Biotechnol.* **2020**, *8*, 1–20. [CrossRef] [PubMed]

15. Operti, M.C.; Bernhardt, A.; Grimm, S.; Engel, A.; Figdor, C.G.; Tagit, O. PLGA-based nanomedicines manufacturing: Technologies overview and challenges in industrial scale-up. *Int. J. Pharm.* **2021**, *605*, 120807. [CrossRef]

16. Cano, A.; Sánchez-López, E.; Ettcheto, M.; López-Machado, A.; Espina, M.; Souto, E.B.; Galindo, R.; Camins, A.; García, M.L.; Turowski, P. Current advances in the development of novel polymeric nanoparticles for the treatment of neurodegenerative diseases. *Nanomedicine* **2020**, *15*, 1239–1261. [CrossRef]

17. Makadia, H.K.; Siegel, S.J. Poly Lactic-co-Glycolic Acid (PLGA) as biodegradable controlled drug delivery carrier. *Polymers* **2011**, *3*, 1377–1397. [CrossRef]

18. Anderson, J.M.; Shive, M.S. Biodegradation and biocompatibility of PLA and PLGA microspheres. *Adv. Drug Deliv. Rev.* **1997**, *28*, 5–24. [CrossRef]

19. Danhier, F.; Ansorena, E.; Silva, J.M.; Coco, R.; Le Breton, A.; Préat, V. PLGA-based nanoparticles: An overview of biomedical applications. *J. Control. Release* **2012**, *161*, 505–522. [CrossRef]

20. Zeng, J.; Martin, A.; Han, X.; Shirihai, O.S.; Grinstaff, M.W. Biodegradable plga nanoparticles restore lysosomal acidity and protect neural pc-12 cells against mitochondrial toxicity. *Ind. Eng. Chem. Res.* **2019**, *58*, 13910–13917. [CrossRef]

21. Zhi, K.; Raji, B.; Nookala, A.R.; Khan, M.M.; Nguyen, X.H.; Sakshi, S.; Pourmotabbed, T.; Yallapu, M.M.; Kochat, H.; Tadrous, E.; et al. Plga nanoparticle-based formulations to cross the blood-brain barrier for drug delivery: From r&d to cgmp. *Pharmaceutics* **2021**, *13*, 500.

22. McCall, R.L.; Sirianni, R.W. PLGA nanoparticles formed by single- or double-emulsion with vitamin E-TPGS. *J. Vis. Exp.* **2013**, *82*, 51015. [CrossRef] [PubMed]

23. Huang, W.; Zhang, C. Tuning the Size of Poly(lactic-co-glycolic Acid) (PLGA) Nanoparticles Fabricated by Nanoprecipitation. *Biotechnol. J.* **2018**, *13*, 1700203. [CrossRef] [PubMed]

24. Ryu, S.; Park, S.; Lee, H.Y.; Lee, H.; Cho, C.W.; Baek, J.S. Biodegradable Nanoparticles-Loaded PLGA Microcapsule for the Enhanced Encapsulation Efficiency and Controlled Release of Hydrophilic Drug. *Int. J. Mol. Sci.* **2021**, *22*, 2792. [CrossRef]

25. Moghimi, S.M.; Moghimi, S.M.; Hunter, A.C.; Murray, J.C. Long-circulating and target-specific nanoparticles: Theory to practice. *Pharmacol. Rev.* **2001**, *53*, 283–318.

26. Hoshyar, N.; Gray, S.; Han, H.; Bao, G. The effect of nanoparticle size on in vivo pharmacokinetics and cellular interaction. *Nanomedicine* **2016**, *11*, 673–692. [CrossRef] [PubMed]

27. Danaei, M.; Dehghankhold, M.; Ataei, S.; Hasanzadeh Davarani, F.; Javanmard, R.; Dokhani, A.; Khorasani, S.; Mozafari, M. Impact of Particle Size and Polydispersity Index on the Clinical Applications of Lipidic Nanocarrier Systems. *Pharmaceutics* **2018**, *10*, 57. [CrossRef]

28. Kumar, A.; Dixit, C.K. Methods for characterization of nanoparticles. In *Advances in Nanomedicine for the Delivery of Therapeutic Nucleic Acids*; Elsevier: Amsterdam, The Netherlands, 2017; pp. 43–58. ISBN 9780081005637.

29. Fröhlich, E. The role of surface charge in cellular uptake and cytotoxicity of medical nanoparticles. *Int. J. Nanomed.* **2012**, *7*, 5577. [CrossRef] [PubMed]

30. Mulvihill, J.J.; Cunnane, E.M.; Ross, A.M.; Duskey, J.T.; Tosi, G.; Grabrucker, A.M. Drug delivery across the blood-brain barrier: Recent advances in the use of nanocarriers. *Nanomedicine* **2020**, *15*, 205–214. [CrossRef]

31. Wünsch, A.; Mulac, D.; Langer, K. Lecithin coating as universal stabilization and functionalization strategy for nanosized drug carriers to overcome the blood–brain barrier. *Int. J. Pharm.* **2021**, *593*, 120146. [CrossRef]

32. Yoo, J.; Park, C.; Yi, G.; Lee, D.; Koo, H. Active targeting strategies using biological ligands for nanoparticle drug delivery systems. *Cancers* **2019**, *11*, 640. [CrossRef]

33. Ray, S.; Sinha, P.; Laha, B.; Maiti, S.; Bhattacharyya, U.K.; Nayak, A.K. Polysorbate 80 coated crosslinked chitosan nanoparticles of ropinirole hydrochloride for brain targeting. *J. Drug Deliv. Sci. Technol.* **2018**, *48*, 21–29. [CrossRef]

34. Joshi, A.S.; Gahane, A.; Thakur, A.K. Deciphering the mechanism and structural features of polysorbate 80 during adsorption on PLGA nanoparticles by attenuated total reflectance-Fourier transform infrared spectroscopy. *RSC Adv.* **2016**, *6*, 108545–108557. [CrossRef]

35. Schwartzberg, L.S.; Navari, R.M. Safety of Polysorbate 80 in the Oncology Setting. *Adv. Ther.* **2018**, *35*, 754–767. [CrossRef] [PubMed]

36. Baskin, J.; Jeon, J.E.; Lewis, S.J.G. Nanoparticles for drug delivery in Parkinson's disease. *J. Neurol.* **2020**. [CrossRef] [PubMed]

37. Gulati, N.M.; Stewart, P.L.; Steinmetz, N.F. Bio inspired shielding strategies for nanoparticle drug delivery applications Graphical Abstract. *Mol. Pharm* **2018**, *15*, 2900–2909. [CrossRef] [PubMed]

38. Suk, J.S.; Xu, Q.; Kim, N.; Hanes, J.; Ensign, L.M. PEGylation as a strategy for improving nanoparticle-based drug and gene delivery. *Adv. Drug Deliv. Rev.* **2016**, *99*, 28–51. [CrossRef]

39. Hu, K.; Shi, Y.; Jiang, W.; Han, J.; Huang, S.; Jiang, X. Lactoferrin conjugated PEG-PLGA nanoparticles for brain delivery: Preparation, characterization and efficacy in Parkinsons disease. *Int. J. Pharm.* **2011**, *415*, 273–283. [CrossRef]

40. Torres-Ortega, P.V.; Saludas, L.; Hanafy, A.S.; Garbayo, E.; Blanco-Prieto, M.J. Micro- and nanotechnology approaches to improve Parkinson's disease therapy. *J. Control. Release* **2019**, *295*, 201–213. [CrossRef] [PubMed]

41. Appelboom, G.; Detappe, A.; LoPresti, M.; Kunjachan, S.; Mitrasinovic, S.; Goldman, S.; Chang, S.D.; Tillement, O. Stereotactic modulation of blood-brain barrier permeability to enhance drug delivery. *Neuro Oncol.* **2016**, *18*, 1601–1609. [CrossRef]

42. Jaiswal, M.; Dudhe, R.; Sharma, P.K. Nanoemulsion: An advanced mode of drug delivery system. *3 Biotech* **2015**, *5*, 123–127. [CrossRef]

43. Gosselet, F.; Loiola, R.A.; Roig, A.; Rosell, A.; Culot, M. Central nervous system delivery of molecules across the blood-brain barrier. *Neurochem. Int.* **2021**, *144*, 104952. [CrossRef]

44. Sharma, D.; Sharma, R.K.; Sharma, N.; Gabrani, R.; Sharma, S.K.; Ali, J.; Dang, S. Nose-To-Brain Delivery of PLGA-Diazepam Nanoparticles. *AAPS Pharmscitech* **2015**, *16*, 1108–1121. [CrossRef] [PubMed]

45. Masserini, M. Nanoparticles for Brain Drug Delivery. *ISRN Biochem.* **2013**, *2013*, 1–18. [CrossRef] [PubMed]

46. Tarawneh, R.; Galvin, J.E. Potential Future Neuroprotective Therapies for Neurodegenerative Disorders and Stroke. *Clin. Geriatr. Med.* **2010**, *26*, 125–147. [CrossRef] [PubMed]

47. Tapias, V. Editorial: Mitochondrial Dysfunction and Neurodegeneration. *Front. Neurosci.* **2019**, *13*, 1372. [CrossRef] [PubMed]

48. Murphy, M.P.; Levine, H. Alzheimer's disease and the amyloid-β peptide. *J. Alzheimer's Dis.* **2010**, *19*, 311–323. [CrossRef] [PubMed]

49. Iqbal, K.; Liu, F.; Gong, C.-X.; Grundke-Iqbal, I. Tau in Alzheimer Disease and Related Tauopathies. *Curr. Alzheimer Res.* **2010**, *7*, 656–664. [CrossRef] [PubMed]

50. Gallagher, D.J.; Mhaolaín, A.N.; Sperling, R.A.; Lawlor, B.A. Alzheimer's Disease. In *Neurodegenerative Disorders*; Springer: London, UK, 2011; pp. 43–64.

51. Armstrong, R.A. Risk factors for Alzheimer's disease. *Folia Neuropathol.* **2019**, *57*, 87–105. [CrossRef]

52. Huang, W.-J.; Zhang, X.; Chen, W.-W. Role of oxidative stress in Alzheimer's disease. *Biomed. Rep.* **2016**, *4*, 519–522. [CrossRef]

53. Yusuf, M.; Khan, M.; Alrobaian, M.M.; Alghamdi, S.A.; Warsi, M.H.; Sultana, S.; Khan, R.A. Brain targeted Polysorbate-80 coated PLGA thymoquinone nanoparticles for the treatment of Alzheimer's disease, with biomechanistic insights. *J. Drug Deliv. Sci. Technol.* **2021**, *61*, 102214. [CrossRef]

54. Elibol, B.; Beker, M.; Terzioglu-Usak, S.; Dalli, T.; Kilic, U. Thymoquinone administration ameliorates Alzheimer's disease-like phenotype by promoting cell survival in the hippocampus of amyloid beta1–42 infused rat model. *Phytomedicine* **2020**, *79*, 153324. [CrossRef]

55. Meng, Q.; Wang, A.; Hua, H.; Jiang, Y.; Wang, Y.; Mu, H.; Wu, Z.; Sun, K. Intranasal delivery of Huperzine A to the brain using lactoferrin-conjugated N-trimethylated chitosan surface-modified PLGA nanoparticles for treatment of Alzheimer's disease. *Int. J. Nanomed.* **2018**, *13*, 705–718. [CrossRef] [PubMed]

56. Ma, T.; Gong, K.; Yan, Y.; Zhang, L.; Tang, P.; Zhang, X.; Gong, Y. Huperzine A promotes hippocampal neurogenesis in vitro and in vivo. *Brain Res.* **2013**, *1506*, 35–43. [CrossRef]

57. Shao, H.; Mi, Z.; Ji, W.G.; Zhang, C.H.; Zhang, T.; Ren, S.C.; Zhu, Z.R. Rhynchophylline Protects Against the Amyloid β-Induced Increase of Spontaneous Discharges in the Hippocampal CA1 Region of Rats. *Neurochem. Res.* **2015**, *40*, 2365–2373. [CrossRef] [PubMed]

58. Xu, R.; Wang, J.; Xu, J.; Song, X.; Huang, H.; Feng, Y.; Fu, C. Rhynchophylline loaded-mPEG-PLGA nanoparticles coated with tween-80 for preliminary study in Alzheimer's disease. *Int. J. Nanomed.* **2020**, *15*, 1149–1160. [CrossRef] [PubMed]

59. Walsh, R.A.; Lynch, T.; Fahn, S. Parkinson's Disease. In *Neurodegenerative Disorders*; Springer: London, UK, 2011; pp. 77–114.

60. Sa, F.; Zhang, L.Q.; Chong, C.M.; Guo, B.J.; Li, S.; Zhang, Z.J.; Zheng, Y.; Hoi, P.M.; Lee, S.M.Y. Discovery of novel anti-parkinsonian effect of schisantherin A in in vitro and in vivo. *Neurosci. Lett.* **2015**, *593*, 7–12. [CrossRef]

61. Huang, J.L.; Jing, X.; Tian, X.; Qin, M.C.; Xu, Z.H.; Wu, D.P.; Zhong, Z.G. Neuroprotective properties of panax notoginseng saponins via preventing oxidative stress injury in SAMP8 mice. *Evid. Based Complement. Altern. Med.* **2017**, *2017*. [CrossRef]

62. Zhang, X.; Xiong, J.; Liu, S.; Wang, L.; Huang, J.; Liu, L.; Yang, J.; Zhang, G.; Guo, K.; Zhang, Z.; et al. Puerarin protects dopaminergic neurons in Parkinson's disease models. *Neuroscience* **2014**, *280*, 88–98. [CrossRef]

63. Chen, T.; Liu, W.; Xiong, S.; Li, D.; Fang, S.; Wu, Z.; Wang, Q.; Chen, X. Nanoparticles mediating the sustained puerarin release facilitate improved brain delivery to treat parkinson's disease. *ACS Appl. Mater. Interfaces* **2019**, *11*, 45276–45289. [CrossRef]

64. Fernandez, H. Malaty Role of rasagiline in treating Parkinson's disease: Effect on disease progression. *Clin. Risk Manag.* **2009**, *5*, 413. [CrossRef]

65. Hattori, N.; Takeda, A.; Takeda, S.; Nishimura, A.; Kitagawa, T.; Mochizuki, H.; Nagai, M.; Takahashi, R. Rasagiline monotherapy in early Parkinson's disease: A phase 3, randomized study in Japan. *Park. Relat. Disord.* **2019**, *60*, 146–152. [CrossRef]

66. Hauser, R.A.; Abler, V.; Eyal, E.; Eliaz, R.E. Efficacy of rasagiline in early Parkinson's disease: A meta-analysis of data from the TEMPO and ADAGIO studies. *Int. J. Neurosci.* **2016**, *126*, 942–946. [CrossRef]

67. Bali, N.R.; Salve, P.S. Impact of rasagiline nanoparticles on brain targeting efficiency via gellan gum based transdermal patch: A nanotheranostic perspective for Parkinsonism. *Int. J. Biol. Macromol.* **2020**, *164*, 1006–1024. [CrossRef]

68. Myers, R.H. Huntington's Disease Genetics. *NeuroRx* **2004**, *1*, 255–262. [CrossRef] [PubMed]

69. Pender, N.P.; Koroshetz, W.J. Huntington's Disease. In *Neurodegenerative Disorders*; Springer: London, UK, 2011; pp. 167–179.

70. Ferrante, R.J.; Kowall, N.W.; Beal, M.F.; Richardson, E.P.; Bird, E.D.; Martin, J.B. Selective sparing of a class of striatal neurons in Huntington's disease. *Science* **1985**, *230*, 561–563. [CrossRef]

71. Rubinsztein, D.C. Lessons from animal models of Huntington's disease. *Trends Genet.* **2002**, *18*, 202–209. [CrossRef]

72. Sung, V.W.; Iyer, R.; Schilling, T.; Buzinec, P.N. Tetrabenazine Use in Patients With Huntington's Disease (HD) Chorea (P2.009). *Neurology* **2017**, *88*, 545–551.

73. Burra, G.; Thakur, A.K. Inhibition of polyglutamine aggregation by SIMILAR huntingtin N-terminal sequences: Prospective molecules for preclinical evaluation in Huntington's disease. *Biopolymers* **2017**, *108*, e23021. [CrossRef] [PubMed]

74. Joshi, A.S.; Singh, V.; Gahane, A.; Thakur, A.K. Biodegradable Nanoparticles Containing Mechanism Based Peptide Inhibitors Reduce Polyglutamine Aggregation in Cell Models and Alleviate Motor Symptoms in a Drosophila Model of Huntington's Disease. *ACS Chem. Neurosci.* **2019**, *10*, 1603–1614. [CrossRef] [PubMed]

75. Leoni, V.; Caccia, C. The impairment of cholesterol metabolism in Huntington disease. *Biochim. Biophys. Acta Mol. Cell Biol. Lipids* **2015**, *1851*, 1095–1105. [CrossRef]

76. Block, R.C.; Dorsey, E.R.; Beck, C.A.; Brenna, J.T.; Shoulson, I. Altered cholesterol and fatty acid metabolism in Huntington disease. *J. Clin. Lipidol.* **2010**, *4*, 17–23. [CrossRef]

77. Valenza, M.; Chen, J.Y.; Di Paolo, E.; Ruozi, B.; Belletti, D.; Ferrari Bardile, C.; Leoni, V.; Caccia, C.; Brilli, E.; Di Donato, S.; et al. Cholesterol-loaded nanoparticles ameliorate synaptic and cognitive function in H untington's disease mice. *EMBO Mol. Med.* **2015**, *7*, 1547–1564. [CrossRef] [PubMed]

78. Hardiman, O. Amyotrophic Lateral Sclerosis. In *Neurodegenerative Disorders*; Springer: London, UK, 2011; Volume 59, pp. 143–166.

79. Andrews, J.A.; Jackson, C.E.; Heiman-Patterson, T.D.; Bettica, P.; Brooks, B.R.; Pioro, E.P. Real-world evidence of riluzole effectiveness in treating amyotrophic lateral sclerosis. *Amyotroph. Lateral Scler. Front. Degener.* **2020**, *21*, 509–518. [CrossRef] [PubMed]

80. Chung, Y.H.; Lin, C.W.; Huang, H.Y.; Chen, S.L.; Huang, H.J.; Sun, Y.C.; Lee, G.C.; Lee-Chen, G.J.; Chang, Y.C.; Hsieh-Li, H.M. Targeting Inflammation, PHA-767491 Shows a Broad Spectrum in Protein Aggregation Diseases. *J. Mol. Neurosci.* **2020**, *70*, 1140–1152. [CrossRef]

81. Rojas-Prats, E.; Tosat-Bitrián, C.; Martínez-González, L.; Nozal, V.; Pérez, D.I.; Martínez, A. Increasing Brain Permeability of PHA-767491, a Cell Division Cycle 7 Kinase Inhibitor, with Biodegradable Polymeric Nanoparticles. *Pharmaceutics* **2021**, *13*, 180. [CrossRef] [PubMed]

82. Dobson, R.; Giovannoni, G. Multiple sclerosis—A review. *Eur. J. Neurol.* **2019**, *26*, 27–40. [CrossRef]

83. Chountoulesi, M.; Demetzos, C. Promising nanotechnology approaches in treatment of autoimmune diseases of central nervous system. *Brain Sci.* **2020**, *10*. [CrossRef] [PubMed]

84. Clerico, M.; Contessa, G.; Durelli, L. Interferon-β 1a for the treatment of multiple sclerosis. *Expert Opin. Biol.* **2007**, *7*, 535–542. [CrossRef] [PubMed]

85. Fodor-Kardos, A.; Kiss, Á.F.; Monostory, K.; Feczkó, T. Sustained: In vitro interferon-beta release and in vivo toxicity of PLGA and PEG-PLGA nanoparticles. *RSC Adv.* **2020**, *10*, 15893–15900. [CrossRef]

86. Yue, X.; Wu, L.; Hu, W. The Regulation of Leukemia Inhibitory Factor. *Cancer Cell Microenviron.* **2015**, *2*, e877.

87. Slaets, H.; Hendriks, J.J.A.; Stinissen, P.; Kilpatrick, T.J.; Hellings, N. Therapeutic potential of LIF in multiple sclerosis. *Trends Mol. Med.* **2010**, *16*, 493–500. [CrossRef]

88. Rittchen, S.; Boyd, A.; Burns, A.; Park, J.; Fahmy, T.M.; Metcalfe, S.; Williams, A. Myelin repair invivo is increased by targeting oligodendrocyte precursor cells with nanoparticles encapsulating leukaemia inhibitory factor (LIF). *Biomaterials* **2015**, *56*, 78–85. [CrossRef]

89. Kuo, Y.C.; Tsai, H.C. Rosmarinic acid- and curcumin-loaded polyacrylamide-cardiolipin-poly(lactide-co-glycolide) nanoparticles with conjugated 83–14 monoclonal antibody to protect β-amyloid-insulted neurons. *Mater. Sci. Eng. C* **2018**, *91*, 445–457. [CrossRef]

90. Mathew, A.; Fukuda, T.; Nagaoka, Y.; Hasumura, T.; Morimoto, H.; Yoshida, Y.; Maekawa, T.; Venugopal, K.; Kumar, D.S. Curcumin loaded-PLGA nanoparticles conjugated with Tet-1 peptide for potential use in Alzheimer's disease. *PLoS ONE* **2012**, *7*, e32616. [CrossRef] [PubMed]

91. Sun, D.; Li, N.; Zhang, W.; Zhao, Z.; Mou, Z.; Huang, D.; Liu, J.; Wang, W. Design of PLGA-functionalized quercetin nanoparticles for potential use in Alzheimer's disease. *Colloids Surf. B Biointerfaces* **2016**, *148*, 116–129. [CrossRef] [PubMed]

92. Loureiro, J.A.; Gomes, B.; Fricker, G.; Coelho, M.A.N.; Rocha, S.; Pereira, M.C. Cellular uptake of PLGA nanoparticles targeted with anti-amyloid and anti-transferrin receptor antibodies for Alzheimer's disease treatment. *Colloids Surf. B Biointerfaces* **2016**, *145*, 8–13. [CrossRef]

93. Sathya, S.; Manogari, B.G.; Thamaraiselvi, K.; Vaidevi, S.; Ruckmani, K.; Devi, K.P. Phytol loaded PLGA nanoparticles ameliorate scopolamine-induced cognitive dysfunction by attenuating cholinesterase activity, oxidative stress and apoptosis in Wistar rat. *Nutr. Neurosci.* **2020**, *8305*, 1–17. [CrossRef] [PubMed]

94. Sathya, S.; Shanmuganathan, B.; Saranya, S.; Vaidevi, S.; Ruckmani, K.; Pandima Devi, K. Phytol-loaded PLGA nanoparticle as a modulator of Alzheimer's toxic Aβ peptide aggregation and fibrillation associated with impaired neuronal cell function. *Artif. CellsNanomed. Biotechnol.* **2018**, *46*, 1719–1730. [CrossRef]

95. Doolaanea, A.A.; Mansor, N.I.; Mohd Nor, N.H.; Mohamed, F. Co-encapsulation of Nigella sativa oil and plasmid DNA for enhanced gene therapy of Alzheimers disease. *J. Microencapsul.* **2016**, *33*, 114–126. [CrossRef]

96. Amin, F.U.; Shah, S.A.; Badshah, H.; Khan, M.; Kim, M.O. Anthocyanins encapsulated by PLGA@PEG nanoparticles potentially improved its free radical scavenging capabilities via p38/JNK pathway against Aβ1-42-induced oxidative stress. *J. Nanobiotechnol.* **2017**, *15*, 1–16. [CrossRef] [PubMed]

97. Baysal, I.; Yabanoglu-Ciftci, S.; Tunc-Sarisozen, Y.; Ulubayram, K.; Ucar, G. Interaction of selegiline-loaded PLGA-b-PEG nanoparticles with beta-amyloid fibrils. *J. Neural Transm.* **2013**, *120*, 903–910. [CrossRef]

98. Chen, T.; Li, C.; Li, Y.; Yi, X.; Wang, R.; Lee, S.M.Y.; Zheng, Y. Small-Sized mPEG-PLGA Nanoparticles of Schisantherin A with Sustained Release for Enhanced Brain Uptake and Anti-Parkinsonian Activity. *Acs Appl. Mater. Interfaces* **2017**, *9*, 9516–9527. [CrossRef] [PubMed]

99. Arisoy, S.; Sayiner, O.; Comoglu, T.; Onal, D.; Atalay, O.; Pehlivanoglu, B. In vitro and in vivo evaluation of levodopa-loaded nanoparticles for nose to brain delivery. *Pharm. Dev. Technol.* **2020**, *25*, 735–747. [CrossRef] [PubMed]

100. Lima, A.F.; Amado, I.R.; Pires, L.R. Poly(D,l-lactide-co-glycolide) (plga) nanoparticles loaded with proteolipid protein (plp)—exploring a new administration route. *Polymer* **2020**, *12*, 1–10.

101. Mohd Sairazi, N.S.; Sirajudeen, K.N.S. Natural Products and Their Bioactive Compounds: Neuroprotective Potentials against Neurodegenerative Diseases. *Evid. Based Complement. Altern. Med.* **2020**, *2020*, 5–7. [CrossRef]

102. Carrera, I.; Cacabelos, R. Current Drugs and Potential Future Neuroprotective Compounds for Parkinson's Disease. *Curr. Neuropharmacol.* **2018**, *17*, 295–306. [CrossRef]

103. Lee, Y.W.; Kim, D.H.; Jeon, S.J.; Park, S.J.; Kim, J.M.; Jung, J.M.; Lee, H.E.; Bae, S.G.; Oh, H.K.; Ho Son, K.H.; et al. Neuroprotective effects of salvianolic acid B on an Aβ25-35 peptide-induced mouse model of Alzheimer's disease. *Eur. J. Pharmacol.* **2013**, *704*, 70–77. [CrossRef]

104. Wang, Z.-Y.; Liu, J.-G.; Li, H.; Yang, H.-M. Pharmacological Effects of Active Components of Chinese Herbal Medicine in the Treatment of Alzheimer's Disease: A Review. *Am. J. Chin. Med.* **2016**, *44*, 1525–1541. [CrossRef] [PubMed]

105. Emanuele, E. Can Trehalose Prevent Neurodegeneration? Insights from Experimental Studies. *Curr. Drug Targets* **2014**, *15*, 551–557. [CrossRef] [PubMed]

106. Liu, R.; Barkhordarian, H.; Emadi, S.; Chan, B.P.; Sierks, M.R. Trehalose differentially inhibits aggregation and neurotoxicity of beta-amyloid 40 and 42. *Neurobiol. Dis.* **2005**, *20*, 74–81. [CrossRef]

107. Sarkar, S.; Davies, J.E.; Huang, Z.; Tunnacliffe, A.; Rubinsztein, D.C. Trehalose, a novel mTOR-independent autophagy enhancer, accelerates the clearance of mutant huntingtin and α-synuclein. *J. Biol. Chem.* **2007**, *282*, 5641–5652. [CrossRef]

108. Khalifeh, M.; Barreto, G.E.; Sahebkar, A. Trehalose as a promising therapeutic candidate for the treatment of Parkinson's disease. *Br. J. Pharmacol.* **2019**, *176*, 1173–1189. [CrossRef]

109. Tanaka, M.; Machida, Y.; Niu, S.; Ikeda, T.; Jana, N.R.; Doi, H.; Kurosawa, M.; Nekooki, M.; Nukina, N. Trehalose alleviates polyglutamine-mediated pathology in a mouse model of Huntington disease. *Nat. Med.* **2004**, *10*, 148–154. [CrossRef] [PubMed]

110. Halbe, L.; Rami, A. Trehalase localization in the cerebral cortex, hippocampus and cerebellum of mouse brains. *J. Adv. Res.* **2019**, *18*, 71–79. [CrossRef]

111. Zhang, X.; Chen, G.; Wen, L.; Yang, F.; Shao, A.L.; Li, X.; Long, W.; Mu, L. Novel multiple agents loaded PLGA nanoparticles for brain delivery via inner ear administration: In vitro and in vivo evaluation. *Eur. J. Pharm. Sci.* **2013**, *48*, 595–603. [CrossRef]

112. Zeng, G.; Tang, T.; Wu, H.J.; You, W.H.; Luo, J.K.; Lin, Y.; Liang, Q.H.; Li, X.Q.; Huang, X.; Yang, Q.D. Salvianolic acid b protects SH-SY5Y neuroblastoma cells from1-methyl-4- phenylpyridinium-induced apoptosis. *Biol. Pharm. Bull.* **2010**, *33*, 1337–1342. [CrossRef] [PubMed]

113. Jiang, P.; Li, C.; Xiang, Z.; Jiao, B. Tanshinone IIA reduces the risk of Alzheimer's disease by inhibiting iNOS, MMP-2 and NF-κBp65 transcription and translation in the temporal lobes of rat models of Alzheimer's disease. *Mol. Med. Rep.* **2014**, *10*, 689–694. [CrossRef]

114. Ren, B.; Zhang, Y.X.; Zhou, H.X.; Sun, F.W.; Zhang, Z.F.; Wei, Z.F.; Zhang, C.Y.; Si, D.W. Tanshinone IIA prevents the loss of nigrostriatal dopaminergic neurons by inhibiting NADPH oxidase and iNOS in the MPTP model of Parkinson's disease. *J. Neurol. Sci.* **2015**, *348*, 142–152. [CrossRef] [PubMed]
115. Wang, Y.; Feng, Y.; Fu, Q.; Li, L. Panax notoginsenoside Rb1 ameliorates Alzheimer's disease by upregulating brain-derived neurotrophic factor and downregulating Tau protein expression. *Exp. Ther. Med.* **2013**, *6*, 826–830. [CrossRef]
116. Anselmo, A.C.; Mitragotri, S. Nanoparticles in the clinic. *Bioeng. Transl. Med.* **2016**, *1*, 10–29. [CrossRef] [PubMed]
117. US FDA CDER. *Current Good Manufacturing Practice (CGMP) Regulations*; FDA: Silver Spring, MD, USA, 2018; p. 5.

pharmaceutics

Review

PLGA Nanoparticle-Based Formulations to Cross the Blood–Brain Barrier for Drug Delivery: From R&D to cGMP

Kaining Zhi [1,*], Babatunde Raji [1], Anantha R. Nookala [2], Mohammad Moshahid Khan [3], Xuyen H. Nguyen [4], Swarna Sakshi [4], Tayebeh Pourmotabbed [5], Murali M. Yallapu [6], Harry Kochat [1], Erene Tadrous [4], Shelby Pernell [4] and Santosh Kumar [4,*]

[1] Plough Center for Sterile Drug Delivery Solutions, University of Tennessee Health Science Center, 208 South Dudley Street, Memphis, TN 38163, USA; braji@uthsc.edu (B.R.); hkochat@uthsc.edu (H.K.)
[2] Covance Inc., Kinsman Blvd, Madison, WI 53704, USA; anfh3@mail.umkc.edu
[3] Department of Neurology, College of Medicine, University of Tennessee Health Science Center, 855 Monroe Avenue, Memphis, TN 38163, USA; mkhan26@uthsc.edu
[4] Department of Pharmaceutical Sciences, University of Tennessee Health Science Center, 881 Madison Ave, Memphis, TN 38163, USA; xnguyen3@uthsc.edu (X.H.N.); ssakshi1@uthsc.edu (S.S.); etadrous@uthsc.edu (E.T.); pernellshelby@gmail.com (S.P.)
[5] Department of Microbiology, Immunology and Biochemistry, College of Medicine, University of Tennessee Health Science Center, 858 Madison Avenue, Memphis, TN 38163, USA; tpourmot@uthsc.edu
[6] Department of Immunology and Microbiology, University of Texas Rio Grande Valley, McAllen, TX 78504, USA; murali.yallapu@utrgv.edu
* Correspondence: kzhi@uthsc.edu (K.Z.); ksantosh@uthsc.edu (S.K.)

Citation: Zhi, K.; Raji, B.; Nookala, A.R.; Khan, M.M.; Nguyen, X.H.; Sakshi, S.; Pourmotabbed, T.; Yallapu, M.M.; Kochat, H.; Tadrous, E.; et al. PLGA Nanoparticle-Based Formulations to Cross the Blood–Brain Barrier for Drug Delivery: From R&D to cGMP. *Pharmaceutics* **2021**, *13*, 500. https://doi.org/10.3390/pharmaceutics13040500

Academic Editor: Oya Tagit

Received: 17 March 2021
Accepted: 5 April 2021
Published: 6 April 2021

Publisher's Note: MDPI stays neutral with regard to jurisdictional claims in published maps and institutional affiliations.

Abstract: The blood–brain barrier (BBB) is a natural obstacle for drug delivery into the human brain, hindering treatment of central nervous system (CNS) disorders such as acute ischemic stroke, brain tumors, and human immunodeficiency virus (HIV)-1-associated neurocognitive disorders. Poly(lactic-*co*-glycolic acid) (PLGA) is a biocompatible polymer that is used in Food and Drug Administration (FDA)-approved pharmaceutical products and medical devices. PLGA nanoparticles (NPs) have been reported to improve drug penetration across the BBB both in vitro and in vivo. Poly(ethylene glycol) (PEG), poly(vinyl alcohol) (PVA), and poloxamer (Pluronic) are widely used as excipients to further improve the stability and effectiveness of PLGA formulations. Peptides and other linkers can be attached on the surface of PLGA to provide targeting delivery. With the newly published guidance from the FDA and the progress of current Good Manufacturing Practice (cGMP) technologies, manufacturing PLGA NP-based drug products can be achieved with higher efficiency, larger quantity, and better quality. The translation from bench to bed is feasible with proper research, concurrent development, quality control, and regulatory assurance.

Keywords: poly(lactic-*co*-glycolic acid) (PLGA); blood–brain barrier (BBB); current Good Manufacturing Practice (cGMP); Food and Drug Administration (FDA); nanotechnology

1. Blood–Brain Barrier (BBB) and Drug Delivery

Compared with other therapeutic areas, drug development is more challenging for brain diseases such as brain cancers, Alzheimer's diseases (AD), acute ischemic stroke, and human immunodeficiency virus (HIV)-1-associated neurocognitive disorders (HAND) [1–4]. Many systemically administered drug products cannot pass the BBB [5]. The BBB restricts the entry of compounds into the central nervous system (CNS) through the presence of brain microvascular endothelial cells, pericytes, perivascular astrocytes, and tight junctions. In addition, the presence of efflux transporters at the BBB has been recognized as a key element to poor drug penetration [6,7]. ATP-binding cassette (ABC) membrane-associated transporters, such as P-glycoprotein (P-gp), breast cancer resistance protein (BCRP), and multidrug resistance-associated protein (MRP1) show significant expressions at the BBB, protecting the brain from potential harmful endogenous and exogenous substances [6,7]. As a result, the BBB only selectively transports molecules such as certain amino acids, sugars, and gaseous molecules (e.g., oxygen and carbon dioxide) into the brain [8]. For example, antiretroviral drugs (ARVs) have shown to be effective in managing HIV-1 [9]. However, due to the inability of ARVs to cross the BBB, they are not highly recommended clinically for the treatment of HAND. Studies showed that, upon boosting with a pharmaco-enhancer, i.e., ritonavir, ARVs including indinavir, elvitegravir, and lopinavir reached therapeutic concentrations in plasma but did not reach therapeutic concentration in the brain, indicating the challenges of delivering drugs to the CNS [10,11].

2. Strategies to Cross BBB

Several strategies have been used to improve drug delivery to the brain. Efforts have been made for the development of inhibitors for ABC transporters due to their high expressions on the BBB [2,12,13]. Studies showed that blocking ABC transporters may significantly improve drug penetrations across the BBB. However, this method has not been used clinically due to the wide distribution of ABC transporters throughout the body, the potential toxicity of inhibitors, and unexpected drug–drug interactions [12]. Another approach is the "BBB opening" approach. Opening the BBB can be achieved by using a hyperosmotic solution to shrink the endothelial cells or using certain cytotoxic agents to disrupt the BBB tight junctions [12,13]. However, opening the tight junctions of the BBB is risky clinically because it may also allow the entry of harmful components into the brain and cause side-effects such as seizures and other long-term neurological complications [12]. Moreover, the development of prodrugs to increase their capacity to penetrate the BBB is another potential delivery approach [12]. Prodrugs can be synthesized with sufficient lipophilicity to facilitate the crossing of the endothelial cell membrane and release the parent ARVs into the brain. However, developing prodrugs as a delivery strategy needs a full evaluation of toxicity, cost, and efficacy, as prodrugs are considered to be a separate chemical entity.

A nanoparticle (NP)-based drug delivery system is considered a promising option to improve drug delivery to the brain [3]. NP-based formulations are usually a colloidal system made of polymers, lipids, or other large macromolecules such as albumin. A therapeutic agent may be released through diffusion or erosion of the matrix [14]. The NP-based delivery system can cross the BBB through membrane transcytosis, bypass efflux transporters, and effectively deliver the therapeutic molecule to the CNS [3]. NPs that have been studied for brain delivery include polymeric NPs such as poly(D,L-lactide-co-glycolide) (PLGA) [15,16] and poly(butyl-cyanoacrylate) (PBCA) NPs [17,18], magnetic NPs (MNPs) composed of an iron oxide core [18], lipid-based nanoformulations such as solid lipid nanoparticles (SLN) and liposomes [12,18], and polymeric micelles-based nanoformulations such as Pluronic micelles [12]. Extracellular vesicles (EVs), liposome-like natural carriers, have drawn attention for delivering drugs into the brain as a potential alternative to NPs [19–21].

3. Introduction of Physical, Chemical, and Biological Characteristics of PLGA Polymer

3.1. Synthesis of PLGA Polymer

PLGA is a synthetic copolymer composed of lactic and glycolic acid polyesters. Synthesis of PLGA is commonly achieved either through ring-opening polymerization reactions of lactide and glycolide [22,23] or through polycondensation reactions of lactic acid and glycolic acid to form PLA and PGA block polymers [24,25]. Ring-opening polymerization processes can be used to generate high-molecular-weight PLGA polymers [26], while polycondensation processes are more suitable for the synthesis of low-molecular-weight polymers [27].

3.2. Physicochemical and Biomolecular Characteristics of PLGA

PLA can exist in D- or L-lactic acid, as well as in D,L-lactic acid, configurations. Homo isomeric PLAs are more crystalline due to the uniform spatial arrangement leading to tighter packing of the polymer chains [28,29]. On the other hand, glycolic acid has no asymmetric carbon; thus, PGA exists only in a highly crystalline form. PLA is more hydrophobic than PGA due to its methyl side groups. As a result, the hydrophobicity and crystallinity of PLGA can be controlled through the ratio of lactide to glycolide. PLGA physicochemical properties, such as mechanical strength, solubility, rate of hydration, rate of hydrolysis, and glass transition temperature (Tg), are heavily influenced by crystallinity and hydrophobicity [30–32]. PLGA with a high degree of crystallinity will have a higher Tg and mechanical strength, as well as a decreased rate of hydration and hydrolysis.

In general, PLGA NPs are susceptible to clearance by the reticuloendothelial system (RES) through opsonin-mediated phagocytosis [33]. RES elimination and biodistribution of PLGA NPs depend on size, hydrophobicity, and surface charge. Cytotoxicity of PLGA NPs was investigated in vitro by monitoring the cell viability of Caco-2 and HeLa cell lines [34,35]. The study indicates that PLGA NPs at the concentration tested were not toxic to cells as both cell lines retained over 75% viability. According to histopathology assays, orally administered PLGA nanoparticles did not elicit adverse effects in mice [35]. Tissue distribution analysis in mice indicated that most of the PLGA NPs were detected in the liver, kidney, heart, and brain, with small amounts detected in plasma [35]. In aqueous conditions, PLGA undergoes hydrolysis of its ester bonds. Hydrolytic biodegradation of PLGA leads to nontoxic byproducts [36,37]. Several studies have investigated the factors that affect PLGA biodegradation, including intrinsic properties such as hydration rate, hydrophobicity/hydrophilicity, polymer chemical composition, molecular weight, crystallinity, and Tg [38]. Moreover, external factors such as pH and chemical additives were also shown to influence hydrolytic degradation [39].

4. PLGA NPs as a Brain Drug Delivery System

PLGA is a highly investigated polymer due to its ability to form NPs, micelles, and microspheres, as it possesses the properties of biocompatibility, biodegradability, and tolerability [40]. As drug delivery systems, PLGA NPs can be used to prepare controlled-release dosage forms of small-molecule drugs, peptides, and nucleic acids [41]. Through proper copolymerization with PEG and surface modifications with linkers, PLGA NPs have been demonstrated as promising carriers for drug delivery across the BBB.

4.1. PLGA NPs Modifications and Mechanisms

PLGA NPs can be prepared through various processing methods including (1) double-emulsion solvent evaporation [42], (2) single-emulsion solvent evaporation [43], (3) phase separation [44], (4) spray-drying [45–47], (5) salting out [48,49], and (6) nanoprecipitation [50,51]. Even though the names are different, processing methods focus on the self-assembly characteristics of PLGA in aqueous solutions to finish the drug encapsulation [52–54]. Details of processing methods and potential scale-up technologies are discussed in Section 5.

PLGA NPs can cross the BBB passively or through active endocytosis mechanisms as shown in Figure 1. Unmodified PLGA NPs cross the BBB primarily through passive internalization based on size, which was found to have low brain uptake. Several strategies have been developed to improve the penetration of NPs into the brain. These strategies modify NPs with components designed to take advantage of BBB endocytosis pathways. Modified PLGA NPs have been designed to cross the BBB through adsorption-mediated transcytosis (AMT) [55], carrier-mediated transport (CMT), and receptor-mediated transcytosis (RMT) [56].

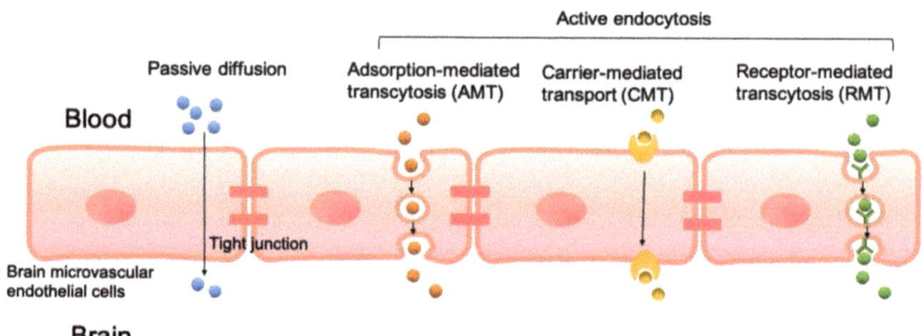

Figure 1. Transport mechanisms of poly(lactic-*co*-glycolic acid) (PLGA) nanoparticles (NPs) to cross the blood–brain barrier (BBB), including passive diffusion, adsorption-mediated transcytosis (AMT), carrier-mediated transport (CMT), and receptor-mediated transcytosis (RMT).

PLGA NP surfaces are modified with positive charges that electrostatically interact with negatively charged regions of the luminal surfaces, which helps PLGA to cross the BBB. Several cationic modifications of PLGA NPs have been demonstrated to utilize the AMT concepts to improve brain uptake. In CMT systems, PLGA NPs are modified with membrane-permeable molecules such as amino acids, nutrients, and membranotropic peptides, and they are able to transport cargo across the BBB endothelium. CMT systems also include designs that take advantage of ABC transporters. With RMT, PLGA NPs are modified or covalently connected with ligands that target specific cell surface receptors known to be BBB transport pathways. A search of the PubMed database with the keywords "PLGA nanoparticles", "BBB", and "drug delivery" from 2015 to date returned 133 publications. From the search results, only research articles that included BBB permeability studies of control "unmodified" PLGA NPs and modified PLGA NPs are summarized in Table 1. Tandem systems that utilized multiple modifications have been reported. Guarnieri et al. [57] demonstrated the cooperative effects of glycoprotein H 625, a CMT modification with iron-mimicking protein CRT and RMT modification, in enhancing PLGA NP permeation of BBB. Liu et al. [58] developed a PLGA NP drug delivery system modified with angiopep-2 (RMT) and 1, 2-Dioleoyl-3-trimethylammonium-propane (AMT), for gefitinib and Golgi phosphoprotein 3 for the treatment of glioblastoma. Intranasal [59–61] or subcutaneous [62] administration of PLGA NP drug systems can bypass the BBB and avoid issues associated with systemic administration.

4.2. PLGA–PEG Co-Polymeric NPs, Modifications, and Applications

To overcome its short half-life, PLGA is combined with polyethylene glycol (PEG) to form PLGA–PEG copolymer NPs [63–65]. The PLGA–PEG copolymer is widely used in pharmaceutical products and devices. Recently, several advancements have been made to modify the surface of the PLGA–PEG NPs to further increase their ability to cross the BBB and deliver drugs into the brain.

The favorable chemistry of the PLGA–PEG NPs makes them amenable for conjugation with various peptides and linkers for use in the treatment of various neurodegenerative diseases and glioma. Memantine is commonly used for the treatment of mild and moderate AD. Encapsulating it in PLGA–PEG NPs using a double-emulsion method increased the delivery to the target tissue with ameliorated pathological markers compared to free memantine [66].

Table 1. Summary of BBB permeability studies of unmodified and modified PLGA NPs.

Ref.	PLGA NPs Modification	Loaded Drug	Proposed Transport Mechanism [1]	Results
[67]	Trimethylated chitosan (TMC)	Coenzyme Q10 6-coumarin	AMT	TMC-modified PLGA NPs loaded with coumarin exhibited increased uptake in mouse brains vs. PLGA-NPs. Neuroprotective effects of Q10 displayed by mice in TMC PLGA NPs were superior to PLGA-NP.
[68]	Angiopep-2	Doxorubicin (DOX), Epidermal growth factor receptor (EGFR) siRNA	RMT	Angiopep-2 modified PLGA NPs improved DOX and siRNA cell uptake. In vivo study showed that the ang-2-PLGA construct can cross BBB.
[69]	8D3 monoclonal antibody	Loperamide	RMT	8D3 functionalized Loperamide-loaded PLGA NPs produced a higher maximal possible antinociceptive effect compared to Loperamide-loaded NPs without 8D3.
[70]	Lactoferrin, folic acid	Etoposide	RMT	BBB permeability coefficient of PLGA NPs increased twofold with Lf-and FA.
[71]	OX26 monoclonal antibody	$1A\beta_5$ peptide	RMT	OX26 increased the uptake of PLGA NPs by BBB endothelial cells which enhanced the peptide transport.
[72]	Lactoferrin	Rotigotine	RMT	Intranasal delivery of rotigotine to the brain was more effective with Lf-PLGA NPs than with PLGA-NPs.
[73]	RVG29	Docetaxel	RMT	RVG29 PLGA NPs showed better BBB penetration in vitro.
[74]	OX26 monoclonal antibody	Temozolomide	RMT	OX26 functionalization enhanced TMZ internalization in glioblastoma cells.
[75]	Dendrimer cationized albumin	Doxorubicin	AMT	Cellular uptake and cell permeability of DOX in dCatAlb-functionalized PLGA NP was improved 1.59-fold and 1.49-fold, respectively, over unmodified PLGA NP.
[76]	D-α-tocopheryl polyethylene glycol succinate (TPGS)	Paclitaxel	AMT	The in vivo evaluation of TPGS–PLGA NPs showed amplified accumulation (>800% after 96 h) of PTX in the brain tissue when compared with bare NPs.
[77]	Polysorbate 80	Rhynchophylline	AMT	In an in vitro BBB model study, functionalized PLGA NPs showed increased transport across bEnd.3 cell monolayers.

[1] Adsorption-mediated transcytosis (AMT), carrier-mediated transport (CMT), and receptor-mediated transcytosis (RMT).

Pioglitazone-loaded PLGA–PEG NPs produced by the solvent displacement technique reduced amyloid burden and decreased memory impairment by increasing the rate of transcytosis across BBB and slowly releasing pioglitazone in the target tissue [78]. Selegiline- or donepezil-loaded PLGA–PEG NPs produced by the solvent evaporation method destabilized the beta-amyloid formation in vitro [79,80]. Various natural compounds and drugs encapsulated in PLGA–PEG NPs were shown to be effective in reducing AD pathology in in vitro models [81]. The antinociceptive effect of loperamide was increased by two- to threefold by encapsulation in a PLGA–PEG–PLGA triblock polymeric NPs coated with poloxamer 188 or polysorbate 80 compared to unmodified NPs alone [82]. With peptides as a modification on PLGA–PEG NPs, Hoyos-Ceballos et al. [35] showed that angiopep-2 conjugated to PLGA–PEG NPs increased their ability to cross the BBB in C57/BL6 mice. PLGA–PEG NPs conjugated with B6 peptide increased the delivery of curcumin into the CNS in an AD mouse model, showing a reduced expression of hallmark AD pathological markers, including amyloid-beta, presenilin-1, phosphorylated tau, and beta-secretase 1, compared to curcumin alone or NPs without B6 peptide [83].

Surnar et al. [84] added a targeting function to PLGA–PEG NPs through conjugating with a lipophilic triphenylphosphonium cation on the surface using a butylene linker. This NP system, when loaded with either coenzyme Q10 or aspirin, was able to cross the mitochondrial double membrane in endothelial cells and astrocytes to reduce the oxidative stress, which is extremely valuable to treat HAND [84]. Yu et al. [85] optimized the development of PLGA–PEG polymersomes conjugated on the surface with lactoferrin as a brain targeted delivery system for peptides. They loaded the NPs with S14G-humanin peptides which exerted a protective effect by decreasing the caspase-3 and bax expression in the rat hippocampus neurons treated with amyloid-beta. Similarly, Bi et al. [72] showed that PLGA–PEG NPs modified with lactoferrin on their surface were able to deliver rotigotine into the striatum by intranasal administration for potential use in Parkinson's disease (PD). Lectin-conjugated PLGA–PEG NPs were able to deliver the basic fibroblast growth factor peptide cargo across the BBB after intranasal administration [86]. Similarly, odrranalectin-conjugated PLGA–PEG NPs were able to efficiently deliver the encapsulated urocortin peptide into the brain as a treatment for PD [86]. Recently, Amanda et al. [87] showed that a modified PLGA–PEG NP system encapsulating epigallocatechin gallate was able to ameliorate neurological deficits induced by 3-nitropropionic acid in Huntington's disease mouse model.

Glioma is an invasive carcinoma of the brain with an average life expectancy of approximately 12–14 months and has poor survival rates [88]. NP systems were used to deliver drug into the brain. Receptor-mediated transcytosis provides a chance to target specific receptors expressed on the surface of cancer cells through ligand–receptor interactions. Cui and coworkers [89] developed a novel dual-targeting PLGA–PEG-based magnetic NP system that crosses the BBB by conjugating transferrin receptor-binding peptide T7 on the surface and encapsulating curcumin and paclitaxel into the NP hydrophobic core. Compared to free drugs, the mice with orthotopic glioma survived with the NP system. Similarly, doxorubicin and tetrahydrocurcumin encapsulated in transferrin-modified PLGA–PEG NPs showed effectiveness in reducing the glioma tumor volume in combination with radiotherapy [90]. Lactoferrin-conjugated PLGA–PEG NPs increased the brain concentrations of shikonin, a naphthoquinone pigment for the potential use in the treatment of glioma [91]. The iNGR-conjugated PLGA–PEG-based NP system designed to target the glioma tumor vessel was effective in delivering the paclitaxel to the glioma parenchyma. Moreover, this NP system was able to travel deeper into the glioma to increase survival rates [92]. Farnesyl thiosalicylic acid, an inhibitor of Ras oncoprotein, was shown to be effective against glioblastoma when administered as PLGA–PEG-based hybrid NPs, which contained 1,2-distearoyl-glycerol-3-phosphoethanolamine and 1,2-dioleoyl-3-trimethylammonium-propane [93].

Another strategy to treat glioma is to target the genes overexpressed in malignant glioma cells. Cyclic hexapeptide-conjugated PLGA–PEG NPs were able to deliver curcumin to the glioma by binding to integrins that are upregulated on the glial cell surface [94]. A nine amino acid linear peptide (Pep-1) targeting the interleukin 13 receptor $\alpha2$ overexpressed on the surface of gliomas was conjugated to PLGA–PEG NPs [95]. Modification with CGKRK peptide, which targets the heparan sulfate expressed on neovascular endothelial cells, resulted in a dual-targeted approach and increased the median survival time in intracranial glioma mice [96]. A biodegradable PLGA–PEG polymer NP system targeting the Fn14 receptor overexpressed on the brain tumor cells was generated by conjugating PLGA–PEG NPs with ITEM4 monoclonal antibody. The half-life of these NPs was more than doubled compared to nontargeted PLGA–PEG NPs [97].

4.3. PLGA NPs for Theranostic Applications

Theranostic represents a novel and powerful emerging platform that integrates targeted therapeutic entities with noninvasive imaging and has the great potential to personalize and advance medicine. Several nanosized delivery vehicles, including gold and iron-oxide nanoparticles, as well as quantum dots, have been extensively studied for theranostic applications [98–100]. Given safety concerns, off-target effects, and slow excretion kinetics from the body, risks may limit its use in the course of diseases. PLGA NP-based theranostic applications can deliver a therapeutic agent while simultaneously monitoring therapy response in real time [101,102]. Contrast agents, such as the radionuclide or fluorophore, play a critical role in enabling visualization of a target with conventional imaging techniques, e.g., magnetic resonance imaging (MRI), optical imaging, and X-ray computed tomography. Contrast agents including superparamagnetic iron oxide (SPIO) and gadolinium were shown to be encapsulated with polymeric nanoparticles [103]. For example, SPIO and chemotherapy drug docetaxel can both be directly encapsulated with PLGA [104]. Similarly, an human epidermal growth factor receptor 2 (HER2)-targeted PLGA–PEG block copolymer nanoparticle, upon encapsulation with $MnFe_2O_4$ and doxorubicin, was designed to target breast cancer in vivo [105]. Encapsulation of anticancer drug N′-(2-Methoxybenzylidene)-3-methyl-1-phenyl-H-Thieno[2,3-c]Pyrazole-5-Carbohyd-razide (MTPC) with plasmonic gold nanorods in PLGA-*b*-PEG polymeric nanospheres enhanced both biodistribution and pharmacokinetics of the MTPC in tumor-bearing mice [106]. Similar to MRI, radionuclide imaging has high sensitivity with no tissue-penetration limitations. Several radionuclide compounds have been extensively studied along with PLGA with the goal of formulating a robust nano delivery system [107]. For instance, Wang and colleagues [108] designed PLGA-lipid hybrid nanocarriers for theranostic therapy, encapsulating an anticancer agent in the matrix, while the lipid shell was chelated with indium-111 or yttrium-90 as radiotherapy agents. Shao et al. [107] demonstrated the therapeutic effect of 32P-CP-PLGA brachytherapy for glioma with the integrin $\alpha v \beta 3$-targeted radiotracer 68Ga-3PRGD2. Press and colleagues [109] demonstrated the cell-type-specific delivery of short interfering RNAs by covalent conjugation of DY-635 fluorescent dye with known hepatobiliary clearance to a PLGA, which allowed them to monitor distribution, uptake, and clearance of short hairpin RNA from the target organ. The major hurdle for the treatment of neurodegenerative diseases is to design therapeutic molecules in a way that it can cross the BBB. Zhang et al. [86] showed that the lectin-modified PLGA nanoparticle encapsulated with basic fibroblast growth factor enhanced drug delivery to the brain, supporting the role of the PLGA NP-based drug delivery system for CNS disorders. Therefore, nanocarriers based on PLGA offer biocompatibility, good stability, and regulated drug release rate and represent an excellent and emerging platform in theranostic medicine.

5. From Research and Development (R&D) to cGMP: Technologies for Scale-Up

To promote promising formulations from the R&D stage to clinical trials, drug products need to be manufactured under certain guidelines. Current Good Manufacturing Practice (cGMP) are regulations enforced by the FDA to ensure the quality of pharmaceutical products [110]. Most PLGA NP-based pharmaceutical products or medical devices are either injectable or implantable [111]. Therefore, sterility and potency are among the top-quality aspects to be considered for PLGA NPs. Per cGMP regulations, sterile products must be manufactured in a registered sterile facility [19]. Furthermore, facilities need to show that a sterile environment is properly maintained with validated sanitization [112]. PLGA NP-based pharmaceutical products have several steps during production: (1) dissolution and mixture of the active pharmaceutical ingredient (API) and PLGA; (2) stabilization of the mixture and removal of organic solvents; (3) separation of free API from PLGA–API product; (4) sterilization; (5) fill-finish. Table 2 presents selected publications showing processing details in the R&D stage.

For scale-up, dissolution of API/PLGA is achieved using stainless-steel tanks with temperature control through double jacketing. Mechanical motors are built into most tanks for light agitation. Since PLGA is not water-soluble, while lipophilic APIs can be dissolved concurrently, hydrophilic ones need to be dissolved separately from PLGA. To stabilize the formulation, PVA is mostly used as a surfactant. Moreover, poloxamer (also named Pluronic), polysorbate, sodium cholate, and D-α-tocopheryl pol-yethylene glycol succinate

(TPGS) are also reported as promising alternatives. Subsequently, aggressive agitation is needed to decrease the mixture droplets' particle size. High-pressure homogenization is widely used in cGMP production, and this technology is also reported with details in Table 2. To remove organic solvents, most publications took advantage of the different boiling points in water and organic solvents through overnight stirring. While this method is useful on the R&D scale, GMP-scale production uses vacuum-assisted rotary evaporation, which is more efficient and easier to validate.

Since PLGA NP-based pharmaceutical products represent a self-assembly drug delivery system, it is critical to separate free APIs from encapsulated ones. The encapsulation rate is one key quality control aspect due to the concern of toxicity. In Table 2, most publications used centrifugation, assisted by filters and membranes. In cGMP production, diafiltration is the technology preferred due to its higher recovery rate and better cost-effectiveness compared to centrifugation. Diafiltration uses a semipermeable membrane to separate the free drug from encapsulated drug based on the difference in cutoff molecular weight. Furthermore, diafiltration can form a closed system to return encapsulated products to the container for reprocessing until a target encapsulation rate is reached with the highest yield [113].

Among popular terminal sterilization methods, steam sterilization (autoclave) and gamma irradiation were reported to cause degradation of PLGA [114]. Vaporized ethylene oxide is commonly used for sterile gowning materials, but its residual is toxic if injected [115]. Electron beam technology is less aggressive than gamma irradiation but can still cause degradation of PLGA [116]. According to the available technologies for cGMP production and the particle size distribution reported from literature, sterile filtration is the best option for PLGA NP-based products. However, a filter compatibility study needs to be performed for validation purposes to minimize API retaining during filtration.

Depending on the concentration of PLGA NP-based pharmaceutical products, the solution may be free-flowing or with high viscosity. High viscosity is a problem for both sterile filtration and automatic filling systems. Filters may be clotted during filtration, the filling accuracy may be compromised, and rejection of vials may occur frequently. Therefore, an engineering fill-finish of placebo is strongly recommended as an industry standard for PLGA NP-based drug products. According to the stability data, lyophilization is recommended for extended shelf life. As presented in Table 2, lyophilization was reported using sugar molecules as cryoprotection. In cGMP production, lyophilization is considered a sterile product and is performed in the last step. The difficult part of lyophilization is the programming of cycles to minimize the moisture content (<5%, *w/w*).

Figure 2 is a concept flowchart for cGMP production of PLGA NP-based pharmaceutical products. The steps in Figure 2 are provided on the basis of the details in Table 2 and the industry standards under cGMP guidelines.

Table 2. Processing details for PLGA NP-based systems from publications.

Ref	API	API Solvent	PLGA Solvent	Primary Aqueous	Secondary Aqueous	Purification
[117]	Doxorubicin	0.001 N HCl	Dichloromethane		1% PVA in PBS	
[118]	Transferrin	RH buffer (pH = 7.4)	Acetone/ethanol	Water		
[66]	Memantine/rhodamine	Water	Ethyl acetate	PVA solution	0.3% PVA	Centrifuge: 15,000 rpm, 20 min
[119]	Rhodamine-6G	PBS	Ethyl acetate		1% PVA	Centrifuge
[80]	Donepezil	Water	Dichloromethane	2% Pluronic F68	0.5% Pluronic F68	Centrifuge: 20,000× g, 20 min
[120]	Doxorubicin	Water	Dichloromethane	Water	1% PVA or 1% HSA PBS with pH 7.2	G2 sintered glass filter
[121]	Elaprase®	Water	Dichloromethane	Water	1% PVA	Centrifuge: 17,000 rpm, 10 min, 5 °C
[122]	Olanzapine	Acetonitrile	Acetonitrile	0.25% Poloxamer 407		Centrifuge: 25,000 rpm, 4 °C
[71]	iAβ5	Chloroform	Chloroform		0.1% Pluronic F127	Centrifuge: 14,300× g, 40 min
[123]	Curcumin	Ethyl acetate	Ethyl acetate	5% PVA	0.3% PVA	Centrifuge: 6000 rpm, 20 min
[124]	Thiazolidinedione	AcOEt/EtOH (80/20)	AcOEt/EtOH (80/20)	PBS with polysorbate 80		
[125]	Bacoside-A	Methanol/dichloromethane (1:2)	Methanol/dichloromethane (1:2)	2% PVA		Centrifuge: 13,000 rpm, 30 min, 4 °C
[126]	Curcumin	Acetonitrile	Acetonitrile	Lipids, EtOH in water		Centrifuge: 10 kDa
[89]	Curcumin, paclitaxel	Chloroform	Chloroform	1% sodium cholate	0.5% sodium cholate	
[127]	3,3′-Diindolylmethane	Ethyl acetate	Ethyl acetate	Didodecyldimethylammonium bromide		Centrifuge: 35,000 rpm, 1 h
[69]	Loperamide	Ethanol/ethyl acetate 20/80	Ethanol/ethyl acetate 20/80	PBS with polysorbate 80		Centrifuge: 3 kDa filter
[82]	Loperamide	Acetone/ethanol	Acetone/ethanol	0.1 wt.% TPGS	1 wt.% polysorbate 80 or 1 wt.% poloxamer 188	Centrifuge: 15,000× g, 1 h, 5 °C
[67]	Coenzyme Q10	Acetone	Acetone	0.02% vitamin E TPGS		Centrifuge: 15,000× g, 15 min, 20 °C

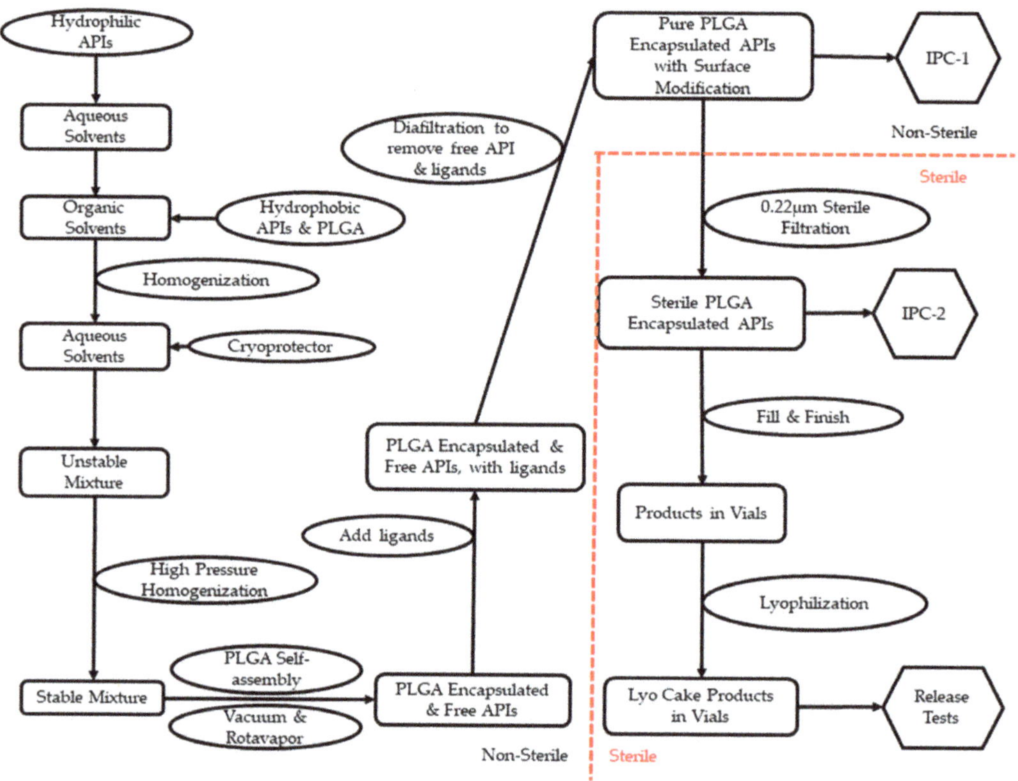

Figure 2. Concept flowchart of cGMP operations for PLGA NP-based drug products.

6. Quality and Regulatory Assurance

The quality of nano-based pharmaceutical products is a learning curve for both manufacturers and regulatory agencies. In 2006, the FDA initialed the Nanotechnology Task Force to improve the service for nano-based products. After its first public report in 2007, FDA published the second report in 2020, emphasizing its commitment to nano-based products. Among the five published guidance for industry regarding nanomaterials, two of them were designed for pharmaceutical and medical products. Another draft guidance is under review for drug products and biological products. On the basis of the above documents, FDA will not accept or reject nanotechnology products according to category. Instead, FDA will conduct a comprehensive review to make a science-focused decision.

Size distribution is the FDA's first consideration among all factors. Dynamic light scattering (DLS) technology is widely used for quality control purposes. Furthermore, a chromatography-based assay of APIs and encapsulation rate is critical to avoid toxicity. United States Pharmacopeia (USP) monograph "Goserelin Implants" also requests HPLC results to provide the retention time of PLGA, which is not rare for quality control of polymers [128]. Table 3 is a summary of the most needed in-process tests and final release tests.

Table 3. In-process and release tests for PLGA NP-based drug products.

Purpose	USP Chapter	Test
IPC-1	N/A	High-performance liquid chromatography
	N/A	Particle size distribution
	N/A	Zeta potential
	<791>	pH
	N/A	Morphology
IPC-2	N/A	Filter integrity test
	<71>	Sterility
	<785>	Osmolality
	<467>	Residual organic solvents
	<281>	Residue on ignition
	<731, 921>	Loss on drying for lyophilized products
Release	<790>	Visible particulate inspection
	<61>	Microbial enumeration
	<791>	pH
	<85>	Bacterial endotoxins
	<788>	Particulate matter for injection
	<1207>	Uniformity of dosages

During the Covid-19 pandemic, several liposome-based vaccines were approved for emergency use [129]. Liposome-based pharmaceutical products have similar self-assembly characteristics compared to PLGA NP-based ones. Therefore, their quality and regulatory focus should be similar. Furthermore, the success of vaccines also revealed the possibility of PLGA NP-based products to be approved through facilitated regulatory pathways, especially for emergency purposes under global regulatory systems [130,131].

7. Conclusions

This review presented recent progress in PLGA NPs as a vehicle to deliver drug to the brain in a controllable and targeted manner. Unlike most NPs, PLGA NPs show a promising future to become a clinically and commercially feasible drug delivery system. PLGA has been approved for both pharmaceutical products and medical devices, which provides clear quality control, quality assurance, and regulatory requirements for future products. Furthermore, the improvement of cGMP technologies, especially those in nanomedicines and sterile injectables, has also removed most obstacles for PLGA NPs. Even though safety is not the major concern, clinical trials are needed to monitor the efficacy and toxicity of PLGA NPs. Uncertainties, such as drug encapsulation rate, assembly stability, particle size distribution stability, and in vivo pharmacokinetics, may be the focus for future research and development.

Author Contributions: All the authors contributed to writing and/or editing the manuscripts.

Funding: The authors acknowledge financial support from the Plough Center for Sterile Drug Delivery Solutions and National Institutes of Health (CA213232).

Institutional Review Board Statement: Not applicable.

Informed Consent Statement: Not applicable.

Data Availability Statement: Not applicable.

Conflicts of Interest: The authors declare no conflict of interest. The funders had no role in the design of the study; in the collection, analyses, or interpretation of data; in the writing of the manuscript, or in the decision to publish the results.

References

1. Pardridge, W.M. Alzheimer's disease drug development and the problem of the blood-brain barrier. *Alzheimers Dement.* **2009**, *5*, 427–432. [CrossRef] [PubMed]
2. Gomez-Zepeda, D.; Taghi, M.; Scherrmann, J.M.; Decleves, X.; Menet, M.C. ABC Transporters at the Blood-Brain Interfaces, Their Study Models, and Drug Delivery Implications in Gliomas. *Pharmaceutics* **2019**, *12*, 20. [CrossRef]
3. Wong, H.L.; Wu, X.Y.; Bendayan, R. Nanotechnological advances for the delivery of CNS therapeutics. *Adv. Drug Deliv. Rev.* **2012**, *64*, 686–700. [CrossRef]
4. Bertrand, L.; Nair, M.; Toborek, M. Solving the Blood-Brain Barrier Challenge for the Effective Treatment of HIV Replication in the Central Nervous System. *Curr. Pharm. Des.* **2016**, *22*, 5477–5486. [CrossRef] [PubMed]
5. Dong, X. Current Strategies for Brain Drug Delivery. *Theranostics* **2018**, *8*, 1481–1493. [CrossRef]
6. Löscher, W.; Potschka, H. Blood-brain barrier active efflux transporters: ATP-binding cassette gene family. *NeuroRx* **2005**, *2*, 86–98. [CrossRef]
7. Mahringer, A.; Ott, M.; Reimold, I.; Reichel, V.; Fricker, G. The ABC of the blood-brain barrier—Regulation of drug efflux pumps. *Curr. Pharm. Des.* **2011**, *17*, 2762–2770. [CrossRef] [PubMed]
8. Kadry, H.; Noorani, B.; Cucullo, L. A blood-brain barrier overview on structure, function, impairment, and biomarkers of integrity. *Fluids Barriers CNS* **2020**, *17*, 69. [CrossRef]
9. Saylor, D.; Dickens, A.M.; Sacktor, N.; Haughey, N.; Slusher, B.; Pletnikov, M.; Mankowski, J.L.; Brown, A.; Volsky, D.J.; McArthur, J.C. HIV-associated neurocognitive disorder–pathogenesis and prospects for treatment. *Nat. Rev. Neurol.* **2016**, *12*, 234–248. [CrossRef]
10. Decloedt, E.H.; Rosenkranz, B.; Maartens, G.; Joska, J. Central nervous system penetration of antiretroviral drugs: Pharmacokinetic, pharmacodynamic and pharmacogenomic considerations. *Clin. Pharm.* **2015**, *54*, 581–598. [CrossRef] [PubMed]
11. Haas, D.W.; Johnson, B.; Nicotera, J.; Bailey, V.L.; Harris, V.L.; Bowles, F.B.; Raffanti, S.; Schranz, J.; Finn, T.S.; Saah, A.J.; et al. Effects of ritonavir on indinavir pharmacokinetics in cerebrospinal fluid and plasma. *Antimicrob. Agents Chemother.* **2003**, *47*, 2131–2137. [CrossRef] [PubMed]
12. Nair, M.; Jayant, R.D.; Kaushik, A.; Sagar, V. Getting into the brain: Potential of nanotechnology in the management of NeuroAIDS. *Adv. Drug Deliv. Rev.* **2016**, *103*, 202–217. [CrossRef]
13. Haluska, M.; Anthony, M.L. Osmotic blood-brain barrier modification for the treatment of malignant brain tumors. *Clin. J. Oncol. Nurs.* **2004**, *8*, 263–267. [CrossRef]
14. Zhi, K.; Lebo, D.B. A preformulation strategy for the selection of controlled-release components to simulate a subcutaneous implant. *Boletín Latinoam. Caribe Plantas Med. Aromáticas* **2020**, *19*, 344–356. [CrossRef]
15. Gong, Y.; Chowdhury, P.; Nagesh, P.K.B.; Rahman, M.A.; Zhi, K.; Yallapu, M.M.; Kumar, S. Novel elvitegravir nanoformulation for drug delivery across the blood-brain barrier to achieve HIV-1 suppression in the CNS macrophages. *Sci. Rep.* **2020**, *10*, 3835. [CrossRef]
16. Gong, Y.; Zhi, K.; Nagesh, P.K.B.; Sinha, N.; Chowdhury, P.; Chen, H.; Gorantla, S.; Yallapu, M.M.; Kumar, S. An Elvitegravir Nanoformulation Crosses the Blood-Brain Barrier and Suppresses HIV-1 Replication in Microglia. *Viruses* **2020**, *12*, 564. [CrossRef]
17. Patel, T.; Zhou, J.; Piepmeier, J.M.; Saltzman, W.M. Polymeric nanoparticles for drug delivery to the central nervous system. *Adv. Drug. Deliv. Rev.* **2012**, *64*, 701–705. [CrossRef]
18. Zhou, Y.; Peng, Z.; Seven, E.S.; Leblanc, R.M. Crossing the blood-brain barrier with nanoparticles. *J. Control Release* **2018**, *270*, 290–303. [CrossRef]
19. Zhi, K.; Kumar, A.; Raji, B.; Kochat, H.; Kumar, S. Formulation, manufacturing and regulatory strategies for extracellular vesicles-based drug products for targeted therapy of central nervous system diseases. *Expert Rev. Precis. Med. Drug Dev.* **2020**, *5*, 469–481. [CrossRef]
20. Kumar, S.; Zhi, K.; Mukherji, A.; Gerth, K. Repurposing antiviral protease inhibitors using extracellular vesicles for potential therapy of COVID-19. *Viruses* **2020**, *12*, 486. [CrossRef] [PubMed]
21. Kumar, A.; Zhou, L.; Zhi, K.; Raji, B.; Pernell, S.; Tadrous, E.; Kodidela, S.; Nookala, A.; Kochat, H.; Kumar, S. Challenges in Biomaterial-Based Drug Delivery Approach for the Treatment of Neurodegenerative Diseases: Opportunities for Extracellular Vesicles. *Int. J. Mol. Sci.* **2020**, *22*, 138. [CrossRef] [PubMed]
22. Gilding, D.K.; Reed, A.M. Biodegradable polymers for use in surgery—polyglycolic/poly(actic acid) homo- and copolymers: 1. *Polymer* **1979**, *20*, 1459–1464. [CrossRef]
23. Deasy, P.B.; Finan, M.P.; Meegan, M.J. Preparation and characterization of lactic/glycolic acid polymers and copolymers. *J. Microencapsul.* **1989**, *6*, 369–378. [CrossRef]
24. Gao, Q.; Lan, P.; Shao, H.; Hu, X. Direct Synthesis with Melt Polycondensation and Microstructure Analysis of Poly(L-lactic acid-co-glycolic acid). *Polym. J.* **2002**, *34*, 786–793. [CrossRef]
25. Fukuzaki, H.; Yoshida, M.; Asano, M.; Kumakura, M. Synthesis of copoly(D,L-lactic acid) with relatively low molecular weight and in vitro degradation. *Eur. Polym. J.* **1989**, *25*, 1019–1026. [CrossRef]
26. Bendix, D. Chemical synthesis of polylactide and its copolymers for medical applications. *Polym. Degrad. Stab.* **1998**, *59*, 129–135. [CrossRef]
27. Lunt, J. Large-scale production, properties and commercial applications of polylactic acid polymers. *Polym. Degrad. Stab.* **1998**, *59*, 145–152. [CrossRef]

28. Sarasua, J.-R.; Prud'homme, R.E.; Wisniewski, M.; Le Borgne, A.; Spassky, N. Crystallization and Melting Behavior of Polylactides. *Macromolecules* **1998**, *31*, 3895–3905. [CrossRef]
29. Sarasua, J.R.; López-Rodríguez, N.; Zuza, E.; Petisco, S.; Castro, B.; del Olmo, M.; Palomares, T.; Alonso-Varona, A. Crystallinity assessment and in vitro cytotoxicity of polylactide scaffolds for biomedical applications. *J. Mater. Sci. Mater. Med.* **2011**, *22*, 2513–2523. [CrossRef]
30. Tsuji, H.; Miyauchi, S. Poly(l-lactide): VI Effects of crystallinity on enzymatic hydrolysis of poly(l-lactide) without free amorphous region. *Polym. Degrad. Stab.* **2001**, *71*, 415–424. [CrossRef]
31. Wang, N.; Wu, X.S.; Li, C.; Feng, M.F. Synthesis, characterization, biodegradation, and drug delivery application of biodegradable lactic/glycolic acid polymers: I. Synthesis and characterization. *J. Biomater. Sci. Polym. Ed.* **2000**, *11*, 301–318. [CrossRef] [PubMed]
32. Kapoor, D.N.; Bhatia, A.; Kaur, R.; Sharma, R.; Kaur, G.; Dhawan, S. PLGA: A unique polymer for drug delivery. *Ther. Deliv.* **2015**, *6*, 41–58. [CrossRef] [PubMed]
33. Owens, D.E., 3rd; Peppas, N.A. Opsonization, biodistribution, and pharmacokinetics of polymeric nanoparticles. *Int. J. Pharm.* **2006**, *307*, 93–102. [CrossRef] [PubMed]
34. Alexis, F.; Pridgen, E.; Molnar, L.K.; Farokhzad, O.C. Factors affecting the clearance and biodistribution of polymeric nanoparticles. *Mol. Pharm.* **2008**, *5*, 505–515. [CrossRef]
35. Hoyos-Ceballos, G.P.; Ruozi, B.; Ottonelli, I.; Da Ros, F.; Vandelli, M.A.; Forni, F.; Daini, E.; Vilella, A.; Zoli, M.; Tosi, G.; et al. PLGA-PEG-ANG-2 Nanoparticles for Blood-Brain Barrier Crossing: Proof-of-Concept Study. *Pharmaceutics* **2020**, *12*, 72. [CrossRef] [PubMed]
36. Erbetta, C.D.A.C.; Alves, R.J.; Magalh, J.; de Souza Freitas, R.F.; de Sousa, R.G. Synthesis and characterization of poly (D, L-lactide-co-glycolide) copolymer. *J. Biomater. Nanobiotechnol.* **2012**, *3*, 208–225. [CrossRef]
37. Gentile, P.; Chiono, V.; Carmagnola, I.; Hatton, P.V. An overview of poly(lactic-co-glycolic) acid (PLGA)-based biomaterials for bone tissue engineering. *Int. J. Mol. Sci.* **2014**, *15*, 3640–3659. [CrossRef] [PubMed]
38. Wu, X.S.; Wang, N. Synthesis, characterization, biodegradation, and drug delivery application of biodegradable lactic/glycolic acid polymers. Part II: Biodegradation. *J. Biomater. Sci. Polym. Ed.* **2001**, *12*, 21–34. [CrossRef]
39. Makadia, H.K.; Siegel, S.J. Poly Lactic-co-Glycolic Acid (PLGA) as Biodegradable Controlled Drug Delivery Carrier. *Polymers* **2011**, *3*, 1377–1397. [CrossRef]
40. Locatelli, E.; Franchini, M.C. Biodegradable PLGA-b-PEG polymeric nanoparticles: Synthesis, properties, and nanomedical applications as drug delivery system. *J. Nanoparticle Res.* **2012**, *14*, 1–17. [CrossRef]
41. Gong, Y.; Chowdhury, P.; Midde, N.M.; Rahman, M.A.; Yallapu, M.M.; Kumar, S. Novel elvitegravir nanoformulation approach to suppress the viral load in HIV-infected macrophages. *Biochem. Biophys. Rep.* **2017**, *12*, 214–219. [CrossRef]
42. Kluge, J.; Fusaro, F.; Casas, N.; Mazzotti, M.; Muhrer, G. Production of PLGA micro- and nanocomposites by supercritical fluid extraction of emulsions: I. Encapsulation of lysozyme. *J. Supercrit. Fluids* **2009**, *50*, 327–335. [CrossRef]
43. Arshady, R. Preparation of biodegradable microspheres and microcapsules: 2. Polyactides and related polyesters. *J. Control. Release* **1991**, *17*, 1–21. [CrossRef]
44. Edelman, R.; Russell, R.G.; Losonsky, G.; Tall, B.D.; Tacket, C.O.; Levine, M.M.; Lewis, D.H. Immunization of rabbits with enterotoxigenic E. coli colonization factor antigen (CFA/I) encapsulated in biodegradable microspheres of poly (lactide-co-glycolide). *Vaccine* **1993**, *11*, 155–158. [CrossRef]
45. Mu, L.; Feng, S.S. Fabrication, characterization and in vitro release of paclitaxel (Taxol) loaded poly (lactic-co-glycolic acid) microspheres prepared by spray drying technique with lipid/cholesterol emulsifiers. *J. Control Release* **2001**, *76*, 239–254. [CrossRef]
46. Jensen, D.M.; Cun, D.; Maltesen, M.J.; Frokjaer, S.; Nielsen, H.M.; Foged, C. Spray drying of siRNA-containing PLGA nanoparticles intended for inhalation. *J. Control Release* **2010**, *142*, 138–145. [CrossRef]
47. Arpagaus, C. PLA/PLGA nanoparticles prepared by nano spray drying. *J. Pharm. Investig.* **2019**, *49*, 405–426. [CrossRef]
48. Ibrahim, H.; Bindschaedler, C.; Doelker, E.; Buri, P.; Gurny, R. Aqueous nanodispersions prepared by a salting-out process. *Int. J. Pharm.* **1992**, *87*, 239–246. [CrossRef]
49. Allémann, E.; Gurny, R.; Doelker, E. Preparation of aqueous polymeric nanodispersions by a reversible salting-out process: Influence of process parameters on particle size. *Int. J. Pharm.* **1992**, *87*, 247–253. [CrossRef]
50. Huang, W.; Zhang, C. Tuning the Size of Poly(lactic-co-glycolic Acid) (PLGA) Nanoparticles Fabricated by Nanoprecipitation. *Biotechnol. J.* **2018**, *13*, 1700203. [CrossRef]
51. Fessi, H.; Puisieux, F.; Devissaguet, J.P.; Ammoury, N.; Benita, S. Nanocapsule formation by interfacial polymer deposition following solvent displacement. *Int. J. Pharm.* **1989**, *55*, R1–R4. [CrossRef]
52. Jain, R.A. The manufacturing techniques of various drug loaded biodegradable poly(lactide-co-glycolide) (PLGA) devices. *Biomaterials* **2000**, *21*, 2475–2490. [CrossRef]
53. Astete, C.E.; Sabliov, C.M. Synthesis and characterization of PLGA nanoparticles. *J. Biomater. Sci. Polym. Ed.* **2006**, *17*, 247–289. [CrossRef]
54. Ding, D.; Zhu, Q. Recent advances of PLGA micro/nanoparticles for the delivery of biomacromolecular therapeutics. *Mater. Sci. Eng. C Mater. Biol. Appl.* **2018**, *92*, 1041–1060. [CrossRef]
55. Hervé, F.; Ghinea, N.; Scherrmann, J.M. CNS delivery via adsorptive transcytosis. *AAPS J.* **2008**, *10*, 455–472. [CrossRef]

56. Pulgar, V.M. Transcytosis to Cross the Blood Brain Barrier, New Advancements and Challenges. *Front. Neurosci.* **2018**, *12*, 1019. [CrossRef]

57. Falanga, A.P.; Melone, P.; Cagliani, R.; Borbone, N.; D'Errico, S.; Piccialli, G.; Netti, P.A.; Guarnieri, D. Design, Synthesis and Characterization of Novel Co-Polymers Decorated with Peptides for the Selective Nanoparticle Transport across the Cerebral Endothelium. *Molecules* **2018**, *23*, 1655. [CrossRef]

58. Ye, C.; Pan, B.; Xu, H.; Zhao, Z.; Shen, J.; Lu, J.; Yu, R.; Liu, H. Co-delivery of GOLPH3 siRNA and gefitinib by cationic lipid-PLGA nanoparticles improves EGFR-targeted therapy for glioma. *J. Mol. Med.* **2019**, *97*, 1575–1588. [CrossRef] [PubMed]

59. Li, X.; Su, J.; Kamal, Z.; Guo, P.; Wu, X.; Lu, L.; Wu, H.; Qiu, M. Odorranalectin modified PEG-PLGA/PEG-PBLG curcumin-loaded nanoparticle for intranasal administration. *Drug Dev. Ind. Pharm.* **2020**, *46*, 899–909. [CrossRef] [PubMed]

60. Shah, P.; Sarolia, J.; Vyas, B.; Wagh, P.; Ankur, K.; Kumar, M.A. PLGA nanoparticles for nose to brain delivery of Clonazepam: Formulation, optimization by 32 Factorial design, in vitro and in vivo evaluation. *Curr. Drug Deliv.* **2020**. [CrossRef]

61. Chatzitaki, A.T.; Jesus, S.; Karavasili, C.; Andreadis, D.; Fatouros, D.G.; Borges, O. Chitosan-coated PLGA nanoparticles for the nasal delivery of ropinirole hydrochloride: In vitro and ex vivo evaluation of efficacy and safety. *Int. J. Pharm.* **2020**, *589*, 119776. [CrossRef]

62. Zhao, P.; Le, Z.; Liu, L.; Chen, Y. Therapeutic Delivery to the Brain via the Lymphatic Vasculature. *Nano Lett.* **2020**, *20*, 5415–5420. [CrossRef]

63. Jusu, S.M.; Obayemi, J.D.; Salifu, A.A.; Nwazojie, C.C.; Uzonwanne, V.; Odusanya, O.S.; Soboyejo, W.O. Drug-encapsulated blend of PLGA-PEG microspheres: In vitro and in vivo study of the effects of localized/targeted drug delivery on the treatment of triple-negative breast cancer. *Sci. Rep.* **2020**, *10*, 14188. [CrossRef]

64. Zhang, K.; Tang, X.; Zhang, J.; Lu, W.; Lin, X.; Zhang, Y.; Tian, B.; Yang, H.; He, H. PEG-PLGA copolymers: Their structure and structure-influenced drug delivery applications. *J. Control Release* **2014**, *183*, 77–86. [CrossRef]

65. Rafiei, P.; Haddadi, A. Docetaxel-loaded PLGA and PLGA-PEG nanoparticles for intravenous application: Pharmacokinetics and biodistribution profile. *Int. J. Nanomed.* **2017**, *12*, 935–947. [CrossRef] [PubMed]

66. Sánchez-López, E.; Ettcheto, M.; Egea, M.A.; Espina, M.; Cano, A.; Calpena, A.C.; Camins, A.; Carmona, N.; Silva, A.M.; Souto, E.B.; et al. Memantine loaded PLGA PEGylated nanoparticles for Alzheimer's disease: In vitro and in vivo characterization. *J. Nanobiotechnol.* **2018**, *16*, 32. [CrossRef]

67. Wang, Z.H.; Wang, Z.Y.; Sun, C.S.; Wang, C.Y.; Jiang, T.Y.; Wang, S.L. Trimethylated chitosan-conjugated PLGA nanoparticles for the delivery of drugs to the brain. *Biomaterials* **2010**, *31*, 908–915. [CrossRef]

68. Wang, L.; Hao, Y.; Li, H.; Zhao, Y.; Meng, D.; Li, D.; Shi, J.; Zhang, H.; Zhang, Z.; Zhang, Y. Co-delivery of doxorubicin and siRNA for glioma therapy by a brain targeting system: Angiopep-2-modified poly(lactic-co-glycolic acid) nanoparticles. *J. Drug Target* **2015**, *23*, 832–846. [CrossRef]

69. Fornaguera, C.; Dols-Perez, A.; Calderó, G.; García-Celma, M.J.; Camarasa, J.; Solans, C. PLGA nanoparticles prepared by nano-emulsion templating using low-energy methods as efficient nanocarriers for drug delivery across the blood-brain barrier. *J. Control Release* **2015**, *211*, 134–143. [CrossRef] [PubMed]

70. Kuo, Y.C.; Chen, Y.C. Targeting delivery of etoposide to inhibit the growth of human glioblastoma multiforme using lactoferrin- and folic acid-grafted poly(lactide-co-glycolide) nanoparticles. *Int. J. Pharm.* **2015**, *479*, 138–149. [CrossRef] [PubMed]

71. Loureiro, J.A.; Gomes, B.; Fricker, G.; Coelho, M.A.N.; Rocha, S.; Pereira, M.C. Cellular uptake of PLGA nanoparticles targeted with anti-amyloid and anti-transferrin receptor antibodies for Alzheimer's disease treatment. *Colloids Surf. B Biointerfaces* **2016**, *145*, 8–13. [CrossRef] [PubMed]

72. Bi, C.; Wang, A.; Chu, Y.; Liu, S.; Mu, H.; Liu, W.; Wu, Z.; Sun, K.; Li, Y. Intranasal delivery of rotigotine to the brain with lactoferrin-modified PEG-PLGA nanoparticles for Parkinson's disease treatment. *Int. J. Nanomed.* **2016**, *11*, 6547–6559. [CrossRef]

73. Hua, H.; Zhang, X.; Mu, H.; Meng, Q.; Jiang, Y.; Wang, Y.; Lu, X.; Wang, A.; Liu, S.; Zhang, Y.; et al. RVG29-modified docetaxel-loaded nanoparticles for brain-targeted glioma therapy. *Int. J. Pharm.* **2018**, *543*, 179–189. [CrossRef]

74. Ramalho, M.J.; Sevin, E.; Gosselet, F.; Lima, J.; Coelho, M.A.N.; Loureiro, J.A.; Pereira, M.C. Receptor-mediated PLGA nanoparticles for glioblastoma multiforme treatment. *Int. J. Pharm.* **2018**, *545*, 84–92. [CrossRef] [PubMed]

75. Muniswamy, V.J.; Raval, N.; Gondaliya, P.; Tambe, V.; Kalia, K.; Tekade, R.K. 'Dendrimer-Cationized-Albumin' encrusted polymeric nanoparticle improves BBB penetration and anticancer activity of doxorubicin. *Int. J. Pharm.* **2019**, *555*, 77–99. [CrossRef] [PubMed]

76. Lei, C.; Davoodi, P.; Zhan, W.; Chow, P.K.; Wang, C.H. Development of Nanoparticles for Drug Delivery to Brain Tumor: The Effect of Surface Materials on Penetration Into Brain Tissue. *J. Pharm. Sci.* **2019**, *108*, 1736–1745. [CrossRef]

77. Xu, R.; Wang, J.; Xu, J.; Song, X.; Huang, H.; Feng, Y.; Fu, C. Rhynchophylline Loaded-mPEG-PLGA Nanoparticles Coated with Tween-80 for Preliminary Study in Alzheimer's Disease. *Int. J. Nanomed.* **2020**, *15*, 1149–1160. [CrossRef]

78. Silva-Abreu, M.; Calpena, A.C.; Andrés-Benito, P.; Aso, E.; Romero, I.A.; Roig-Carles, D.; Gromnicova, R.; Espina, M.; Ferrer, I.; García, M.L.; et al. PPARγ agonist-loaded PLGA-PEG nanocarriers as a potential treatment for Alzheimer's disease: In vitro and in vivo studies. *Int. J. Nanomed.* **2018**, *13*, 5577–5590. [CrossRef]

79. Baysal, I.; Yabanoglu-Ciftci, S.; Tunc-Sarisozen, Y.; Ulubayram, K.; Ucar, G. Interaction of selegiline-loaded PLGA-b-PEG nanoparticles with beta-amyloid fibrils. *J. Neural Transm.* **2013**, *120*, 903–910. [CrossRef]

80. Baysal, I.; Ucar, G.; Gultekinoglu, M.; Ulubayram, K.; Yabanoglu-Ciftci, S. Donepezil loaded PLGA-b-PEG nanoparticles: Their ability to induce destabilization of amyloid fibrils and to cross blood brain barrier in vitro. *J. Neural Transm.* **2017**, *124*, 33–45. [CrossRef] [PubMed]

81. Amin, F.U.; Shah, S.A.; Badshah, H.; Khan, M.; Kim, M.O. Anthocyanins encapsulated by PLGA@PEG nanoparticles potentially improved its free radical scavenging capabilities via p38/JNK pathway against Aβ(1-42)-induced oxidative stress. *J. Nanobiotechnology* **2017**, *15*, 12. [CrossRef]

82. Chen, Y.C.; Hsieh, W.Y.; Lee, W.F.; Zeng, D.T. Effects of surface modification of PLGA-PEG-PLGA nanoparticles on loperamide delivery efficiency across the blood-brain barrier. *J. Biomater. Appl.* **2013**, *27*, 909–922. [CrossRef] [PubMed]

83. Fan, S.; Zheng, Y.; Liu, X.; Fang, W.; Chen, X.; Liao, W.; Jing, X.; Lei, M.; Tao, E.; Ma, Q.; et al. Curcumin-loaded PLGA-PEG nanoparticles conjugated with B6 peptide for potential use in Alzheimer's disease. *Drug Deliv.* **2018**, *25*, 1091–1102. [CrossRef] [PubMed]

84. Surnar, B.; Basu, U.; Banik, B.; Ahmad, A.; Marples, B.; Kolishetti, N.; Dhar, S. Nanotechnology-mediated crossing of two impermeable membranes to modulate the stars of the neurovascular unit for neuroprotection. *Proc. Natl. Acad. Sci. USA* **2018**, *115*, E12333–E12342. [CrossRef]

85. Yu, Y.; Pang, Z.; Lu, W.; Yin, Q.; Gao, H.; Jiang, X. Self-assembled polymersomes conjugated with lactoferrin as novel drug carrier for brain delivery. *Pharm. Res.* **2012**, *29*, 83–96. [CrossRef]

86. Zhang, C.; Chen, J.; Feng, C.; Shao, X.; Liu, Q.; Zhang, Q.; Pang, Z.; Jiang, X. Intranasal nanoparticles of basic fibroblast growth factor for brain delivery to treat Alzheimer's disease. *Int. J. Pharm.* **2014**, *461*, 192–202. [CrossRef] [PubMed]

87. Cano, A.; Ettcheto, M.; Espina, M.; Auladell, C.; Folch, J.; Kühne, B.A.; Barenys, M.; Sánchez-López, E.; Souto, E.B.; García, M.L.; et al. Epigallocatechin-3-gallate PEGylated poly(lactic-co-glycolic) acid nanoparticles mitigate striatal pathology and motor deficits in 3-nitropropionic acid intoxicated mice. *Nanomedicine* **2021**, *16*, 19–35. [CrossRef]

88. Price, R.L.; Chiocca, E.A. Evolution of malignant glioma treatment: From chemotherapy to vaccines to viruses. *Neurosurgery* **2014**, *61* (Suppl. 1), 74–83. [CrossRef]

89. Cui, Y.; Zhang, M.; Zeng, F.; Jin, H.; Xu, Q.; Huang, Y. Dual-Targeting Magnetic PLGA Nanoparticles for Codelivery of Paclitaxel and Curcumin for Brain Tumor Therapy. *Acs Appl. Mater. Interfaces* **2016**, *8*, 32159–32169. [CrossRef]

90. Zhang, X.; Zhao, L.; Zhai, G.; Ji, J.; Liu, A. Multifunctional Polyethylene Glycol (PEG)-Poly (Lactic-Co-Glycolic Acid) (PLGA)-Based Nanoparticles Loading Doxorubicin and Tetrahydrocurcumin for Combined Chemoradiotherapy of Glioma. *Med. Sci. Monit.* **2020**, *25*, 9737. [CrossRef]

91. Li, H.; Tong, Y.; Bai, L.; Ye, L.; Zhong, L.; Duan, X.; Zhu, Y. Lactoferrin functionalized PEG-PLGA nanoparticles of shikonin for brain targeting therapy of glioma. *Int. J. Biol. Macromol.* **2018**, *107*, 204–211. [CrossRef] [PubMed]

92. Kang, T.; Gao, X.; Hu, Q.; Jiang, D.; Feng, X.; Zhang, X.; Song, Q.; Yao, L.; Huang, M.; Jiang, X.; et al. iNGR-modified PEG-PLGA nanoparticles that recognize tumor vasculature and penetrate gliomas. *Biomaterials* **2014**, *35*, 4319–4332. [CrossRef]

93. Kaffashi, A.; Lüle, S.; Bozdağ Pehlivan, S.; Sarısözen, C.; Vural, İ.; Koşucu, H.; Demir, T.; Buğdaycı, K.E.; Söylemezoğlu, F.; Karlı Oğuz, K.; et al. Farnesylthiosalicylic acid-loaded lipid-polyethylene glycol-polymer hybrid nanoparticles for treatment of glioblastoma. *J. Pharm. Pharm.* **2017**, *69*, 1010–1021. [CrossRef]

94. Zhang, X.; Li, X.; Hua, H.; Wang, A.; Liu, W.; Li, Y.; Fu, F.; Shi, Y.; Sun, K. Cyclic hexapeptide-conjugated nanoparticles enhance curcumin delivery to glioma tumor cells and tissue. *Int. J. Nanomed.* **2017**, *12*, 5717–5732. [CrossRef]

95. Wang, B.; Lv, L.; Wang, Z.; Zhao, Y.; Wu, L.; Fang, X.; Xu, Q.; Xin, H. Nanoparticles functionalized with Pep-1 as potential glioma targeting delivery system via interleukin 13 receptor α2-mediated endocytosis. *Biomaterials* **2014**, *35*, 5897–5907. [CrossRef]

96. Lv, L.; Jiang, Y.; Liu, X.; Wang, B.; Lv, W.; Zhao, Y.; Shi, H.; Hu, Q.; Xin, H.; Xu, Q.; et al. Enhanced Antiglioblastoma Efficacy of Neovasculature and Glioma Cells Dual Targeted Nanoparticles. *Mol. Pharm.* **2016**, *13*, 3506–3517. [CrossRef]

97. Wadajkar, A.S.; Dancy, J.G.; Roberts, N.B.; Connolly, N.P.; Strickland, D.K.; Winkles, J.A.; Woodworth, G.F.; Kim, A.J. Decreased non-specific adhesivity, receptor targeted (DART) nanoparticles exhibit improved dispersion, cellular uptake, and tumor retention in invasive gliomas. *J. Control Release* **2017**, *267*, 144–153. [CrossRef]

98. Elgqvist, J. Nanoparticles as Theranostic Vehicles in Experimental and Clinical Applications-Focus on Prostate and Breast Cancer. *Int. J. Mol. Sci.* **2017**, *18*, 1102. [CrossRef] [PubMed]

99. Sharma, R.; Mody, N.; Agrawal, U.; Vyas, S.P. Theranostic Nanomedicine; A Next Generation Platform for Cancer Diagnosis and Therapy. *Mini Rev. Med. Chem.* **2017**, *17*, 1746–1757. [CrossRef]

100. Chen, F.; Ehlerding, E.B.; Cai, W. Theranostic nanoparticles. *J. Nucl. Med.* **2014**, *55*, 1919–1922. [CrossRef] [PubMed]

101. Mieszawska, A.J.; Kim, Y.; Gianella, A.; van Rooy, I.; Priem, B.; Labarre, M.P.; Ozcan, C.; Cormode, D.P.; Petrov, A.; Langer, R.; et al. Synthesis of polymer-lipid nanoparticles for image-guided delivery of dual modality therapy. *Bioconjug. Chem.* **2013**, *24*, 1429–1434. [CrossRef] [PubMed]

102. Xin, Y.; Liu, T.; Yang, C. Development of PLGA-lipid nanoparticles with covalently conjugated indocyanine green as a versatile nanoplatform for tumor-targeted imaging and drug delivery. *Int. J. Nanomed.* **2016**, *11*, 5807–5821. [CrossRef] [PubMed]

103. Zhou, S.; Sun, J.; Sun, L.; Dai, Y.; Liu, L.; Li, X.; Wang, J.; Weng, J.; Jia, W.; Zhang, Z. Preparation and characterization of interferon-loaded magnetic biodegradable microspheres. *J. Biomed. Mater. Res. B Appl. Biomater.* **2008**, *87*, 189–196. [CrossRef]

104. Ling, Y.; Wei, K.; Luo, Y.; Gao, X.; Zhong, S. Dual docetaxel/superparamagnetic iron oxide loaded nanoparticles for both targeting magnetic resonance imaging and cancer therapy. *Biomaterials* **2011**, *32*, 7139–7150. [CrossRef] [PubMed]

105. Yang, J.; Lee, C.H.; Ko, H.J.; Suh, J.S.; Yoon, H.G.; Lee, K.; Huh, Y.M.; Haam, S. Multifunctional magneto-polymeric nanohybrids for targeted detection and synergistic therapeutic effects on breast cancer. *Angew. Chem. Int. Ed. Engl.* **2007**, *46*, 8836–8839. [CrossRef]

106. Darwish, W.M.A.; Bayoumi, N.A. Gold nanorod-loaded (PLGA-PEG) nanocapsules as near-infrared controlled release model of anticancer therapeutics. *Lasers Med. Sci.* **2020**, *35*, 1729–1740. [CrossRef] [PubMed]

107. Shao, G.; Wang, Y.; Liu, X.; Zhao, M.; Song, J.; Huang, P.; Wang, F.; Wang, Z. Investigation of Newly Prepared Biodegradable (32)P-chromic Phosphate-polylactide-co-glycolide Seeds and Their Therapeutic Response Evaluation for Glioma Brachytherapy. *Contrast Media Mol. Imaging* **2018**, *2018*, 2630480. [CrossRef]

108. Wang, A.Z.; Yuet, K.; Zhang, L.; Gu, F.X.; Huynh-Le, M.; Radovic-Moreno, A.F.; Kantoff, P.W.; Bander, N.H.; Langer, R.; Farokhzad, O.C. ChemoRad nanoparticles: A novel multifunctional nanoparticle platform for targeted delivery of concurrent chemoradiation. *Nanomedicine* **2010**, *5*, 361–368. [CrossRef] [PubMed]

109. Press, A.T.; Traeger, A.; Pietsch, C.; Mosig, A.; Wagner, M.; Clemens, M.G.; Jbeily, N.; Koch, N.; Gottschaldt, M.; Bézière, N.; et al. Cell type-specific delivery of short interfering RNAs by dye-functionalised theranostic nanoparticles. *Nat. Commun.* **2014**, *5*, 5565. [CrossRef]

110. US-FDA. Current Good Manufacturing Practice (CGMP) Regulations. Available online: https://www.fda.gov/drugs/pharmaceutical-quality-resources/current-good-manufacturing-practice-cgmp-regulations (accessed on 18 February 2021).

111. Park, K.; Skidmore, S.; Hadar, J.; Garner, J.; Park, H.; Otte, A.; Soh, B.K.; Yoon, G.; Yu, D.; Yun, Y.; et al. Injectable, long-acting PLGA formulations: Analyzing PLGA and understanding microparticle formation. *J. Control. Release* **2019**, *304*, 125–134. [CrossRef]

112. Polarine, J.; Chai, R.; Kochat, H.; Pulliam, P.J.; Zhi, K.; Brooks, K. In-Situ Disinfectant Validation Case Study. *Am. Pharm. Rev.* **2021**, 1–6. [CrossRef]

113. Wasalathanthri, D.P.; Feroz, H.; Puri, N.; Hung, J.; Lane, G.; Holstein, M.; Chemmalil, L.; Both, D.; Ghose, S.; Ding, J. Real-time monitoring of quality attributes by inline Fourier transform infrared spectroscopic sensors at ultrafiltration and diafiltration of bioprocess. *Biotechnol. Bioeng.* **2020**, *117*, 3766–3774. [CrossRef] [PubMed]

114. Yaman, A. Alternative methods of terminal sterilization for biologically active macromolecules. *Curr. Opin. Drug Discov. Devel.* **2001**, *4*, 760–763. [PubMed]

115. Lü, J.M.; Wang, X.; Marin-Muller, C.; Wang, H.; Lin, P.H.; Yao, Q.; Chen, C. Current advances in research and clinical applications of PLGA-based nanotechnology. *Expert Rev. Mol. Diagn.* **2009**, *9*, 325–341. [CrossRef]

116. Loo, J.; Ooi, C.; Boey, F. Degradation of poly (lactide-co-glycolide)(PLGA) and poly (L-lactide)(PLLA) by electron beam radiation. *Biomaterials* **2005**, *26*, 1359–1367. [CrossRef]

117. Malinovskaya, Y.; Melnikov, P.; Baklaushev, V.; Gabashvili, A.; Osipova, N.; Mantrov, S.; Ermolenko, Y.; Maksimenko, O.; Gorshkova, M.; Balabanyan, V.; et al. Delivery of doxorubicin-loaded PLGA nanoparticles into U87 human glioblastoma cells. *Int. J. Pharm.* **2017**, *524*, 77–90. [CrossRef]

118. Chang, J.; Paillard, A.; Passirani, C.; Morille, M.; Benoit, J.P.; Betbeder, D.; Garcion, E. Transferrin adsorption onto PLGA nanoparticles governs their interaction with biological systems from blood circulation to brain cancer cells. *Pharm. Res.* **2012**, *29*, 1495–1505. [CrossRef] [PubMed]

119. Semete, B.; Booysen, L.; Lemmer, Y.; Kalombo, L.; Katata, L.; Verschoor, J.; Swai, H.S. In vivo evaluation of the biodistribution and safety of PLGA nanoparticles as drug delivery systems. *Nanomedicine* **2010**, *6*, 662–671. [CrossRef] [PubMed]

120. Wohlfart, S.; Khalansky, A.S.; Gelperina, S.; Maksimenko, O.; Bernreuther, C.; Glatzel, M.; Kreuter, J. Efficient chemotherapy of rat glioblastoma using doxorubicin-loaded PLGA nanoparticles with different stabilizers. *PloS ONE* **2011**, *6*, e19121. [CrossRef]

121. Rigon, L.; Salvalaio, M.; Pederzoli, F.; Legnini, E.; Duskey, J.T.; D'Avanzo, F.; De Filippis, C.; Ruozi, B.; Marin, O.; Vandelli, M.A.; et al. Targeting Brain Disease in MPSII: Preclinical Evaluation of IDS-Loaded PLGA Nanoparticles. *Int. J. Mol. Sci.* **2019**, *20*, 2014. [CrossRef] [PubMed]

122. Seju, U.; Kumar, A.; Sawant, K.K. Development and evaluation of olanzapine-loaded PLGA nanoparticles for nose-to-brain delivery: In vitro and in vivo studies. *Acta Biomater.* **2011**, *7*, 4169–4176. [CrossRef] [PubMed]

123. Mathew, A.; Fukuda, T.; Nagaoka, Y.; Hasumura, T.; Morimoto, H.; Yoshida, Y.; Maekawa, T.; Venugopal, K.; Kumar, D.S. Curcumin loaded-PLGA nanoparticles conjugated with Tet-1 peptide for potential use in Alzheimer's disease. *PloS ONE* **2012**, *7*, e32616. [CrossRef] [PubMed]

124. Monge, M.; Fornaguera, C.; Quero, C.; Dols-Perez, A.; Calderó, G.; Grijalvo, S.; García-Celma, M.J.; Rodríguez-Abreu, C.; Solans, C. Functionalized PLGA nanoparticles prepared by nano-emulsion templating interact selectively with proteins involved in the transport through the blood-brain barrier. *Eur. J. Pharm. Biopharm.* **2020**, *156*, 155–164. [CrossRef] [PubMed]

125. Jose, S.; Sowmya, S.; Cinu, T.A.; Aleykutty, N.A.; Thomas, S.; Souto, E.B. Surface modified PLGA nanoparticles for brain targeting of Bacoside-A. *Eur. J. Pharm. Sci.* **2014**, *63*, 29–35. [CrossRef]

126. Orunoğlu, M.; Kaffashi, A.; Pehlivan, S.B.; Şahin, S.; Söylemezoğlu, F.; Oğuz, K.K.; Mut, M. Effects of curcumin-loaded PLGA nanoparticles on the RG2 rat glioma model. *Mater. Sci. Eng. C Mater. Biol. Appl.* **2017**, *78*, 32–38. [CrossRef] [PubMed]

127. Bhowmik, A.; Chakravarti, S.; Ghosh, A.; Shaw, R.; Bhandary, S.; Bhattacharyya, S.; Sen, P.C.; Ghosh, M.K. Anti-SSTR2 peptide based targeted delivery of potent PLGA encapsulated 3,3'-diindolylmethane nanoparticles through blood brain barrier prevents glioma progression. *Oncotarget* **2017**, *8*, 65339–65358. [CrossRef]

128. USP-NF. Goserelin Implants. Available online: https://online.uspnf.com/uspnf/document/1_GUID-556307CA-0AAB-4ED2-9 4D6-6878DCEAF168_1_en-US?source=Search%20Results&highlight=PLGA (accessed on 21 February 2021).

129. US-FDA. COVID-19 Vaccines. Available online: https://www.fda.gov/emergency-preparedness-and-response/coronavirus-disease-2019-covid-19/covid-19-vaccines (accessed on 23 February 2021).
130. Liberti, L.; Breckenridge, A.; Hoekman, J.; Leufkens, H.; Lumpkin, M.; McAuslane, N.; Stolk, P.; Zhi, K.; Rägo, L. Accelerating access to new medicines: Current status of facilitated regulatory pathways used by emerging regulatory authorities. *J. Public Health Policy* **2016**, *37*, 315–333. [CrossRef] [PubMed]
131. Liberti, L.; Breckenridge, A.; Hoekman, J.; Leufkens, H.; Lumpkin, M.; McAuslane, N.; Stolk, P.; Zhi, K.; Rägo, L. Practical aspects of developing, implementing and using facilitated regulatory pathways in the emerging markets. In Proceedings of the Poster Drug Information Association Annual Meeting, Philadelphia, PA, USA, 26–30 June 2016.

MDPI
St. Alban-Anlage 66
4052 Basel
Switzerland
Tel. +41 61 683 77 34
Fax +41 61 302 89 18
www.mdpi.com

Pharmaceutics Editorial Office
E-mail: pharmaceutics@mdpi.com
www.mdpi.com/journal/pharmaceutics